D0641146

Advances in

ECOLOGICAL RESEARCH

VOLUME 17

Advances in

ECOLOGICAL RESEARCH

Edited by

A. MACFADYEN

23 Mountsandel Road, Coleraine, Northern Ireland

E. D. FORD

Center for Quantitative Science, University of Washington, 3737 15th Avenue Seattle, Washington 98195, USA

VOLUME 17

1987

ACADEMIC PRESS

Harcourt Brace Jovanovich, Publishers

London Orlando

San Diego New York Austin Boston

Sydney Tokyo Toronto

ACADEMIC PRESS INC. (LONDON) LTD.
24/28 Oval Road
London NW1

United States Edition published by
ACADEMIC PRESS INC.
Orlando, Florida 32887

British Library Cataloguing in Publication Data

Advances in ecological research
Vol. 17
1. Ecology
I. Macfadyen, A. II. Ford, E. D.
574.5 QH541

ISBN 0-12-013917-0

Typeset by Paston Press, Loddon, Norfolk
and printed in Great Britain
by St Edmundsbury Press
Bury St Edmunds, Suffolk

Contributors to Volume 17

P. H. CROWLEY, *T. H. Morgan School of Biological Sciences, University of Kentucky, Lexington, Kentucky 40506, USA.*
P. K. DAYTON, *A-001 Scripps Institution of Oceanography, University of California, San Diego, La Jolla, California 92093, USA.*
W. S. C. GURNEY, *Department of Applied Physics, University of Strathclyde, Glasgow G4 0NG, Scotland.*
M. A. JERVIS, *Department of Zoology, University College, P.O. Box 78, Cardiff CF1 1XL, Wales.*
J. A. KITCHING, *School of Biological Sciences, University of East Anglia, Norwich NR4 7TJ, England.*
J. H. LAWTON, *Department of Biology, University of York, Heslington, York YO1 5DD, England.*
R. M. NISBET, *Department of Applied Physics, University of Strathclyde, Glasgow G4 0NG, Scotland.*
M. J. TEGNER, *A-001 Scripps Institution of Oceanography, University of California, San Diego, La Jolla, California 92093, USA.*
N. WALOFF, Department of Pure and Applied Biology, Imperial College at Silwood Park, Ascot, Berks SL5 7PY, England.
M. R. WARBURG, *Department of Biology, Technion—Israel Institute of Technology, Haifa 32000, Israel.*
G. W. YEATES, *New Zealand Soil Bureau, Private Bag, Lower Hutt, New Zealand.*

Preface

This very full volume of six reviews contains two follow-on papers by previous contributors to the series. Professor J. A. Kitching's earlier paper (in collaboration with F. J. Ebling, Vol. 4, pp. 198–291, 1967) on Lough Ine summarised work up to that date on factors determining the distributions of marine organisms in a unique locality which, although sheltered and with limited tidal range enjoys normal marine salinity conditions. The current paper, using the more correct name of L. Hyne, covers progress over two further decades and extends to further localities within and outside the Lough which provide a greater range of combinations of environmental factors. Field observations have now been supplemented by experimental studies including the transplantation of organisms between habitats and some fascinating studies on migratory movements. Insight into recent geological history has been gained by bottom mud sampling.

Dr Waloff's first paper (Volume 11, pp. 82–215, 1980) brought together studies, mainly at Silwood Park, on grassland leafhoppers (Auchenorrhyncha) which affect grasses both directly through feeding and also by disease transmission. The new paper with Dr M. A. Jervis, extends the field to Auchenorrhyncha throughout Europe and concentrates on parasite/parasitoid complexes. The practical importance of this work for biological control, especially in the context of "green revolution cereals" cannot be over-emphasised. The paper will be of general interest to ecologists also for its account of the subtle ways in which resources are divided between closely related species apparently exploiting the same resources.

The first paper in the volume by P. H. Crowley and others builds on nearly two decades' work on Damselflies at York and elsewhere and develops a most ambitious and successful multi-factor model of population regulation in these freshwater insects. They exhibit particularly complex life histories with quite different regulating factors in the different stages, and factors promoting or suppressing asynchrony of sub-populations receive special attention. Analogies emerge from this work which remind one of the non-linear and threshold functions which are well known in epidemiology.

Insect-plant relationships have been the subject of much attention over the past two decades and more. The nematodes, which rival insects in terms of the number of individuals on the Earth if not in their diversity of species have had far less attention, especially as regards the effects of plants on their populations. The practical importance of such effects is demonstrated by the limited success of campaigns against potato-root eelworm, for instance. Dr G. W. Yeates' paper helps to rectify the situation by reviewing the many kinds of effects which plants have on vital nematode functions and life stages and the great range of nematode reponses in return. Parallels are drawn with plant-insect relationships to the advantage of workers in both fields.

The Terrestrial Isopod Crustacea seem to the outsider to be quite anomalous. Members of a primarily aquatic group, they nevertheless thrive under surprisingly dry conditions, and are of major importance as decomposers of certain soils. Dr Warburg's article demonstrates how their success on land is ensured by combinations of features of anatomy, physiology and behaviour and life history strategy. Moisture, however, proves to remain the ultimate constraint to their range. This is a good example of the growing recrudescence of functional studies in ecology and the contribution they can make to population ecology.

Before dismissing the paper by M. J. Tegner and P. K. Dayton as yet another on the El Niño phenomenon, the reader must realize that their paper concerns effects far North of the Equator, in fact the specific effect on Kelp beds in the Bay of California. Although the ultimate cause of the effects described is oceanographic these are modified and amplified in a most revealing way by biological factors which act sometimes reversibly, sometimes in an irreversible manner. A point of interest is that both here and in Professor Kitching's studies a critical role is played by Sea Urchins and by the effects of temperature on seaweed physiology and development.

Once again the papers in this volume demonstrate the vital importance of longer term studies in ecology. Few of the principles which emerge here could have been unearthed by studies lasting less than a decade: which is a very sobering thought at a time of curtailment of funding both to individual projects and to whole institutions which are not solely committed to short term research.

AMYAN MACFADYEN

Contents

Population Regulation in Animals with Complex Life-histories: Formulation and Analysis of a Damselfly Model

P. H. CROWLEY, R. M. NISBET, W. S. C. GURNEY and J. H. LAWTON

How Plants Affect Nematodes

G. W. YEATES

Ecological Studies at Lough Hyne

J. A. KITCHING

Isopods and Their Terrestrial Environment

M. R. WARBURG

El Niño Effects on Southern California Kelp Forest Communities

MIA. J. TEGNER AND PAUL. K. DAYTON

Communities of Parasitoids Associated with Leafhoppers and Planthoppers in Europe

N. WALOFF and M. A. JERVIS

Population Regulation in Animals with Complex Life-histories: Formulation and Analysis of a Damselfly Model

P. H. CROWLEY, R. M. NISBET, W. S. C. GURNEY, and J. H. LAWTON

I. SUMMARY

The complex life-histories of taxa such as damselflies (Odonata: Zygoptera) obscure the mechanisms of population regulation. Most of the available data are for the family Coenagrionidae, and its best-known member is the British and European species *Ischnura elegans*. This information suggests four

ADVANCES IN ECOLOGICAL RESEARCH Vol. 17
ISBN 0-12-013917-0

plausible mechanisms of damselfly population regulation: food availability; feeding-related intraspecific interference; mortality-related intraspecific interference; and density-dependent predation.

We derive a mathematical model of a damselfly population and obtain parameter values largely based on *I. elegans*. The model represents six damselfly life-stages and their interactions with a population of aquatic prey, using coupled ordinary and delay-differential equations, which are solved numerically. Also incorporated are seasonal driving functions, one modifying feeding and mortality parameters according to temperate-zone temperature oscillations, and the other controlling emergence as if by photoperiod or temperature cues. We analyze the model's behavior both in steady state and dynamically with our literature-derived parameter values, and perform sensitivity analyses.

The resulting larval densities, larval stage durations, emergence rates, and general emergence pattern for the standard parameter values are in good agreement with those in the literature: the generation time slightly exceeds one year, and the emergence pattern is strongly bimodal, as observed for some *I. elegans* populations in the British Midlands. Varying the size needed to achieve emergence strongly influences these patterns and densities, emphasizing the need for more data on the body sizes of emerging damselflies.

Varying the carrying capacity of the prey assemblage demonstrates a threshold below which damselflies are unable to persist, and a general increase in densities and decrease in stage durations of larval damselflies with increasing carrying capacity. Stage durations rapidly approach their minimum at and above intermediate prey levels, but larval densities continue to rise even at high carrying capacities. Despite this apparent food-limitation of damselfly larvae, they are generally unable to deplete their prey substantially, and are thus seemingly unable to compete with each other for food. They should nevertheless be susceptible to such competition from other animals such as fish that may be capable of substantially reducing prey densities. Feeding-related interference has essentially no effect on the damselfly population.

The damselfly population is sensitive to changes in the larval and adult mortality parameters: particularly at low prey levels, threefold increases in one of the density-independent mortality parameters generally resulted in extinction of the damselflies. The effectiveness of mortality-related interference and density-dependent predation in regulating the model population is clearly indicated in the stage-by-stage damping of shifts in fecundity: small larvae responded strongly, large larvae weakly, and subsequent emergence rates hardly at all.

Emergence patterns produced by the model seem to reflect the balance between forces promoting and opposing the coexistence of the asynchronous subpopulations that produce separate emergence peaks; promoting coexistence are density-dependent predation and intra-stage, mortality-related larval interference, and opposing it is interstage interference.

II. INTRODUCTION

A. Overview

Some populations exhibit wild density fluctuations; others are more tranquil. Population ecologists face the daunting task of explaining how severely fluctuating populations often manage to persist, and how other populations can remain within relatively narrow ranges of density. The typical magnitude and temporal pattern of density in a population are determined by aspects of the biotic environment (e.g., predators, food supply), the abiotic environment (e.g., seasonal cycles), the population itself (e.g., intraspecific competition), and interactions among them. The magnitudes of excursions from typical densities, and the tendency to return, depend strongly on an important subset of these density-determining factors, referred to herein as *regulatory* and characterized by *direct density-dependence*.

Identifying the factors that determine and regulate densities of natural populations becomes even more formidable for organisms with complex life-histories. For these, the relevant physiological and ecological constraints may shift dramatically during post-embryonic development. Thorough investigations of such organisms encompassing all life-stages may encounter serious methodological and logistical problems, which may largely account for the rarity of these studies in the literature. An alternative approach attempts to combine, within a single conceptual framework, results from separate investigations of different life-stages. This at least offers the hope of generating some testable hypotheses about how population densities of organisms with complex life-histories are regulated.

Here we follow the latter approach, synthesizing information on a fascinating group of organisms—the damselflies—in an attempt to understand how their population densities are regulated. Damselflies (Odonata: Zygoptera) are classical examples of animals with complex life-histories. During the relatively long-lived aquatic larval stage, damselflies experience a continuously changing spectrum of enemies, food and environmental stresses; they then metamorphose directly into flying adults, a transformation in morphology, habitat, and behavior as dramatic as any seen in the animal kingdom. The adults live in a world totally unlike that of the larvae. Indeed, the transition from aquatic larva to terrestrial adult can be viewed as a seasonal "migration" by damselflies between two quite different habitats. The population dynamic consequences of such behavior are poorly understood (Fretwell, 1972; Kot and Schaffer, 1984).

Long-term population dynamic studies of damselflies are rare, though extensive natural history observations attest that populations of many species persist within relatively narrow bounds for several years, provided that habitats remain unaltered. Macan (1974) presents data on the number of large larvae of two species (*Pyrrhosoma nymphula* and *Enallagma cyathigerum*) in Hodson's Tarn in northern England, for 17 and 16 years respectively (1955– and 1956–1971). Populations fluctuated rather little at first, with a ratio of maximum to minimum numbers in *Pyrrhosoma* of 2·02 over the 11 years 1955–1965, and 2·51 over the corresponding 10 years for *Enallagma* (coefficients of variation in log-transformed counts of large larvae were 0·054 and 0·067 respectively). Thereafter, numbers of both species declined to lower levels, apparently because of deterioration in *Littorella* and *Myriophyllum* weed beds in the tarn. These data suggest that damselfly populations are relatively stable and well regulated, at least without severe habitat modification (Williamson, 1972; Hassell *et al.*, 1976; Connell and Sousa, 1983).

We begin our attempt to identify the ecological mechanisms responsible for these relatively tranquil population dynamics with a brief description of damselfly life-histories. We then review relevant published observations, hypotheses, and experimental studies, proceeding through the life stages sequentially. Much of this material necessarily focuses on the relatively long larval (or naiad or nymph) stage, for which the most quantitative information is available. Using these ideas and data, we develop and analyze a model incorporating major facets of damselfly life-histories. This exercise illuminates the more glaring deficiencies in our current state of knowledge, and helps us to evaluate some ways that damselfly population densities may be regulated in nature.

B. Damselfly Life-histories

For readers unfamiliar with damselfly biology, this section sketches the bare outline of a much more detailed and informative picture presented by Corbet (1962, 1980). Insects in the ancient hemimetabolous order Odonata, damselflies in the suborder Zygoptera and dragonflies in the suborder Anisoptera, have both aquatic and terrestrial stages. The aquatic larval stage usually lasts considerably longer than the terrestrial adult stage, particularly towards higher latitudes. Following hatching, the aquatic larvae pass through 9–15 instars, the duration of each instar depending on feeding rate and temperature. Severe environmental changes (such as the onset of winter) may halt or drastically slow development. In many species late instar larvae enter diapause in response to temperature or photoperiod cues, or both; but in other species, a simple winter quiescence is apparently induced by low ambient temperatures. The larvae are generalized predators on invertebrates and occasionally on vertebrates, especially cladocerans and midge larvae; damselfly larvae are themselves common prey of fish and other aquatic vertebrates, as well as bugs, dragonflies, and other damselflies. The larvae are usually ambush predators in shallow water (<2 m deep) and are known to interact aggressively with each other.

Once a larva has developed sufficiently within the final instar, and subject to other constraints imposed by water temperature and photoperiod, it undergoes metamorphosis. Feeding ceases; mouthparts degenerate; and eyes, wings, new mouthparts, and other structures develop rapidly within the larval exoskeleton over several days. Then weather permitting, the larva emerges by climbing up onto the shore or a stalk of vegetation (or some comparable object) and pulls itself out of the exuvia. After a few minutes, when its wings have dried, the new adult can fly. At temperate latitudes, a population may emerge more synchronously early in the year (usually spring); or less synchronously later in the year (summer or autumn); or in multimodal seasonal patterns. Typically, immature adults remain away from water for 1–3 weeks, a period of dispersal, feeding, and reproductive development. Adults eat mainly smaller insects and are frequent prey of birds, fish, frogs, and spiders. Mature adult damselflies return to water to mate, which generally features aggressive interactions among conspecific males. Mated females oviposit in stalks of aquatic or overhanging terrestrial vegetation, or in floating leaves, algal mats, or debris; or, rarely, directly into water. Egg development may proceed immediately to completion and hatching, or there may be a delay of variable duration in response to photoperiod, temperature, or other cues.

III. REGULATORY FACTORS AND PROCESSES

The following sections review many components that impinge upon the complex life-cycles of damselflies, and that might conceivably contribute to population regulation.

A. Larval Growth

1. Prey Availability and Dynamics

Larvae of damselflies and other odonates are generalist predators in the field, consuming cladocerans, larval dipterans, copepods, larval ephemeropterans, ostracods, oligochaetes, larval odonates, and other invertebrates in proportions largely determined by the relative abundances of each kind of prey (e.g., Chutter, 1961; Pritchard, 1964; Pearlstone, 1973; Thompson, 1978b,c). These prey differ considerably in morphology, behavior, density, distribution, and population growth rates, but relatively little in nutritional content or in the efficiency with which they are assimilated (Lawton, 1970, 1971). Damselflies in intermediate and later instars eat mostly cladocerans and midges (usually the most abundant prey—see above references), and there is good evidence from the work of Macan (1964, 1974), Lawton (1971), Folsom (1980), Pickup *et al.* (1984), and Baker (1986a,b) that these larvae can be food-limited in the field, but Thompson (1982) and Folsom and Collins (1982a,b), were unable to detect food limitation. In contrast, the earliest instars feed on the small end of the prey size-spectrum, including large protozoans, copepod nauplii, rotifers, small cladocerans, and first-instar chironomids (Corbet, 1962; Walker, 1953); these prey may generally be abundant and productive enough to make food limitation less likely.

There is some evidence that anisopteran larvae at natural densities can reduce prey densities in the field (Folsom, 1980; Jeffries, 1984; D. M. Johnson *et al.*, 1987; see also Peckarsky, 1984), but this has not been unequivocally demonstrated for damselflies (Jeffries, 1984). Predators that consume damselflies may also deplete damselfly prey (e.g., fish: Hayne and Ball, 1956; Macan, 1966; Bohanan and Johnson, 1983; Johnson *et al.*, 1983; Morin, 1984b) or otherwise make them unavailable by inducing shifts in prey or damselfly behavior (Macan, 1966). These possibilities for prey depletion and food limitation, except perhaps in the earliest instars, argue that the density and dynamics of prey may be important in damselfly population regulation. Unfortunately, diversity in the damselfly prey assemblage makes evaluation of this view difficult.

2. Functional Response to Prey Density

Most damselfly larvae, with the exception of some in the family Lestidae, are classic sit-and-wait predators, moving seldom and slowly, and depending primarily on prey movement to bring them into striking range (Corbet, 1962). In laboratory experiments in structurally simple containers with a single prey type, the relation between killing rate per larva and prey density (i.e., the functional response to prey density) has consistently followed a decelerating rise to a plateau with increasing prey density (e.g., Thompson, 1978a,b), generally known as type II response (Holling, 1959, 1966). It is possible that physical structure (Benke, 1978), density-dependent prey behavior (Crowley, 1975), or presence of alternative prey (Lawton *et al.*, 1974; Akre and Johnson, 1979) could tend to make the functional response sigmoid (i.e., type III) in the field. However, there is evidence that damselfly larvae at higher prey densities hunt less actively or less frequently for prey (Savan, 1979; Thompson, 1975; Wilson, 1982) and kill and consume fewer prey captured (Johnson *et al.*, 1975), which may tend to reverse this effect.

Thus a type II functional response may also apply, at least very roughly, in the field.

The parameters of the type II functional response to prey density, the attack coefficient and the handling time (Hassell, 1978), have been estimated in a number of laboratory studies, though almost always with a single daphnid prey type in aquaria containing minimal structure. These parameters are known to depend on larval size (Thompson, 1978b) and hunger level (Wilson, 1982), temperature (Thompson, 1978a; Gresens *et al.*, 1982), prey characteristics (i.e., morphology, behavior, distribution: Johnson and Crowley, 1980), and on the presence or absence of damselfly predators (Heads, 1985). They may also be influenced by physical structure, illumination (Crowley, 1979), and the time of day (Cloarec, 1975).

In a number of predator–prey interactions, nonrandom spatial distribution of prey, and predator "aggregation" in response to patches or clumps of prey are known to markedly influence the effective feeding rate (i.e., functional response) of individual predators, and the stability of predator–prey interactions (e.g., Hassell, 1978). Though laboratory data suggest that some damselflies may show aggregative responses to patchily distributed prey (e.g., *Coenagrion resolutum*: Baker, 1980, 1982), other laboratory studies (e.g., *Lestes disjunctus*; Baker, 1981b), and a field experiment (Baker, 1986b) find no such effects. Accordingly, we have made no systematic attempt to incorporate nonrandom distributions of prey and predators into our model.

3. Feeding-related Interference

At high larval densities in the laboratory, odonate larvae may interfere with each other's feeding by interactions featuring distraction or overt aggression (Ross, 1971; Machado, 1977; Rowe, 1980; Uttley, 1980; Baker, 1980, 1981a, 1983, in preparation; McPeek and Crowley, 1987; also see Crowley, 1984). It has been suggested that such behavior may influence larval growth rates and act to regulate the number of larvae emerging as adults (Macan, 1977). This interference effect on feeding has been incorporated into mathematical representations of the type II functional response by two slightly different approaches (Beddington, 1975; Rogers and Hassell, 1974). Results of laboratory experiments with larval *Ischnura elegans* as predators and *Daphnia magna* as prey (Uttley, 1980) agree more closely with Beddington's model, in which encounters between larvae interrupt searching in a way closely analogous to the handling time expended during predation.

Recent field experiments demonstrate interference both by feeding inhibition (Johnson *et al.*, 1984; Pierce *et al.*, 1985) and mortality (Johnson *et al.*, 1985; Machado, 1977) in natural systems. But accurately predicting interference intensities in the field from laboratory observations remains a major challenge. This laboratory-to-field extrapolation depends on understanding and quantifying the structure of the substrate on which the interactions occur. In particular, the surface area (e.g., leaves, stems, detrital surfaces) above a standard area of bottom must strongly influence the intensity of interference among damselfly larvae distributed on that surface (see Ross, 1971; Baker, 1986b). Since different aquatic systems differ widely in effective area, this factor may account for much of the variance in interference intensity among these systems.

4. Relation Between Growth Rate and Feeding Rate

Little information is available on the relation between the rates of growth and feeding in invertebrate predators (see Beddington *et al.*, 1976). Lawton *et al.* (1980) have established this relationship in the last three instars of the damselfly *I. elegans* under laboratory conditions. It is strongly nonlinear (concave downward) and approaches an asymptotic straight line of positive slope at high feeding rates. Thus it does not fit either the linear or logarithmic relationships proposed by Beddington *et al.* (1976). More data to quantify this relationship for other instars and species, and mechanistic models predicting the functional form, are desirable because, as we show later, our model is particularly sensitive to the precise form of this relationship.

5. Interspecific Competition

Interspecific competition among larval odonates, manifested both as growth inhibition and as mortality, has been detected experimentally in the field (Johnson *et al.*, 1984, 1985; Pierce *et al.*, 1985), and there are other field data consistent with this interpretation (Crowley and Johnson, 1982a,b). It has been proposed that there may be more potential for interspecific competition in Zygoptera (damselflies) than in Anisoptera, because damselfly species tend to have more broadly overlapping emergence patterns and thus less opportunity for temporal segregation (Johannsson, 1978). This view assumes that populations with displaced emergence intervals interact less intensively (see also Paulson and Jenner, 1971); Ingram and Jenner, 1976; and laboratory observations in Ingram, 1971; Carchini and Nicolai, 1984) but there is some experimental evidence that greater size differences lead to more intense interactions among anisopterans in the field (Benke, 1978; Johnson *et al.*, 1985). In any case, fish may often prey heavily enough on odonate larvae to reduce their densities sufficiently to suppress competition among them (Morin, 1984b; Johnson and Crowley, 1980). Instead, or in addition, fish and other vertebrates may function as competitors, severely depleting the odonates' food supply (See Sections IIIA1 and IIIB2).

More field experiments are needed before any general importance is ascribed to interspecific competition, and our model makes no attempt to incorporate this feature.

B. Larval Mortality

1. Interference Mortality

Cannibalism in odonate populations has been demonstrated by the results of both gut dissections (L. Martin, unpublished manuscript) and fecal studies (Fischer, 1961b; Lawton, 1970; Benke, 1972; Baker and Clifford, 1981; Merrill and Johnson, 1984; Bohanan and Johnson, 1983; but see Thompson, 1978c). Individual larvae have been observed being killed by equal or larger instars in both seminatural laboratory systems (e.g., Fischer, 1961b; McPeek, 1984) and in field experiments (Benke, 1978; Johnson *et al.*, 1985; Crowley *et al.*, 1987). Every larva/larva encounter thus has a small but finite probability of resulting directly in the death of at least one of the participants, and may also expose both participants to additional risks of death by predation (e.g., if movement associated with encounters tends to attract fish). These small but frequently incurred risks can contribute significantly to overall mortality (McPeek, 1984; see also Benke, 1978). In subsequent sections, we shall refer to this effect as interference mortality.

2. Predation by Other Invertebrates and by Vertebrates

Many invertebrates are known to prey upon damselflies (see Corbet *et al.*, 1960), including particularly other odonate species as noted above, Hemiptera (Heads, 1985), and Diptera. Very few field experiments have been performed to detect a density-related mortality effect of other invertebrate predators on odonate populations, and these few seem to be restricted to the intra-odonate studies already cited.

Fish commonly prey heavily on damselfly populations, as indicated by laboratory observations and experiments (Pierce *et al.*, 1985) and by numerous field gut analyses (e.g., Wright, 1946; Gerking, 1962; Martin, 1986) and field experiments (e.g., Hall *et al.*, 1970; Mittelbach, 1981; Morin, 1984a,b). Often this predation is size-selective, focusing on the last several instars (e.g., Faragher, 1980; Mittelbach, 1981; see Benke, 1972; Morin, 1984a,b for anisopteran examples). In a few cases, field experiments have demonstrated reduced densities of damselfly larvae in the presence of fish (e.g., Macan, 1966; Hall *et al.*, 1970). Other vertebrates have also been shown to consume damselflies (Corbet, 1962; Strohmeier *et al.*, in preparation), but field density reductions at natural predator densities remain to be demonstrated.

Though the potential importance of nonodonate predators of damselfly larvae is clear, neither the general forms of functional responses nor the values of the parameters in particular cases have been firmly established. Accordingly, it is impossible, given the data currently available, to be sure what role such predators might play in regulating damselfly larval populations. Much the same conclusion was reached by Ubukata (1981), who discusses the role of predators in the population dynamics of Anisoptera.

3. Other Sources

Other possible mortality sources for damselfly larvae in the field include disease (bacterial, fungal, and protozoan), parasitoid attack, parasites, starvation, freezing, desiccation, and asphyxiation. There is no strong evidence that any of these is consistently and directly important in the field; nevertheless, some of them may interact with other factors in significant ways (e.g., with growth—see below) that remain as yet undocumented.

4. Interaction with Growth

Growth and mortality are closely linked. Results of a field study suggest that higher mortality (e.g., from fish predation: Macan, 1964, 1966) may alleviate intra-odonate competition sufficiently to speed growth and shorten the larval period. On the other hand, reduced growth rates may indirectly increase larval mortality: slowing growth increases the duration of the larval period, exposing the animal for a longer time to the above-mentioned mortality sources. Slower biomass accumulation generally implies less frequent molting (Lawton *et al.*, 1980), which may increase the risk that the integument will be penetrated by microbial pathogens. Poorly fed, and thus slowly growing, larvae may tend to take bigger predation risks while foraging (Heads, 1985) and may even be somewhat less able to escape from predators. Larvae with a high endoparasite burden may move around more and thus also be more susceptible to predators (cf. immature adults in Corbet, 1980). Thus the overall potential for increased mortality resulting from reduced growth appears high; and more rapid growth resulting from increased mortality may also be possible.

C. Emergence

1. Size Constraints and Environmental Cues

Termination of the larval stage depends on both physiological and environmental signals. Physiologically, emergence requires the completion of development and accumulation of sufficient biomass, which could differ between sexes. Size at emergence in length or dry mass for a given population and sex probably varies little at any particular site and sampling date, but may vary more between sites (Pickup *et al.*, 1984) and dates. In several studies (Penn, 1951; Cothran and Thorp, 1982; Banks and Thompson, 1985a; Harvey and Corbet, 1985; Crowley, in preparation) it has tended to decline through the emergence season. More information is needed about size at emergence, including the possibility of density dependence (see Ross, 1967; Pickup *et al.*, 1984; Harvey and Corbet, 1985), since this will be correlated with the larval stage duration and through-stage survival (see Section IIIB4).

Seasonal variations in photoperiod, temperature, or both may generate seasonal patterns in larval growth or emergence. The mechanisms here apparently vary widely among zygopteran species, and generalizations are elusive. However, temperate-zone damselflies will often spend their last winter in one or more late instars in response to a photoperiod cue (Corbet, 1980; but see Procter, 1973) or temperature cues (see Lutz, 1968), which helps to partially synchronize the subsequent emergence.

2. Mortality

There is some evidence that emergence is a time of particularly high mortality, as a result of aquatic and terrestrial predation, competition for emergence sites, and risk of damage from desiccation, water, or heat. Some of this mortality could be density-dependent, as for example in synchronized emergences of anisopterans, which can attract birds (Corbet, 1957), fish (Corbet, 1961; Macan, 1966), or crocodiles (Corbet, 1959). Alternatively, highly synchronized emergences could satiate predators and yield *inversely* density-dependent mortality (Baker, pers. comm.). On available evidence, emergence mortality could be a contributing regulatory factor in some cases (Corbet, pers. comm.), but its generality is not established. By estimating emergence mortality over several years, Ubukata (1981) inferred that populations of the anisopteran *Cordulia aenea* were not regulated by density-dependent mortality at emergence.

3. Seasonal Pattern

Many different damselfly emergence patterns have been observed in the temperate zone, including single, double, or multimodal peaks of emergence within one season; sharp spikes to broad or irregular distributions; and maximal emergence in spring, summer, or autumn (Corbet, 1962, 1980). One pertinent example will suffice: several populations of *Ischnura elegans* studied in west-central England (latitude 53–54° N) appear to have bimodal emergence patterns, with a sharp, early peak in June, followed by a broader, lower peak in mid- or late summer (Parr, 1969b, 1970, 1973).

The characteristics of a population's seasonal emergence distribution generally correlate well with latitude and seasonal temperature (Johannsson, 1978; Thompson, 1978a; Norling, 1976, 1984). However, emergence patterns for particular damselfly species may vary considerably between years at a given site (Kormondy and Gower, 1965) or between sites in a given year (Gower and Kormondy, 1963). These patterns may to a large extent reflect the regulatory (i.e., density-dependent) factors operating in the larval stages of a population (see Paulson and Jenner, 1971) and thus merit close scrutiny, particularly for comparing the mechanisms and extent of regulation in different populations. Unfortunately, data on numbers emerging, and patterns of emergence at sites where predators have been manipulated (e.g., Macan, 1964, 1966, 1974) are not sufficiently detailed to justify analysis.

D. The Immature Adult Stage

1. Feeding and Mortality

Newly emerged adult damselflies remain immature for about 1–3 weeks (Corbet, 1980). During this period, the adults feed mainly on small insects (Corbet, 1962) and are preyed upon by birds, frogs, spiders, and large dragonflies (Walker, 1953; Corbet, 1962; Fincke, 1982). They may be weakened by parasites (Imamura and Mitchell, 1973; Abro, 1971, 1982) and killed or seriously injured by severe weather. In the near absence of quantitative data, and particularly of experimental data, there currently appears little reason to suspect either significant food limitation or density-dependent mortality (but see Section IIID2) of immature adults. The few available data indicate that the *per capita* mortality rate is relatively constant during this stage (e.g., Parr (1965) for *I. elegans*, but see Parr (1973)).

2. Dispersal

Adult odonates generally spend most of the immature stage away from water (Corbet, 1980), though *I. elegans* (Parr, 1973) and *I. verticalis* (R. L. Baker, pers. comm.) sometimes provide exceptions. Apparently, no studies have managed to completely disentangle dispersal and mortality of adults. The immature stage appears to be the primary dispersal stage (see Banks and Thompson, 1985a), and the common failure to recover individuals marked at emergence (e.g., Bick and Bick, 1961) suggests little or no site fidelity between emergence and reproduction, provided that suitable breeding habitats are not widely scattered. Though it seems plausible that dispersal of immature adults is density-dependent, there is little empirical evidence to confirm this, except possibly by inference from the data of Moore (1964). Dispersal could increase the rate of mortality from predators and severe weather in this stage, resulting in a local density reduction even where exchanges of migrants with adjacent aquatic systems would otherwise equilibrate.

E. The Mature Adult Stage

1. Territoriality and Inter-male Aggression

Some mature adult male damselflies are territorial, and essentially all populations studied show aggressive interactions among males that help determine access to females (e.g., see Bick and Bick, 1963; Bick, 1972). It has thus been suggested that territoriality of adults may play a role in production regulation of odonates (Moore, 1953, 1957; Ito, 1978). However, it has not yet been shown that female fecundity is limited by aggressive encounters among males, and the generally high percentage of females mated (Fincke, 1982; Parr and Palmer, 1971) is evidence to the contrary. We conclude that an unequivocal link between territoriality, or male–male aggression in general, and population regulation in adult damselflies remains to be established.

2. Feeding and Mortality

Average longevities of mature adult damselflies range from about 2 days to 2 weeks in the field, where predators or the gradual accumulation of physical damage may generally set the upper limit (Moore, 1952; Parr, 1965). Essentially the same predators attack mature adults as attack immatures, though aquatic predators should be much more important for mature adults: ovipositing females and fighting males may be especially susceptible to predation by fish and amphibians. Generally, there may be more reason to suspect density-dependent predation on mature adults than on immature adults, simply because mature adults spend more time aggregated at aquatic sites; dense populations there may be disproportionately likely to attract predators, as noted above at emergence. This possibility merits further empirical study. The scanty data suggest *per capita* mortality rates roughly constant through time and generally similar to those experienced by immatures (Parr, 1965, 1973; Ueda and Iwasaki, 1982; Banks and Thompson, 1985b). As for immature adults, there is little reason to expect that mature adults, which feed on small insects, are food-limited. However, we note that extended periods of inclement weather or low temperatures could seriously hamper feeding, ultimately reducing fecundity, and perhaps influence survival as well (P. S. Corbet, pers. comm.).

3. Fecundity

The proportion of females successfully mated appears to be generally high and unrelated to the densities of males or females or to their ratio (see above). This, with the lack of evidence for oviposition-site or food limitation, suggests that fecundity is essentially independent of adult density. However, it remains possible that *larval* density could influence adult fecundity. For example, if high larval densities result in the emergence of small adults (see Ross, 1967), and if small adults mate less frequently or lay fewer eggs than large adults, then a delayed density-dependence would result (see also Harvey and Corbet, 1985). We have not incorporated such effects into our model, but they deserve further study.

4. Dispersal

In the mature adult stage, dispersal appears to be negligible in the best-studied cases, even between nearby subpopulations (Mitchell, 1962; Ueda, 1976; Banks and Thompson, 1985b). Though damselflies can disperse in the mature adult stage, any relation of such dispersal to the frequency of aggressive encounters among males and thus to density remains controversial (Corbet, 1980; see references cited therein).

F. The Egg Stage

1. Viability

Within and between individuals and species, the proportion of eggs laid that are viable varies considerably (see Rivard *et al.*, 1975; Sonehara, 1979; Masseau and Pilon, 1982), but there is apparently no evidence that this variance can be attributed to differences in population density.

2. Development Time

For damselflies in which eggs begin developing immediately after oviposition and continue developing without interruption until hatching or death (all except the Lestidae), development times depend strongly on temperature (Masseau and Pilon, 1982). Other factors have apparently not been shown to influence egg development time consistently.

3. Mortality

There are a few documented cases of damselfly egg predation and parasitism (see Corbet, 1962). Infestation by fungi and protozoa are also well known, but the principal source of odonate egg mortality is probably desiccation (Corbet, 1962). Though these factors may well be important in nature, there is currently no evidence for density-dependence.

IV. THE MODEL

The foregoing literature review indicates an extensive array of mechanisms and factors that may contribute to the regulation of damselfly populations. Though no clear consensus has emerged from the empirical evidence, four of these mechanisms appear particularly likely to contribute to population regulation via density-dependent effects operating on the larvae: interference via feeding inhibition; interference via mortality; density-dependent predation; and food limitation. We now investigate the implications of these mechanisms using a mathematical model that is capable of simulating the ecological and life-history characteristics of many damselflies. Parameter values used in the model have been estimated or measured primarily for the damselfly *Ischnura elegans*, but data from other species in the same family (Coenagrionidae) have been used where necessary. This approach does not ignore the possibility that future work may identify other sources of density-dependence in damselfly populations. Rather, we view the result as a "minimal model" or as a "working hypothesis" to be improved and extended in future studies.

A. Derivation

1. Model Structure

Our model exploits recent theoretical methods (Gurney *et al.*, 1983; Nisbet and Gurney, 1983; Gurney *et al.*, 1985) of formulating population models for species whose life-history is made up of a number of well defined physiological stages, within which all individuals are assumed to be identical in feeding behavior and probability of death. In the present analysis we depict the damselfly life-history by the scheme shown in Fig. 1, which we believe to contain the minimal resolution of damselfly life histories needed to incorporate and evaluate the four mechanisms. The four aquatic (eggs and larvae) and two terrestrial (adults) life-stages we use are defined as follows:

"eggs"	egg and first 3 or 4 larval instars
small larvae	middle 3 or 4 larval instars
large larvae	final 3 or 4 larval instars
"holding bin"	fully developed larvae prevented from emerging by seasonal constraints
immature adults	adults incapable of reproductive activity
mature adults	adults capable of reproductive activity

The progress of an individual through a particular stage is quantified by a *development index* (*DI*: Gurney *et al.*, 1985) representing the state of development or the "physiological age" of the individual within that stage at a particular time. The *DI* increases at the same instantaneous rate for all individuals in the stage at a given time. Development within a stage is thus akin to passage along a conveyer belt moving at a variable speed (Oster and Takahashi, 1974; Gurney and Nisbet, 1983). For all stages other than the holding bin, maturation out of a stage occurs on achieving a fixed value of the

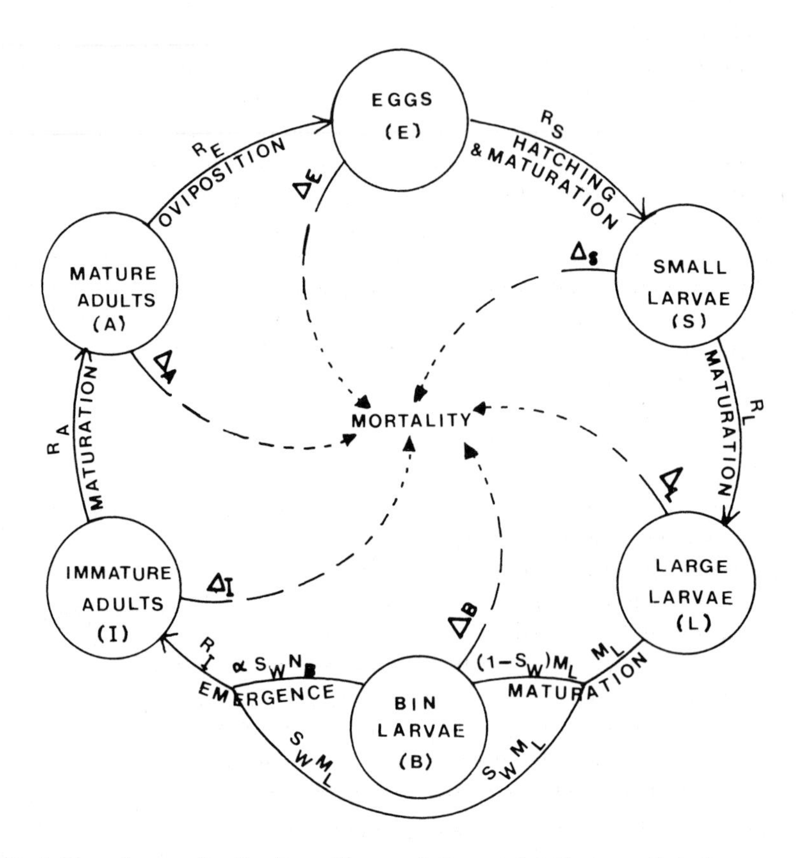

Fig. 1. Flow diagram for the damselfly population model. Each small circle represents a state variable having dynamics specified by coupled ordinary differential and delay-differential equations. The arrows indicate flow rates of individuals into and out of these state-variable circles; the population balance terms corresponding to each arrow are indicated, where R is recruitment rate, M is maturation rate, Δ is mortality rate, N is population density, and α is a constant. Seasonality is introduced by sinusoidal variation in some of the rate parameters and by opening and closing an "emergence window". The operation of the window is controlled by the approximately square-wave seasonal parameter S_W. When $S_W = 0$, larvae large enough to emerge are recruited to the holding bin, otherwise continuing to function as large larvae. When the season is appropriate for emergence, S_W increases quickly from 0 to 1 (i.e., the window is open), and both holding-bin larvae and other larvae of sufficient size can emerge. An aquatic prey assemblage (not shown) is coupled to damselfly larval growth dynamics as an additional ordinary differential equation, including terms representing size-specific functional responses of the larvae to prey density.

development index, analogous to reaching the end of the conveyer belt; holding-bin larvae emerge at a rate determined by seasonal factors and not by their state of development.

Mathematically, the model is a system of coupled ordinary differential and delay-differential equations (Table 1: equations (1–12)), using variables (Table 2) and parameters (Table 3) of clear biological significance. In principle, the population in each stage is described by three differential equations, expressing changes in total population numbers, through-stage survival, and stage duration; but for all stages other than small and large larvae, the assumptions of the model result in one or more of these quantities remaining constant. Evaluation of the various quantities that appear on the right-hand sides of the differential equations requires assumptions about development rates and *per capita* death rates for all stages, and about fecundity for the mature adult stage. We discuss these assumptions in some detail in the next subsection.

2. Model Detail

We select different development indices (*DI*) for each stage. Since there is no evidence that food limitation (or other factors) influences rates of development in early instar larvae (hence our lumping of these stages with the eggs), we assume that development in this stage is determined solely by temperature, and use degree-days as *DI*. For the small and large larval stages, the selection of a suitable *DI* is more difficult, as both temperature and feeding are implicated in development. We assume that development rates for these stages are affected by temperature only via the corresponding feeding rates, and we base our description of development on curve fits (equations (40) and (41)) to the data of Lawton *et al.* (1980), relating stage durations for later instars of *I. elegans* to feeding rates. The most complete data set relates to the eleventh or penultimate instar, and we therefore require two parameters (i_S and i_L) characterizing the durations of our "small" and "large" larval stages relative to instar 11 of this particular species. It is assumed that no development takes place in the holding bin, though holding-bin larvae are assumed to feed at the same rate as large larvae, and thus no *DI* need be defined for this stage. The two adult phases are short relative to the others, and we merely use age, or more precisely, time in stage, as *DI* for these stages.

Having chosen to let development in certain stages depend on feeding, and to let both feeding and mortality be influenced by intraspecific interactions among larvae, we require descriptions of "food," "feeding," and "interference". We adopt an "encounter frequency" approach (Koopman 1956a,b; Gerritsen and Strickler 1977; and Gerritsen 1980; see Jervey (1982) for an application to predation by damselflies) to larva–larva and to larva–prey interactions; this involves defining a physical surface over which

Table 1
Equations of the damselfly population model: symbols are defined in Tables 2 and 3.

Differential equations

(1)	$\dot{H}(t) = R_H(t) - f_S(t)N_S(t)/w - f_L(t)[N_L(t) + N_B(t)]/w$	Prey population density
(2)	$\dot{N}_E(t) = R_E(t) - R_S(t) - \Delta_E N_E(t)$	Densities of damselfly life-stages
(3)	$\dot{N}_S(t) = R_S(t) - R_L(t) - \Delta_S(t)N_S(t)$	
(4)	$\dot{N}_L(t) = R_L(t) - M_L(t) - \Delta_L(t)N_L(t)$	
(5)	$\dot{N}_B(t) = R_B(t) - E_B(t) - \Delta_L(t)N_B(t)$	
(6)	$\dot{N}_I(t) = R_I(t) - R_A(t) - \Delta_I N_I(t)$	
(7)	$\dot{N}_A(t) = R_A(t) - \Delta_A N_A(t)$	
(8)	$\dot{\tau}_E(t) = 1 - g_E(t)/g_E(t - \tau_E(t))$	Egg and larval stage durations
(9)	$\dot{\tau}_S(t) = 1 - g_S(t)/g_S(t - \tau_S(t))$	
(10)	$\dot{\tau}_L(t) = 1 - g_L(t)/g_L(t - \tau_L(t))$	
(11)	$\dot{P}_S(t) = P_S(t)[g_S(t)\Delta_S(t - \tau_S(t))/g_S(t - \tau_S(t)) - \Delta_S(t)]$	Larval through-stage survival
(12)	$\dot{P}_L(t) = P_L(t)[g_L(t)\Delta_L(t - \tau_L(t))/g_L(t - \tau_L(t)) - \Delta_L(t)]$	

Initial conditions

(13)	$H(0) = K$
(14)	$N_E(0) = 0$
(15)	$N_S(0) = 0$
(16)	$N_L(0) = 0$
(17)	$N_B(0) = N_{B0}$
(18)	$N_I(0) = 0$
(19)	$N_A(0) = 0$
(20)	$\tau_E(0)$: implicit solution of $\tau_E(0) - \beta \sin(\omega\tau_E(0))/\omega = \eta$
(21)	$\tau_S(0)$: implicit solution of $\int_{-\tau_S(0)}^{0} g_S(t)\,dt = i_S$
(22)	$\tau_L(0)$: implicit solution of $\int_{-\tau_L(0)}^{0} g_L(t)\,dt = i_L$
(23)	$P_S(0) = \exp\{-\delta_S\tau_S(0)\}$
(24)	$P_L(0) = \exp\{-\delta_L\tau_L(0)\}$

Auxiliary equations

(25)	$R_H(t) = rS_T(t)H(t)[1 - H(t)/K]$	Population balance
(26)	$R_E(t) = \varrho\Delta_A N_A(t)$	
(27)	$R_S(t) = R_E(t - \tau_E(t)) \exp\{-\Delta_E\tau_E(t)\}g_E(t - \tau_E(t))$	
(28)	$R_L(t) = R_S(t - \tau_S(t))P_S(t)g_S(t)/g_S(t - \tau_S(t))$	
(29)	$M_L(t) = R_L(t - \tau_L(t))P_L(t)g_L(t)/g_L(t - \tau_L(t))$	
(30)	$E_L(t) = S_W(t)M_L(t)$	
(31)	$R_B(t) = [1 - S_W(t)]M_L(t)$	
(32)	$E_B(t) = \alpha S_W(t)N_B(t)$	
(33)	$R_I(t) = S_W(t)[M_L(t) + \alpha N_B(t)]$	
(34)	$R_A(t) = R_I(t - \tau_I) \exp\{-\Delta_I\tau_I\}$	
(35)	$f_S(t) = a_S S_T(t)H(t)/[1 + a_S b_S H(t) + \gamma_{SS}N_S(t) + \gamma_{SL}[N_L(t) + N_B(t)]]$	Vital rates
(36)	$f_L(t) = a_L S_T(t)H(t)/[1 + a_L b_L H(t) + \gamma_{LS}N_S(t) + \gamma_{LL}[N_L(t) + N_B(t)]]$	
(37)	$\Delta_S(t) = \delta_S + \delta_{SP}N_S(t) + [\delta_{SS}N_S(t) + \delta_{SL}[N_L(t) + N_B(t)]]S_T(t)$	
(38)	$\Delta_L(t) = \delta_L + \delta_{LP}[N_L(t) + N_B(t)] + [\delta_{LS}N_S(t) + \delta_{LL} \times [N_L(t) + N_B(t)]]S_T(t)$	
(39)	$g_E(t) = S_T(t)/\eta$	
(40)	$g_S(t) = g_{S0}[1 - \exp\{-f_S(t)/f_{S0}\}] + e_S f_S(t)$	
(41)	$g_L(t) = g_{L0}[1 - \exp\{-f_L(t)/f_{L0}\}] + e_L f_L(t)$	

Table 1 *continued.*

(42) $S_T(t) = 1 - \beta \cos(\omega t)$ (43) $S_W(t) = [t - n - x_0]/x_w$, when $n + x_0 \leqslant t \leqslant n + x_0 + x_w$, for some integer n; $= 1$, when $n + x_0 + x_w \leqslant t \leqslant n + x_s - x_w$; $= [n + x_s - t]/x_w$, when $n + x_s - x_w \leqslant t \leqslant n + x_s$; $= 0$, when $n \leqslant t \leqslant n + x_0$ or $n + x_s \leqslant t \leqslant n + 1$.	Seasonal variation

Table 2
Variables of the damselfly population model.

Symbol	Definition	Units
t	Time	yr
$H(t)$	Density of prey	prey liter^{-1} of water
$N_E(t)$	Density of eggs	eggs m^{-2} of bottom
$N_S(t)$	Density of small larvae	larvae m^{-2} of bottom
$N_L(t)$	Density of large larvae	larvae m^{-2} of bottom
$N_B(t)$	Density of larvae in holding bin	larvae m^{-2} of bottom
$N_I(t)$	Density of immature adults	adults m^{-2} of bottom
$N_A(t)$	Density of mature adults	adults m^{-2} of bottom
$\tau_E(t)$	Duration of egg stage	yr
$\tau_S(t)$	Duration of small larval stage	yr
$\tau_L(t)$	Duration of large larval stage	yr
$P_S(t)$	Through-stage survival of small larvae	dimensionless
$P_L(t)$	Through-stage survival of large larvae	dimensionless
$R_H(t)$	Recruitment rate of prey	prey liter^{-1} yr^{-1}
$R_E(t)$	Recruitment rate of eggs	eggs m^{-2} yr^{-1}
$R_S(t)$	Recruitment rate of small larvae	larvae m^{-2} yr^{-1}
$R_L(t)$	Recruitment rate of large larvae	larvae m^{-2} yr^{-1}
$M_L(t)$	Maturation rate of large larvae	larvae m^{-2} yr^{-1}
$E_L(t)$	Emergence rate of large larvae	larvae m^{-2} yr^{-1}
$R_B(t)$	Recruitment rate of holding-bin larvae	larvae m^{-2} yr^{-1}
$E_B(t)$	Emergence rate of holding-bin larvae	larvae m^{-2} yr^{-1}
$R_I(t)$	Recruitment rate of immature adults	adults m^{-2} yr^{-1}
$R_A(t)$	Recruitment rate of mature adults	adults m^{-2} yr^{-1}
$f_S(t)$	Feeding rate of small larvae	prey larva^{-1} yr^{-1}
$f_L(t)$	Feeding rate of large larvae	prey larva^{-1} yr^{-1}
$\Delta_S(t)$	*Per capita* mortality rate of small larvae	yr^{-1}
$\Delta_L(t)$	*Per capita* mortality rate of large larvae	yr^{-1}
$g_E(t)$	Development rate of eggs	yr^{-1}
$g_S(t)$	Development rate of small larvae	yr^{-1}
$g_L(t)$	Development rate of large larvae	yr^{-1}
$S_T(t)$	Sinusoidal temperature driving function	dimensionless
$S_W(t)$	Trapezoidal emergence driving function	dimensionless

Table 3

Parameters of the damselfly population model: notes explain the choice of these parameter values, which are referred to in the text as "default values".

Symbol	Definition	Magnitude	Units
r	Mean prey intrinsic growth rate[a]	18	yr^{-1}
K	Prey carrying capacity[b]	20	prey $liter^{-1}$
a_S	Mean attack coefficient of small larvae[c]	16	liter yr^{-1} $larva^{-1}$
a_L	Mean attack coefficient of large larvae[d]	62	liter yr^{-1} $larva^{-1}$
b_S	Mean handling time of small larvae[c]	$1 \cdot 1 \times 10^{-3}$	larva-yr $prey^{-1}$
b_L	Mean handling time of large larvae[d]	$1 \cdot 4 \times 10^{-4}$	larva-yr $prey^{-1}$
w	Water depth[e]	1×10^{3}	mm
g_{S0}	Characteristic development rate of small larvae[f]	10·6	yr^{-1}
g_{L0}	Characteristic development rate of large larvae[f]	10·6	yr^{-1}
f_{S0}	Characteristic feeding rate of small larvae[f]	730	prey $larva^{-1}$ yr^{-1}
f_{L0}	Characteristic feeding rate of large larvae[f]	730	prey $larva^{-1}$ yr^{-1}
e_S	Incremental development efficiency of small larvae[f]	0·0015	larvae $prey^{-1}$
e_L	Incremental development efficiency of large larvae[f]	0·0015	larvae $prey^{-1}$
i_S	Development index at maturation of small larvae[f]	1·7	dimensionless
i_L	Development index at maturation of large larvae[f]	3·3	dimensionless
V_S	Mean movement speed of small larva[g]	35	m yr^{-1}
V_L	Mean movement speed of large larva[g]	105	m yr^{-1}
D_S	Effective encounter diameter of small larvae[h]	0·02	m $larva^{-1}$
D_L	Effective encounter diameter of large larvae[h]	0·04	m $larva^{-1}$
ϕ_{SS}	Mean area clearance rate for small encountering small[i]	1·98	m^2 yr^{-1} $larva^{-1}$
ϕ_{SL}	Mean area clearance rate for small encountering large[i]	6·64	m^2 yr^{-1} $larva^{-1}$
ϕ_{LS}	Mean area clearance rate for large encountering small[i]	6·64	m^2 yr^{-1} $larva^{-1}$
ϕ_{LL}	Mean area clearance rate for large encountering large[i]	11·9	m^2 yr^{-1} $larva^{-1}$
λ	Interference area index[j]	5	dimensionless
σ_{SS}	Mean time wasted by small larva encountering small[k]	$0 \cdot 46 \times 10^{-4}$	yr
σ_{SL}	Mean time wasted by small larva encountering large[k]	$0 \cdot 23 \times 10^{-4}$	yr
σ_{LS}	Mean time wasted by large larva encountering small[k]	$0 \cdot 23 \times 10^{-4}$	yr

σ_{LL}	Mean time wasted by large larva encountering large[k]	$2{\cdot}28 \times 10^{-4}$	yr
ν_{SS}	Probability that small larva dies encountering small[l]	0·005	dimensionless
ν_{SL}	Probability that small larva dies encountering large[l]	0·2	dimensionless
ν_{LS}	Probability that large larva dies encountering small[l]	0·002	dimensionless
ν_{LL}	Probability that large larva dies encountering large[l]	0·02	dimensionless
γ_{SS}	Feeding interference coefficient, small on small[m]	9×10^{-5}	m^2 larva^{-1}
γ_{SL}	Feeding interference coefficient, large on small[m]	3×10^{-5}	m^2 larva^{-1}
γ_{LS}	Feeding interference coefficient, small on large[m]	3×10^{-5}	m^2 larva^{-1}
γ_{LL}	Feeding interference coefficient, large on large[m]	11×10^{-5}	m^2 larva^{-1}
δ_{SS}	Mean interference mortality coefficient, small on small[n]	0·002	m^2 yr^{-1} larva^{-1}
δ_{SL}	Mean interference mortality coefficient, large on small[n]	0·27	m^2 yr^{-1} larva^{-1}
δ_{LS}	Mean interference mortality coefficient, small on large[n]	0·003	m^2 yr^{-1} larva^{-1}
δ_{LL}	Mean interference mortality coefficient, large on large[n]	0·047	m^2 yr^{-1} larva^{-1}
Δ_E	*Per capita* mortality rate of eggs[o]	1·8	yr^{-1}
δ_S	"Background" *per capita* mortality rate of small larvae[p]	0·9	yr^{-1}
δ_L	"Background" *per capita* mortality rate of large larvae[p]	0·9	yr^{-1}
Δ_I	*Per capita* mortality rate of immature adults[q]	74	yr^{-1}
Δ_A	*Per capita* mortality rate of mature adults[q]	174	yr^{-1}
δ_{SP}	Density-dependent predation mortality rate, small larvae[r]	0·002	m^2 yr^{-1} larva^{-1}
δ_{LP}	Density-dependent predation mortality rate, large larvae[r]	0·047	m^2 yr^{-1} larva^{-1}
α	*Per capita* emergence rate by holding-bin larvae[s]	35	yr^{-1}
x_0	Fractional time of year at which emergence begins[t]	0·4	yr
x_s	Fractional time of year at which emergence ends[t]	0·6	yr
x_w	Time spent opening or closing emergence window[t]	0·02	yr
τ_I	Immature adult stage duration[u]	0·0211	yr
ϱ	Total oviposition per mature adult[v]	120	eggs adult^{-1}
η	Duration of "egg" stage at the equinox[w]	0·2	yr
β	Seasonal amplitude[x]	0·9	dimensionless
ω	Seasonal frequency coefficient[x]	2π	radians yr^{-1}
N_{B0}	Initial density of holding-bin larvae[y]	0·1	larvae m^{-2}

Notes to Table 3 overleaf.

Notes to Table 3.

[a] Estimates for daphnids (used in the laboratory studies from which the larval feeding and development parameters were obtained) are about 100–150 yr^{-1} at 20°C (e.g., Smith, 1963; Frank *et al.*, 1957; Hall 1964); these rates are relatively insensitive to food levels but decrease sharply at lower temperatures (to 25–40 yr^{-1} at 11°C in Hall, 1964). Since we are more interested in the natural prey of damselfly larvae, we have used data on prey turnover for an assemblage of odonate prey, which imply an approximate $r = 40\ yr^{-1}$ at 16°C mean annual temperature (see Benke, 1976). Adjusting this to a mean annual temperature of 10°C yields $r = 18\ yr^{-1}$ (see Johnson and Brinkhurst, 1971), a result reasonably consistent with the laboratory daphnid data. This and all other parameters that vary seasonally with temperature are set at estimated annual mean values characteristic of annual mean water temperatures in central Britain (10–12°C).

[b] The carrying capacity for laboratory daphnid populations is roughly proportional to the rate at which food is supplied (Slobodkin, 1954); similarly, for natural damselfly prey assemblages, K should vary widely with primary production, detrital influx, and other fluxes of energy and nutrients. As a rough baseline, we note that invertebrate standing stock (excluding odonates) in ponds often approximates 1 g m^{-2} of bottom (see, e.g., Benke, 1976; Hall *et al.*, 1970). Since the daphnids used in laboratory studies to obtain the feeding and development parameters averaged 44 μg (Thompson, 1978a; Lawton *et al.*, 1980), the invertebrate standing stock in daphnid equivalents is about 20 000 m^{-2}. Distributing these over a 1 m water column yields 20 prey per liter. We assume that prey are randomly distributed.

[c] $a_S = a_L/4$; $b_S = 8b_L$: these parameter adjustments for smaller larvae are suggested by the data of Thompson (1975).

[d] Thompson (1978a), 12°C values (*Ischnura elegans vs.* 1·7 mm *Daphnia magna*).

[e] Needed to put larval and prey densities in compatible units. A depth of 1 m is quite typical for larval damselfly habitat (see Corbet, 1962).

[f] Estimated from Fig. 5 of Lawton *et al.* (1980), for instar 11. Neither the linear nor the logarithmic relationships proposed by Beddington *et al.* (1976) for the relation between development rate and feeding rate appear to fit these data adequately. The combination of linear and exponential terms used here (equations (40) and (41)) fits very well, even without explicitly incorporating maintenance costs. (In any case, maintenance requirements are probably low even at summer temperatures in Britain: see Lawton (1970, 1971) on *Pyrrhosoma nymphula* at 16°C.)

[g] P. A. Heads (Department of Biology, University of York, pers. comm.), based on long-term (>12 h) observations of *Ischnura elegans* in the laboratory. Shorter-term observations of coenagrionid damselfly larvae indicate considerably more movement (e.g., McPeek, 1984; P. H. Crowley, personal observations of *I. verticalis*), but this may reflect an initial adjustment to new surroundings.

[h] Approximate movement velocities and distances apart at which damselfly larvae orient toward each other with good visibility in the laboratory (P. H. Crowley, personal observations of *Ischnura verticalis*). Though these distances are probably reduced in the dark, larvae may move around more in the dark (Pierce, 1982; but see Crowley, 1979); thus area clearance rates ϕ are assumed to be independent of illumination.

[i] $\phi_{LL} = 2\sqrt{2}D_L V_L$; $\phi_{LS} = \phi_{SL} = [D_L + D_S]\sqrt{V_L^2 + V_S^2}$; $\phi_{SS} = 2\sqrt{2}D_S V_S$. These relationships are obtained by assuming that larval encounters are adequately represented as collisions resulting from random two-dimensional motion (Koopman 1965a,b). In this case, the area clearance rate ϕ (analogous to a predator's attack coefficient a) is the product of the relative velocity of the two larvae and the sum of their encounter diameters.

[j] The total area of surface that larvae move around on (e.g., submerged surface area of vegetation in an aquatic weed bed), per unit area in which larval densities are expressed (e.g., area of pond bottom). This index is conceptually equivalent to the "leaf area index" widely used by botanists.

[k] Estimated from data of Uttley (1980) for *I. elegans*, instars 9–11, 8°C.

[l] Very rough estimates of interference mortality based on laboratory observations of encounters (McPeek, 1984; P. H. Crowley, unpublished) and examination of fecal pellets from field-collected larvae (Merrill and Johnson 1984; D. M. Johnson, unpublished data). ν_{LS} is nonzero primarily because interactions with small larvae may cause larger ones to move or disperse, thereby increasing their vulnerability to other predators, an effect which also contributes to the other values.

[m] Calculated by expressing the interference effect as a product of an intensity term (σ) and an encounter rate term (ϕ), per unit of available substrate (λ). $\gamma_{LL} = \phi_{LL}\sigma_{LL}/\lambda; \gamma_{LS} = \phi_{LS}\sigma_{LS}/\lambda; \gamma_{SL} = \phi_{SL}\sigma_{SL}/\lambda; \gamma_{SS} = \phi_{SS}\sigma_{SS}/\lambda$. These calculated values are in reasonable agreement with those observed in the laboratory by Uttley (1980).

[n] Calculated as for γ (see note *m*), except with the intensity term ν instead of σ. $\delta_{LL} = \phi_{LL}\nu_{LL}/\lambda$; $\delta_{LS} = \phi_{LS}\nu_{LS}/\lambda$; $\delta_{SL} = \phi_{SL}\nu_{SL}/\lambda$; $\delta_{SS} = \phi_{SS}\nu_{SS}/\lambda$.

[o] Since the earliest three or four instars of larvae generally eat different food, are distributed differently, are probably susceptible to different predators, and are very poorly studied (Corbet, 1962), we combine them with eggs as the initial or "egg" stage in the model. The only available data on survival in this initial stage (Ubukata (1981), on the Japanese dragonfly *Cordulia aenea amurensis*) suggest higher mortality than for later instars; we therefore take $\Delta_E = 2\delta_S = 2\delta_L$.

[p] Mortality unrelated to predation or interference: data inferred from information in Lawton (1970) for the damselfly *Pyrrhosoma nymphula*.

[q] Fincke (1982) for a field population of the damselfly *Enallagma hageni* in northern Michigan, USA. See also similar values in Parr (1965) and Lord (1961) for *Ischnura elegans*.

[r] Since the magnitudes of these surely vary across orders of magnitude and are poorly documented, we arbitrarily set $\delta_{SP} = \delta_{SS}$ *and* $\delta_{LP} = \delta_{LL}$ as a baseline. Note that positive δ_{SP} and δ_{LP} imply direct density-dependence; we do not consider inversely density-dependent predation in this analysis.

[s] Chosen to be representative of well synchronized damselfly emergences—a negative exponential distribution in which about 63% of the larvae in the holding bin emerge in just over 10 days after the onset of emergence (see Corbet, 1962).

[t] Consistent with data for a population of *Ischnura elegans* in the Pocklington Canal near York, England (P. H. Crowley, in preparation).

[u] Fincke (1982) for *E. hageni* in northern Michigan; see also Parr (1969b) for *I. elegans*.

[v] From Fincke (1982); adult female *Enallagma hageni* averaged 1·22 bouts of oviposition during an average adult lifetime of 2·1 days. This is 0·58 bouts per day. Female coenagrionid damselflies average about 200 eggs per bout (P. S. Corbet, pers. comm.; see Corbet, 1980). Since damselfly sex ratios at emergence generally approximate 1:1 (see Lawton, 1972; Parr, 1969), this is 58 eggs per adult per day; multiplied by the mean lifetime of 2·1 days, this yields $\varrho = 120$ eggs per adult.

[w] When water temperatures reach 16–20°C, development time for nondiapausing eggs is about 2–3 weeks, with a comparable additional time required to complete the first four instars. Mean development rate over the whole year (i.e., the rate at around 10°C) should be about half this value, yielding a rough estimate of $\eta = 0{\cdot}2$ yr.

[x] Determines the amplitude and frequency of oscillation for the temperature-sensitive parameters that fluctuate seasonally according to the sinusoidal equation (42). When $\beta = 1$, the maximum is twice the mean for these parameters, but growth rate falls to zero whenever t is an integer (i.e., at the beginning of each year). Since the stage durations become extremely long as development rate approaches zero (see equations (8–10)), we instead set $\beta = 0{\cdot}9$ to obtain a similar seasonal trend without numerical problems. So that each seasonal period corresponds to one year, we take $\omega = 2\pi$ radians yr^{-1}.

[y] Chosen arbitrarily.

random movements by larvae determine the frequency of interlarval encounters, and a volume within which random movements by prey determine the frequency of larva–prey encounters. Since most of the experiments on larval feeding use *Daphnia* as prey, we imagine *Daphnia* uniformly distributed through the water column as the archetypal prey for both larval stages. (Differences in distribution between uniformly distributed *Daphnia* and the diverse natural prey assemblage can be partially compensated for by adjusting the effective water depth parameter relative to the actual water depth. This is because prey density in the model is inversely related to effective water depth. If, for example, prey spend most of their time near substrate frequented by damselfly larvae, then they are effectively compressed into a shallower layer around the substrate, increasing the density of prey immediately surrounding the larvae.) But because the natural prey assemblage is sufficiently diverse that any biologically reasonable representation of a particular species would lack generality, we crudely represent this assemblage as a pseudo-population with logistic dynamics: the carrying capacity is estimated from the standing stock of invertebrate biomass for ponds containing damselflies, and the intrinsic growth rate is intermediate among those of the most common damselfly prey. On coarse timescales, the logistic equation provides a qualitatively plausible description of population growth, at least in some stable *Daphnia* populations (Frank *et al.*, 1957), but see Murdoch and McCauley (1985) on the diversity of dynamics observed in natural *Daphnia* populations. We assume a type II functional response (e.g. see Thompson, 1978a) incorporating interference among larvae (Beddington, 1975). This formulation follows from the random movement assumptions, together with assumed constant reductions in the time available to search out prey for the time spent handling each prey captured and interacting with each other larva encountered.

We assume density-independent mortality in the "egg" stage and in both adult stages, but partition the mortality of small, large, and holding-bin larvae into a density-independent term of "background" mortality, including some predation-associated mortality, and three density-dependent terms. We argue, in the absence of empirical evidence for strong shifts in larval behavior with larval density, that encounter frequencies and interference-associated mortality will be roughly proportional to the product of the relevant densities, as implied here. While the density-dependence of predation-associated mortality is potentially rather more complex, we use the same algebraic form for this component as a general, parsimonious, first approximation.

Seasonality is introduced into the model in two ways. First, seasonal changes in water temperature influence several of the model's physiological and behavioral parameters. For example, increasing temperature has been-shown to increase the attack coefficients and decrease the handling times of damselflies (Thompson, 1978a). Analogously, encounter coefficients should increase larval movement, and time wasted per encounter should decrease. As a rough approximation, we represent these seasonal trends by multiplying or dividing, for direct and inverse relationships to temperature respectively, by the sinusoidal function $S_T(t)$ in equation (42). Thus the seasonal trends for products of terms oppositely related to temperature are simply assumed to cancel, such as for the product *ab* in equations (35) and (36) and for the γ component of the interference coefficient in the same equations.

The second way in which seasonality enters the model is in the description of emergence. We introduce the concept of an "emergence window," that part of the year during which emergence is possible. Large larvae that complete development while the emergence window is open emerge immediately (except during an extremely short interval when the window is opening or closing); those maturing at other times enter the holding bin, and the subsequent survivors emerge when the window re-opens.

B. Steady States

1. Static Environment Variant

A normal first step in analyzing any complex model is to examine the equilibrium or "steady state" conditions; but our full model described above, which postulates regular seasonal changes in environment, clearly cannot exhibit states in which all the subpopulations remain strictly constant. However, we might reasonably expect to find such states in a "static environment variant" of the model, in which we assume constant environmental temperature and photoperiod. This requires setting to unity the two seasonal driving functions, S_T and S_W, to be described below.

The steady state abundances predicted by such a caricature may differ markedly from the average abundances predicted by the full model. Nevertheless, a careful analysis of the sensitivity of the steady state behavior of the static-environment variant to changes in parameter values provides, at negligible computational expense, an instructive overview of the relative importance of the various model parameters in determining abundances of various life-history stages, stage durations, and total annual emergence.

2. Default Behavior

We begin by examining the model's behavior with all nonseasonal parameters assigned the values for *Ischnura elegans* set out in Table 3 and hereafter referred to as their "default" values. However, with such a complex model it is unsafe to blindly calculate a particular steady state solution and, instead, we must first check for the existence of multiple steady states. These are not only of interest in their own right (May, 1977), but are indicative of potentially complex dynamics in a seasonal environment. With some algebra it is possible to show that there are circumstances in which, for specified values of the model parameters, there *can* be two sets of steady states. The cause of this is the combination of logistic prey growth and density-dependent larval mortality that permits equilibria with prey levels close to or well below carrying capacity or both (cf. Steele and Henderson, 1981). Fortunately, the region of parameter space permitting multiple steady states is well away from the neighborhood of our default parameter set; to date we have located the other steady states only by artificially setting the effective water depth to 1 mm and assuming excessively large values of carrying capacity.

With the default parameter set of Table 3, the only nontrivial steady state has:

prey density = 18 *Daphnia* per liter
small larval density = 76 larvae per m^2 bottom
large laval density = 10 larvae per m^2 bottom
small larval stage duration = 0·55 yr
large larval stage duration = 0·35 yr
through-stage survival (small larvae) = 0·12
through-stage survival (large larvae) = 0·48
annual emergence = 19 damselflies per yr per m^2 bottom

Since the default parameter set has a carrying capacity of 20 *Daphnia* per liter, we note that the damselflies only cause a small depression of the prey and that the larval densities, stage durations, and annual emergence are within the ranges observed in the field (see below). We further note that with the above stage durations, the through-stage survival probabilities would increase sharply to 0·61 (small larvae) and 0·73 (large larvae) if the density-dependent mortality terms were absent, one indication of the impact of density-dependent mortality.

3. Dependence on Prey Dynamics

One of the more egregious oversimplifications in our model is the representation of a heterogeneous prey assemblage by a logistic pseudo-population, having parameters r (the instantaneous *per capita* growth rate of the prey assemblage) and K (the carrying capacity of the prey assemblage). Without this or some similar assumption, any modeling of a generalist predator would be a hopeless task; nevertheless, it is not clear *a priori* that the simplification is reasonable, nor is it certain that our *Daphnia*-based parameters are appropriate. Furthermore, even if this model structure is adequate, we would expect habitats to differ in effective carrying capacity by as much as two or three orders of magnitude.

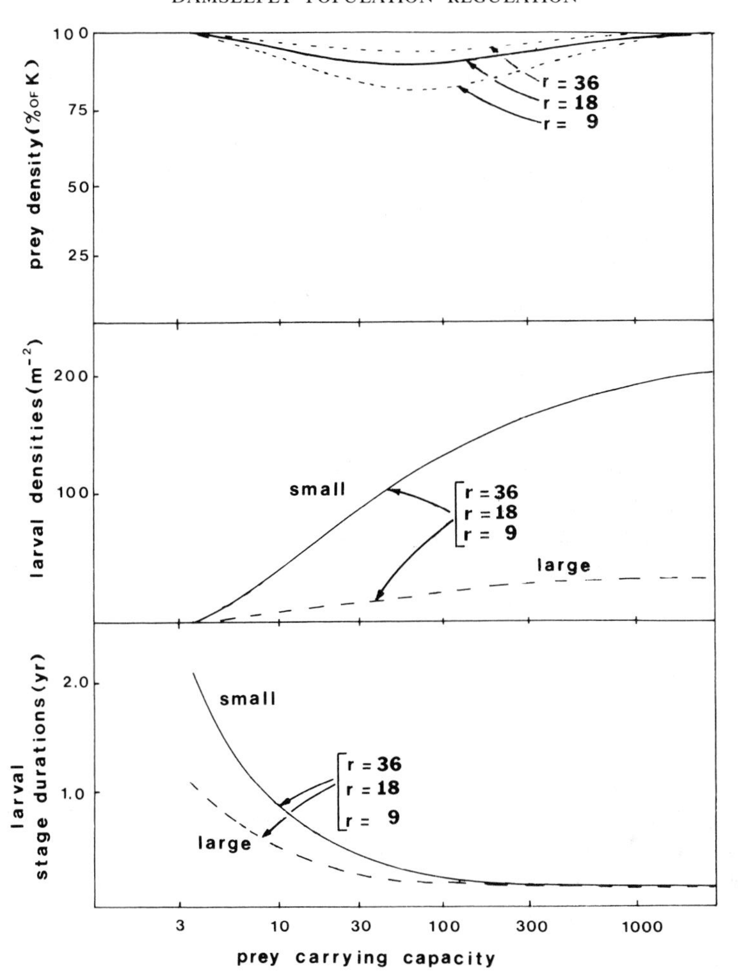

Fig. 2. Steady states for the default parameter set: prey density, larval densities and larval stage durations for three prey *per capita* growth rates, and across a range of prey carrying capacities. Note that the relations of both larval density and stage duration with prey carrying capacity are essentially unaffected by prey *per capita* growth rates in the relevant range.

We have therefore studied the steady states with all parameters other than the carrying capacity K assigned their default values from Table 3, but with a wide range of K values. The results are plotted in Fig. 2, from which we see that:

(a) there is a threshold value of K (3 *Daphnia* l^{-1}) necessary to support a larval population;

(b) prey depression is never large (maximum around 10%);
(c) at high K values, the larval populations saturate, and the prey density approaches carrying capacity;
(d) there is considerable variation in stage durations at low food levels, but for high carrying capacities the development times are essentially constant.

Some laborious algebra (available from R. M. Nisbet) indicates that the threshold carrying capacity for the support of a damselfly population depends on the larval life-history parameters that are effective at low densities, but not on r, the intrinsic growth rate of the prey. It can also be shown that the saturation values of the larval densities are set primarily by the density-dependent mortality terms. Since there was considerable arbitrariness in the parameter values selected for these terms, it is reassuring that we have arrived at a span of possible larval densities covering the same orders of magnitude as observed for peak larval populations in a variety of habitats (ones to thousands per m^2: Uttley, 1980).

We have investigated the effect of r on densities away from the threshold and again find that it is small. The effects of halving and doubling the value of r are shown in Fig. 2 (dotted lines); the maximal prey depression is increased to 22%, but the effects on larval populations or stage durations are scarcely perceptible.

4. Larval Feeding and Development

Table 4 shows the sensitivity of the steady states to changes in the parameters related to feeding in two different model environments, $K = 20$, the default value, and $K = 200$, where interlarval interference is strongly affecting the steady states. In essence, we measure the *resistance*, in the sense of Pimm (1984), to changes in the parameters.

Probably the most important, unequivocal conclusion to emerge from the calculations is that the feeding interference is of no significance whatever in setting the equilibrium densities. This follows from the manner in which the interference terms enter the functional response equations (35 and 36): feeding rates can only be influenced by interference at larval densities greater than the reciprocal of the largest feeding interference coefficient γ, i.e., around 10 000 larvae per m^2, well in excess of the saturation values shown in Fig. 2. This conclusion does not depend upon whether Beddington's (1975) or Rogers and Hassell's (1974) interference equations are used to derive equations (35) and (36).

The other feeding-related parameters affect the steady states in different ways, depending on the value of K. With $K = 20$, the larval populations have lower resistance to changes in attack rates than to changes in handling times. However, with $K = 200$, which produces prey densities well above the half-saturation constants for feeding, 115 and 57 *Daphnia* per liter for

Table 4

Percentage shifts in steady state magnitudes of six variables, from the magnitudes observed in the standard or "default" run, in response to alterations in feeding-related parameters of the model.[a]

		Variables					
Parameter(s)	Alteration	H, prey density	N_S, density of small larvae	N_L, density of large larvae	τ_S, duration of small-larval stage	τ_L, duration of large-larval stage	E, total annual emergence
Attack coefficients	$a_S, a_L \times 3$	−24(−4)	+50(+16)	+76(+13)	−37(−10)	−26(−12)	+135(+29)
	$a_S, a_L \div 3$	+8(+4)	−69(−20)	−73(−25)	+117(+30)	+88(+22)	−86(−38)
Handling times	$b_S, b_L \times 3$	+4(+6)	−20(−28)	−36(−54)	+40(+101)	+13(+33)	−42(−62)
	$b_S, b_L \div 3$	−3(−12)	+6(+21)	+17(+45)	−12(−30)	−4(−18)	+20(+70)
Feeding interference coefficients	$\gamma_{SS}, \gamma_{SL}, \gamma_{LS}, \gamma_{LL} \times 3$	0(0)	0(0)	−1(0)	+1(0)	0(0)	−1(−1)
	$\gamma_{SS}, \gamma_{SL}, \gamma_{LS}, \gamma_{LL} \div 3$	0(0)	0(0)	0(0)	(0)	0(0)	0(0)

[a] Parenthetical percentages were obtained with a carrying capacity (K) for the prey population of 200 prey per liter, in contrast to the nonparenthetical percentages obtained with the "default" carrying capacity of 20 prey per liter.

large and small larvae respectively, the situation is reversed: the larval densities are insensitive to changes in the attack coefficients but responsive to shifts in handling time.

Feeding, of course, influences the larval dynamics *indirectly* through the dependence of development rate on feeding rate (equations (40) and (41)), and we therefore studied the sensitivity of the steady states to changes in the parameters of the development function (Table 5). The most important parameters here are the critical development indices (i_S and i_L), which scale the stage durations for small and large larvae relative to the duration of instar 11 in *I. elegans*, from which the other parameters were obtained (Lawton *et al.*, 1980). The parameters f_{S0} and f_{L0}, which set the point at which development becomes less sensitive to feeding rate, are clearly also of importance, especially at low prey carrying capacities.

5. Mortality in Aquatic Stages

The effects of selected changes in the 11 parameters associated with larval mortality are summarized in Table 6. Our main aim is to assess the importance of the various density effects; to this end we have studied the sensitivity of the steady states to the density-independent mortality coefficients, and to those parameters related to predation and interference.

A population in the low-food environment ($K = 20$) is clearly vulnerable to large shifts in baseline mortality: a threefold increase in either the aquatic or the terrestrial components increases the threshold prey level above 20 and produces extinction. With $K = 200$, a situation where the larval populations are higher and density effects stronger, the population can resist the same shifts in the background larval death rates, but still cannot withstand a threefold increase in post-emergence mortality.

In view of our arbitrary assignment of equal values to the interference mortality and predation parameters δ_{SS} and δ_{SP} (also δ_{LL} and δ_{LP}), it is perhaps initially surprising that the larval populations are significantly more sensitive to interference mortality than to density-dependent predation. The greater sensitivity to interference clearly must result from the between-class interference mortality terms δ_{LS} and δ_{SL}, a deduction that is confirmed by setting these terms to zero and observing that the density of emerging adults is thereby increased by almost an order of magnitude.

6. Terrestrial Stages

The effect of adult female fecundity is embodied in two of the model parameters, the mean egg production per lifetime (ϱ), and the mature adult *per capita* death rate (Δ_A). Table 7 shows that although changes in ϱ significantly affect the density of small larvae, this change is damped as development proceeds, and the end effect on annual emergence is relatively small. A similar phenomenon accompanies a reduction in adult death rates,

Table 5

Percentage shifts in steady state magnitudes of six variables, from the magnitudes observed in the standard or "default" run, in response to alterations in development-related parameters of the model.[a]

		Variables					
Parameter(s)	Alteration	H, prey density	N_S, density of small larvae	N_L, density of large larvae	τ_S, duration of small-larval stage	τ_L, duration of large-larval stage	E, total annual emergence
"Egg"-stage duration	$\eta \times 3$	+4(+3)	−55(−52)	−29(−26)	−3(−1)	−2(−1)	−21(−18)
	$\eta \div 3$	−2(−2)	+26(+25)	+7(+8)	+1(0)	+1(0)	+4(+3)
Development indices at larval maturation	$i_S, i_L \times 2$	+6(+4)	−69(−58)	−67(−57)	+92(+98)	+94(+99)	−84(−79)
	$i_S, i_L \div 2$	−11(−10)	+127(+113)	+117(+111)	−46(−49)	−47(−49)	+328(+323)
Characteristic larval feeding rates	$f_{S0}, f_{L0} \times 3$	+6(+2)	−64(−7)	−72(−50)	+119(+92)	+68(+8)	−84(−48)
	$f_{S0}, f_{L0} \div 3$	−6(−2)	+31(−8)	+126(+40)	−52(−32)	−21(0)	+155(+25)
Immature adult stage duration	$\tau_I \times 3$	+9(+8)	−100(−100)	−100(−100)	Damselfly extinction		−100(−100)
	$\tau_I \div 3$	−9(−7)	+145(+143)	+24(+28)	+7(+2)	+5(+1)	+3(+7)

[a] Parenthetical percentages were obtained with a carrying capacity (K) for the prey population of 200 prey per liter, in contrast to the nonparenthetical percentages obtained with the "default" carrying capacity of 20 prey per liter.

Table 6

Percentage shifts in steady state magnitudes of six variables, from the magnitudes observed in the standard or "default" run, in response to alterations in aquatic-stage mortality parameters of the model.[a]

Parameter(s)	Alteration	Variables					
		H, prey density	N_S, density of small larvae	N_L, density of large larvae	τ_S, duration of small-larval stage	τ_L, duration of large-larval stage	E, total annual emergence
Predation (all)	$\delta_S, \delta_L \times 2$ *and* $\delta_{SP}, \delta_{LP} \times 3$	+4(+3)	−51(−37)	−48(−32)	−3(−1)	−2(−1)	−54(−43)
	$\delta_S, \delta_L \div 2$ *and* $\delta_{SP}, \delta_{LP}, = 0$	−3(−2)	+40(+30)	+32(+23)	+2(0)	+2(0)	+47(+41)
Predation (density dependent)	$\delta_{SP}, \delta_{LP} \times 3$	+2(+2)	−21(−25)	−18(−20)	−1(0)	−1(0)	−26(−32)
	$\delta_{SP}, \delta_{LP} = 0$	−1(−2)	+16(+20)	+12(+15)	+1(0)	+1(0)	+20(+28)
Interference mortality	$\delta_{SS}, \delta_{SL}, \delta_{LS}, \delta_{LL} \times 3$	+6(+4)	−62(−62)	−62(−63)	−4(−1)	−2(−1)	−59(−60)
	$\delta_{SS}, \delta_{SL}, \delta_{LS}, \delta_{LL}, \div 3$	−10(−11)	+115(+119)	+119(+133)	+8(+2)	+6(+2)	+87(+94)
	$\delta_{SL}, \delta_{LS} = 0$	−20(−22)	+199(+175)	+257(+282)	+18(+5)	+13(+5)	+78(+49)
"Background" mortality (aquatic stages)	$\delta_S, \delta_L, \Delta_E \times 3$	+9(+5)	−100(−73)	−100(−57)	—[b](−1)	—[b](−1)	−100(−56)
	$\delta_S, \delta_L, \Delta_E \div 3$	−4(−2)	+57(+38)	+30(+17)	+3(0)	+2(0)	+32(+16)

[a] Parenthetical percentages were obtained with a carrying capacity (K) for the prey population of 200 prey per liter, in contrast to the nonparenthetical percentages obtained with the "default" carrying capacity of 20 prey per liter.
[b] Undefined because of damselfly extinction.

Table 7

Percentage shifts in steady state magnitudes of six variables, from the magnitudes observed in the standard or "default" run, in response to alterations in terrestrial-stage and environmental parameters of the model.[a]

Parameter(s)	Alteration	Variables					
		H, prey density	N_S, density of small larvae	N_L, density of large larvae	τ_S, duration of small-larval stage	τ_L, duration of large-larval stage	E, total annual emergence
Oviposition per mature adult	$\varrho \times 3$	−10(−7)	+156(+155)	+24(+30)	+8(+2)	+6(+2)	+2(+6)
	$\varrho \div 3$	+6(+4)	−73(−69)	−47(−41)	−4(−1)	−3(−1)	−39(−31)
Mortality (terrestrial stages)	$\Delta_I, \Delta_A \times 3$	+9(+8)	−100(−100)	−100(−100)	Damselfly extinction		−100(−100)
	$\Delta_I, \Delta_A \div 3$	−9(−7)	+145(+143)	+24(+28)	+7(+2)	+5(+2)	+3(+7)
Water depth	$w \times 2$	+4(+4)	+3(+1)	+3(+1)	−3(−1)	−2(−1)	+6(+1)
	$w \div 2$	−8(−9)	−6(−2)	−7(−1)	+9(+2)	+4(+1)	−11(−3)
	$w \div 5$	−27(−40)	−22(−10)	−26(−11)	+24(+11)	+18(+9)	−37(−18)
Temperature index[b]	$S_T \times 1{\cdot}5$	+3(+2)	−16(−13)	−18(−12)	−9(−17)	−4(−10)	−14(−3)
	$S_T \times 0{\cdot}7$	−4(−2)	+6(+5)	+10(+4)	+12(+22)	+5(+9)	+5(−4)

[a] Parenthetical percentages were obtained with a carrying capacity (K) for the prey population of 200 prey per liter, in contrast to the nonparenthetical percentages obtained with the "default" carrying capacity of 20 prey per liter.

[b] S_T is defined in Table 2.

where a threefold decrease, which produces a threefold increase in total oviposition and a reduction in mortality in the immature adult stage, increases annual emergence by less than 10%.

Immature adult mortality can eventually become significant for population persistence if the parameter Δ_I, representing total losses during and after emergence up to reproductive maturity, becomes too large; this is demonstrated in Table 7, covering changes in *per capita* death rates, and in Table 5, where a similar result is achieved by varying the duration of this stage.

7. Environmental Parameters

The interference area index (λ; see Table 3), which expresses the area of vegetation available to larvae per unit area of substrate, significantly influences steady state larval abundances. This can be seen from the sensitivity of larval densities to alterations of the mortality interference coefficients (δ_{SS}, δ_{SL}, δ_{LS}, δ_{LL}: see Table 6), which are inversely related to the interference area index. Thus greater abundances of larvae would generally be expected in denser vegetation via reductions in the intensity of interference mortality.

One of the less precisely defined parameters in the model is the "effective water depth", w, whose exact value is in part determined by the extent to which the prey assemblage resembles a homogeneous population of *Daphnia*. It is thus reassuring to discover (Table 7) that the sensitivity of the densities, stage durations, and annual emergence to this parameter is very small unless this depth is very small. For example, with an effective depth of only 20 cm, the damselflies cause significant depletion of their prey, with a consequent increase in stage durations. Though this comparison may be reasonable in some cases, as for example when w is reduced for an assemblage of active prey that remains in close association with vegetation containing the damselfly larvae, in others it may be more reasonable to vary w and λ in concert. If reducing water depth implies proportionally less vegetation or other substrate per unit area of bottom, then the interference–area–index effect noted above tends to dominate, and the resulting lower larval densities are insufficient to deplete prey very extensively.

The final environmental quantity represented in the model is the water temperature, having an effect represented by the variable $S_T(t)$. By arbitrarily setting its average value to unity in the steady state calculations, we have assumed a temperature equal to that (10°C) for which the default parameter set was calculated. To assess the general implications of temperature changes, we recognized that population parameters are likely to be linearly related to temperature only over a rather small dynamic range, and we thus calculated (in Table 7) the sensitivity to rather small changes in temperature. The most important conclusion is that, as one would expect, stage durations are significantly affected by temperature.

C. Dynamics

We have already indicated the limitations of steady state analyses in this particular model. Although the calculations reported in the previous section shed considerable light on the significance for population regulation of the various parameters in our model, they all relate to a hypothetical stationary population in a static environment. Further progress requires us to consider numerical solutions of the delay-differential equations. These have been calculated using a predictor–corrector method (see e.g., Wylie, 1975) implemented in the program template SOLVER (Mass *et al.*, 1984); the program is written in PASCAL and designed for the UCSD p-System, which is available for many microcomputers.

Because we have assumed that adults have short life-spans and that maturation from larval stages occurs at a fixed value of development index, our model has the potential to exhibit irregular fluctuations over relatively long periods. Transient emergence bursts and declines are poorly damped by the relatively synchronous oviposition and undamped by the subsequent lock-step development of eggs and larvae, often persisting for tens of years before yielding to a steady pattern. We believe the extent of this transient behavior to be a biologically uninteresting artifact of our assumption of fixed stage end-points, rather than an indicator of low resilience (in the sense of Pimm, 1984) in natural populations, which would exhibit some variability in stage end-points. In the absence of a defensible representation in our model of this natural end-point variability, we have based our analysis and interpretation exclusively on the model's long-term dynamic behavior.

1. Stationary Environment

In Fig. 3 we display the effect of a stationary environment on our model's behavior after all transients have elapsed. All nonseasonal parameters are assigned their default (Table 3) values, with temperature held constant and emergence window held permanently open. Several points deserve comment:

(a) The steady state is clearly *unstable*; the population is executing large-amplitude limit cycles with a period almost exactly equal to the equilibrium value of the average generation time (1·1 yr).

(b) As predicted by the steady state analysis, the prey abundance is depressed only a little below saturation; its mean value (18·0 l^{-1}) is identical with the steady state.

(c) The mean abundances for both small and large larvae (99·5 m^{-2} and 11·3 m^{-2} respectively) are increased above their steady state values, because the large cycles allow the two age classes to "time-share" the environment and hence avoid the heavy interference mortality which occurs when both classes are present simultaneously.

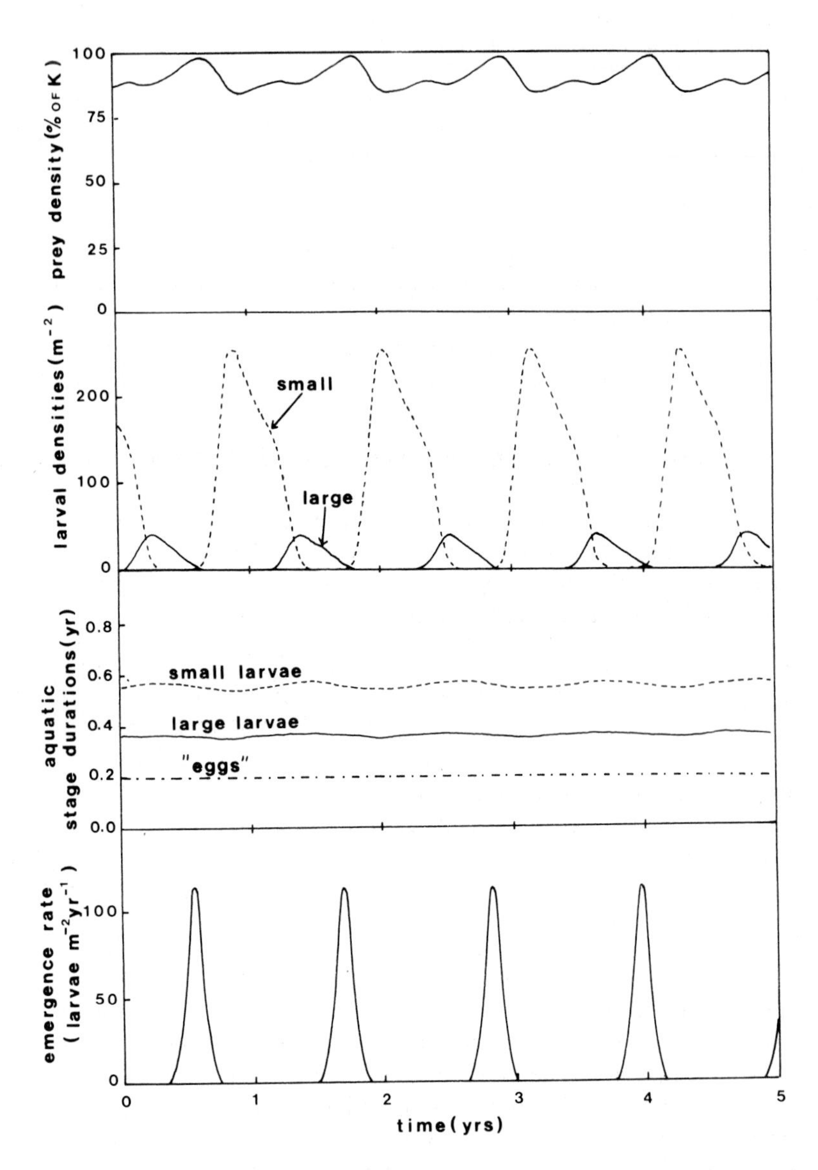

Fig. 3. Dynamic bahavior of the static environment variant of the default model: prey density, larval densities, aquatic stage durations, and emergence rate across 5 years.

(d) There is very little variation in stage durations through the cycles, and their mean values (0·56 yr for larges and 0·36 yr for smalls) are very close to the steady state.

The model is thus exhibiting the "single-generation cycles" that have been shown to occur in simpler models with larval competition and a short adult phase (Nisbet and Gurney, 1984; Gurney and Nisbet, 1985).

2. Seasonality

As noted earlier in the text, we have included representations of two aspects of seasonality in the model: temperature dependence of many damselfly attributes through sinusoidal variation of certain parameters, and seasonal constraints on emergence through the opening and closing of the emergence window. The long-term dynamics when both are included are summarized in Fig. 4, hereafter referred to as the "default run", which shows:

(a) significant depression of both the peak and mean larval populations from their values in Fig. 3;
(b) a pattern of annual emergences which, while formally irregular, varies from year to year to an extent that would be empirically undetectable;
(c) a characteristic bimodality in emergence pattern; and
(d) substantial cyclic fluctuations in the stage durations.

In order to elucidate the mechanisms responsible for these phenomena, we performed two further runs: one with strong sinusoidal variation in parameter values, but with the window fully open at all times (Fig. 5); and the other with constant parameters, but with the window constraining the timing of emergences (Fig. 6). *A priori*, we might have expected that sinusoidal parameter variation alone would be sufficient to cause the single generation cycles of Fig. 3 to synchronize to some subharmonic of the forcing frequency (as occurs in some other models with time delays: Nisbet and Gurney 1976, 1982 (Ch. 2); Stokes, 1985). That this does not happen with our default parameter set is obvious from Fig. 5, which also illustrates (by having a burst of adults emerging in mid-winter!) the need for some seasonal constraint on emergence.

The effect of the emergence window is to break the "default" population into three out-of-phase subpopulations, each of which has essentially discrete generations of adults. The successive generations of any one of the subpopulations appear in a sequence with a repeat period of three years and alternating short (just over one year) and long (just under two years) generations, as follows:

(1) *Short generation*. Eggs laid by adults emerging *early* during the window in *year 0* reach maturity *late* during the window in *year 1*, emerge, and lay eggs.

(2) *Long generation*. Eggs laid by adults emerging *late* during the window in *year 1* reach maturity after the window in year 2 is closed, are forced to over-winter in the holding bin, and emerge as soon as the window opens (*early*) in *year 3*.

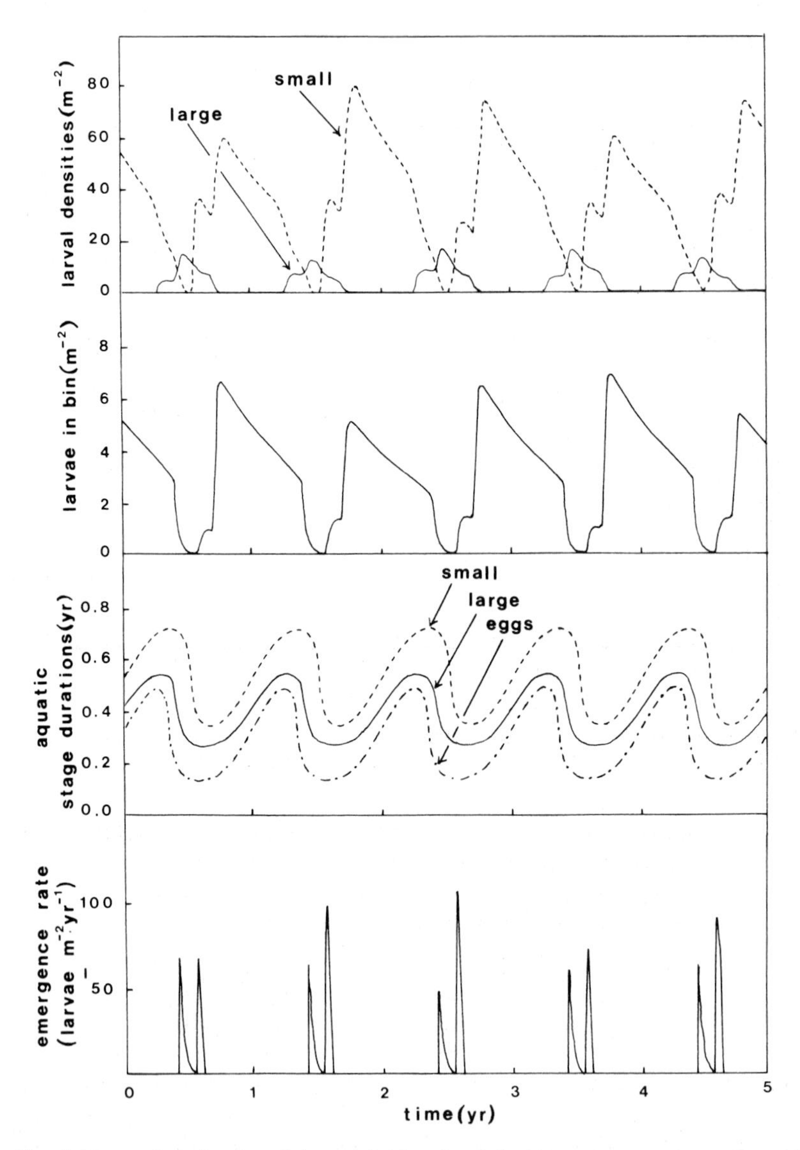

Fig. 4. Dynamic behavior of the model for the default parameter set: small and large larval densities, bin larval density, aquatic stage durations, and emergence rate across 5 years.

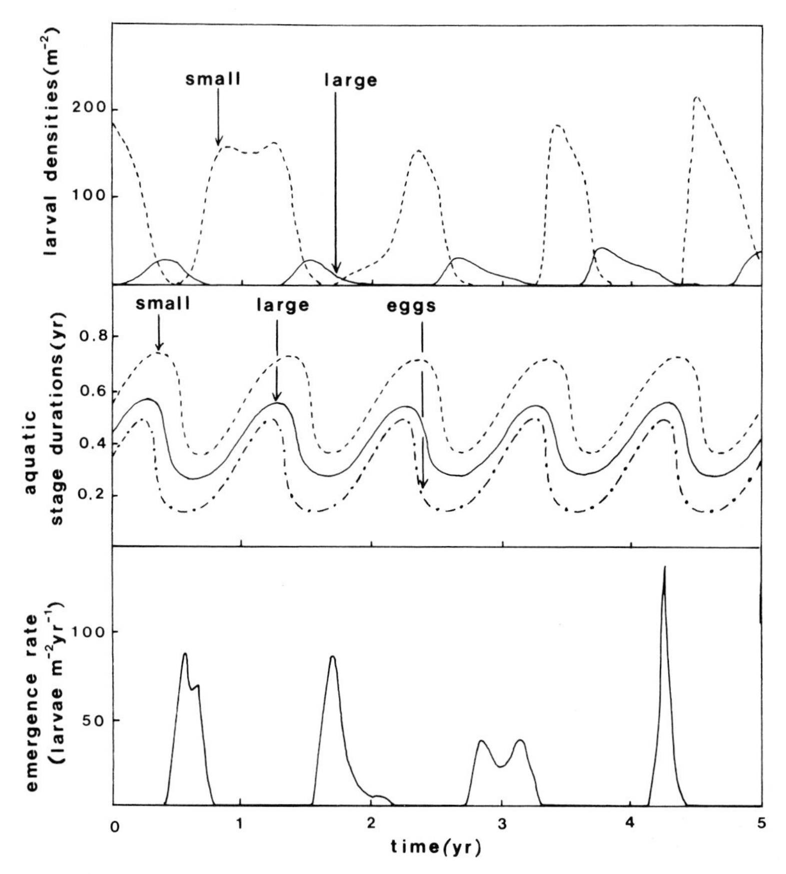

Fig. 5. Dynamic behavior of the model with no seasonal restriction of emergence: larval densities, aquatic stage durations, and emergence rate across 5 years.

The emergence patterns of each of the three separate subpopulations then merge into the overall emergence pattern of Fig. 4, as follows:

Year	Early emergence peak	Late emergence peak
0	Subpopulation 1	Subpopulation 2
1	Subpopulation 3	Subpopulation 1
2	Subpopulation 2	Subpopulation 3

In the default run, these three subpopulations coexist with, on average, approximately equal densities, the seasonally driven variations in development time allowing the necessary flexibility in phase. With constant parameters (Fig. 6), the relative magnitudes of the subpopulations are no longer comparable, and one eventually dominates the other two.

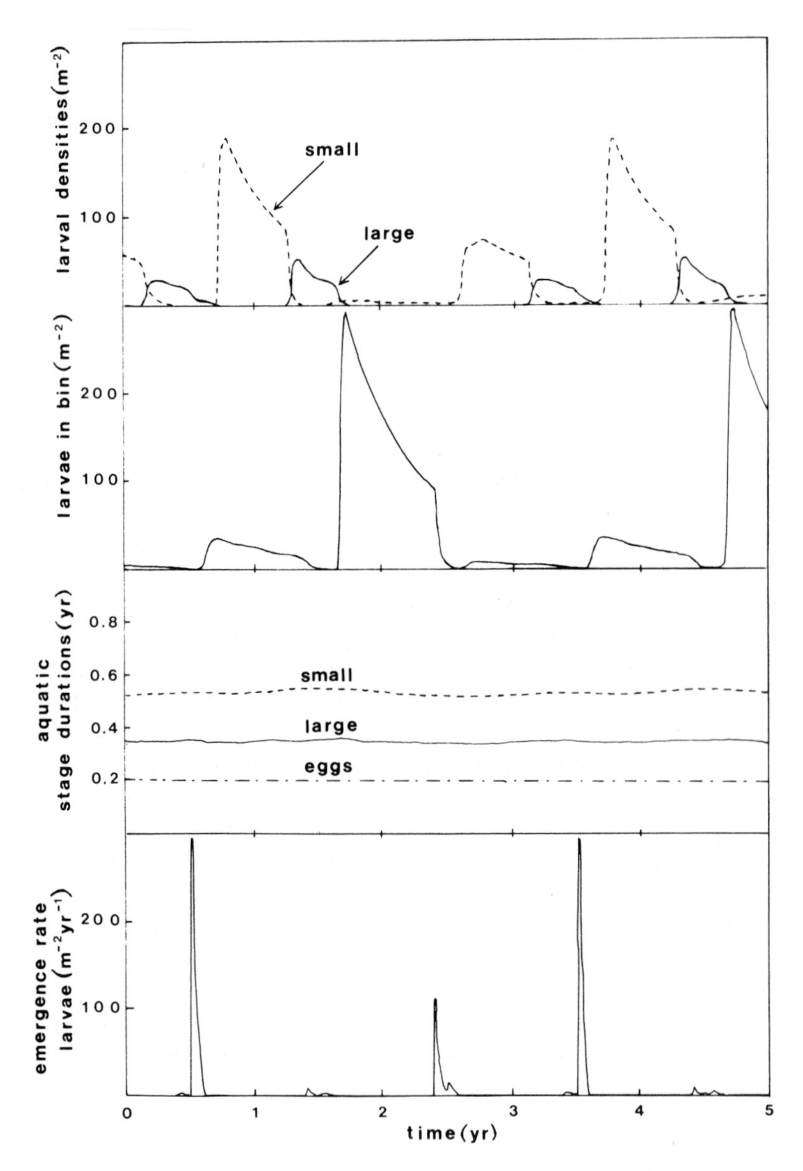

Fig. 6. Dynamic behavior of the model with no seasonal oscillations of aquatic-stage parameters: small and large larval densities, bin larval density, aquatic stage durations, and emergence rate across 5 years.

3. *Prey Dynamics*

Our investigation of steady states indicated that the intrinsic growth rate (r) of the prey population plays very little part in regulating larval densities, but that the carrying capacity (K) is significant in determining both abundances and stage durations. Intrinsic prey growth rate has just as little influence on the dynamic patterns in the default system; a run with $r = 9\ \mathrm{yr}^{-1}$ (not illustrated) produced results scarcely distinguishable from those of Fig. 4, except that the year-to-year emergences were even more regular.

The importance of prey carrying capacity to the dynamics of the model was evaluated by performing a run (not illustrated) identical to the default run in all respects, except with $K = 200$ *Daphnia* per liter. This parameter set produces a straightforward univoltine life-history, with a sharply unimodal early emergence pattern. We already know that the steady state generation time is thereby reduced to less than one year, and this change is reflected in the dynamics. Emergence begins as soon as the window opens, with egg production shortly thereafter, and the subsequent completion of larval development early the following year. These mature larvae then enter the holding bin to await the cue for emergence.

4. *Larval Mortality*

The steady state analysis pointed to density-dependent larval mortality, deriving from both predation and intraspecific interference, as playing a vital role in population regulation (cf. Table 6). We therefore performed two simulations with the aim of determining whether these phenomena have a similarly dramatic effect on the dynamic pattern. Figure 7 illustrates the long-term behavior in a run where all parameters had their default values, except that there was *no predation* on the larval populations. Clearly the default pattern of essentially constant year-to-year emergences is broken; the three subpopulations evident in Fig. 6 reappear, with one of these subpopulations clearly dominating the other two.

Figure 8 contains results from a run in which all parameters had their default values, but there was no *cross-class* interference mortality, because δ_{SL} and δ_{LS} are set to zero. A robust, stable pattern of bimodal yearly emergences is established, with the generation time of an individual always exceeding one year. Furthermore, the mean densities are much greater than in the default system, a result consistent with the expectations based on Table 6.

5. *Environmental Parameters*

To assess the effects of variation in the *mean* temperature, as opposed to fluctuations about the default mean, we performed two simulations in which the mean values of the function $S_T(t)$ were 0·7 and 1·5. The latter change

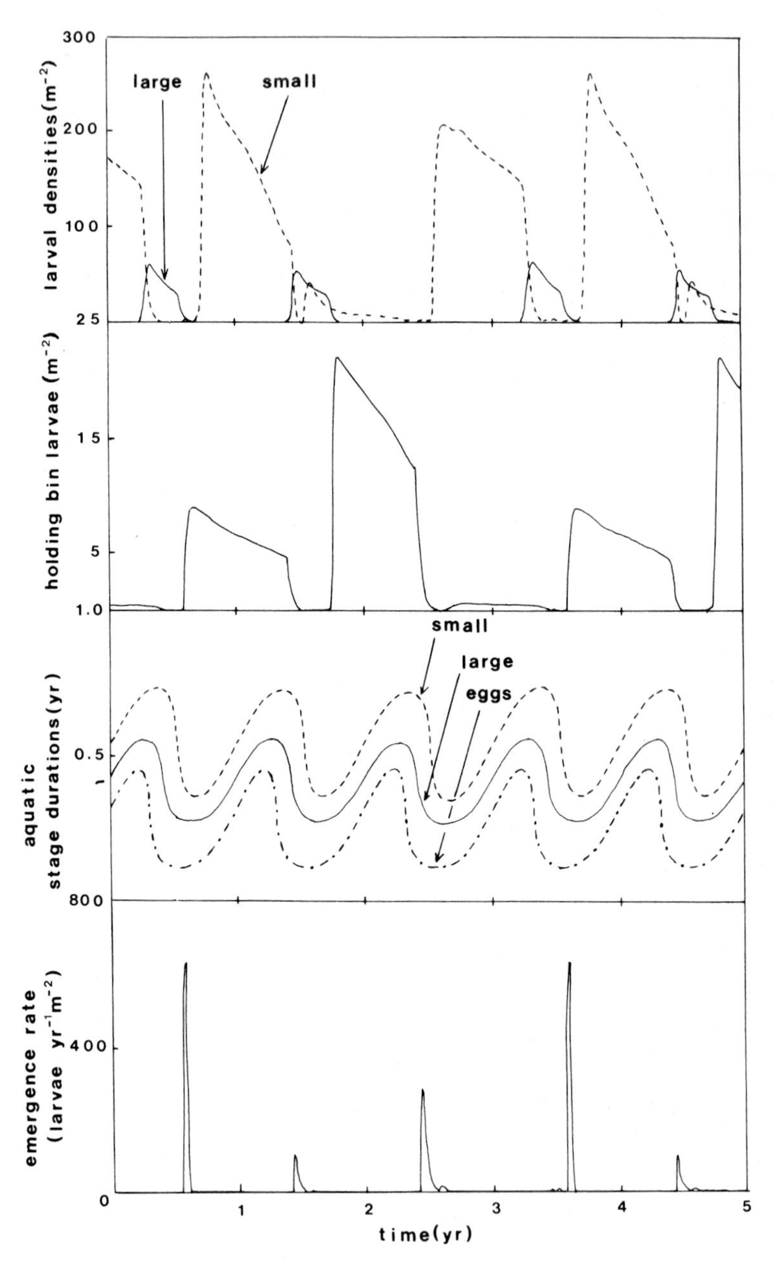

Fig. 7. Dynamic behavior of the model without predation on larvae: small and large larval densities, bin larval density, aquatic stage durations, and emergence rate across 5 years.

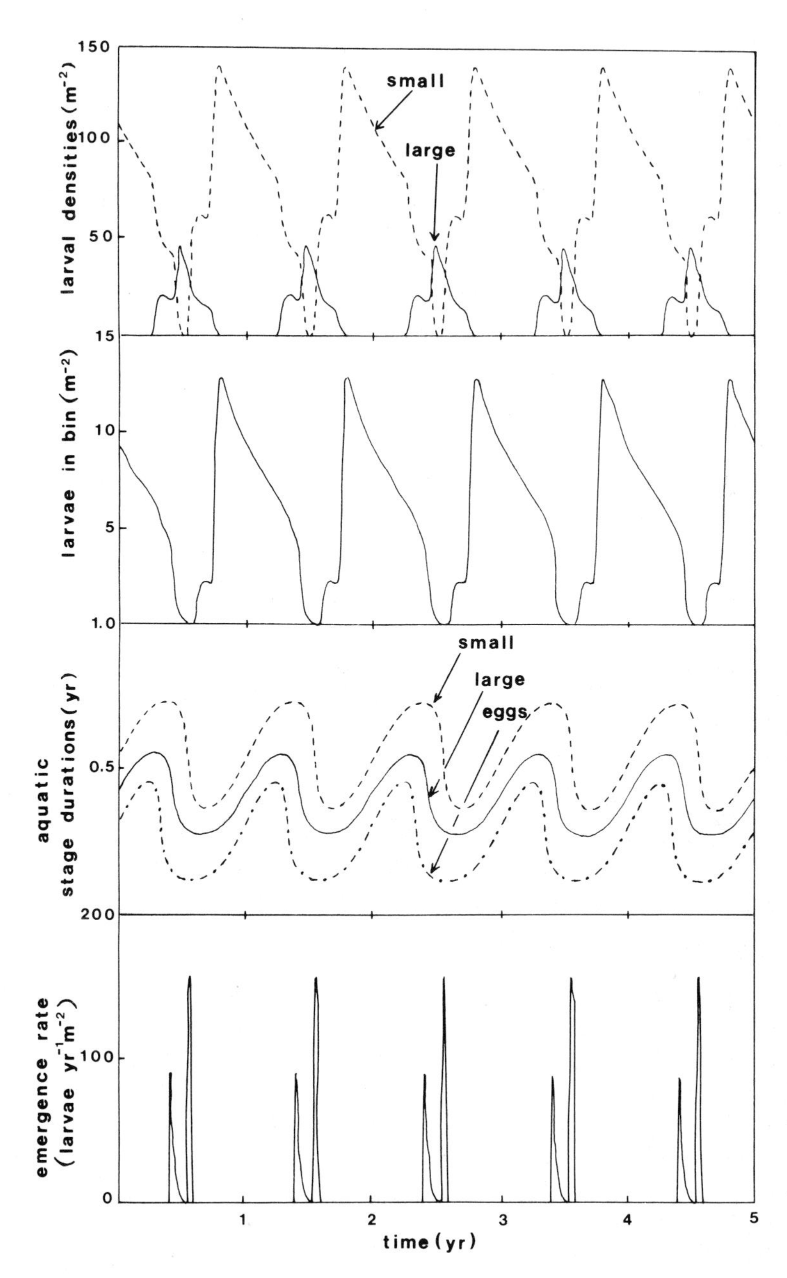

Fig. 8. Dynamic behavior of the model without asymmetric interference mortality of larvae: small and large larval densities, bin larval density, aquatic stage durations, and emergence rate across 5 years.

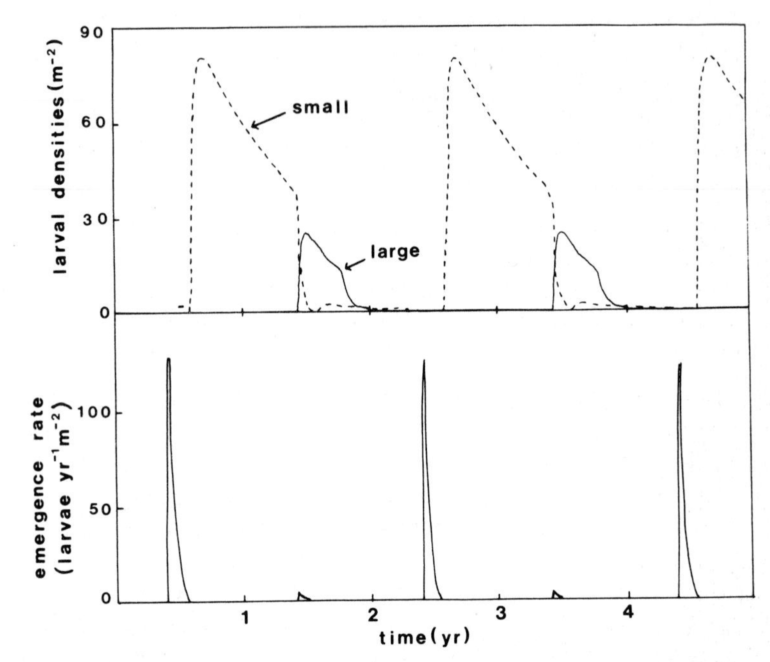

Fig. 9. Dynamic behavior of the model with modified ("high latitude") seasonal oscillations and mean values of some aquatic-stage parameters: larval densities and emergence rate across 5 years.

produced the stable pattern of unimodal year-to-year emergences expected for a life-cycle shorter than one year. The case where the mean value of S_T drops to 0·7 is illustrated in Fig. 9; there is a strong 2-year periodicity evident, with very low emergences in the "off" years.

6. *Coexistence of Subpopulations*

When the average life-cycle of an individual is less than a year, the model's dynamics appears to be relatively straightforward (but see Kot and Schaffer (1984) on the dynamic implications of multivoltinism, which can arise with life-cycles much shorter than one year). When life-cycles exceed one year, the key to disentangling the more complex underlying pattern is the identification of essentially discrete subpopulations. These subpopulations may coexist at comparable densities (as in Figs 4 and 8), or one may dominate the other(s) (as in Figs 6, 7, and 9).

In this situation, interference between large and small larvae becomes analogous to "interspecific" competition between subpopulations, while the remaining interference mortality terms and the density-dependent predation terms are analogous to "intraspecific" competition. There is an extensive literature on discrete models of both interspecific (e.g., Haussman, 1974; Hassell, 1976; Comins and Hassell, 1976; Hassell and Comins, 1976; Fisher and Goh, 1977; Gilpin *et al.*, 1982) and inter-age-class (e.g., May *et al.*, 1974; Hoppensteadt, 1982, pp. 12–16) competition; from these we note the following results, which appear to be independent of the precise number of competing species or cohorts:

(a) Stable coexistence requires that intraspecific interactions be more intense than interspecific interactions.
(b) An unstable community of competitors may persist when exposed to certain types and intensities of predation (predator-mediated coexistence).

By chance, our default parameter set has yielded three subpopulations perched at the margin of stable coexistence. This fragile *ménage à trois* is sustained in part by predation, since removal of predators destroys it (Fig. 7), but it is jeopardized by cross-class competition (cf. Fig. 8).

We conclude that the timing and duration of emergence largely determine the number of subpopulations into which a damselfly population subdivides. The relative intensities of (1) interference *within* and density-dependent predation *on* subpopulations and (2) interference *between* subpopulations then determine the relative densities, including the possibility that one or more subpopulations may be completely eliminated.

V. DISCUSSION

Undoubtedly one of the most interesting features of the present model is that by simplifying the life-history of damselflies to six stages, and by estimating parameter values from a wide variety of sources (Table 3), we generate predictions about population behavior, larval densities, and adult emergence patterns similar to those seen in real damselfly populations.

Under a wide range of environmental conditions, larval predation depresses prey populations by relatively small amounts, typically 10% to 20%. Such levels of depression will be extremely difficult to detect in field experiments. Jeffries (1984) performed a series of experiments with different densities of final and penultimate instars of *Ischnura elegans*, confined at realistic densities in field enclosures. He was unable to detect statistically significant reductions in density in prey populations, except in one experiment where chironomid larvae may have been slightly depleted (see also Johnson *et al.*, 1987).

The small prey depressions indicated by our model also imply that manipulations of larval density should produce no detectable alteration in larval growth rates or development times. However, for cool, unproductive ecosystems (i.e., low $\bar{S}_T$ and low r) such as those studied by Macan (1966, 1974), the model does predict prey depression that is apparently sufficient to slow development (e.g., note the trend with decreasing r in Fig. 2(a)). This implies that, under these circumstances, development time should indeed decrease at lower larval densities, as Macan observed. It is nonetheless important to realize that the generally weak prey depression does not exempt the larvae from food limitation: the model clearly predicts that an increase in prey density should produce a marked increase both in larval growth rates and ultimately in larval densities (see Section VI).

Another interesting prediction to emerge from our model is that we would generally expect greater abundances of larvae in dense vegetation, because mortality rates from larval interference are reduced when there is more habitat surface area. Systematic field data on this point are rare, but it is interesting to note that Macan (1974) found major long-term reductions in large larvae of both *Enallagma* and *Pyrrhosoma* in Hodson's Tarn, as weed beds deteriorated after 1966. Macan attributed the decline in damselfly numbers to "greater exposure to predation". Our model suggests that high densities of larvae on reduced areas of weed lead to greater larval interference, and hence to higher risks of predation.

One other feature of our model also appears to mirror particularly well what actually happens in the field. *Ischnura elegans*, the species that provides most of the parameter estimates, has a life-history that varies with latitude. It is semivoltine in the north of its range in the Outer Hebrides (Parr, 1969a,b) and multivoltine in southern France (Aguesse, 1961). In Lancashire and Yorkshire (central England), populations contain mixtures of univoltine and semivoltine individuals (Parr, 1970; Thompson, 1977). We find it encouraging that our models span the full range of these possibilities, using parameter values appropriate to warmer or cooler climates respectively. Moreover, Parr's (1970) interesting hypothesis that semivoltine and univoltine cohorts in *Ischnura, Coenagrion*, and other odonates may be the product of a balanced genetic polymorphism is shown to be unnecessary by our model. Mixed voltinism is easily generated when odonate larvae experience food limitation in a cool-temperate seasonal climate.

The details of emergence curves generated by our model certainly differ from those seen in real damselfly populations. In central England, for example, *Ischnura elegans* populations have a sharp early emergence and a lower, later emergence, in contrast to the curves of Fig. 4. Such discrepancies are unsurprising, in light of the simplifying assumptions from which the model is derived; nor do they appear to undercut the qualitative conclusions drawn in the present analysis.

Climates in cool, temperate regions are, of course, rather unpredictable from year to year, while our model has a completely deterministic seasonal cycle. Two points are worth making. First, density-independent effects on adult reproductive performance, such as cold, wet summers that reduce egg inputs, should have relatively little impact on subsequent adult numbers. Second, though it is tempting and plausible to attribute both early and late emergence of adults (e.g., Macan, 1974; Gower and Kormondy, 1963; Kormondy and Gower, 1965) and good and bad years for total adult numbers (e.g., Gower and Kormondy, 1963; Kormondy and Gower, 1965) to the effects of climate, our model, with entirely deterministic weather patterns, can produce very similar results. We conclude that it may be dangerous to attribute poor or late adult emergence of dragonflies and damselflies to climate, when only short runs of data are available. Different emergence patterns within different populations of the same species, or between species, are not likely to be solely due to photoperiod and temperature cues (see Corbet (1962, 1980) for full discussions of such physiological effects). Larval population dynamics may also play an important part.

Our model has no density-dependent processes acting at emergence or in the adult stage that might contribute to population stability. For workers interested in the role of territoriality in Odonata (e.g., see Heymer, 1972; Corbet, 1980; Bick and Bick, 1980; Baker, 1983; Parr, 1983; and references therein), this may seem remarkable. It undoubtedly greatly simplifies the population dynamics, and may be one reason why we have only a single plausible equilibrium rather than multiple equilibria (e.g., Fretwell, 1972) in our model.

VI. TESTS AND HYPOTHESES

Formulating and analyzing mathematical models of biological processes can often aid understanding by establishing and clarifying the implications of explicit assumptions, as has been a particular goal of the present study. Modeling can also be valuable in focusing empirical work on the most important missing information and the most significant hypotheses; this is the primary concern in the remainder of this section.

A. In Search of Data

The following list indicates some needed information that would be particularly helpful in sharpening the analysis attempted in the first four major sections of this chapter:

(1) Field experiments on density-dependent predation by vertebrates, especially fish—such as replicated, short-term enclosure studies.
(2) More field experiments capable of detecting and measuring both feeding- and mortality-related competition—including interspecific competition with fish and Anisoptera, as well as with other damsel-

flies. Density-dependent predation and interference mortality are of central importance because they supply the *regulation* in the model, though other factors also help *determine* densities.

(3) More field experiments on the potential for prey depletion by damselfly larvae. This could revive or lay to rest the possibility of exploitation competition among the larvae.

(4) Additional laboratory studies of the relation between growth rate and feeding rate for larval damselflies. We found that the nonlinearity (concave downward) of this relationship (observed in *Ischnura elegans* by Lawton *et al.*, 1980) is critical in reducing the sensitivity of the stage durations to the feeding parameters. The possible temperature-sensitivity of the growth parameters should be investigated. A more mechanistic model of this relationship would be particularly useful.

(5) Quantitative data on the relative spatial distributions of larval instars in the field (cf. Johannsson, 1978). This is essential to a better assessment of the potential for interference between larval stages, a key determinant of the model's behavior. Moreover, existing predator–prey models highlight predator aggregations in regions of high prey density as a key stabilizing mechanism (Hassell, 1978). The possibility that damselfly larvae exhibit aggregative responses to nonrandomly distributed prey also deserves more attention. Aggregation enhances interference, but also increases individual feeding rates above levels assumed by models with random encounters between predators, and between predators and prey.

(6) Measurements of emergence mortality in relation to density. This could be attempted in several shoreline pens, some with larval densities augmented by net-removals from others, and stocked with appropriate numbers of fish. Though difficult to obtain, the results could indicate another source of density-dependence.

(7) Means, variances, and seasonal trends (by sex) in size at emergence. These need to be better documented, since the population densities yielded by the model are particularly sensitive to the magnitude of the development index required for emergence.

(8) We also need to know whether small adults are produced by density-dependent effects operating on the larvae. If fecundity depends on adult size, the possibility then exists for an additional, delayed density-dependent component in the life-history.

(9) Estimates of adult dispersal in relation to density—perhaps long-term studies at individual sites of annual emergence *vs.* mature adult density, ideally with marking of emergers, will help evaluate this elusive phenomenon.

(10) Finally, measurements of fecundity in relation to the densities of adult males and females could perhaps be done in conjunction with long-term

studies suggested above, or in relatively intensive within-season observations of isolated sites. Though the subsequent emergence rate in the model is quite insensitive to the fecundity of the previous generation, density-dependence in the adult stages could become more important in relatively unproductive systems (i.e., the expected lower larval densities imply less interference and perhaps less predation, reducing overall larval density-dependence).

B. Some Testable Hypotheses

The results of this study summarized in Section IV suggest a number of hypotheses that could be empirically tested. Some of the more interesting and significant of these are noted here.

Larval damselfly densities and emergence rates are directly related to overall prey abundance but generally unrelated to prey productivity. Larvae are thus food-limited but unable to deplete prey enough that prey productivity becomes important. Experimental augmentation of prey of larval damselflies in the field should increase growth rates.

Thus, other animals capable of reducing prey densities in the field are also at least indirectly capable of reducing damselfly densities as well. To the extent that such animals are important predators on damselflies too, they will have a doubly inhibitory effect on larval densities and emergence rates. This may be further compounded by any tendency for damselflies to feed less efficiently in the presence of predators. These effects could perhaps be sorted out in the field in large fish enclosures and exclosures, with some of the damselflies further sequestered in smaller prey-permeable containers.

Only a relatively depauperate and especially unproductive prey assemblage could be depleted sufficiently by damselfly predation to raise the possibility of lengthened larval stage durations via exploitation competition. Combining these first three hypotheses suggests that different fish predators in certain aquatic systems could have opposing effects on larval stage durations. In an unproductive system, size-selective insectivores such as many trout may shorten the life-cycle by depleting damselflies but allowing some of their prey to increase (see Macan, 1966); more typically, generalized predators such as many sunfish may lengthen the life-cycle via prey depletion.

Interference mortality among damselfly larvae is especially important in determining and regulating larval densities and emergence rates, but interference that reduces feeding rates is unimportant. We note that there is some field evidence cited in Section IIIA3 indicating that feeding-related interference may indeed be important in nature; this apparent inconsistency deserves attention both in additional field experiments and in reconsidering the structure and parameter values of the model. Small, replicated enclosures with different densities of larvae (see, e.g., Pierce *et al.*, 1985) should prove generally useful here.

Factors that slow larval development within a generation (e.g., an unusually cool summer) *will tend to reduce survival through the larval stages by exposing the larvae to mortality sources for a longer time* (assuming that the mortality rate is not substantially reduced at the lower temperatures). *But factors that slow larval development across all generations* (e.g., cool temperatures associated with higher latitude) *will yield lower larval densities rather than lower survival*. This follows from the lack of density-dependence in the adult and egg stages, because any reduction in larval survival is necessarily translated into a subsequent reduction in larval density, which raises survival again via density-dependent mortality. The converse of the within-generation prediction can be tested by food augmentation in large field enclosures, comparing subsequent emergence against controls. Comparative studies of similar systems among latitudes should test the across-generation prediction.

Emergence patterns of damselflies having life-cycles exceeding a year are more likely to be multimodal in the presence of fish and other strongly density-dependent predators than in their absence; in the (near) absence of density-dependent predation, off-years for emergence may be frequent and regular. Density-dependent predation opposes the tendency for competition between larval stages to reduce the number of coexisting subpopulations that are seasonally out of phase; where present, particularly in the absence of heavy predation on larvae, these subpopulations tend to appear as separate modes in emergence patterns. Comparative studies may also provide the test here.

The density of small larvae is directly related to total oviposition by females of the previous generation; but the density of large larvae is only weakly related, and the total emergence is essentially unrelated to this previous egg production, because of density-dependent larval mortality. This could possibly be tested in small, replicated experimental ponds, or by using netting to subdivide individual ponds. Stems of vegetation could be covered with mesh to reduce oviposition in some ponds (or pond subdivisions) and subsequent larval development and emergence observed (cf. Benke, 1972).

ACKNOWLEDGEMENTS

We thank the following people for valuable discussions and many helpful suggestions on the manuscript: Rob Baker, Jim Bence, Joy Bergelson, Philip Corbet, Pat Dillon, Phil Heads, Dan Johnson, Mary Linton, Ed McCauley, Mike Parr, Andy Sih, Kevin Strohmeier, and Dave Thompson.

Rob Baker and Lee Martin supplied useful unpublished material. Grant support was provided by the National Science Foundation (DEB 8104424 and BSR 8400377 to PHC).

REFERENCES

Abro, A. (1971). Gregarines: their effects on damselflies. *Entomol. Scand.* **2**, 294–300.

Abro, A. (1982). The effects of parasitic water mite larvae (*Arrenurus* spp.) on zygopteran imagoes (Odonata). *J. Invert. Pathol.* **39**, 373–381.

Aguesse, P. (1961). Contribution à l'étude ecologique des Zygopteres de Camargue. Ph.D. thesis, University of Paris.

Akre, B. G. and Johnson, D. M. (1979). Switching and sigmoid functional response curves by damselfly naiads with alternative prey available. *J. Anim. Ecol.* **48**, 703–720.

Baker, R. L. (1980). Use of space in relation to feeding areas by zygopteran nymphs in captivity. *Can. J. Zool.* **58**, 1060–1065.

Baker, R. L. (1981a). Behavioural interactions and use of feeding areas by nymphs of *Coenagrion resolutum* (Coenagrionidae: Odonata). *Oecologia* (*Berl.*) **49**, 353–358.

Baker, R. L. (1981b). Use of space in relation to areas of food concentration by nymphs of *Lestes disjunctus* (Lestidae, Odonata) in captivity. *Can. J. Zool.* **59**, 134–135.

Baker, R. L. (1982). Effects of food abundance on growth, survival, and use of space by nymphs of *Coenagrion resolutum* (Zygoptera). *Oikos* **38**, 47–51.

Baker, R. L. (1983). Spacing behaviour by larval *Ischnura cervula* Selys: effects of hunger, previous interactions, and familiarity with an area (Zygoptera: Coenagrionidae). *Odonatologica* **12**, 201–207.

Baker, R. L. (1986a). Estimating food availability for larval dragonflies: a cautionary note. *Can. J. Zool.* **64**, 1036–1038.

Baker, R. L. (1986b). Food limitation and population density of larval dragonflies: a field test of spacing behaviour. *Can. J. Fish. Aquat. Sci.* **43**, 1720–1725.

Baker, R. L. and Clifford, H. F. (1981). Life cycles and food of *Coenagrion resolutum* (Coenagrionidae: Odonata) and *Lestes disjunctus disjunctus* (Lestidae: Odonata) populations from the Boreal Forest of Alberta, Canada. *Aquatic Insects* **3**, 179–191.

Banks, M. J. and Thompson, D. J. (1985a). Lifetime mating success in the damselfly *Coenagrion puella. Anim. Behav.* **33**, 1175–1183.

Banks, M. J. and Thompson, D. J. (1985b). Emergence, longevity and breeding area fidelity in *Coenagrion puella* (L.) (Zygoptera: Coenagrionidae). *Odonatologica* **14**, 279–286.

Beddington, J. R. (1975). Mutuai interference between parasites and predators and its effect on searching efficiency. *J. Anim. Ecol.* **44**, 331–340.

Beddington, J. R., Hassell, M. P. and Lawton, J. H. (1976). The components of arthropod predation. II. The predator rate of increase. *J. Anim. Ecol.* **45**, 165–186.

Benke, A. C. (1972). An experimental field study on the ecology of coexisting larval odonates. Ph.D. dissertation, University of Georgia, Athens, Georgia, USA. 112 pp.

Benke, A. C. (1976). Dragonfly production and prey turnover. *Ecology* **57**, 915–927.
Benke, A. C. (1978). Interactions among coexisting predators—a field experiment with dragonfly larvae. *J. Anim. Ecol.* **47**, 335–350.
Bick, G. H. (1972). A review of territorial and reproductive behavior in Zygoptera. *Contactbrief Nederlandse Libellenonderzockers, Utrecht* No. 10, Suppl., 14 pp.
Bick, G. H. and Bick, J. C. (1961). An adult population of *Lestes disjunctus australis* (Walker). *Southwest. Nat.* **6**, 111–137.
Bick, G. H. and Bick, J. C. (1963). Behavior and population structure of the damselfly, *Enallagma civile* (Hagen). *Southwest. Nat.* **8**, 57–84.
Bick, G. H. and Bick, J. C. (1980). A biobliography of reproductive behavior of Zygoptera of Canada and coterminous United States. *Odonatologica* **9**, 5–18.
Bohanan, R. E. and Johnson, D. M. (1983). Response of littoral invertebrate populations to a spring fish exclusion experiment. *Freshwat. Invertebr. Biol.* **2**, 28–40.
Carchini, G. and Nicolai, P. (1984). Food and time resource partitioning in two coexisting *Lestes* species (Zygoptera: Lestidae). *Odonatologica* **13**, 461–466.
Chutter, F. M. (1961). Certain aspects of the morphology and ecology of the nymphs of several species of *Pseudagrion* Selys (Odonata). *Arch. Hydrobiol.* **57**, 430–463.
Cloarec, A. (1975). Variations quantitatives circadiennes de la prise alimentaire des larves d'*Anax imperator* Leach (Anisoptera: Aeshnidae). *Odonatologica* **4**, 125–206.
Comins, H. L. and Hassell, M. P. (1976). Predation in multi-prey communities. *J. theor. Biol.* **62**, 93–114.
Connell, J. H. and Sousa, W. P. (1983). On the evidence needed to judge ecological stability and persistence. *Am. Nat.* **121**, 789–824.
Corbet, P. S. (1957). The life-history of the Emperor Dragonfly *Anax imperator* Leach (Odonata: Aeshnidae). *J. Anim. Ecol.* **26**, 1–69.
Corbet, P. S. (1959). Notes on the insect food of the Nile crocodile. *Proc. R. ent. Soc., Lond. A* **34**, 17–22.
Corbet, P. S. (1961). The food of non-cichlid fishes in the Lake Victoria basin, with remarks on their evolution and adaptation to lacustrine conditions. *Proc. Zool. Soc., Lond.* **136**, 1–101.
Corbet, P. S. (1962). *A Biology of Dragonflies*. Chicago, Ill.: Quadrangle Books, 247 pp.
Corbet, P. S. (1980). Biology of Odonata. *Ann. Rev. Entomol.* **25**, 189–217.
Corbet, P. S., Longfield, C. and Moore, N. W. (1960). *Dragonflies*. London: Collins.
Cothran, M. L. and Thorp, J. H. (1982). Emergence patterns and size variation of Odonata in a thermal reservoir. *Freshwat. Invertebr. Biol.* **1**, 30–39.
Crowley, P. H. (1975). Spatial heterogeneity and the stability of a predator–prey link. Ph.D. dissertation, Michigan State University, East Lansing, Michigan.
Crowley, P. H. (1979). Behavior of zygopteran nymphs in a simulated weed bed. *Odonatologica* **8**, 91–101.
Crowley, P. H. (1984). Evolutionarily stable strategies for larval dragonflies. In *Mathematical Ecology: Proceedings, Trieste* (Ed. by S. I. Levin and T. G. Hallam), pp. 55–74. Lecture notes in biomathematics, 54. Berlin: Springer-Verlag.
Crowley, P. H. and Johnson, D. M. (1982a). Habitat and seasonality as niche axes in an odonate community. *Ecology* **63**, 1064–1077.
Crowley, P. H. and Johnson, D. M. (1982b). Co-occurrence of Odonata in the eastern United States. *Adv. Odonatol.* **1**, 15–37.
Crowley, P. H., Dillon, P. M., Johnson, D. M. and Watson, C. N. (1987).

Intraspecific interference among larvae in a semivoltine dragonfly population. *Oecologia* (*Berl.*) **71**, 447–456.

Faragher, R. A. (1980). Life cycle of *Hemicordulia tau* Selys (Odonata: Corduliidae) in Lake Eucumbene, N.S.W., with notes on predation on it by two trout species. *J. Aust. ent. Soc.* **19**, 269–276.

Fincke, O. M. (1982). Lifetime mating success in a natural population of the damselfly, *Enallagma hageni* (Walsh) (Odonata: Coenagrionidae). *Behav. Ecol. Sociobiol.* **10**, 293–302.

Fischer, Z. (1961a). Some data on the Odonata larvae of small pools. *Int. Rev. ges. Hydrobiol.* **46**, 269–275.

Fischer, Z. (1961b). Cannibalism among larvae of the dragonfly *Lestes nympha* Selys. *Ekol. Polsk.* (*ser. B*) **7**, 33–39.

Fisher, M. E. and Goh, B. S. (1977). Stability in a class of discrete time models of interacting populations. *J. math. Biol.* **4**, 265–274.

Folsom, T. C. (1980). Predation ecology and food limitation of the larval dragonfly *Anax junius* (Aeshnidae). Ph.D. dissertation, University of Toronto. 138 pp.

Folsom, T. C. and Collins, N. C. (1982a). An index of food limitation in the field for the larval dragonfly *Anax junius* (Odonata: Aeshnidae). *Freshwat. Invertebr. Biol.* **1**, 25–32.

Folsom, T. C. and Collins, N. C. (1982b). Food availability in nature for the larval dragonfly *Anax junius* (Odonata: Aeshnidae). *Freshwat. Invertebr. Biol.* **1**, 33–40.

Frank, P. W., Boll, C. D. and Kelley, R. W. (1957). Vital statistics of laboratory cultures of *Daphnia pulex* DeGeer as related to density. *Physiol. Zool.* **30**, 287–305.

Fretwell, S. D. (1972). *Populations in a Seasonal Environment*. Princeton: Princeton University Press, 217 pp.

Gerking, S. D. (1962). Production and food utilization in a population of bluegill sunfish. *Ecol. Monogr.* **32**, 31–78.

Gerritsen, J. (1980). Adaptive responses to encounter problems. In *Ecology and Evolution of Zooplankton Communities* (Ed. by W. C. Kerfoot), pp. 52–62. Hanover, NH: University Press of New England.

Gerritsen, J. and Strickler, J. R. (1977). Encounter probabilities and community structure in zooplankton: a mathematical model. *J. Fish. Res. Bd Can.* **34**, 73–82.

Gilpin, M. E. Case, T. J. and Bender, E. A. (1982). Counterintuitive oscillations in systems of competition and mutualism. *Am. Nat.* **119**, 584–588.

Gower, J. L. and Kormondy, E. J. 1963. Life history of the damselfly *Lestes rectangularis* with special reference to seasonal regulation. *Ecology* **44**, 398–402.

Gresens, S. E., Cothran, M. L. and Thorp, J. H. (1982). The influence of temperature on the functional response of the dragonfly *Celithemis fasciata* (Odonata, Libellulidae). *Oecologia* **53**, 281–284.

Gurney, W. S. C. and Nisbet, R. M. (1983). The systematic formulation of delay-differential models of age or size structured populations. In *Proceedings of International Conference on Population Biology, Edmonton, Alberta, Canada* (Ed. by H. I. Freeman and C. Strobeck) pp. 163–172. Berlin: Springer-Verlag.

Gurney, W. S. C. and Nisbet, R. M. (1985). Fluctuation periodicity, generation separation and the expression of larval competition. *Theor. Pop. Biol.* **28**, 150–180.

Gurney, W. S. C., Nisbet, R. M. and Lawton, J. H. (1983). The systematic formulation of tractable single-species population models incorporating age-structure. *J. Anim. Ecol.* **52**, 479–495.

Gurney, W. S. C., Nisbet, R. M. and Blythe, S. P. (1986). The systematic formulation of models of stage-structured populations. In *The Dynamics of*

Physiologically Structured Populations (Ed. by J. A. J. Metz and O. Diekmann), Heidelberg: Springer-Verlag, in press.
Hall, D. J. (1964). An experimental approach to the dynamics of a natural population of *Daphnia galeata mendotae*. *Ecology* **45**, 94–112.
Hall, D. J., Cooper, W. E. and Werner, E. E. (1970). An experimental approach to the production dynamics and structure of freshwater animal communities. *Limnol. Oceanogr.* **15**, 839–928.
Harvey, I. F. and Corbet, P. S. (1985). Territorial behavior of larvae enhances mating success of male dragonflies. *Anim. Behav.* **33**, 561–565.
Hassell, M. P. (1976). *The Dynamics of Competition and Predation*. London: Edward Arnold, 68 pp.
Hassell, M. P. (1978). *The Dynamics of Arthropod Predator–Prey Systems*. Princeton: Princeton University Press, 237 pp.
Hassell, M. P. and Comins, H. N. (1976). Discrete time models for two-species competition. *Theor. Pop. Biol.* **9**, 202–221.
Hassell, M. P., Lawton, J. H. and May, R. M. (1976). Patterns of dynamical behaviour in single species populations. *J. Anim. Ecol.* **45**, 471–486.
Haussman, U. E. (1974). Coexistence of species in a discrete system. In *Mathematical Problems in Biology* (Ed. by P. van den Driessche), Berlin: Springer-Verlag.
Hayne, D. W. and Ball, R. C. (1956). Benthic productivity as influenced by fish predation. *Limnol. Oceanogr.* **1**, 162–175.
Heads, P. A. (1985). The effect of invertebrate and vertebrate predators on the foraging movements of *Ischnura elegans* larvae (Odonata: Zygoptera). *Freshwat. Biol.* **15**, 559–571.
Heymer, A. 1972. Comportements social et territorial des Calopterygidae (Odon. Zygoptera). *Annales Société Entomologique de France* (N.S.) **8**, 3–53.
Holling, C. S. (1959). Some characteristics of simple types of predation and parasitism. *Can. Ent.* **91**, 385–398.
Holling, C. S. (1966). The functional response of invertebrate predators to prey density. *Mem. Entomol. Soc., Can.* **48**, 1–86.
Hoppensteadt, F. C. (1982). *Mathematical Methods of Population Biology*. Cambridge: Cambridge University Press, 150 pp.
Imamura, T. and Mitchell, R. (1973). The water mites parasitic on the damselfly, *Cercion hieroglyphicum* Brauer. I. Systematics and life history. *Annot. Zool. Jpn* **40**, 28–36.
Ingram, B. R. (1971). The seasonal ecology of two species of damselflies with special reference to the effects of photoperiod and temperature on nymphal development. Ph.D. dissertation, University of North Carolina, Chapel Hill. 232 pp.
Ingram, B. R. and Jenner, C. E. (1976). Life histories of *Enallagma hageni* (Walsh) and *E. aspersum* (Hagen) (Zygoptera: Coenagrionidae). *Odonatologica* **5**, 331–345.
Ito, Y. (1978). *Hikaku Seitaigaku* [Comparative Ecology], 2nd edn. Tokyo: Iwanami, 421 pp.
Jeffries, M. J. (1984). The impact of freshwater invertebrate predators upon the structure of ecological communities. D.Phil. dissertation, University of York. 300 pp.
Jervey, T. O. (1982). Prey encounter frequency, encounter surface, and attack rate of *Ischnura verticalis* (Odonata: Coenagrionidae). M.S. thesis, University of Kentucky. 66 pp.
Johannsson, O. E. (1978). Co-existence of larval Zygoptera (Odonata) common to the Norfolk Broads (U.K.). *Oecologia* (*Berl.*) **32**, 303–321.

Johnson, D. M., Akre, B. G. and Crowley, P. H. (1975). Modelling arthropod predation: wasteful killing by damselfly naiads. *Ecology* **56**, 1081–1093.
Johnson, D. M., Bohanan, R. E., Watson, C. N. and Martin, T. H. (1983). Exploitation competition between dragonfly larvae and small sunfish. *Bull. Ecol. Soc. Am.* **64**, 90.
Johnson, D. M., Bohanan, R. E., Watson, C. N. and Martin, T. H. (1984). Coexistence of *Enallagma divagans* and *Enallagma traviatum* (Zygoptera: Coenagrionidae) in Bays Mountain Lake: an *in situ* enclosure experiment. *Adv. Odonatol.* **2**, 57–70.
Johnson, D. M. and Crowley, P. H. (1980). Odonate "hide and seek": habitat specific rules? In *The Evolution and Ecology of Zooplankton Communities* (Ed. by W. C. Kerfoot), pp. 569–579. Hanover, NH: University Press of New England.
Johnson, D. M., Crowley, P. H., Bohanan, R. E., Watson, C. N. and Martin, T. H. (1985). Competition among larval dragonflies: a field enclosure experiment with *Tetragoneuria cynosura* and *Celithemis elisa* (Odonata: Anisoptera). *Ecology* **66**, 119–128.
Johnson, D. M., Pierce, C. L., Martin, T. H., Bohanan, R. E., Watson, C. N. and Crowley, P. H. (1987). Prey depletion by odonate larvae: combining evidence from multiple field experiments. *Ecology*, in press.
Johnson, M. G. and Brinkhurst, R. O. (1971). Production of benthic macroinvertebrates of Bay of Quinte and Lake Ontario. *J. Fish. Res. Bd Can.* **28**, 1699–1714.
Koopman, B. O. (1956a). The theory of search. I. Kinematic bases. *Opr. Res.* **4**, 324–346.
Koopman, B. O. (1956b). The theory of search. II. Target detection. *Opr. Res.* **4**, 503–531.
Kormondy, E. J. and Gower, J. L. (1965). Life history variations in an association of Odonata. *Ecology* **46**, 882–886.
Kot, M. and Schaffer, W. M. (1984). The effects of seasonality on discrete models of population growth. *Theor. Pop. Biol.* **26**, 340–360.
Lawton, J. H. (1970). A population study on larvae of the damselfly *Pyrrhosoma nymphula* (Sulzer) (Odonata: Zygoptera). *Hydrobiologia* **36**, 33–52.
Lawton, J. H. (1971). Maximum and actual field feeding-rates in larvae of the damselfly *Pyrrhosoma nymphula* (Sulzer) (Odonata: Zygoptera). *Freshwat. Biol.* **1**, 99–111.
Lawton, J. H. (1972). Sex ratios in Odonata larvae, with particular reference to the Zygoptera. *Odontologica* **1**, 209–219.
Lawton, J. H., Beddington, J. R. and Bonser, R. (1974). In *Ecological Stability* (Ed. by M. B. Usher and M. H. Williamson), pp. 141–158. London: Chapman & Hall.
Lawton, J. H., Thompson, B. A. and Thompson, D. J. (1980). The effects of prey density on survival and growth of damselfly larvae. *Ecol. Entomol.* **5**, 39–51.
Lord, P. M. (1961). A study of the colour varieties of some damselflies. Ph.D. dissertation, University of Wales.
Lutz, P. E. (1968). Life-history studies on *Lestes eurinus* Say (Odonata). *Ecology* **49**, 576–579.
Maas, P., Gurney, W. S. C. and Nisbet, R. M. (1984). SOLVER Rev 2.0: An adaptable program template for initial value problem solving. Applied Physics Industrial Consultants, University of Strathclyde, Glasgow, U.K.
Macan, T. T. (1964). The Odonata of a moorland fishpond. *Int. Rev. ges. Hydrobiol.* **49**, 325–360.
Macan, T. T. (1966). The influence of predation on the fauna of a moorland fishpond. *Arch. Hydrobiol.* **61**, 432–452.
Macan, T. T. (1974). Twenty generations of *Pyrrhosoma nymphula* (Sulzer) and

Enallagma cyathigerum (Charpentier) (Zygoptera: Coenangrionidae). *Odontologica* **3**, 107–119.
Macan, T. T. (1977). The influence of predation on the composition of fresh-water animal communities. *Biol. Rev.* **52**, 45–70.
Machado, A. B. M. (1977). Ecological studies on the larva of the plant breeding damselfly *Roppaneura beckeri* Santos (Zygoptera: Protoneuridae). *Abstr. Pap. 4th Int. Symp. Odonatol., Gainesville*, p. 11.
Martin, T. H. (1986). The diets of redear and bluegill sunfish in Bays Mountain Lake, M.S. thesis, East Tennessee State University, Johnson City, Tennessee.
Masseau, M. J. and Pilon, J. G. (1982). Action de la temperature sur le developpement embryonnaire de *Enallagma hageni* (Walsh) (Odonata: Coenagrionidae). *Adv. Odonatol.* **1**, 117–150.
May, R. M. (1977). Thresholds and breakpoints in ecosystems with a multiplicity of steady states. *Nature* **269**, 471–477.
May, R. M., Conway, G. R., Hassell, M. P. and Southwood, T. R. E. (1974). Time delays, density dependence, and single-species oscillations. *J. Anim. Ecol.* **43**, 747–770.
McPeek, M. A. (1984). A laboratory study of intraspecific interference in the larvae of the damselfly *Ischnura verticalis* (Say). M.S. thesis, University of Kentucky. 168 pp.
McPeek, M. A. and Crowley, P. H. (1987). The effects of density and relative size on the aggressive behaviour, movement, and feeding of damselfly larvae (Odonata: Coenagrionidae). *Anim. Behav.*, in press.
Merrill, R. J. and Johnson, D. M. (1984). Dietary niche overlap and mutual predation among coexisting larval Anisoptera. *Odonatologica* **13**, 387–406.
Mitchell, R. (1962). Storm-induced dispersal in the damselfly *Ischnura verticalis* (Say). *Am. Midl. Nat.* **82**, 359–366.
Mittelbach, G. G. (1981). Foraging efficiency and body size: a study of optimal diet and habitat use by bluegills. *Ecology* **62**, 1370–1386.
Moore, N. W. (1952). On the length of life of adult dragonflies (Odonata—Anisoptera) in the field. *Proc. Bristol Nat. Soc.* **28**, 267–272.
Moore, N. W. (1953). Population density in adult dragonflies (Odonata—Anisoptera). *J. Anim. Ecol.* **22**, 344–359.
Moore, N. W. (1957). Territory in dragonflies and birds. *Bird Study, Oxford* **4**, 125–130.
Moore, N. W. (1964). Intra- and interspecific competition among dragonflies. *J. Anim. Ecol.* **33**, 49–71.
Morin, P. J. (1984a). The impact of fish exclusion on the abundance and species composition of larval odonates: results of short-term experiments in a North Carolina farm pond. *Ecology* **65**, 53–60.
Morin, P. J. (1984b). Odonate guild compositions: experiments with colonization history and fish predation. *Ecology* **65**, 1866–1873.
Murdoch, W. W. and McCauley, E. (1985). Stability and cycles in planktonic systems: implications for ecological theory. *Nature* **316**, 628–630.
Nisbet, R. M. and Gurney, W. S. C. (1976). Population dynamics in a periodically varying environment. *J. theor. Biol.* **56**, 459–476.
Nisbet, R. M. and Gurney, W. S. C. (1982). *Modelling Fluctuating Populations.* Chichester: Wiley.
Nisbet, R. M. and Gurney, W. S. C. (1983). The systematic formulation of population models for insects with dynamically varying instar duration. *Theor. Pop. Biol.* **23**, 114–135.
Nisbet, R. M. and Gurney, W. S. C. (1984). Models of uniform larval competition.

In *Mathematical Ecology: Proceedings, Trieste* (Ed. by S. I. Levin and T. Hallam), pp. 97–113. Lecture notes in biomathematics, 54. Berlin: Springer-Verlag.

Norling, U. (1976). Seasonal regulation in *Leucorrhinia dubia* (Vander Linden). *Odonatologica Utrecht* **5**, 245–263.

Norling: U. (1984). The life cycle and larval photoperiodic responses of *Coenagrion hastulatum* (Charpentier) in two climatically different areas (Zygoptera: Coenagrionidae). *Odonatologica* **13**, 429–449.

Oster, G. F. and Takahashi, Y. (1974). Models for age-specific interactions in a periodic environment. *Ecol. Monogr.* **44**, 483–501.

Parr, M. J. (1965). A population study of a colony of imaginal *Ischnura elegans* (Vander Linden) (Odonata: Coenagrionidae) at Dale, Pembrokeshire. *Fld Stud.* **2**, 237–282.

Parr, M. J. (1969a). On the ecology of zygopteran dragonfly populations. *Entomologist* **102**, 114–116.

Parr, M. J. (1969b). Comparative notes on the distribution, ecology and behaviour of some damselflies (Odonata: Coenagridae). *Entomologist* **102**, 151–161.

Parr, M. J. (1970). The life histories of *Ischnura elegans* (Vander Linden) and *Coenagrion puella* (L.) (Odonata) in south Lancashire. *Proc. R. ent. Soc., Lond. A* **45**, 172–181.

Parr, M. J. (1973). Ecological studies of *Ischnura elegans* (Vander Linden). II. Survivorship, local movements and dispersal. *Odonatologica Utrecht* **2**, 159–174.

Parr, M. J. (1983). An analysis of territoriality in libellulid dragonflies (Anisoptera: Libellulidae). *Odonatologica* **12**, 39–57.

Parr, M. J. and Palmer, M. (1971). The sex ratios, mating frequencies and mating expectancies of three coenagriids in northern England. *Entomol. Scand.* **2**, 191–204.

Paulson, D. R. and Jenner, C. E. (1971). Population structure in overwintering larval Odonata in North Carolina in relation to adult flight season. *Ecology* **52**, 96–107.

Pearlstone, P. S. M. (1973). The food of damselfly larvae in Marion Lake, British Columbia. *Syesis* **6**, 33–39.

Peckarsky, B. L. (1984). Predator–prey interactions among aquatic prey. In *The Ecology of Aquatic Insects* (Ed. by V. H. Resh and D. M. Rosenberg), pp. 196–254. New York: Praeger.

Penn, G. H. (1951). Seasonal variation in the adult size of *Pachydiplax longipennis* (Burmeister) (Odonata, Libellulidae). *Ann. ent. Soc. Am.* **44**, 193–197.

Pickup, J., Thompson, D. J. and Lawton, J. H. (1984). The life history of *Lestes sponsa* (Hansemann): larval growth (Zygoptera: Lestidae). *Odonatologica* **13**, 451–459.

Pierce, C. L. (1982). The relationship of behavior to competition and predation in two larval odonate populations. M.S. thesis, University of Kentucky.

Pierce, C. L., Crowley, P. H. and Johnson, D. M. (1985). Behavior and ecological interactions of larval Odonata. *Ecology* **66**, 1504–1512.

Pimm, S. L. (1984). The complexity and stability of ecosystems. *Nature, Lond.* **307**, 321–326.

Pritchard, G. (1964). The prey of dragonfly larvae in ponds in northern Alberta. *Can. J. Zool.* **42**, 785–800.

Procter, D. L. C. (1973). The effect of temperature and photoperiod on larval development in Odonata. *Can. J. Zool.* **51**, 1165–1170.

Rivard, D., Pilon, J.-G. and Thiphrakesone, S. (1975). Effect of constant temperature environments on egg development of *Enallagma boreale* Selys (Zygoptera: Coenagrionidae). *Odonatologica* **4**, 271–276.

Rogers, D. J. and Hassell, M. P. (1974). General models for insect parasite and predator searching behavior: interference. *J. Anim. Ecol.* **43**, 239–253.

Ross, Q. E. (1967). The effect of different naiad and prey densities on the feeding behavior of *Anax junius* (Drury) naiads. M.S. thesis, Cornell University, Ithaca.

Ross, Q. E. (1971). The effect of intraspecific interactions on the growth and feeding behavior of *Anax junius* (Drury) naiads. Ph.D. dissertation, Michigan State University, East Lansing.

Rowe, R. J. (1980). Territorial behavior of a larval dragonfly *Xanthocnemis zealandica* (McLachlan) (Zygoptera: Coenagrionidae). *Odonatologica* **9**, 285–292.

Savan, B. I. (1979). Studies on the foraging behavior of damselfly larvae (Odonata: Zygoptera). Ph.D. dissertation, University of London, Imperial College. 339 pp.

Slobodkin, L. B. (1954). *Growth and Regulation of Animal Populations.* New York: Holt, Rinehart and Winston, 184 pp.

Smith, F. E. (1963). Population dynamics in *Daphnia magna* and a new model for population growth. *Ecology* **44**, 651–662.

Sonehara, I. (1979). The number of eggs in the egg-string of *Epitheca bimaculata sibirica* and their hatching ratio. *Tombo* **22**, 27.

Steele, J. H. and Henderson, E. W. (1981). A simple plankton model. *Am. Nat.* **117**, 676–691.

Stokes, T. K. (1985). Long term changes in a laboratory insect population. Ph.D. dissertation, University of Strathclyde, Glasgow.

Thompson, D. J. (1975). Towards a predator–prey model incorporating age structure: the effects of predator and prey size on the predation of *Daphnia magna* by *Ischnura elegans. J. Anim. Ecol.* **44**, 907–916.

Thompson, D. J. (1977). Field and laboratory studies on the feeding ecology of damselfly larvae. D. Phil. dissertation. University of York. 132 pp.

Thompson, D. J. (1978a). Towards a realistic predator–prey model: the effect of temperature on the functional response and life history of larvae of the damselfly, *Ischnura elegans. J. Anim. Ecol.* **47**, 757–767.

Thompson, D. J. (1978b). Prey size selection by larvae of the damselfly, *Ischnura elegans* (Odonata). *J. Anim. Ecol.* **47**, 769–785.

Thompson, D. J. (1978c). The natural prey of larvae of the damselfly, *Ischnura elegans* (Odonata: Zygoptera). *Freshwat. Biol.* **8**, 377–384.

Thompson, D. J. (1982). Prey density and survival in damselfly larvae: field and laboratory studies. In *Advances in Odonatology*, Vol. 1 (Ed. by R. M. Gambles), pp. 267–280. Proceedings of the Sixth International Symposium of Odonatology, Chur, Switzerland.

Thompson, D. J., Banks, M. J., Cowley, S. E. and Pickup, J. (1985). Horses as a major cause of mortality in *Coenagrion puella* (L.) *Noctulae Odontologica,* **2**, 104–105.

Ubukata, H. (1981). Survivorship curve and annual fluctuation in the size of emerging population of *Cordulia aenea amurensis* Selys (Odonata: Corduliidae). *Jap. J. Ecol.* **31**, 335–346.

Ueda, T. (1976). The breeding population of damselfly, *Cercion calamorum* Ris. I. Daily movements and spatial structure. *Physiol. Ecol. Jpn* **17**, 303–312.

Ueda, T. and Iwasaki, M. (1982). Changes in the survivorship, distribution, and movement pattern during the adult life of a damselfly, *Lestes temporalis* (Zygoptera: Odonata). In *Advances in Odonatology*, Vol. 1 (Ed. by R. M. Gambles), pp. 281–291. Proceedings of the Sixth International Symposium of Odonatology, Chur, Switzerland.

Uttley, M. G. (1980). A laboratory study of mutual interference between freshwater invertebrate predators. D. Phil. dissertation, University of York. 173 pp.

Walker, E. M. (1953). *The Odonata of Canada and Alaska*. Toronto: University of Toronto Press.
Williamson, M. (1972). *Analysis of Biological Populations*. London: Edward Arnold, 180 pp.
Wilson, A. D. (1982). Handling time and the functional response of damselfly larvae (Odonata: Zygoptera). M.S. thesis, University of Kentucky.
Wright, M. (1946). The economic importance of dragonflies (Odonata). *J. Tenn. Acad. Sci.* **21**, 60–71.
Wylie, C. R. (1975). *Advanced Engineering Mathematics*, 4th edn. New York: McGraw-Hill, 937 pp.

How Plants Affect Nematodes

G. W. YEATES

I. SUMMARY

The pathogenic effects of plant-parasitic nematodes are well known, but as heterotrophic organisms all nematodes are dependent on autotrophs for their energy supply. If total soil and plant nematodes are considered their abundance may be positively correlated with ecosystem productivity, in contrast to the negative correlation for plant-parasitic forms.

Not only does the quantity of plant material influence specific nematode populations, but seasonal variation in its quality is important. This may be

ADVANCES IN ECOLOGICAL RESEARCH Vol. 17
ISBN 0-12-013917-0

manifest through the need for a root-derived factor to stimulate hatching or moulting or, apparently, through the nutrient content of the plant material itself.

A wide range of published data on fertilizer application, pruning, grazing, root growth, root diameter and root exudates is re-assessed in terms of the quantity and quality of plant material available to nematodes.

Nematode : nematode interactions within roots reflect the total resource available; "poor" host-plants may result in smaller nematodes; crowding may lead to a greater proportion of males which utilize less resource. Reproductive strategies often correlate with pulses of food availability.

Plant resistance mechanisms to migratory nematodes are poorly understood, but it appears that fertilizer may increase the ability of resistant plants to support nematodes. Due to the abundance of nematodes, selection for races among the genetic range may occur frequently and resistant plant cultivars should be used strategically.

Once parallels are drawn, nematode ecology is very similar to insect ecology. The relation of nematode populations to their resources offers wide scope for ecological advance.

II. INTRODUCTION

Nematodes are the most numerous multicellular animals on Earth. As a result of their economic and social effects, plant and animal parasitic nematodes dominate both scientific literature and popular attitudes. As heterotrophic organisms, nematodes are ultimately dependent on autotrophic organisms for their energy supply, but it is the plant disease and crop loss aspects of the relationship that have developed into "plant nematology" (e.g., Dropkin, 1980; Southey, 1978). The past decade has seen growing understanding of the role of "free-living" bacterial and fungal feeding nematodes in stimulating mineralization of plant nutrients (e.g., Coleman *et al.*, 1983; Ingham *et al.*, 1985) and this has brought "nematologists" and "ecologists" closer (e.g., Freckman, 1982). This review strives to present an ecological view of some aspects of plant–nematode interactions, while avoiding the pathogenic aspects of plant disease.

Measures of nematode dependence on higher plants which I wish to discuss are complementary to the work of Jones (1983) on "Weather and plant-parasitic nematodes"; Jones would certainly acknowledge the need for "potato root exudate" to stimulate emergence of *Globodera rostochiensis* (Woll.) juveniles before moisture, rainfall and temperature play their direct role on nematode development, in addition to acting through the plant. Similarly, the statistical view, "Interactions between nematodes and other factors on plants", espoused by Wallace (1983) implies that more information is needed on the effects of plants on nematodes.

Two orders of nematodes have evolved dependence on plants, the Tylenchida (Class Secernentia or Phasmidia), containing some 2000 plant dependent species, and the Dorylaimida (Class Adenophorea or Aphasmidia), some 200 species according to Maggenti (1981). As this review will show, the degree of association between plant and nematodes is very diverse and I maintain the view that while a few genera with sedentary, saccate females are "plant-parasites" in the strict sense, most nematodes have a looser, grazing or browsing association and are better termed "plant-feeding" (Yeates, 1971). Many of the Dorylaimida have their food described as "miscellaneous" or "unknown". They and many of the less pathogenic Tylenchida are probably subclinical pathogens or "root pruners" and may have an important role in nutrient cycling through their influence on the microbial community surrounding the roots (Ingham and Coleman, 1983).

Ecological relationships and population dynamics of plant and soil nematodes are poorly known and while references are made to niches, interactions and density dependence these cannot compare with the knowledge of plant–insect relations (e.g., Strong, Lawton and Southwood, 1984). However, given the poorly defined feeding habits of nematodes, the absence of distinct trophic levels, the increase in "quality" or "information content" of food (whether energy, phosphorus or nitrogen) with progression from the autotroph, and stimulation of prey species by grazers or predators, their populations are worthy of study.

III. ECOSYSTEM PRODUCTIVITY

The food of plant and soil nematodes is living plant material or the bacteria and fungi associated with decaying plant material. Thus the finding that *total* abundance of plant and soil nematodes and grassland productivity or root weight (Fig. 1(a), (b)) are, within a sampling series, positively correlated is not surprising, although it conflicts with the more traditional negative correlation between *plant–parasitic* nematodes and productivity (Fig. 1(c), (d)). Forest and cropping systems differ from grasslands in architecture, turnover time and physical disturbance; without a close root : shoot relation the nematode : productivity relation is more difficult to analyze, and strategies of plant resistance (Coley *et al.*, 1985) may differ.

Under cropping regimes the availability of suitable plant material provides a resource which may be exploited by a plant-feeding nematode. In such agro-ecosystems the regular cultivation and the sequence of crops are keys to the build-up of plant-feeding nematode populations. The relationship

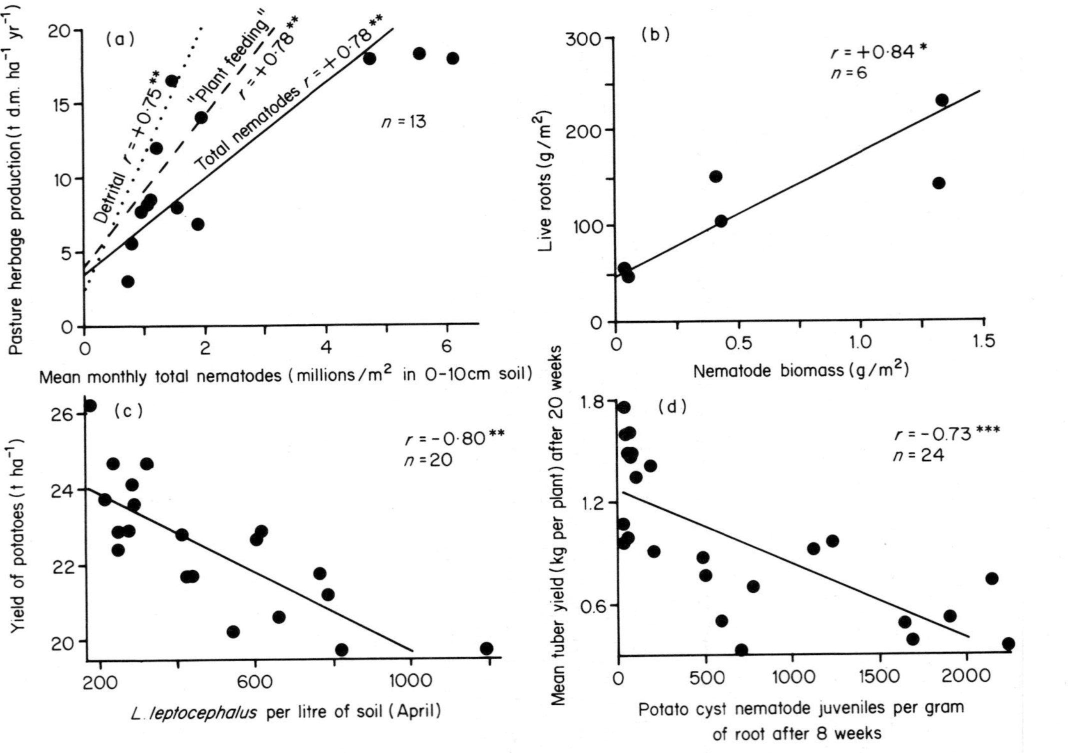

Fig. 1. Relationship between nematode populations and measures of ecosystem productivity. (a) Mean monthly total nematode populations and annual herbage dry matter production in New Zealand pastures; regression lines are given for total nematodes, "plant feeding" and detrital components (data of Yeates, 1984b). (b) Total nematode biomass and live root weight of six US grassland sites (mean of three values for 1972) (data from French, 1979). (c) Root infestation by *Globodera* spp. and yield of potato tubers (data from Trudgill *et al.*, 1975). (d) Population of *Longidorus leptocephalus* and yield of potatoes (redrawn from Sykes, 1979).

between an increase in plant-parasitic nematode population density and a decline in net plant productivity has been described in mathematical equations (Jones and Kempton, 1978) and it is not intended to review them here; nor are effects of nematode damage on plant succession considered. However, the interaction of fertilizer and cultivation regimes, crop sequences, plant cultivars, nematode species, nematode race and climate are dealt with below.

In the absence of simple data on plant-feeding nematodes, Table 1 shows how the pre-reproductive period of the bacterial-feeding nematode *Caenorhabditis briggsae* decreases as available food increases, and that total egg-production increases with food density. Similar results can be expected for plant-feeding nematodes and response to applied potassium (Section V) may be an example. Trials with soybean (*Glycine max* (L.)) by Weaver *et al.* (1985) show a positive correlation ($r = +0{\cdot}5651^{**}$, $n = 21$) between seed yield and soil populations of *Meloidogyne incognita* in the absence of soil fumigation, indicating that even in "pathogenic" situations nematode populations may reflect plant yield.

Many plants appear to be able to support abundant populations of *Helicotylenchus* without apparent yield loss (e.g., 3000 *H. dihystera* per ml soil on cotton (*Gossypium hirsutum* L.) (Bernard and Hussey, 1979), 2 149 000 *H. pseudorobustus* per m^2 pasture (Yeates, 1984a)). In pot trials with sunflower (*Helianthus annuus* L.) an inoculum of 1500 *H. dihystera* increased to an average of 43 970 over 18 weeks, while *Pratylenchus alleni* increased from 1000 to only 6532 over 75 days (Bernard and Keyserling,

Table 1

Effect of bacterial cell concentration on developmental time and fecundity of two bacterial feeding nematodes cultured at 20°C.[a]

Caenorhabditis briggsae			*Plectus palustris*		
E. coli cells per ml	Prereproductive period (days)	Total egg production	*Acinetobacter* cells per ml	Pereproductive period (days)	Daily egg production
			6–9 × 10^7	∞	0
2 × 10^8	10·2	17·7			
5 × 10^8	6·0	53·0			
			6–9 × 10^8	18·5	12·6
1 × 10^9	4·2	90·2			
5 × 10^9	3·6	134·0			
			6–9 × 10^9	12·5	37·7
1 × 10^{10}	3·3	144·9			
5 × 10^{10}	3·1	153·6			

[a] Adapted from Scheimer *et al.* (1980) and Scheimer (1982).

1985). The ability of plants to support significant populations of plant-feeding nematodes without yield loss is central to a discussion of plant tolerance, resistance and susceptibility to nematodes. For the purposes of this review, a plant is regarded as *tolerant* of a nematode population if under the particular conditions there is not significant loss of net plant production; *resistant* plants lead to a reduction in the nematode population ($P_f < P_i$) over the growing season, and *susceptible* plants lead to nematode multiplication ($P_f > P_i$). Wallace (1973) has critically discussed the circumstances under which parasitic nematodes become pathogenic and contribute to plant disease, and loss of plant production. Stress alleviation, through manipulation of the plant and environment to avoid disease, can be important in maintaining root health (Cook, 1984) and plant productivity (Lyda, 1981).

IV. PLANT QUALITY

The carbon content of plant nematodes has been assessed as 52·5% dry weight by Persson (1983) and 12·4% wet weight (or 49·6% assuming 25% dry matter content) by Jensen (1984). The N content is given as 11·5% dry weight (Persson, 1983) and P as 4·51% dry weight (McKercher *et al.*, 1979). However, the complementary knowledge on how nematode populations are limited by restricted ingestion of elements in their food is less well established and we must rely largely on symposia such as *Nitrogen as an ecological factor* (Lee *et al.*, 1983) and various entomological papers (e.g., Pierce, 1985; Room and Thomas, 1985). However, White (1978) has clearly proposed that "for many if not most animals—both herbivore and carnivore, vertebrate and invertebrate—the single most important factor limiting their abundance is a relative shortage of nitrogenous food for the very young". In examining the changing populations of post-hatching stages of *Helicotylenchus pseudorobustus* (Steiner) (Tylenchida) and *Pungentus maorium* (Clark) (Dorylaimida) over 36 months, Yeates (1982, 1984a) found significant, positive correlations between their average monthly reproductive activity and previous measurements of herbage nitrogen levels at the two pasture sites at which the species occurred (Table 2). Bird (1970) found that increasing the nitrogen status of the host plant increased the growth rate of developing *Meloidogyne javanica*. While these data are indicative of a situation similar to that enunciated by White (1978), only when periodicity of reproduction of a range of nematodes has been correlated with that of ingestible/assimilatable amino acids in plant roots will knowledge approach that for sap-sucking insects.

Plant growth after germination is dependent on nitrogen transported from the roots through the xylem in various organic forms; the phloem sap has a much greater amino-acid content than the xylem sap (Lea and Miflin, 1980).

Table 2
Correlation coefficients for monthly populations of each stage of two nematode species and herbage nitrogen content.[a]

Nematode species	*Pungentus maorium*		*Helicotylenchus pseudorobustus*	
Site climate:	Summer dry		Summer moist	
Stage	Grass	Clover	Grass	Clover
I	−0·35[b]	−0·44[b]	(−0·40)[c]	(−0·28)[c]
II	+0·53†	+0·52†	−0·52[b]	−0·47[b]
III	+0·77**	+0·65*	+0·52	+0·54
IV	+0·77**	+0·72*	+0·29	+0·30
♀	+0·61*	+0·54*	+0·01	+0·29
Total	+0·59*	+0·55†	+0·37	+0·44

[a] From Yeates (1982, 1984a).
[b] First post-hatching stage.
[c] Parentheses signify that this stage is normally within egg.
† Denotes $p < 0{\cdot}1$; * $p < 0{\cdot}05$; ** $p < 0{\cdot}01$.

There appears to be a paucity of data on nitrogen and amino-acid content of roots of many crop plants (Krauss, 1980), although ecologists have documented seasonal patterns in plant species growing in mixed swards (Gay *et al.*, 1982). For several nematodes, however, root factors have recently been invoked to aid interpretation of nematode reproductive activity:

> Plant age, condition of the root system and depth of feed roots were some of the more important factors which appeared to affect vertical distribution in the soil of larvae and males of citrus nematode. (Husain *et al.*, 1981)

> . . . plant root production is probably a factor limiting reproduction by the nematodes under natural conditions . . . (*Xiphinema diversicaudatum*). (Brown and Coiro, 1983)

> The production of new females is seen to be correlated with total root production . . . (*Globodera rostochiensis*). (Storey, 1982)

This area offers great potential for advance and to obtain adequate results it will be necessary to:

(a) monitor qualitative and quantitative changes in roots;
(b) study migratory nematodes whose reproductive cycle is not as closely bound to plant activity as the Heteroderoidea;
(c) study perennial plants where roots are potentially available for nematode feeding throughout the year;

(d) relate the timing of root growth to "resistance" to nematodes (Spaull, 1980);
(e) determine how stylet length determines feeding site/niche within the root (Yeates, 1986) and if this is similar to the xylem/phloem segregation in foliar feeding insects (Raven, 1983).

V. FERTILIZER RESPONSE

Plant quality influences nematodes, so modification of the plant's growth by fertilizer application would be expected to affect any associated nematode populations. Plant response to applied fertilizer is influenced by climate, pre-existing nutrient levels, trace elements, soil chemistry, plant cultivar, pathogens, mycorrhizae and other rhizosphere organisms. Any effect on the plant–nematode relationship will be confounded upon these responses. In addition to change in root quality the mass of roots, the root:shoot ratio (e.g., Steen, 1985), the resistance of the plant to invasion, and the proportions of nematodes in roots and soil may vary. While increased plant growth or nutrient uptake may lead to fertilizer offsetting nematode damage, the increased plant growth potentially provides additional food for nematodes.

Although the manner in which a given amount of an element is supplied greatly affects soil biological activity (e.g., Edwards and Lofty, 1982) there is little comparative data for nematodes. However, it has been shown clearly by Juhl (1981) that addition of nitrogen as $Ca(NO_3)_2$ significantly depresses populations of cereal cyst nematode (*Heterodera avenae*). Nematode infestation may increase translocation of elements from the roots (Price and Sanderson, 1984) or lead to accumulation in the roots (Spiegel *et al.*, 1982); both these consequences vary with the age of the infestation.

Under some conditions a ceiling in nematode populations may be reached reflecting all the factors affecting nematodes in the particular soil–plant system. Figure 2 shows how, in a pot trial with *Heterodera trifolii*, the number of full cysts some 16 weeks after sowing was a function of white clover production (reflecting root mass) but that there was a maximum value beyond which additional roots did not permit a larger population. Similar results can be obtained in the field (e.g., Jones, 1956). The "carrying capacity" of the roots exceeds the "carrying capacity" of the soil fabric in some way; this may be an artifact of monoculture and not applicable to natural ecosystems. These may be the conditions under which "natural" population crashes due to biocontrol pathogens occur. In some trials (e.g. Dolliver, 1961) results are of dubious value because final populations in many treatments are below inoculation levels.

Because of the wide range of possible plant response only some major plant nutrients are considered here. Kirkpatrick *et al.* (1964) have tabulated

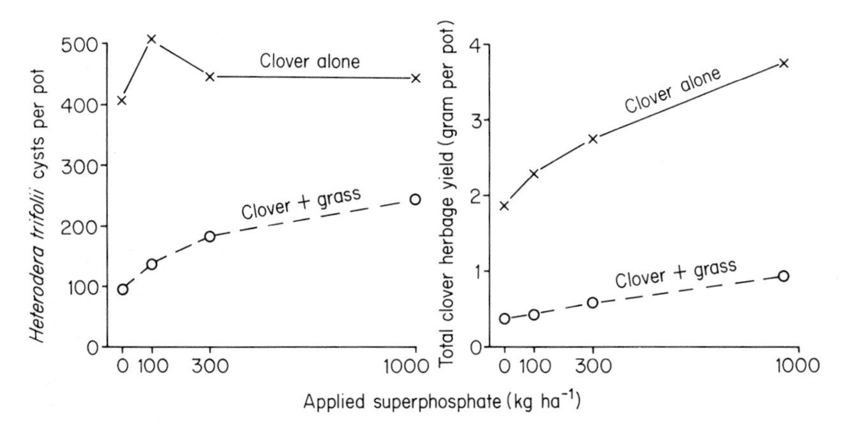

Fig. 2. Effect of supplementing white clover (*Trifolium repens* L.) with ryegrass (*Lolium perenne* L.) on the final *Heterodera trifolii* cyst number in pots and clover herbage production, at four levels of applied superphosphate (from data of Yeates, 1974b).

some nematode responses to applied NPK. Nematicides and other pesticides have potential direct effects on plant growth; therefore experiments in which they have been used to manipulate nematode populations should be avoided in simpler analyses. Further, as the size of nematode populations is significantly influenced by the amount of root available to them, experiments omitting plant growth or yield data are of limited value.

In a study of nitrogen application to eggplant (*Solanum melongana* L.) Mahmood and Saxena (1980) showed that increasing rates of nitrogen increased root and soil populations of *Rotylenchulus reniformis* proportionally. However, recalculation of their results on a per gram basis shows that populations decline with increasing nitrogen (Table 3). This may reflect the increasing phenolic content of the total plant (Section XIV), but free amino acids also increased (Table 3).

In wheat (*Triticum aestivum* L.) Simon and Rovira (1985) found that as superphosphate application (*c.* 10% P) increased from 0 to 400 kg ha^{-1}, female *Heterodera avenae*/root system increased from 141 to 324, but both ears per m^2 and grain yield also increased (148 to 236 ears per m^2; 100 to 199 g m^{-2}). Grazed pastures which received 125 or 500 kg ha^{-1} superphosphate (*c.* 10% P) annually gave only a 3% increase in herbage dry matter production, but Yeates (1976) considered the effect of the increase in *Pratylenchus* from 27 800 to 106 000 per m^2 effectively offset fertilizer response.

Potassium was shown to speed the maturation of *Meloidogyne incognita* by Otiefa (1953) (Table 4), and subsequently *Tylenchorhynchus, Pratylenchus, Tylenchulus* and *Rotylenchulus* have been found in greater abundance

Table 3

Effect of ammonium sulphate and urea on *Rotylenchulus reniformis* populations and plant chemistry in pot-grown eggplant.[a]

		R. reniformis in roots		*R. reniformis* in soil			
Fertilizer	g N kg^{-1} soil	Total in root system	per g fresh root	Total in pot	per g fresh root	Total phenols (%)	Total free amino acids (%)
Control	0·0	105	34	6 503	2 132	0·328	0·138
Ammonium sulphate	0·5	122	25	7 203	1 500	0·338	0·142
	1·0	130	22	8 350	1 415	0·369	0·147
	2·0	146	21	9 500	1 387	0·394	0·160
Urea	0·5	133	26	8 020	1 542	0·362	0·145
	1·0	138	23	9 203	1 521	0·385	0·155
	2·0	155	20	10 050	1 317	0·419	0·168

[a] Adapted from Mahmood and Saxena (1980).

after the addition of potassium (Otiefa and Diab, 1961; Badra and Yousif, 1979). Physiologically, potassium is an important ion, and while there must be a question as to whether additional potassium offsets additional damage to the roots, the additional potassium does not always reduce plant loss. However, additional potassium increases nematode multiplication. The composition of the root or rhizosphere affects the proportion of nematodes in the root at a particular potassium level (e.g., Badra and Yousif, 1979) and this, coupled with the absence of a control (no potassium) treatment, may have influenced assessment of *Pratylenchus loosi* populations in tea (*Camellia sinensis* L.) by Gnanapragasam (1982), who showed depression of *P. loosi* per root system and per gram of root at a higher potassium level.

Table 4

Influence of potassium supply on development of *Meloidogyne incognita* in lima beans (*Phaseolus lunatus* L.).[a]

	Potassium level[b]		
	Deficient	Optimum	Excessive
Days from inoculation			
all moulted to females	36	26	20
first egg production	40	24	16
75% laying eggs	54	28	22

[a] Adapted from Otiefa (1953).
[b] Potassium levels maintained by applying nutrient solution every 2 days.

Table 5
Effect of fertilization schemes on nematode populations and crop yield.[a]

Fertilization[b] scheme	*Helicotylenchus dihystera* per 450 cm^3 soil[c]	*Paratrichodorus christiei* per 450 cm^3 soil[c]	Yield ($kg\ ha^{-1}$)[c]
LPK+	563·3	58·3	2933
LPK−	7·2	77·5	1426
LNPK−	11·7	147·5	2868
LNK+	122·2	95·0	1613
LNP+	22·5	95·8	1524
NPK+	273·3	351·8	1781
LNPK+	$3{\cdot}3 \times 10^6$	105·8	3150
LNPK+ m.e.	$6{\cdot}5 \times 10^6$	155·0	3059

[a] Calculated from Rodriguez-Kabana and Collins (1979).
[b] L, lime; N, nitrogen; P, phosphorus; K, potassium at crop-specific rates; m.e., minor elements; ±, with or without winter legumes.
[c] Averages for corn, soybean and cotton in 1970 and 1971.

A fertilizer trial of at least 15 years' standing, on a loamy sand in Alabama, was sampled by Rodriguez-Kabana and Collins (1979) with special reference to *Helicotylenchus dihystera* and *Paratrichodorus christiei*. The authors reported virtual elimination of *H. dihystera* from plots deficient in nutrients, with lesser depression of *P. christiei*, although they reported that lack of lime resulted in significant increases in this species. For each fertilizer treatment the mean population of each nematode over three crops and two years, and the mean crop yield have been derived (Table 5). There is a marked increase in *H. dihystera* populations with increasing "soil fertility", as noted by Rodriguez-Kabana and Collins (1979), with a correlation between *H. dihystera* and yield in Table 5 of +0·5984. Over all crops liming actually decreased the *P. christiei* population (NPK+ *vs.* LNPK+; Table 5); this is considered to illustrate the impression obtained from the literature that Trichodoridae are poor competitors as the same treatment comparison shows a great increase in *H. dihystera*. Growth of winter legumes (common vetch and crimson clover) also leads to a marked increase in *H. dihystera* and depression of *P. christiei* (treatments LPK− *vs.* LPK+ and LNPK− *vs.* LNPK+). These and other comparisons in Table 5 indicate that with greater fertility the total nematode population feeding on the roots increased, but in the same situation there was a change in the proportion of the two species populations counted.

The relation between applied fertilizer and nematode populations reflects plant growth, and analysis of results must account for this and for possible shifts in specific composition of the nematode infestation.

VI. GRAZING OR PRUNING RESPONSES

Seasonal patterns of plant activity may be significantly altered, particularly in managed systems, by grazing or pruning. The relationship between grasses and grazers has been debated (e.g., Owen and Wiegert 1981; Thompson and Uttley, 1982) and the dynamic aspects of herbivory have been reviewed by Crawley (1983). Moderate defoliation may increase carbohydrate contents of roots and severe defoliation may decrease it (Kinsinger and Hopkins, 1961). Petelle (1982) suggested that herbivore activity may stimulate phyllosphere and rhizosphere micro-organisms, enhance plant nitrogen relations and increase soil organic matter. Such influences, which may persist for more than a year (Pease *et al.*, 1979), are significant for nematode populations.

In a six-year-old ryegrass (*Lolium perenne* L.)/white clover (*Trifolium repens* L.) pasture in New Zealand there were significant increases in several nematode populations in response to increased grazing pressure (Table 6); herbage production was 2% greater in those plots with greater grazing pressure which also had a greater percentage of ryegrass cover. Although site data are lacking, increases in total plant-feeding nematode populations in cattle-grazing areas have been reported in a Washington shrub–steppe ecosystem (1 751 000 to 2 238 000 per m^2 in 0–40 cm soil for three 1974 samplings; Smolik and Rogers, 1976) and in a Colorado shortgrass prairie (1 767 000 to 3 480 000 per m^2 in 0–60 cm soil in August 1973; Smolik and Dodd, 1983). Grazing of a South Dakota prairie by the rodent *Cynomys ludovicianus* significantly increased populations of Rhabditida, Tylenchida

Table 6
Effect of stocking rate on nematode populations in a six-year-old pasture.[a]

Nematode population	Sheep per ha 14·8	Sheep per ha 22·2
Pratylenchus per m^2	53 400	80 100**
Pratylenchus per 5 g root	23.3	32·6*
Paratylenchus per m^2	92 800	95 700
Helicotylenchus per m^2	13 900	15 500
Heterodera trifolii juveniles per m^2	163 000	234 000**
H. trifolii cysts per m^2	43 400	44 800
Mononchidae per m^2	10 800	16 600**
Dorylaimida per m^2	206 000	230 000
Total nematodes per m^2	1 610 000	1 950 000**

[a] Adapted from Yeates (1976).
* Denotes $p < 0{\cdot}05$; ** $p < 0{\cdot}01$.

Table 7
Nematode populations, root and herbage weights in improved pasture under differing stocking regimes.[a]

Sheep per ha	Total nematodes per m^2 in 0–25 cm soil	Washed roots (kg D.M. ha^{-1})	Green herbage (kg D.M. ha^{-1})
10	270 000	6 800	3 125
20	191 000	4 263	1 959
30	101 000	2 516	620

[a] Adapted from King and Hutchinson (1976).

and plant-parasitic Dorylaimida (Ingham and Detling, 1984). Changes in vegetative composition, soil temperature and increase in total soil nitrogen were associated with grazing and explain many nematode population changes. It also appears that in the rodent-grazed areas nematodes consumed proportionately twice as much root production as in ungrazed areas, reducing root standing crop and plant ability to withstand stress.

Conversely, King and Hutchinson (1976, 1983) have found in New South Wales that increases in stocking rate lead to decreased total nematode populations in both natural ($P < 0{\cdot}2$) and improved ($P < 0{\cdot}05$) pastures. Their data suggest these decreases were due largely to changes in the habitat, as evidenced by the root and herbage mass (Table 7). With decreased food availability nematode populations would be expected to decline.

The removal of aerial parts of banana (*Musa acuminata* Colla) suckers was found by Mateille *et al.* (1984) to influence population dynamics of *Radopholus similis* and *Helicotylenchus multicinctus*. Root-infesting *H. multicinctus* preferentially multiplied in organs weakened by pruning, while the population of *R. similis* in both roots and cortex was inhibited (Table 8). Application of the growth hormone 2,4-D to cut surfaces restored more natural populations.

The defoliation of pot-grown peas (*Pisium sativum* L.) by Dolliver (1961) did not significantly affect levels of root infestation by *Pratylenchus penetrans*.

A pot trial by Stanton (1983) used three trimming regimes to influence the relation between blue grama grass (*Bouteloua gracilis* (H.B.K.)) and *Helicotylenchus exallus*. It was found that, in addition to net production lost as a result of grazing (trimming or *H. exallus*), the intensity of above-ground grazing affected populations of both *H. exallus* and microbial-feeding

Table 8
Percentages of sectors of banana corms containing nematode species.[a]

Nematode species	Upper sectors	Lower sectors
Radopholus similis	39·4	44·4
Meloidogyne spp.	15	15·6
Hoplolaimus pararobustus	18·1	49·4
R. similis + *Meloidogyne* spp.	6 (6)[b]	5·4 (7)
R. similis + *H. pararobustus*	27·8 (7)	45·6 (22)
H. pararobustus + *Meloidogyne* spp.	13 (3)	21·5 (8)

[a] From Quénéhervé and Cadet (1985).
[b] Values in parentheses are percentages of this mixture if mixing is random.

nematodes by reducing root biomass, and probably by changing the root nutrient status or the chemical defence system (i.e., via root quantity and quality).

Despite limited data it is clear that through its effects on root quantity and quality, soil physical and chemical conditions, and perhaps by altering the vegetative composition of the plant community, grazing or pruning of plants influences nematode populations.

VII. ROOTING PATTERNS

Plant roots provide the resources of plant-feeding nematodes and root distribution determines nematode distribution. Generally, both root and nematode abundance decline with depth (Rossner, 1972) but some root-feeding genera may have their greatest population well below the surface, for example *Xiphinema* below cultivation depth in potato fields (Evans, 1979) and *Paratylenchus* in some grazed pastures (Yeates *et al.*, 1983). Perhaps the most dramatic effect of roots on nematodes are the root chamber observations from southern England. In *Trichodorus viruliferus* a greater proportion of females in the rhizosphere had oocytes than in root-free soil (Pitcher and McNamara, 1970). The population was associated with the extending root tip of apple trees (*Malus sylvestris* (L.)) and migrated with the fast-growing root tip (Pitcher, 1967). Such migration with host roots also occurs in *Radopholus similis* but generally self-dispersal of root feeding nematodes is of much shorter range (*c.* 10 cm yr^{-1}; Jones, 1980; Thomas, 1981) to just behind the zone of root extension. Caged aphid populations have similarly been found to move to the part of the plant offering the best phloem sap (Larsson, 1985).

Plant root distribution, even in perennial plants, varies seasonally and is not always synchronous with shoot growth (Montenegro *et al.*, 1982). Apart from specific stimuli such as hatching factors, roots are generally considered most suitable for nematode feeding in spring or when seeds germinate (e.g., Olthof, 1982). Older, tougher leaves may cause wear of insect mouth parts (Raupp, 1985); the importance of root toughness and thickening with age on nematodes is unknown, but it may be reflected in more robust stylets. Not only is nutrient content important, but thickening of cell walls, development of defence mechanisms and accumulation of possibly toxic plant metabolites may also restrict subsequent nematode feeding. Some defence mechanisms may translocate compounds primarily initially evolved to protect leaves from insect grazing. The pattern of root-hair development varies markedly between species (Itoh and Barber, 1983) and perhaps between cultivars; their abundance will influence those Tylenchida with delicate stylets better suited to such tissue (e.g., *Tylenchus, Cephalenchus*). In addition to the roots and root-hairs, hyphae of mycorrhizal fungi may extend the range of roots and provide selective exploitation of the heterogeneous soil (St John *et al.*, 1983). However, nematode grazing on such hyphae probably reduces their effectiveness and leaves roots more vulnerable to attack by pathogens (Ruehle, 1972).

Just as the architecture of above-ground vegetation may influence insect herbivore diversity through seasonally changing resource availability (Stinson and Brown, 1983) migratory root-feeding nematodes may move from one plant species to another during the year. However, the sedentary cyst-nematodes (*Heterodera* and *Globodera*) have their seasonal activity strictly limited to the period in which their hosts have suitable roots (*Globodera rostochiensis* once a year in potatoes, *Heterodera trifolii* twice a year in white clover; Williams, 1978; Yeates, 1973); the balance of the year is spent in resistant cysts. In root-knot nematodes (*Meloidogyne*), however, there is some evidence that the roots of the host plant may provide a sanctuary for the nematode in periods when migratory juveniles would suffer serious mortality in the soil (Yeates *et al.*, 1985), although work in a cropping regime in Africa suggests vertical migration of second-stage juveniles thus avoiding moisture stress (de Guiran and Germani, 1980).

In addition to the general stress imposed on the plant by nematode feeding, the activities of root-feeding nematodes may have a significant effect on root-growth patterns and may reduce the efficiency of the root network and thus the competitiveness of the particular plant. Gross effects include the inducing of root-knots by *Meloidogyne*, stubby roots by *Trichodorus*, root galls by *Anguina*, and the necrosis and root death which infection with secondary pathogens may cause following *Pratylenchus* invasions. Reduced competitiveness of the plant may result in a change in the structure of the plant community, just as insect (McBrien *et al.*, 1983) or mammallian (Crawley, 1983) grazing may affect plant succession. If a root-feeding nematode is host-specific, extinction of its host would-be serious; strategies in the plant–nematode balance are discussed in Section XI.

Roots of a given plant may be used as a resource by several nematodes during the year. Many agricultural plants can be used by *Heterodera* and *Meloidogyne* (sedentary endoparasites), *Pratylenchus* (migratory endoparasite), one or more migratory ectoparasites (e.g., *Criconemoides, Xiphinema, Trichodorus*) and external root-browsers of lower pathogenicity (e.g., *Tylenchus, Helicotylenchus, Pungentus*). The interpretation of interactions between such infections is difficult (Section IX), and recent results for infestations in banana roots (Table 8) are the first to suggest "conditioning" of roots by one nematode for invasion by another.

The use of several plant species by a nematode species is poorly documented—apart from the importance of "weeds" maintaining pathogenic nematodes during a rotation (Hooper and Stone, 1981)—but as co-existing grasses differ in temporal and vertical patterns of growth (Veresoglou and Fitter, 1984) so a nematode may move between the intertwined roots of potential hosts, selecting the most favourable resource at a particular time. Scandinavian work on nematodes in coniferous forests (e.g., Magnusson, 1983) is a developing avenue in this regard. Selection may involve shifting between monocotyledonous and dicotyledonous plants whose rooting patterns may be complementary as the former tend to take up more divalent cations and the latter monovalent cations (Woodward *et al.*, 1984). While the narrow host range of some *Heterodera* spp. and *Anguina* spp. limits them to either monocotyledons or dicotyledons there appears to be no general preference for either class.

Seasonal pulses of root growth are often associated with the production of root exudate (Section VIII) and there is commonly an assumed correlation between root growth and nematode population increase. The migration of *Trichodorus* with growing apple-tree roots has already been discussed. In grapes McKenry (1984) has demonstrated the occurrence of a flush of root initiation just prior to bloom and a second flush just after harvest; while nematode populations have yet to be related to these flushes, valuable data is available on the abundance of root primordia and implications for chemical control strategies are considerable. In lucerne (*Medicago sativa* (L)) Olthof (1982) found that penetration by *Pratylenchus penetrans* decreased with increasing age of root tissue; again control of the nematode should recognize the root age factor. Growth rates of *Meloidogyne javanica* in bean (*Vicia faba* L.), tomato (*Lycopersicon esculentum* Mill.) and cabbage (*Brassica oleracea* var. *capitata*) roots were found by Bird (1972) to decrease in that order, just as root weight gain decreased.

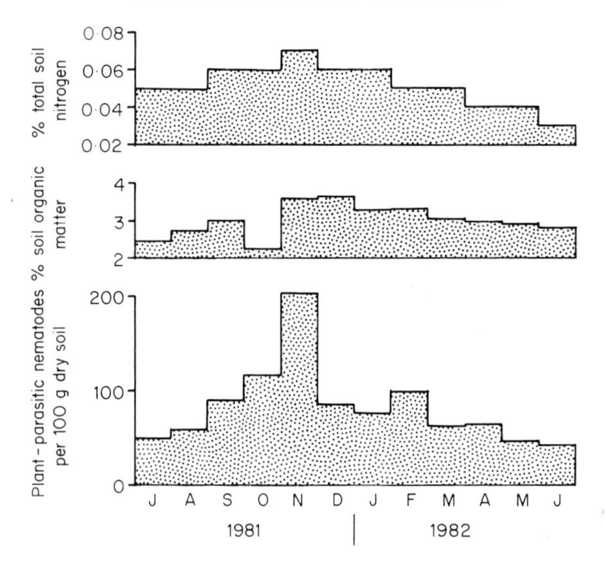

Fig. 3. Monthly changes in total plant-parasitic nematode populations, soil organic matter (by muffle furnace) and soil nitrogen (micro-Kjeldahl) in the rhizosphere of the grass *Andropogon pumilus* Roxb. in a savanna-type tropical hill ecosystem in Orissa, India (from data of Pradhan and Dash, 1984).

Nematode dynamics in the rhizosphere of the grass *Andropogon pumilus* Roxb. were studied by Pradhan and Dash (1984) (Fig. 3). Plant-parasitic nematode abundance in the rhizosphere was better correlated with total nitrogen ($r = +0{\cdot}7688^{**}$) than with percent organic matter ($r = +0{\cdot}4247$), while abundance in root-free soil was poorly correlated with these factors ($r = -0{\cdot}2056$ N; $r = +0{\cdot}1250$ OM). Differential population increase of plant-parasitic nematodes in the rhizosphere was also significantly correlated with both total nitrogen ($r = +0{\cdot}7843^{**}$) and organic matter ($r = +0{\cdot}8713^{***}$). While these data are compatible with results for *Helicotylenchus* and *Pungentus* (Section III) the nitrogen measurement includes root exudates and components that are nematode defecation products, such as amino acids and carbohydrates (Ingham and Coleman, 1983). Clearly direct, concurrent estimations of plant-root N and plant-feeding nematode populations are needed.

The root diameters of the two tall fescue (*Festuca arundinacea* Schreb.) cultivars AF-81 and AF-7 are 1·0 and 2·2 mm respectively. Williams *et al.* (1983) found that AF-7, in addition to having superior root penetration, herbage yield, persistence and sward cover, supported more plant-parasitic nematodes per unit volume of soil and weight of root. Root architecture is apparently important in this, as a pot trial showed; AF-7 yielded 450

Hoplolaimus galeatus (L = 1·2–1·9 mm) per 10 g root compared with 220 per 10 g in AF-81, and Elkins *et al.* (1979) considered root diameter *per se* to be an important factor; the two genotypes also differed in herbage composition (Section III). Differences in root diameter will influence the length of stylet required to reach vascular tissue (Section IX), effective nematode density per unit length of root (Section XII), and the annular space available to nematodes between the epidermis and vascular tissue.

Root growth is affected by soil moisture. However, as soil moisture affects nematode hatching, migration and root penetration (Wallace, 1966) the relation between root growth and nematode population increase is complex. Figure 4(a) shows that as moisture and soybean root weight decrease, so does the *Heterodera glycines* cyst level after 81 days; the abundance per gram of root is greatest at 0·3–0·4 bar moisture tension, presumably reflecting optimum conditions for nematode activity. More field-related moisture regimes were used when *Pratylenchus penetrans* multiplication on cherry (*Prunus mahaleb* L.) seedlings was assessed over 120 days (Fig. 4(b)); infestations per plant and per gram of root both fell markedly with increasing moisture tension.

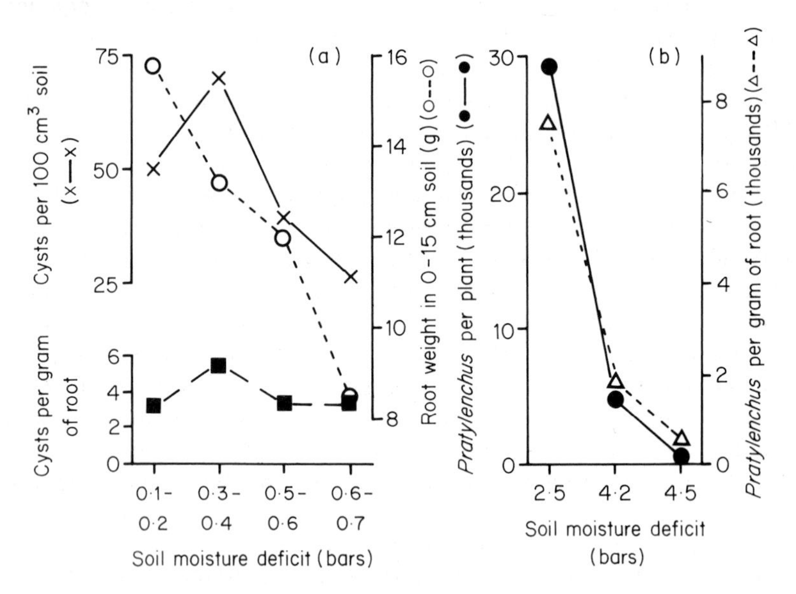

Fig. 4. Influence of soil moisture on nematode populations. (a) *Heterodera glycines* cysts and soybean (*Glycine max.* L.) root weight in 0–15 cm soil after 81 days' growth under greenhouse conditions (from data of Heatherly *et al.*, 1982). (b) *Pratylenchus penetrans* infestations on a root weight and root system basis 120 days after inoculating pots each containing a single cherry (*Prunus mahaleb* L.) seedling with 325 *P. penetrans* (from data of Kable and Mai, 1968).

Embedding techniques for studying the distribution of micro-arthropods in litter (Anderson and Healey, 1970) and of nematodes in undisturbed soil (Boag and Robertson, 1983) have been described. However, such techniques are not practical if one wishes to study the relationship between rooting pattern and mixed nematode infestations, or specific vermiform populations in which stadial determination is required.

VIII. ROOT EXUDATES AND NEMATODE ACTIVITY

Various authors have referred to those chemicals which originate from roots and which stimulate nematode activity as diffusates, exudates and secretions. The physiological origin of these chemicals is unknown and for consistency will be referred to here as *exudates*; Rovira *et al.* (1979) discuss the nomenclature of organic materials in the rhizosphere.

A. Hatching

The role of root exudates in stimulating the hatch of infective, second-stage juveniles of *Heterodera* and *Globodera* and their subsequent emergence from the cyst has been the subject of extensive research (Shepherd, 1962). More recently, work has studied the changing permeability of the egg-shell, trehalose levels and distinctions between quiescence and diapause (Clarke and Perry, 1977). Some emergence occurs in water, some in artificial hatching agents (e.g., zinc chloride, zinc sulphate, picric acid, flavianic acid and picrolonic acid; Shepherd, 1970) but root exudates have attracted most attention. Only for *Heterodera glycines* Ichinoche has a natural hatching stimulus, glycinoclepin A, been identified (Masamune *et al.*, 1982) but, for that species, Tefft and Bone (1984) suggest that zinc may be the cation mediating hatching and not calcium as in other species (Perry and Clarke, 1981; Atkinson and Taylor, 1983). Change in egg-shell permeability and loss of solutes (including trehalose in *G. rostochiensis, Heterodera goettingiana* and *Ascaris suum*; Perry *et al.*, 1983) appears to remove osmotic stress and permit the activation of unhatched juveniles. Change in egg-shell permeability and loss of solutes may constitute the breaking of diapause; chilling breaks diapause in *Meloidogyne naasi* and in *Heterodera avenae*. There is evidence of a winter diapause in other *Heterodera* spp. (Clarke and Perry, 1977), and in *M. incognita* a genetically predetermined proportion of eggs enter diapause (de Guiran, 1979). The termination of diapause (as a consequence of changed egg-shell permeability) and the stimulation of hatching (if it occurs in *M. naasi*) may represent separate, interlocking mechanisms (the latter not being active until after the former). Increased hatch of *Rotylenchulus reniformis* in the presence of host root exudate has been demonstrated (Khan, 1985). All the claimed effects of host root exudate, viz.

(a) stimulation of hatching (Clarke and Perry, 1977),
(b) attraction to plant roots (Prot, 1980),
(c) enhanced juvenile susceptibility to nematicides (Evans and Wright, 1982), and
(d) stimulated juvenile movement and migration (Clarke and Hennessy, 1984),

may reflect enhanced nematode activity in the presence of host root diffusate, increasing with exudate concentration and thus proximity to the emitting root. The plant clearly affects the nematode and, as a result of the evolutionary process, the nematode has an increased opportunity of invading a suitable host plant; intraspecific variability in response to exudates would give a species greater resilience in unpredictable, heterogeneous environments.

B. Egg Hatch

Results on the influence of root exudates on egg hatch in *Meloidogyne* have been reviewed by Hamlen *et al.* (1973), who found that the neutral carbohydrate fraction of lucerne (*Medicago sativa* L.) root exudates significantly increased hatch of *M. incognita* eggs. However, the water hatch exceeded 50% and so any treatment increase was not practically significant. Their method differed from that of most other workers in wetting soil with the exudate instead of using simple aqueous suspensions. However, Hamlen *et al.* (1973) did not consider that their greatest increase in hatch (5·14% to 70·2%) following immersion in exudate from lucerne seedlings could represent a mechanism capable of stimulating the residual eggs in a soil following water hatch, and death, of the vast majority.

C. Moulting

Among plant and soil nematodes, only *Paratylenchus* has been shown to be dependent on host root exudates for moulting stimulus, and the most comprehensive work is that of Fisher (1966) (Table 9) and Ishibashi *et al.* (1975). Although much of the work on *Paratylenchus* has been related to woody plants (e.g., Fisher, 1966; Solov'eva, 1975), great concentrations of resistant sub-adult *Paratylenchus* have been reported in grasslands (Townshend and Potter, 1973; Yeates, 1978). However, the extremely aggregated nature of their populations makes investigation of population dynamics difficult. In a grassland climosequence, Yeates (1974a) reported significant *Paratylenchus* populations at the drier sites (annual precipitation up to 850

Table 9
Effect of apricot root secretion on moulting of fourth-stage juveniles of *Paratylenchus nanus*.[a]

	Percent assigned to condition			Percentage
	Male	Female	Unmoulted	moulted
Trial A: five replicates of 20 juveniles				
Unboiled secretion	29	52	15	84[b]
Boiled secretion	18	27	52	47[b]
Distilled water (control)	23	28	35	59[b]
Trial B: 50 juveniles per treatment				
Secretion from growing seedlings	28	44	28	72
Secretion from dormant seedlings	8	26	64	35
Distilled water (control)	8	20	64	30

[a] Calculated from Fisher (1966).
[b] Least significant difference at $P < 0{\cdot}05 = 14$.

mm) with *Pratylenchus, Radopholus* and *Macroposthonia* being the plant-feeding Tylenchida at wetter, colder sites.

D. Toxic Exudates

Root exudates of marigolds (*Tagetes* spp.) contain alphaterthienyl and bi-thienyl which have nematostatic or nematotoxic effects on a range of nematodes (*Meloidogyne, Pratylenchus, Helicotylenchus, Rotylenchulus, Tylenchorhynchus* and *Tylenchus*). A nematocidal glycoside in the root exudates of asparagus (*Asparagus officinalis* L.) affects *Trichodorus* and other genera (Rohde and Jenkins, 1958). Gommers and voor in T'Holt (1976) reported compounds toxic to *Pratylenchus* in many Compositae, and Saiki and Yoneda (1981) considered an ester produced by goldenrod (*Solidago altissima* L., Compositae) to have both allelopathic and nematocidal effects.

While the physiological basis for production of specific exudates by plant roots is largely unknown, such exudates may have apparent stimulatory or nematostatic effects on nematodes in their rhizosphere. Much remains to be done.

IX. NEMATODE : NEMATODE INTERACTIONS WITHIN ROOTS

The analysis of mixed nematode infestations in plants has been considered to be a difficult area, and nematode–pathogen interactions are more easily analysed (e.g., Dropkin, 1980), although Norton's (1978) discussion of "cohabitation by nematodes" was an advance in considering heterogeneity of niches. Occasional sampling of mixed nematode infestations gives a series of effectively unrelated points. The roots of a given plant vary in quantity and quality during the year, and thus their "carrying capacity" for plant-feeding nematodes also varies. Sequential sampling should indicate the population dynamics of the various nematode species associated with the roots, indicate how much their populations are complementary (partitioning the resource the root offers) and how much they compete. All types of root-feeding nematodes must be involved in the analysis of interactions around the roots of a given plant, while different feeding types may utilize differing niches the total resource is limited.

Agricultural crops, in which plant presence and growth are grossly managed, in which soil is regularly cultivated and in which vegetation is a restricted cultivar rather than a community of potential alternative food sources, do not provide a suitable regime in which to find generalizations. However, they do offer avenues for "experimentation".

Soil physical aspects of the nematode habitat were discussed by Jones *et al.* (1969) and soil factors are important in determining the composition of the nematode fauna (Ferris *et al.*, 1971: Yeates 1984b). Jones *et al.* noted that tillage limits the distribution of *Xiphinema diversicaudatum* (adult length *c.* 4 mm), and tillage leads to a marked reduction in *Helicotylenchus dihystera* below *Sorghum halepense* (L.) (Barker *et al.*, 1969; Johnson *et al.*, 1974). Dropping or tumbling of soil can result in the significant reduction of many nematode populations, with Trichodoridae being particularly susceptible (Spaull and Murphy, 1983; Suatmadji, 1984).

Reduction in tillage has been used to decrease energy inputs to crops and to reduce the risk of erosion. In two cases, specific nematode populations have been studied. Caveness (1979) showed that after 7 years of *Zea mays* L., *Helicotylenchus pseudorobustus* and *Meloidogyne incognita* in both soil and roots were decreased by annual disc ploughing to 20 cm, but *Pratylenchus* spp. increased in response to the treatment (Table 10). As the treatment would influence maize growth, soil organic matter, soil structure, moisture and a range of pathogens, the mode of action is uncertain. However, a compensatory increase in *Pratylenchus* to exploit the available root resource represents a remarkable, yet totally predictable, phenomenon.

Thomas (1978) presented graphs of four nematode populations in the sixth year of a maize trial which has been subjected to seven tillage regimes. Soil populations of *Helicotylenchus pseudorobustus, Xiphinema americanum* and both soil and root populations of *Pratylenchus hexincisus* + *scribneri* were all greatest in the "no-till ridge" treatment, as were plant yields. Whether tillage effects *per se* plant growth were responsible for

Table 10
Effect of seven years' tillage on nematode populations in maize (*Zea mays* L.) crops.[a]

	Nematodes per litre soil		Nematodes per gram of root	
	Tilled	No-tillage	Tilled	No-tillage
Helicotylenchus pseudorobustus	171	1 075	0	0·08
Meloidogyne incognita	235	825	0	124
Pratylenchus sefaensis + *brachyurus*	21 480	4 078	5 111	1 629

[a] Adapted from Caveness (1979).

the observed differences is unclear, but most nematodes occurred in the treatment yielding most maize.

The addition of *Meloidogyne incognita* to tomato (*Lycopersicon esculentum* Mill.) significantly decreased the *Pratylenchus penetrans* population in the roots, and while Estores and Chen (1972) used the results of split-root plants to suggest plant metabolic mediation in the effect, basic food/space availability would provide a simpler explanation. The addition of *Pratylenchus coffeae* to yam (*Dioscorea rotundata* Poir) significantly decreased *Scutellonema bradys* populations according to Acosta and Ayala (1976); *P. coffeae* was apparently the dominant species in their pots. Results in Table 6 show that banana roots contain a greater frequency of mixed infestations of *Radopholus similis* + *Hoplolaimus pararobustus* and *H. pararobustus* + *Meloidogyne* spp. than expected from random occurrence and suggest some form of synergy, but data on how the relative numbers changed is not given. Results in Fig. 5 indicate a remarkably uniform level of total nematode populations in pots of Bermuda grass whether infected with one, two or three species of migratory plant-feeding nematodes; reductions in plant weight were also similar. These pots provided an environment which produced a uniform amount of plant material which, in turn, could support a uniform nematode population.

In Section V there was an indication of *Paratrichodorus christiei* populations being suppressed by *Helicotylenchus dihystera* (Table 5). In soybeans (*Glycine max*) Weaver *et al.* (1985) significantly reduced the soil *Meloidogyne incognita* population by fumigation, reduction occurring in 20 out of 21 plots, but in the same plots *P. christiei* was reduced only once, and it increased in all other plots with ten of the increases being significant ($P < 0{\cdot}05$). A field study by Castaner (1963) of nematodes attacking corn (*Zea mays* L.) is summarized in Table 11 and indicates that for most fertilizer additions *Helicotylenchus microlobus* decreased at the expense of *Pratylen-*

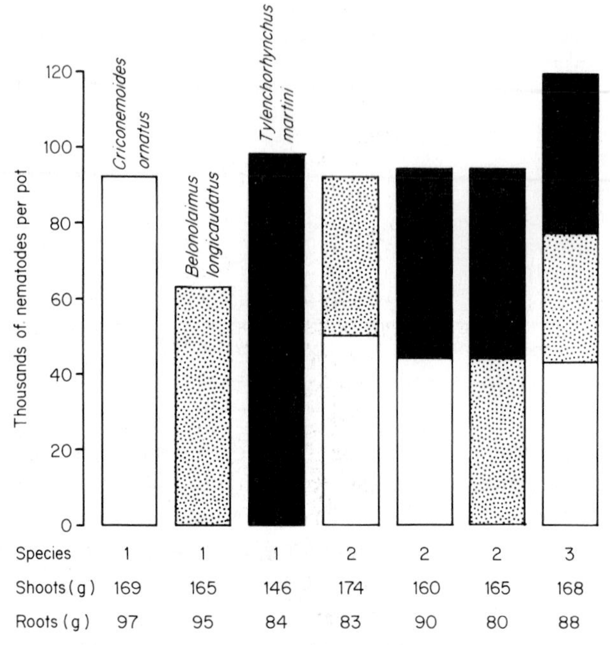

Fig. 5. Total nematodes recovered from pots of Bermudagrass (*Cynodon dactylon* L.) cv. Tufcote 155 days after inoculating with one, two or three species of migratory plant-feeding nematodes. Cumulative leaf weight and final root weight are given with an indication of whether these differ significantly ($P < 0{\cdot}05$) from uninoculated controls which were 209 g and 123 g respectively (from data of Johnson, 1970).

Table 11

Abundance of three nematode species per 500 ml soil in corn (*Zea mays* L.) plots receiving different soil amendments.[a]

Amendment[b]	*Helicotylenchus microlobus*[c]	*Pratylenchus* spp.[c]	*Xiphinema americanum*[c]
none	760	230	240
NPK	225	310	420
m	575	220	645
NPK, m	20	1260	415
lime	520	195	510
lime, m	20	875	1315
NPK, lime	1115	145	315
NPK, lime, m	60	1270	570

[a] Calculated from Castaner (1963).
[b] NPK—these elements annually: m, manure each 4 years 1915–1952; lime, limestone to adjust soil to pH 6·0.
[c] Average of June and September 1962 population estimates.

chus spp. ($r = -0{\cdot}8024^*$); *H. microlobus* showed a similar, but weaker ($r = -0{\cdot}5251$) relation with *Xiphinema americanum*.

These examples have generally indicated some degree of competition between nematode species for the limited resource offered by the plant roots. Intraspecific competition may play a role in sex determination (Section XII). Three-weekly sampling of nematodes in roots of white clover (*Trifolium repens* L.) in pasture plots by Yeates *et al.* (1985) has shown an apparent sequential, complementary distribution of *Meloidogyne hapla, Heterodera trifolii* and *Pratylenchus* over three sites (Fig. 6(a)). While climatic or edaphic factors may exclude one or more genera from a given site, the phenology of *Heterodera* was remarkably consistent (reflecting the close association with *T. repens* growth), *Meloidogyne* had the longest duration of activity where infestation was highest, and *Pratylenchus* showed greatest diversity in phenology, perhaps benefiting from being able to use grass as an alternative host. Monthly sampling of two successive cotton crops also suggests a sequential, complementary distribution of three nematode genera (Fig. 6(b)).

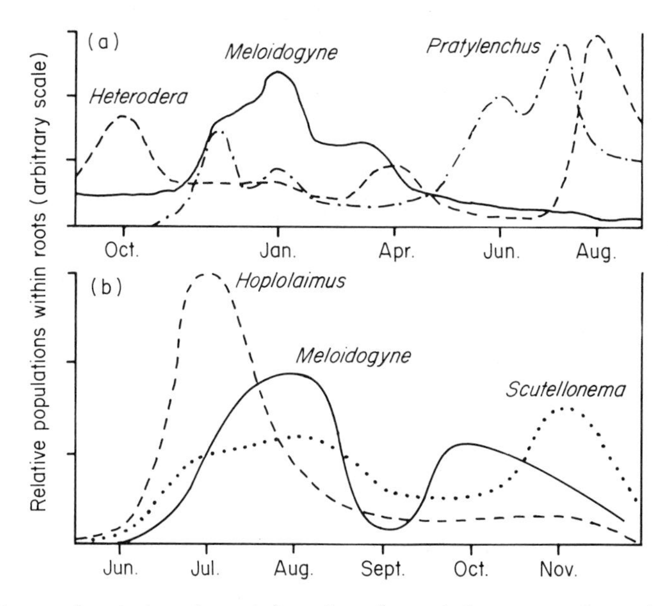

Fig. 6. Seasonal variation of root infestation of co-existing nematode species; in each case the area under each species graph is the same to give populations equal weight. (a) *Meloidogyne hapla, Heterodera trifolii* and *Pratylenchus* sp. in roots of white clover (Trifolium repens L.) averaged over three pasture sites in Waikato, New Zealand (redrawn from Yeates *et al.*, 1985). (b) *Meloidogyne incognita, Hoplolaimus columbus* and *Scutellonema brachyurum* in roots of cotton (*Gossypium hirsutum* L.) in South Carolina (from data in Kraus-Schmidt and Lewis, 1979).

Shorrocks and Rosewell (1986) have shown that many guilds may contain about seven species. Data in Thomas (1978) and Yeates (1974a, 1978, 1984b) suggest that seven Tylenchida often occur in soil samples, while the root/fungal feeding guild described by Magnusson (1983) comprised 6–9 abundant forms.

The understanding of interspecific relations within plant roots will progress significantly when such mixed infestations are monitored and populations of each species and stage are related to the quantity and quality of the root resource available. The use of species associations to detect the presence of small populations of virus-vector species (Boag and Topham, 1985) indicates a positive use for our knowledge of population composition and interaction.

X. HOST PLANT AND NEMATODE MORPHOMETRICS

In the absence of gross morphological diversity, nematologists tend to characterize nematode populations in terms of morphometrics. While marked differences exist in the size of some nominal species, in an increasing number of cases biologically valid species with similar dimensions are being distinguished on biological/biochemical grounds (e.g., *Globodera rostochiensis* and *pallida*; *Radopholus similis* and *citrophilus*).

Fungal-feeding nematodes may be cultured under controlled conditions in the laboratory and Evans and Fisher (1970a,b) showed that by changing the temperature the size of *Ditylenchus myceliophagus, D. destructor* and *Aphelenchus avenae* could be altered. At constant temperatures Pillai and Taylor (1967) found such variation in host preference, host suitability, and morphometrics between five species reared on ten fungi to suggest that each nematode–fungus association produces characteristic effects. Such results must be borne in mind when considering plant-feeding nematode dimensions from various plant/soil/climate regimes.

Several workers have found that progeny of a single egg-mass of *Meloidogyne* species when raised on a variety of plants show a range of size, both in juveniles and females. Nutritional factors appear to be important and Davide (1980) concluded that adult nematodes in less susceptible crops tend to be relatively smaller than those in more susceptible crops, i.e. plants which are "better" hosts provide "better" conditions for nematode growth. Results of Singh *et al.* (1985) (Table 12) illustrate typical results; environmentally influenced body dimensions clearly vary more than genetically determined stylet length. Differing NPK fertilizer treatments of *L. esculentum* cv. Marglobe were found by Pant *et al.* (1983) to give mean body lengths of 449–890 μm, body widths of 456–548 μm and neck lengths of 225–326 μm.

Table 12
Effect of host plant on dimensions[a] of *Meloidogyne incognita* females.[b]

Host	Body length	Body width	Neck length	Stylet length
Artemisia sp. (initial inoculum)	725	420	235	17
Lycopersicon pimpinellifolium (red fruit)	600 ± 4	429 ± 3	217 ± 4	16 ± 0·06
L. esculentum cv. Meeruti	708 ± 8	440 ± 5	253 ± 5	17 ± 0·03
L. esculentum cv. Pusa Ruby	717 ± 6	443 ± 4	239 ± 4	17 ± 0·06
Medicago sativa	723 ± 6	454 ± 5	247 ± 4	17 ± 0·03
Phaseolus vulgaris cv. Royal Red	734 ± 5	462 ± 5	252 ± 5	17 ± 0·06
Vigna sinensis cv. Russian Jaint	766 ± 6	493 ± 3	253 ± 4	17 ± 0·06
Least significant difference ($p < 0{\cdot}05$)	30	28	18	1·3

[a] Means of 50 ♀♀ ± standard errors (μm).
[b] Adapted from Singh *et al.* (1985).

Populations of *Xiphinema diversicaudatum* from a range of hosts in European, North American and New Zealand biotopes, examined morphometrically by Brown and Topham (1985), were considered to represent a single, amphimictic biological species. However, when seven of the populations were cultured for 4 years under uniform greenhouse conditions most showed significant changes in body length, maximum and anal body width, odontostyle and odontophore length (Brown, 1985). However, at the end of this period of uniform conditions significant phenotypic differences remained in all the dimensions reported.

Thus while host plant and its nutrition can significantly influence nematode dimensions, populations may have genetic differences which result in phenotypic differences even under uniform plant/soil/climate conditions.

XI. NEMATODE REPRODUCTIVE STRATEGIES

Being dependent on living roots for their existence, plant-feeding nematodes must balance their life-cycle between

(1) the episodic availability of suitable plant material,
(2) potentially lethal physical conditions,
(3) biological pathogens, and
(4) problems of locating, penetrating and establishing in a host.

If any parasite causes the death of its host or parasitizes an ephemeral or annual organism, that parasite must have a dispersal or survival mechanism so that sufficient of its propagules infect another host to perpetuate the gene pool. Because of the overall need to complete its life-cycle, I prefer the view that selection operates on the entire life-history (Tuomi *et al.*, 1983) rather than individually on each stage of the life-history (Wilbur, 1980).

In agricultural systems which operate on an annual basis many common crops have both host-specific nematodes and facultative nematode pests. These nematodes show a range of modifications to basic nematode morphology and stadial development which reflect the nematode–plant relationship. All post-hatching stages of the migratory, facultative nematodes are able to grow and disperse. Parasitic life-cycles require large numbers of infective agents because of the hazards of transfer from host to host, and this commonly means a relative increase in reproductive effort (Rogers, 1962). This may be accomplished through greater gamete production or uniparental reproduction, but in Heteroderidae the non-feeding nature of the infective second-stage means that egg size is determined by the physical requirements for the size of this stage (in *Heterodera* and *Meloidogyne* these infective juveniles apparently approach the minimum size for functional efficiency; Yeates, 1986). The reproductive effort in sedentary plant-parasitic nematodes is said to be channelled into somewhat increased egg production, coupled with development of the female body wall into a protective cyst in *Heterodera*, and with the production of the protective gelatinous egg-sac in Heteroderoidea as a whole. Data on nematode fecundity are relatively few and, because of the range of experimental conditions under which it has been collected, difficult to assess; the field data on population changes over a cropping cycle (P_f:P_i) not only reflects net population change but, for many species, reflects the compound effect of two or more generations.

As the ability to withstand drought in an anhydrobiotic condition is apparently not confined to any one life-stage or trophic group (Demeure and Freckman, 1981; Glazer and Orion, 1983) and as nematode eggs are, until stimulated in some manner, also rather resistant, we must assume that for unspecialized nematodes all stages are equally susceptible to mortality (this is confirmed by the mixture of stages recovered from soil). The exceptions to this generalization can be enumerated as follows:

(1) *Ditylenchus* stage 4 resistance more extreme ("eelworm wool");
(2) *Paratylenchus* stage 4 resistant;
(3) Heteroderoidea stage 2 resistant (embryonated eggs in cyst or gelatinous matrix);
(4) *Anguina* stage 2 resistant (in seed-gall).

Jones (1980), using life-history correlates of the type given by Pianka (1970), divided plant nematodes by their population strategies into

(1) exploiters—unstable populations, *r*-strategists (*Ditylenchus, Aphelenchoides, Anguina, Rhadinaphelenchus*); and
(2) persisters—relatively stable populations, *K*-strategists (*Heterodera, Longidorus, Helicotylenchus, Pratylenchus, Paratylenchus, Trichodorus* etc.).

While these concepts are useful, a range of meanings have been applied to them (Parry, 1971; Boyce, 1984), and some additional strategies proposed (Andrews and Rouse, 1982; Greenslade, 1983). Their general utility has yet to be demonstrated. The strict interpretation of *r* and *K* implies acceptance of the overall concept of density dependence so that in resource-limited populations *K*-selection will increase resource utilization efficiency (e.g., increased body size, decreased reproductive effort) and *r*-selection will result in maximum resource use to enhance fitness (e.g., decreased body size, increased reproductive effort). How well do these concepts fit plant nematodes?

Of the four variables which plant nematodes must balance in their life-histories, only the episodic availability of food appears truly independent of nematode abundance (density) in all cases. Lethal physical conditions, while generally density independent, may be partly ameliorated through aggregation (as in desiccation resistant "eelworm wool" of *Ditylenchus dipsaci*) or non-dispersal (second stages of *Anguina* and Heteroderidae). Biological pathogens are strongly density dependent as "prey" availability is critical for their success; increasing awareness of the fungal parasites of the egg masses of Heteroderidae (Kerry, 1984) and the prey–predator cycles (Hallas and Yeates, 1972) support this. The establishment of sedentary nematodes in plant roots has been shown to be density dependent, a discrete amount of root being required by each female (e.g., Trudgill, 1972); in migratory forms no data is available. In seed-galls one female *Anguina* dominates the infection in each floret. Thus it appears that only for the sedentary Heteroderoidea and *Anguina* is there a distinct, density-dependent resource limitation; only *within* such groups will the strict *r*-: *K*- strategy be meaningful.

The occurrence of a necessary "diapause" during the development of many plant nematodes has been discussed by Clarke and Perry (1977); progression beyond such a stage can only proceed after a suitable stimulus (e.g., temperature regime, root exudate). Such diapause is widely documented in insects and the similarity of insect diapause to arrested development in zooparasitic nematodes has been noted (Horak, 1981). It is unclear whether plants, through their root exudates, actually contribute to breaking such diapause (Section VIII).

XII. SEX RATIOS

A plant root system has a limited capacity to support nematodes, but nematodes need to produce a sufficient abundance and variety of progeny to locate, invade and reproduce in the root system of another plant. Nematode sex ratios play a part in this balance. While sexual reproduction is normal in nematodes, uniparental reproduction has developed independently in many taxa (Poinar and Hansen, 1983). Although the cytogenetic situation has been resolved in some nematode groups (Triantophyllou and Hirschmann, 1980), overall reproductive strategies are poorly known in ecological terms. Nematodes may show cell constancy, and if chromosomal DNA can be reduced through haploidy this may be advantageous (Lewis, 1985). Most plant nematodes breed once each generation and each generation lasts up to a year; they are thus uniparous (Kirkendall and Stenseth, 1985). Longidoridae may not breed until their second year (Flegg, 1966); they are also uniparous.

Adoption of uniparental reproduction, an adaption both to a parasitic way of life and to life in a heterogeneous, discontinuous environment, is common. The potential increase in population derived from a given resource by adopting uniparental reproduction has long been considered to be at the cost of genetic variability (such reduced variability may in fact be an adaptive advantage in preventing significant departure from specific parasite–host interactions (Law and Lewis, 1983)). However, Slobodchikoff (1982) has shown phenotypic variation in the parthenogenetic wasp *Venturia canescens*, and clonal niche structure has been documented in the parthenogenetic earthworm *Octolasion tyrtaeum* (Jaenike *et al.*, 1980). In the parthenogenetic *Heterodera trifolii* complex Hirschmann and Triantophyllou (1979) found statistically significant differences in stylet length, body length and tail length of second-stage juveniles and in female stylet length between six populations (chromosome number 24–34), distinguishing each chromosomal form. Quantitative differences in multiplication on seven host plants also occurred. This degree of variability is similar to that among the various species and pathotypes of the bisexual *Globodera* complex analysed by Stone (1983). In the bisexual *Radopholus similis/citrophilus* complex morphologically identical biotypes have differing host preferences (Kaplan and O'Bannon, 1985). Clearly the mode of reproduction does not affect the ability of nematodes to adapt to differing species or cultivars of host plant.

Even in obligatory bisexual nematodes the occurrence of specific attractants or pheremones enables contact to be made and may produce aggregations of males and multiple mating of females (Green, 1980; Green *et al.*, 1970). Thus a 1♂ : 1♀ sex ratio is not essential. Some aspects of departure from the 1:1 ratio are discussed below.

From time to time, males are reported in normally uniparental plant nematode species. They may:

(a) indicate the presence of a second species (e.g. differentiation of *Heterodera daverti* from *H. trifolii* by Wouts and Sturhan (1978));
(b) be non-functional or intersexes;
(c) be produced as a population control measure (Section XI).

Unbalanced sex ratios are common among plant-parasitic nematodes. Kerstan (1969) and Ross and Trudgill (1969) independently postulated that in *Globodera rostochiensis* the size of the giant cell group initiated by the second-stage juvenile influences its sex. They contended that under crowded conditions a greater proportion become males, resulting in the residual females being more fecund. This suggestion of "environmental sex determination" has been questioned (Charnov, 1982; Triantophyllou, 1973) and differential mortality of prospective males and females proposed as an alternative mechanism. Bridgeman and Kerry (1980) assessed the fate of juveniles individually placed near seedlings and for susceptible hosts found that *Heterodera cruciferae, H. schactii, Globodera rostochienis* and *G. pallida* all produced a sex ratio of 1 : 1 when sexually differentiated juveniles were counted. On plants with some degree of resistance sex ratios of 3 : 1, 7 : 1 and 8 : 1 were recorded, but these were a result of decreased recovery of female nematodes and were not associated with an increase in the number of males. Fungal infection of roots may increase the proportion of males in *G. rostochiensis* populations (Ketudat, 1969) and an "environmental factor" seems important here. While such results do not apply to all populations of cyst-nematodes (see especially Mugniery and Fayet, 1984) they suggest that for some populations in unfavourable conditions "potential female" juveniles may be unable to establish syncytial transfer-cells and develop to maturity. Sex appears genetically determined, but differential maturation depends on "environmental conditions". Other populations apparently show almost total environmental control of sex determination.

In *Meloidogyne*, which like *Heterodera* lacks sex chromosomes (Goldstein, 1981), sex reversal may lead to increased male production. The number of testes in such males depends on whether conditions were initially adverse (one testis) or favourable, in which case female-destined juveniles with a double gonad-primordium with onset of adverse conditions produce males with two testes (Triantophyllou, 1960; Papadopoulou and Triantophyllou, 1982). Incomplete sex reversal in *M. javanica* may produce intersexes. However, in *M. arenaria, M. graminis* and *M. hapla* the situation is complicated by some populations being normally sexually reproducing and other populations being uniparental (Triantophyllou, 1973). In *M. graminicola* Prasad and Rao (1984) found that foliar applications of nematicides increased the proportion of males in populations in pots, and

Bird (1970) found most *M. javanica* males in nitrogen-deficient plants with high levels of inoculation. Thus sex ratios are influenced by species, populations and "environmental conditions". Triantophyllou (1973) includes temperature, plant growth regulators, infection density, host resistance, age of host plant and host nutrition among these conditions, clearly illustrating the importance of the plant.

Information on sex ratios in vermiform plant nematodes is rarely given and in many species studied males are absent or rare. Co-existing bisexual populations of *Rotylenchus robustus, Trichodorus primitivus, T. velatus* and *Paratrichodorus pachydermus* in forest nurseries were sampled over 36 months by Boag (1981, 1982). In *R. robustus*, females were 1·0–2·2 times more abundant than males, while in the trichodorids the factor was 1·0–1·6. In contrast (Yeates (1982, 1984a) studied *Pungentus maorium* and *Helicotylenchus pseudorobostus* under grazed pastures for the same period and found a single male *P. maorium*. Clearly these two sets of populations are operating under contrasting regimes in terms of reproductive strategies and resource utilization. Females normally predominate in *Paratylenchus*, although the non-feeding males may be equally numerous in *P. elachistus* and *P. peperpotti* (Solov'eva, 1975). A population study of *P. dianthus* gave sex ratios of 1 : 3·3, 1 : 1·5 and 1 : 1·5 in a 3-year study (D'Errico and Di Maio, 1979). Viglierchio and Croll (1968) report that the usual sex ratio of the garlic race of *Ditylenchus dipsaci* is 1:2, but in callus culture they found ratios of 1:3 to 3:1, the proportion of males increasing as host suitability decreased. In the field Yeates (1972) found the ratio to vary, the maximum being 1:24 in July (when the total nematode fauna had been subjected to drought-induced mortality). Unfortunately pot data for *Pratylenchus penetrans* which suggests a greater proportion of males under drier conditions (Norton and Burns, 1971) is based on only that part of the population within roots, and the treatments used are such as to suggest differential migration into the soil.

The occasional males reported in species such as *Pratylenchus zeae, Helicotylenchus dihystera, Pungentus maorium, Longidorus leptocephalus, Paralongidorus maximus, Xiphinema index* and *Paratrichodorus minor* represent anomalies just as do intersexes (e.g. Goseco and Ferris, 1973) and forms with two vulvae (e.g. Jairajpuri *et al.*, 1980) etc. Their rarity means that their genetic basis cannot be studied and that they have little effect on population dynamics. However, information that they occur in response to some form of environmental stress on the population (as in *P. maorium*; Yeates, 1982) would confirm the picture given by species with more abundant males.

There is an overall trend for populations to contain a greater proportion of males when they are subjected to some form of stress, commonly acting through the plant host but including temperature, crowding, plant resistance, plant nutrition and plant age. The response may be seen as optimizing the use of the available resource and population survival while maintaining genetic diversity.

Among the nematodes as a whole, male investment in reproductive activities is, as in many Lepidoptera (Marshall, 1982), minimal and spermatophores are only known in the marine *Prorhynchonema warwicki* (Monhysterida) (Gourbault and Renaud-Mornant, 1983). Effective sex ratios in plant nematodes are determined by both genetic and environmental factors and, in addition to understanding these effects, it would be worthwhile to understand how evolutionary, distributional, geograpical and environmental conditions have interacted to produce the current mix of bisexual and uniparental nematode populations. Theories of sex allocation, and breeding systems as a whole, are still being developed (Charnov, 1982).

The relatively sedentary nature of plant-feeding nematodes, in contrast to the mobility of many phytophagous insects (Thompson, 1983), restricts nematode response to lack of food. Modification of the sex ratio, however achieved, and thus altered population growth can be seen as a nematode response to the limited resources available in the plant.

XIII. NEMATODE RACES AND PLANT CULTIVARS

Only in the past 20 years have the traditional morphological nematode species been critically assessed in terms of the "biological species" concept (Mayr, 1969) which combines the genetic, reproductive and ecological properties of populations. Experiments have demonstrated the "biological validity" of *Rhabditis elegans* and *R. briggsae* (Nigon and Dougherty, 1949), *Heterodera carotae, cruciferae* and *goettingiana* (Yeates, 1970), *Globodera rostochiensis* and *G. pallida* (Parrott, 1972), *Ditylenchus dipsaci* and *D. destructor* (Ladygina, 1975), and *Pratylenchus penetrans* and *P. fallax* (Perry *et al.*, 1980). However, from as long ago as 1888 there was awareness of "physiological races" of the polyphagous stem nematode (*Ditylenchus dipsaci*); these races have yet to be adequately defined on a worldwide basis. I support Sturhan (1969) who considered that monoculture of a particular host favours certain gene combinations and suppresses others, hence changing the gene frequency and the pathogenic potential of a population. *D. dipsaci* does not have "biological races" (non-interbreeding sympatric populations; *sensu* Mayr, 1969) and while some plant nematodes may be found to have such "biological races" which are reproductively isolated by host-plant preference and are a forerunner to "sibling species" (morphologically identical), "physiological race" seems most appropriate. Nematologists use "pathotype" to reflect pathogenic specificity on a restricted range of test plants (e.g., Jones *et al.*, 1981); several pathotypes may co-exist within a population.

As discussed in Section XII, many plant-feeding nematodes have permanent uniparental reproduction and thus produce a "clone" of genetically similar progeny (variability in such clones is discussed above). In such species the morphological species concept must be applied (Mayr, 1969), although physiological and genetic characteristics of populations of such species can be used for subspecific ranking (e.g., in the *Heterodera trifolii* complex).

Vegetative reproduction means that "clones" of genetically uniform plant material can be established. In those plants solely sexually reproducing in nature, breeding can give "lines" with a narrow genetic basis. In agricultural terms these are both termed "cultivars". Genetic shift occurs naturally in clonal breeding plants (e.g., Burdon, 1983; Handel, 1985) and there is recent evidence that each meristem in a plant can be part of a mosaic genetic pattern (Antolin and Strobeck, 1985).

Thus, under natural conditions, each population of plant-feeding nematode, whether reproducing sexually or uniparentally, is feeding on a plant population whose genetic composition may slowly change. The diversity of present nematode faunae reflects the complementary changes in plants and nematodes since the Devonian Period. The examples given above illustrate some biologically valid species, while the gene flow between *Deladenus* spp. (Akhurst, 1975), *Haemonchus* spp. (Le Jambre, 1981), four races of *Heterodera glycines* (Price *et al.*, 1978) and three *Aphelenchoides* spp. (Cayrol and Dalmasso, 1975) show varying degrees of species separation.

Under agricultural conditions economic crop loss due to plant–nematode interactions reflects a series of changes imposed on the biota by man, such as the following:

(1) reduced plant diversity in a field so that dispersal mortality is reduced;
(2) the sequential cultivation of a host cultivar in conditions where dispersal mortality is also reduced;
(3) the exposure of plant species to nematode species to which they may not have previously been exposed and thus not been able to develop immunity;
(4) the exposure of cultivars of a plant species to races of a nematode species to which they may not have previously been exposed and thus not able to develop immunity;
(5) the restriction of the plant or nematode gene pool as a result of establishing populations from few colonizers ("founder effect");
(6) the use of a plant rotation which effectively selects for a previously rare form in the nematode population, and apparent emergence of a new race;
(7) the use of a pesticide regime which effectively selects for a particular genotype of plant nematode;

(8) the possible loss or gain of biological control agents for the nematode population.

The prime example of such changes is given by the introduction from the Andes to Europe of potato (*Solanum tuberosum* L.) and potato cyst nematode (*Globodera rostochiensis* and *G. pallida*): they have subsequently been widely dispersed. Resistance of *S. tuberosum* to potato cyst nematode was first reported in 1948 (Ellenby, 1948), *Solanum tuberosum* spp. *andigena* being used to produce resistant cultivars, but in 1957 resistance-breaking biotypes of potato cyst nematode were found (Jones, 1957). Within five years the initially rare resistance-breaking biotype became dominant in some plots (Cole and Howard, 1962; Jones and Kempton, 1978). The resistance-breaking form is now recognized as a distinct species (*G. pallida*), but the two species each have several races (Jones *et al.*, 1981). However, within each species cropping practice can lead to selection (Hominick, 1982; Turner *et al.*, 1983), possibly through selection acting on the proportions of the pathotypes within the progeny of a given female (Kort and Jaspers, 1973). Glasshouse and field trials may give contrasting results (Forrest and Phillips, 1984). Caligari and Phillips (1984) have questioned the method of shift in *G. pallida* populations. The genetic basis for resistance to potato cyst nematode has been reviewed by Jones *et al.* (1981), but whether the European races of potato cyst nematode represent "founder effects" or more recent selection may never be established.

In the migratory endoparasite *Radopholus similis* a "banana race" and a "citrus race" were recognized in 1956 and the latter has now been described as *R. citrophilus*, a sibling species. Further, two morphologically and karyologically identical races of *R. citrophilus*, differing in their ability to reproduce in various citrus rootstocks, have now been reported (Kaplan and O'Bannon, 1985). Problems occur in using resistant plant cultivars to effectively manage populations of cereal cyst nematode (*Heterodera avenae*), soybean cyst nematode (*H. glycines*), *Meloidogyne incognita* and *M. arenaria*, due to the occurrence of nematode races (Brown, 1984; Riggs *et al.*, 1977; Taylor and Sasser, 1978). Races in many other species undoubtedly await recording, and identification of such races is economically important (Mai, 1984).

While mechanisms are not always understood, grasses such as digitgrass (*Digitaria decumbens* Stent.), *Andropogon gayanus, Brachiaria* spp., *Panicum maximum* Jacq. and *Hemarthria altissima* may suppress *Meloidogyne* spp. populations. However, it is clear from the work of Haroon and Smart (1983) and Lenne (1981) that only certain cultivars of such grasses are effective. Thus not only is host plant cultivar specifically important but specificity of allelopathic action must also be considered at the cultivar level. With the pine wood nematode (*Bursaphelenchus xylophilus*) the interaction of nematode race and plant cultivar is compounded by the presence of a fungal component in the disease syndrome (Mamiya, 1983).

The plant–nematode interaction reflects the genetic make-up of the host and parasite populations; evolutionary aspects of this have been considered by Jones (1973) and Stone (1979). It is the local populations which interact and, perhaps, change. While plant breeders are aware of the problems (e.g., Kinloch and Hinson, 1973), experimental workers who habitually culture their nematode population on an "easy" plant (e.g., *Meloidogyne* on tomato) rather than the usual field host for the population may be significantly "selecting" their results. It is also significant that Johnson (1982) reported that the resistance of available resistant varieties often does not hold up in the presence of large nematode populations. Diversity in the large nematode populations must be a factor; if sufficient reproduction of resistance-breaking individuals occurs so may the development of a new race and, given time, a new species of nematode.

XIV. PLANT RESISTANCE MECHANISMS

Most plants are neither invaded by nor hosts to most plant-feeding nematodes: ". . . in nature most plants are resistant to most nematodes. Generally that resistance is manifested at the preinfection stage" (Veech, 1981). The situation is reflected in the preparation of lists of resistant plants (e.g., Franklin and Hooper, 1959) in contrast to the more normal "host lists".

The suitability of an invaded plant as a host for a given nematode is a poorly explored area except for the sedentary endoparasitic forms. The feeding of these nematodes is dependent on a precise biochemical interaction resulting in the development of a metabolically hyperactive transfer cell (syncytia in *Heterodera* spp., giant cells in *Meloidogyne* spp.). Failure to establish a functional transfer cell results in the death of such nematodes and is thus a mechanism of resistance. To be convincing a resistance mechanism, whether physical or chemical, must occur at the site of infection, at the time of infection and have a detrimental effect on the infective agent. While host suitability and resistance mechanisms are complementary, plant reaction (gall or knot formation) may be a secondary effect and not directly related to the mechanism of infection establishment (Kiraly *et al.*, 1972).

In many cases substances that promote plant growth will promote nematode population growth provided that the plant has prior susceptibility. Adverse conditions which may limit both plant and nematode growth may be removed by growth regulators (and fertilizers—Section V) and lead to increased nematode population growth. Chemical treatment to actually cause a breakdown in resistance does not seem to have been recorded.

High levels of phenolics and tannins reduce browsing by plant-feeding insects (Zucker, 1982; Tempel, 1981) and both feeding deterrents (Higgins and Pedigo, 1979; Russell *et al.*, 1978) and organo-tin antifeedants (Ascher, 1979) are widespread in plants. Stimulation of insect feeding also occurs (Bristow *et al.*, 1979) and some insects metabolize plant phenolics (Bernays and Woodhead, 1982). Production of defence mechanisms by plants has a metabolic cost and in annual agricultural crops mobile agents such as alkaloids, phenolic and cyanogenic glycosides seem more likely than the metabolically more costly polyphenols and fibre (see Coley *et al.*, 1985). What then is the situation with plant nematodes?

Levels of phenols in roots have been correlated with cultivar resistance to nematodes (Ponin *et al.*, 1977; Singh and Choudhury, 1973) but this is not always the case (Bruseke and Dropkin, 1973; Feldman and Hanks, 1971). Table 3 shows the increase in phenol content of eggplant with increasing fertilizer application and, while this is correlated with decreasing *Rotylenchus* root populations per gram ($r = -0{\cdot}8718^{*}$) amino-acid content also increases ($r = +0{\cdot}9788^{***}$). The balance between the adverse effects of phenolic compounds on nematodes and parallel changes in beneficial aspects of root quality need to be assessed. There have apparently been no critical analyses of active phenolic levels near the site of resistance, and no correlation made of the presence of such compounds in roots and in foliage as arthropod deterrents. Simple phenols modifying IAA oxidase activity form the basis of Giebel's (1974) "Holistic hypothesis of incompatibility" which has not found acceptance (Kaplan and Keen, 1980; Veech, 1981) in either the *Globodera*–potato interaction or as a general resistance mechanism.

The role of high levels of plant growth regulators in favouring susceptibility to nematode establishment is debatable. While susceptible plants may have higher levels of IAA or NAA, or kinetin, ethylene or tryptophan (Veech, 1981) and treatment of susceptible plants may increase susceptibility, treatment of resistant plants may not break resistance (Sawhney and Webster, 1975). Adjustment of auxin (Setty and Wheeler, 1968) or ethylene (Akitt *et al.*, 1980) levels to an amount per unit weight of tissue (i.e., adjusting for plant size) may eliminate differences between susceptible and resistant plants or reverse the postulated trend. Growth inhibitors, such as maleic hydrazide (Davide and Triantophyllou, 1968), inhibit nematode population development. Inhibition of ethylene action in tomatoes by silver thiosulphate or aminoethoxyvivyl-glycine offset the pathogenic symptoms of *Meloidogyne javanica* without significantly altering nematode populations (Glazer *et al.*, 1984).

Allelochemics are plant chemicals which affect other organisms and may be beneficial or deleterious to those organisms. Beneficial effects as we assess them may be the product of co-evolution of the plant and the nematode. The stimulation of emergence and migration of Heteroderidae by root exudates (Section VIII) is the best known example. Deleterious effects of allelochemics are manifest as resistance. Terthienyls and terpenoids fall into this category (Veech, 1981). Marigolds (*Tagetes* spp.) and margosa (*Azadirachata indica*) are well established as plants whose allelopathic effects on nematodes may be used in the cropping situation (Alam *et al.*, 1975; Suatmadji, 1968), while asparagus (*Asparagus officinalis* L.) is of less practical value (Rohde and Jenkins, 1958).

If plant reaction to invasion produces a specific chemical detrimental to the invader, that chemical is termed a phytoalexin; such compounds have been reported in over 75 species from 20 plant families, with legumes being very reactive (Yoshikawa, 1978). Anti-nematode phytoalexins have been reported in lima beans (*Phaseolus lunatus* L.) in response to *Pratylenchus scribneri* (Rich *et al.*, 1977), in soybeans in response to *Meloidogyne incognita* but not to *M. javanica* (Kaplan *et al.*, 1980), and in resistant cotton cultivars following invasion by *M. incognita* (Veech, 1978). These phytoalexins (coumestrol, glyceollin and terpenoids respectively), are apparently reactions within roots as a response to infection and which have an adverse effect on the invading organism. While the mechanism presumably developed in a mixed vegetation, in agriculture it offers a very effective mechanism of resistance to specific nematodes. However, phytoalexins can have complex effects in ecosystems. Ryegrass (*Lolium perenne* L.) infected with an endophytic fungus (*Acremonium loliae* Latch, Christensen and Samuels) produces a phytoalexin; while the effect on the fungus is unknown the compound affects the nervous system of grazing mammals and significantly reduces feeding by insect herbivores (Clay *et al.*, 1985; Prestidge *et al.*, 1985).

As phytoalexins are produced in response to infection it is not possible to simply screen plants for their presence. Plants must be "challenged" by the nematode before the presence is assessed.

Given specific rejection/acceptance of transfer cell formation, phytoalexin and allelochemic production, and developing knowledge of the genetics of the plant–nematode system (Sidhu and Webster, 1981) the future for the breeding of nematode-resistant plants should lie in manipulating the frequency of these genes. This attitude was also adopted by Harris and Frederiksen (1984) after reviewing host plant resistance to fungi and arthropods. Attempts to modify the physical or chemical parameters of plants, although perhaps contributing to our knowledge of biochemical mechanisms, may not tell us much more than we already know—if a susceptible plant can grow better it will support a greater nematode population.

XV. THE PLANT–BACTERIAL–NEMATODE PATHWAY

A review which shows plant quantity and quality to be critical for plant-feeding nematode populations would not be complete without commenting on the mineralization of plant material. Biological strategies of nutrient cycling have been reviewed, in this series, by Coleman *et al.* (1983) and an understanding of the importance of mobilization of nutrients from soil microbial biomass is developing (e.g., Marumoto *et al.*, 1982; Sparling *et al.*, 1985). While much of the work has been carried out in microcosms, through modelling (Ingham *et al.*, 1985), scaling (Shirazi *et al.*, 1984) and field studies (Parker *et al.*, 1984) such results can be demonstrated as representing real effects in the field. Not only nematodes but all microbial-feeding animals contribute significantly to the mineralization of nutrients in the soil, and this contribution is greater than can be deduced from the proportion these animals contribute to soil biomass or respiration. Data presented in Fig. 1(a,b) shows how total nematode abundance correlates with measures of plant production; this positive correlation is the result of some of the negative effects of the plant-feeding forms being exceeded by the positive effects of the grazing forms of the detritus food web. The distribution of the total nematode fauna is primarily related to root distribution (Yeates *et al.*, 1983; Freckman and Mankau, 1986).

Excretory products of nematodes may contain amino acids, amines, ammonia, proteins and aldehydes (Myers and Krusberg, 1965). While excretory products of bacterial feeding *Mesodiplogaster* significantly influence the forms of nitrogen available to plants in experimental microcosms (Anderson *et al.*, 1983), root feeding by *Tylenchorhynchus claytoni* leads to increased bacterial populations in soil but unchanged nitrogen and phosphorus levels (Ingham and Coleman, 1983).

For an annual desert plant Parker *et al.* (1984) consider that the timing of rain events, ephemeral plant production, plant decomposition, and perennial production are related. Micro-arthropods are important regulators of decomposition and increase the turnover of nitrogen immobilized in fungi. In the absence of micro-arthropods, nematodes become the system regulators, and thus the decomposition process becomes closely associated with soil water content.

In work based on a conceptual model, using *Bouteloua gracilis* (H.B.K.) in gnotobiotic microcosms and studying various trophic interactions, including the nematodes *Pelodera, Acrobeloides* and *Aphelenchus*, Ingham *et al.* (1985) found that microbial grazers may significantly increase mineralization and thus plant growth, at times when mineralization by the microflora is insufficient to meet plant requirements. Although they found mineralization by microbial grazers to have short-term significance, its occurrence in many ecosystems may be at those times ideal for plant growth.

Thus, while microbial-feeding nematodes appear to modify limits to primary production in some situations, in others micro-arthropods also play a significant role.

XVI. CONCLUSIONS

Early naturalists observed insects feeding on plants, and those relationships have been investigated in great detail. Plant-feeding nematodes have been recognized as pathogens for less than 200 years and their effects have been studied over the past 50 years or so. This chapter has brought together information to show that plants, in a dozen or more ways, through their quality and quantity, have a significant effect on nematodes. Detailed analysis of these interactions, although more difficult because of the opaque milieu and microscopic techniques required, should be even more fruitful than studying foliar-feeding insects. Not only are crop loss, development of nematode-resistant varieties and stress alleviation involved, but the nematode-induced flow of plant metabolites into the rhizosphere may be significant in nutrient cycling.

As potato cyst nematode (*Globodera* spp.) has been a basis for modelling population dynamics, and bacterial-feeding *Caenorhabditis briggsae* a tool in genetic studies, so the effects of plants on nematodes offer wide scope for ecological advance.

ACKNOWLEDGEMENTS

D. C. Coleman and M. Luxton have influenced the development of ideas presented here. I am grateful to I. A. E. Atkinson and K. R. Tate for constructive comment on the draft manuscript, and Ngaire McLean for typing.

REFERENCES

Acosta, N. and Ayala, A. (1976). Effects of *Pratylenchus coffeae* and *Scutellonema bradys* alone and in combination on Guinea yam (*Dioscorea rotundata*). *J. Nematol.* **8**, 315–317.

Akhurst, R. J. (1975). Cross-breeding to facilitate the identification of *Deladenus* spp., nematode parasites of woodwasps. *Nematologica* **21**, 267–272.

Akitt, D. B., Bown, A. W. and Potter, J. W. (1980). Role of ethylene in the response of tomato plants susceptible and resistant to *Meloidogyne incognita*. *Phytopathology* **70**, 94–97.

Alam, M. M., Masood, A. and Husain, I. (1975). Effect of margosa and marigold root-exudates on mortality and larval hatch of certain nematodes. *Indian J. exp. Biol.* **13**, 412–414.

Anderson, J. M. and Healey, I. N. (1970). Improvements to the gelatin-embedding technique for woodland soil and litter samples. *Pedobiologia* **10**, 108–120.

Anderson, R. V., Gould, W. D., Woods, L. E., Cambardella, C., Ingham, R. E. and Coleman, D. C. (1983). Organic and inorganic nitrogenous losses by microbivorous nematodes in soil. *Oikos* **40**, 75–80.

Andrews, J. H. and Rouse, D. I. (1982). Plant pathogens and the theory of *r*- and *K*-selection. *Am. Nat.* **120**, 283–296.

Antolin, M. F. and Strobeck, G. (1985). The population genetics of somatic mutation in plants. *Am. Nat.* **126**, 52–62.

Ascher, K. R. S. (1979). Fifteen years (1963–1978) of organotin antifeedants—a chronological bibliography. *Phytoparasitica* **7**, 117–137.

Atkinson, H. J. and Taylor, J. D. (1983). A calcium-binding sialoglycoprotein associated with an apparent eggshell membrane of *Globodera rostochiensis*. *Ann. appl. Biol.* **102**, 345–354.

Badra, T. and Yousif, G. M. (1979). Comparative effects of potassium levels on growth and mineral composition of intact and nematised cowpea and sour orange seedlings. *Nematol. Medit.* **7**, 21–27.

Barker, K. R., Nusbaum, C. J. and Nelson, L. A. (1969). Seasonal population dynamics of selected plant-parasitic nematodes as measured by three extraction procedures. *J. Nematol.* **1**, 232–239.

Bernard, E. C. and Hussey, R. S. (1979). Population development and effects of the spiral nematode, *Helicotylenchus dihystera*, on cotton in microplots. *Pl. Dis. Reptr.* **63**, 807–810.

Bernard, E. C. and Keyserling, M. L. (1985). Reproduction of root-knot, lesion, spiral, and soybean cyst nematodes on sunflower. *Pl. Dis.* **69**, 103–105.

Bernays, E. A. and Woodhead, S. (1982). Plant phenols utilized as nutrients by a phytophagous insect. *Science, N.Y.* **216**, 201–203.

Bird, A. F. (1970). The effect of nitrogen deficiency on the growth of *Meloidogyne javanica* at different population levels. *Nematologica* **16**, 13–21.

Bird, A. F. (1972). Quantitative studies on the growth of syncytia induced in plants by root knot nematodes. *Int. J. Parasit.* **2**, 157–170.

Boag, B. (1981). Observations on the population dynamics and vertical distribution of trichodorid nematodes in a Scottish forest nursery. *Ann. appl. Biol.* **98**, 463–469.

Boag, B. (1982). Observations on the population dynamics, life cycle and ecology of the plant parasitic nematode *Rotylenchus robustus*. *Ann. appl. Biol.* **100**, 157–165.

Boag, B. and Robertson, L. (1983). A technique for studying micro distribution of nematodes in undisturbed soil. *Revue Nématol.* **6**, 146–148.

Boag, B. and Topham, P. B. (1985). The use of associations of nematode species to aid the detection of small numbers of virus-vector nematodes. *Pl. Path.* **34**, 20–24.

Boyce, M. S. (1984). Restitution of *r*- and *K*-selection as a model of density-dependent natural selection. *Ann. Rev. ecol. Syst.* **15**, 427–447.

Bridgeman, M. R. and Kerry, B. R. (1980). The sex ratios of cyst-nematodes produced by adding single second-stage juveniles to host roots. *Nematologica* **26**, 209–213.

Bristow, P. R., Doss, R. P. and Campbell, R. L. (1979). A membrane filter bioassay for studying phagostimulatory materials in leaf extracts. *Ann. ent. Soc. Am.* **72**, 16–18.

Brown, D. J. F. (1985). The effect, after four years, of a change in biotope on the morphometrics of populations of *Xiphinema diversicaudatum* (Nematoda: Dorylaimoidea). *Nematol. Medit.* **13**, 7–13.

Brown, D. J. F. and Coiro, M. I. (1983). The total reproductive capacity and longevity of individual female *Xiphinema diversicaudatum* (Nematoda: Dorylaimida). *Nematol. Medit.* **11**, 87–92.

Brown, D. J. F. and Topham, P. B. (1985). Morphometric variability between populations of *Xiphinema diversicaudatum* (Nematoda: Dorylaimoidea). *Revue Nématol.* **8**, 15–26.

Brown, R. H. (1984). Ecology and control of cereal cyst nematode (*Heterodera avenae*) in southern Australia. *J. Nematol.* **16**, 216–222.

Brueske, C. H. and Dropkin, V. H. (1973). Free phenols and root necrosis in Nematex tomato infected with root knot nematode. *Phytopathology* **63**, 319–334.

Burdon, J. J. (1983). Biological flora of the British Isles. *Trifolium repens* L. *J. Ecol.* **71**, 307–330.

Caligari, P. D. S. and Phillips, M. S. (1984). A re-examination of apparent selection in *Globodera pallida* on *Solanum vernei* hybrids. *Euphytica* **33**, 583–586.

Castaner, D. (1963). Nematode populations in corn plots receiving different soil amendments. *Proc. Iowa Acad. Sci.* **70**, 107–113.

Caveness, F. E. (1979). Nematode populations under a no-tillage soil management regime. In *Soil Tillage and Crop Production* (Ed. by R. Lal), pp. 133–145. Ibadan: International Institute of Tropical Agriculture.

Cayrol, J. C. and Dalmasso, A. (1975). Affinités interspécifiques entre trois nématodes de feuilles (*A. fragariae, A. ritzemabosi* et *A. besseyi*). *Cah. O.R.S.T.O.M. Sér. Biol.* **10**, 215–225.

Charnov, E. L. (1982). *The Theory of Sex Allocation*. Princeton University Press. 355 pp.

Clarke, A. J. and Hennessy, J. (1984). Movement of *Globodera rostochiensis* (Wollenweber) juveniles stimulated by potato-root exudate. *Nematologica* **30**, 206–212.

Clarke, A. J. and Perry, R. N. (1977). Hatching of cyst-nematodes. *Nematologica* **23**, 350–368.

Clay, K., Hardy, T. N. and Hammond, A. M. (1985). Fungal endophytes of grass and their effects on an insect herbivore. *Oecologia* **66**, 1–5.

Cole, C. S. and Howard, H. W. (1962). Further results from a field experiment on the effect of growing resistant potatoes on a potato root eelworm (*Heterodera rostochiensis*) population. *Nematologica* **7**, 57–61.

Coleman, D. C., Reid, C. P. P. and Cole, C. V. (1983). Biological strategies of nutrient cycling in soil systems. *Adv. ecol. Res.* **13**, 1–55.

Coley, P. D., Bryant, J. P. and Chapin, F. S. (1985). Plant availability and plant antiherbivore defence. *Science, N.Y.* **230**, 895–899.

Cook R. J. (1984). Root health: importance and relationship to farming practices. *Am. soc. Agron. sp. Publ.* **46**, 111–127.

Crawley, M. J. (1983). *Herbivory: The Dynamics of Animal–Plant Interactions.* Oxford University Press. 437 pp.

Davide, R. G. (1980). Influence of different crops on the dimensions of *Meloidogyne arenaria* isolated from fig. *Proc. helminth. Soc. Wash.* **47**, 80–84.

Davide, R. G. and Triantophyllou, A. C. (1968). Influence of the environment on development and sex differentiation of root-knot nematodes. III. Effect of foliar application of maleic hydrazide. *Nematologica* **14**, 37–46.

Demeure, Y. and Freckman, D. W. (1981). Recent advances in the study of anhydrobiotic nematodes. In *Plant Parasitic Nematodes* (Ed. by B. M. Zuckerman and R. A. Rohde), Vol. 3, pp. 205–226. New York: Academic Press.

D'Errico, F. P. and Di Maio, F. (1979). [A 3-year study of seasonal dynamics and population density of *Paratylenctus dianthus* on carnation.] In *Nematodi delle colture ortofloricole e industriali Atti Giornate Nematologiche 1979, Firenze, Italy*, pp. 177–190. Societa' Italia a di Nematologia, Italy.
Dolliver, J. S. (1961). Population levels of *Pratylenchus penetrans* as influenced by treatments affecting dry weight of Wando pea plants. *Phytopathology* **51**, 364–367.
Dropkin, V. H. (1980). *Introduction to Plant Nematology*. New York: Wiley, 292 pp.
Edwards, C. A. and Lofty, J. R. (1982). Nitrogenous fertilizers and earthworm populations in agricultural soils. *Soil Biol. Biochem.* **14**, 515–521.
Elkins, C. B., Haalund, R. L., Rodriguez-Kabana, R. and Hoveland, C. S. (1979). Plant-parasitic nematode effects on water use and nutrient uptake of small- and large-rooted tall fescue genotype. *Agron. J.* **71**, 497–500.
Ellenby, C. (1948). Resistance to the potato-root eelworm. *Nature, Lond.* **162**, 704.
Estores, R. A. and Chen, T. A. (1972). Interactions of *Pratylenchus penetrans* and *Meloidogyne incornita* as coinhabitants in tomato. *J. Nematol.* **4**, 170–174.
Evans, A. A. F. and Fisher, J. M. (1970a). The effect of environment on nematode morphometrics. Comparison of *Ditylenchus myceliophagus* and *D. destructor*. *Nematologica* **16**, 113–120.
Evans, A. A. F. and Fisher, J. M. (1970b). Some factors affecting the number and size of nematodes in populations of *Aphelenchus avenae*. Nematologica **16**, 295–304.
Evans, K. (1979). Nematode problems in the Woburn ley-arable experiment, and changes in *Longidorus leptocephalus* population density associated with time, depth, cropping and soil type. *Rep. Rothamsted exptl Stn* 1978(2), 27–45.
Evans, S. G. and Wright, D. J. (1982). Effects of the nematicide oxamyl on life cycle stages of *Globodera rostochiensis*. *Ann. appl. Biol.* **100**, 511–519.
Feldman, A. W. and Hanks, R. W. (1971). Attempts to increase tolerance of grapefruit seedlings to the burrowing nematode (*Radopholus similis*) by application of phenolics. *Phytochemistry* **10**, 701–709.
Ferris, V. R., Ferris, J. M., Bernard, R. L. and Probst, A. H. (1971). Community structure of plant-parasitic nematode related to soil types in Illinois and Indiana soybean fields. *J. Nematol.* **3**, 399–408.
Fisher, J. M. (1966). Observations on moulting of fourth-stage larvae of *Paratylenchus nanus*. *Aust. J. biol. Sci.* **19**, 1073–1079.
Flegg, J. J. M. (1966). Once-yearly reproduction in *Xiphinema vuittenezi*. *Nature, Lond.* **212**, 741.
Forrest, J. M. S. and Phillips, M. S. (1984). The effect of continuous rearing of a population of *Globodera pallida* (Pa 2) on susceptible or partially resistant potatoes. *Pl. Path.* **33**, 53–56.
Franklin, M. T. and Hooper, D. J. (1959). Plants recorded as resistant to root-knot nematodes (*Meloidogyne* spp.). *Commonw. Bur. helminth. Tech. Communs* **31**, 1–33.
Freckman, D. W., Ed. (1982). *Nematodes in Soil Ecosystems*. Austin, Texas: University of Texas Press, 206 pp.
Freckman, D. W. and Mankau, R. (1986). Abundance, distribution, biomass and energetics of soil nematodes in a northern Mojave desert ecosystem. *Pedobiologia* **29**, 129–142.
French, N., Ed. (1979). *Perspectives in Grassland Ecology*. New York: Springer-Verlag, 204 pp.
Gay, P. E., Grubb, P. J. and Hudson, H. J. (1982). Seasonal changes in the concentrations of nitrogen, phosphorus and potassium, and in the density of

mycorrhiza, in biennial and matrix-forming perennial species of closed chalkland turf. *J. Ecol.* **70**, 571–593.

Giebel, J. (1974). Biochemical mechanisms of plant resistance to nematodes: a review. *J. Nematol.* **6**, 175–184.

Glazer, I., Apelbaum, A. and Orion, D. (1984). Reversal of nematode-induced growth retardation in tomato plants by inhibition of ethylene action. *J. Am. Soc. hort. Sci.* **109**, 886–889.

Glazer, I. and Orion, D. (1983). Studies on anhydrobiosis of *Pratylenchus thornei*. *J. Nematol.* **15**, 333–338.

Gnanapragasam, N. C. (1982). Effect of potassium fertilization and of soil temperature on the incidence and pathogenicity of the root-lesion nematode, *Pratylenchus loosi* Loof, on tea (*Camellia sinensis* L.). *Tea Q.* **5**, 169–174.

Goldstein, P. (1981). Sex determination in nematodes. In *Plant Parasitic Nematodes* (Ed. by B. M. Zuckerman and R. A. Rohde), Vol. 3, pp. 37–60. New York: Academic Press.

Gommers, F. J. and voor in 'T Holt, D. J. M. (1976). Chemotaxonomy of Compositae related to their host suitability for *Pratylenchus penetrans*. *Neth. J. Pl. Path.* **82**, 1–8.

Goseco, C. G. and Ferris, V. R. (1973). Intersexes of *Leptonchus obtusus* Thorne. *J. Nematol.* **5**, 226–228.

Gourbault, N. and Renaud-Mornant, J. (1983). Système reproducteur d'un nématode marin à fécondation par spermatophore. *Revue Nématol.* **6**, 51–56.

Green, C. D. (1980). Nematode sex attractants. *Helminth. Abst.* **49B**, 81–93.

Green, C. D., Greet, D. N. and Jones, F. G. W. (1970). The influence of multiple mating on the reproduction and genetics of *Heterodera rostochiensis* and *H. schachtii*. *Nematologica* **16**, 309–326.

Greenslade, P. J. M. (1983). Adversity selection and the habitat templet. *Am. Nat.* **122**, 352–365.

de Guiran, G. (1979). A necessary diapause in root-knot nematodes. Observations on its distribution and inheritance in *Meloidogyne incognita*. *Revue Nématol.* **2**, 223–231.

de Guiran, G. and Germani, G. (1980). Fluctuations des populations de *Meloidogyne incognita* dans un sol cultivé en climat tropical humide. *Revue Nématol.* **3**, 51–60.

Hallas, T. E. and Yeates, G. W. (1972). Tardigrada of the soil and litter of a Danish beech forest. *Pedobiologia* **12**, 287–304.

Hamlen, R. A., Bloom, J. R. and Lukesic, F. L. (1973). Hatching of *Meloidogyne incognita* eggs in the neutral carbohydrate fraction of root exudates of gnotobiotically grown alfalfa. *J. Nematol.* **5**, 142–146.

Handel, S. N. (1985). The intrusion of clonal growth patterns on plant breeding systems. *Am. Nat.* **125**, 367–384.

Haroon, S. and Smart, G. C. (1983). Development of *Meloidogyne incognita* inhibited by *Digitaria decumbens* cv. Pangola. *J. Nematol.* **15**, 102–105.

Harris, M. K. and Frederiksen, R. A. (1984). Concepts and methods regarding host plant resistance to arthropods and pathogens. *Ann. Rev. Phytopath.* **22**, 247–272.

Heatherly, L. G., Young, L. D., Epps, J. M. and Hartwig, E. E. (1982). Effect of upper-profile soil water potential on numbers of cysts of *Heterodera glycines* on soybeans. *Crop Sci.* **22**, 833–835.

Higgins, R. A. and Pedigo, L. P. (1979). A laboratory antifeedant simulation bioassay for phytophagous insects. *J. Econ. Ent.* **72**, 238–244.

Hirschmann, H. and Triantophyllou, A. C. (1979). Morphological comparison of members of the *Heterodera trifolii* species complex. *Nematologica* **25**, 458–481.

Hominick, W. M. (1982). Selection of a rapidly maturing population of *Globodera rostochiensis* by continuous cultivation of early potatoes in Ayrshire, Scotland. *Ann. appl. Biol.* **100**, 345–351.

Hooper, D. J. and Stone, A. R. (1981). Role of wild plants and weeds in the ecology of plant-parasitic nematodes. In *Pests Pathogens and Vegetation* (Ed. by J. M. Thresh), pp. 199–215. Boston: Pitman.

Horak, I. G. (1981). The similarity between arrested development in parasitic nematodes and diapause in insects. *J. S. Afr. vet. Assoc.* **52**, 299–303.

Husain, S. I., Mohammad, H. Y. and Al-Zarari, A. J. (1981). Studies on the vertical distribution and seasonal fluctuation of the citrus nematode in Iraq. *Nematol. Medit.* **9**, 7–19.

Ingham, R. E. and Coleman, D. C. (1983). Effects of an ectoparasitic nematode on bacterial growth in gnotobiotic soil. *Oikos* **41**, 227–232.

Ingham, R. E. and Detling, J. K. (1984). Plant–herbivore interactions in a North American mixed-grass prairie. III. Soil nematode populations and root biomass on *Cynomys ludovicianus* colonies and adjacent uncolonised areas. *Oecologia* **63**, 307–313.

Ingham, R. E., Trofymow, J. A., Ingham, E. R. and Coleman, D. C. (1985). Interactions of bacteria, fungi, and their nematode grazers: effects on nutrient cycling and plant growth. *Ecol. Monogr.* **55**, 119–140.

Ishibashi, N., Kondo, E. and Kashio, T. (1975). The induced molting of 4th-stage larvae of pin nematode, *Paratylenchus aciculus* Brown (Nematoda; Paratylenchidae) by root exudate of host plant. *Appl. Ent. Zool.* **10**, 275–283.

Itoh, S. and Barber, S. A. (1983). Phosphorus uptake by six plant species as related to root hairs. *Agron. J.* **75**, 457–461.

Jaenike, J., Parker, E. D. and Selander, R. K. (1980). Clonal niche structure in the parthenogenetic earthworm *Octolasion tyrtaeum*. *Am. Nat.* **116**, 196–205.

Jairajpuri, M. S., Ahmad, W. and Dhanachand, C. (1980). Double set of reproductive organs in *Discolaimus tenex*. *Ind. J. Nematol.* **10**, 83–86.

Le Jambre, L. F. (1981). Hybridization of Australian *Haemonchus placei* (Place, 1893), *Haemonchus contortus cayugensis* (Das & Whitlock, 1960) and *Haemonchus contortus* (Rudolphi, 1903) from Louisiana. *Int. J. Parasit.* **11**, 323–330.

Jensen, P. (1984). Measuring carbon content in nematodes. *Helgolander Meeresunters.* **38**, 83–86.

Johnson, A. W. (1970). Pathogenicity and interaction of three nematode species on six Bermudagrasses. *J. Nematol.* **2**, 36–41.

Johnson, A. W. (1982). Managing nematode populations in crop production. In *Nematology in the Southern Region of the United States* (Ed. by R. D. Riggs), pp. 193–203. Southern Cooperative Series bulletin 276. Fayetteville: Southern Regional Research Committee.

Johnson, A. W., Dowler, C. C. and Hauser, E. W. (1974). Seasonal population dynamics of selected plant-parasitic nematodes on four monocultured crops. *J. Nematol.* **6**, 187–190.

Jones, D. A. (1973). Co-evolution and cyanogenesis. In *Taxonomy and Ecology* (Ed. by V. H. Heywood), pp. 213–242. London: Academic Press.

Jones, F. G. W. (1956). Soil populations of beet eelworm (*Heterodera schachtii* Schm.) in relation to cropping. II. Microplot and field plot results. *Ann. appl. Biol.* **44**, 25–56.

Jones, F. G. W. (1957). Resistance-breaking biotypes of the potato root eelworm (*Heterodera rostochiensis* Woll.). *Nematologica* **2**, 185–192.

Jones, F. G. W. (1980). Some aspects of the epidemiology of plant parasitic nematodes. In: *Comparative Epidemiology: a Tool for Better Disease Management* (Ed. by J. Polti and J. Kranz), pp. 71–92. Wageningen: Pudoc.

Jones, F. G. W. (1983). Weather and plant parasitic nematodes. *Bull. Eur. Pl. Prot. Orgn* **13**, 103–110.

Jones, F. G. W. and Kempton, R. A. (1978). Population dynamics, population models and integrated control. In *Plant Nematology* (Ed. by J. F. Southey), pp. 333–361. London: HMSO.

Jones, F. G. W., Larbey, D. W. and Parrott, D. M. (1969). The influence of soil structure and moisture on nematodes, especially *Xiphinema, Longidorus, Trichodorus* and *Heterodera* spp. *Soil Biol. Biochem* **1**, 153–165.

Jones, F. G. W., Parrott, D. M. and Perry, J. N. (1981). The gene-for-gene relationship and its significance for potato cyst nematodes and their solanaceous hosts. In *Plant Parasitic Nematodes* (Ed. by B. M. Zuckerman and R. A. Rohde), Vol. 3, pp. 23–36. New York: Academic Press.

Juhl, M. (1981). The influence of increasing amounts of nitrogen on the propagation of the cereal cyst nematode (*Heterodera avenae* Woll.). *Tidsskr. PlAvl.* **85**, 281–289.

Kable, P. F. and Mai, W. F. (1968). Influence of soil moisture on *Pratylenchus penetrans*. *Nematologica* **14**, 101–122.

Kaplan, D. T. and Keen, N. T. (1980). Mechanisms conferring plant incompatibility to nematodes. *Revue Nématol.* **3**, 123–134.

Kaplan, D. T., Keen, N. T. and Thomason, I. J. (1980). Studies on the mode of action of glyceollin in soybean incompatibility to the root knot nematode *Meloidogyne incognita*. *Physiol. Pl. Path.* **16**, 319–325.

Kaplan, D. T. and O'Bannon, J. H. (1985). Occurrence of biotypes in *Radopholus citrophilus*. *J. Nematol.* **17**, 158–162.

Kerry, B. R. (1984). Nematophagous fungi and the regulations of nematode populations in soil. *Helminth. Abst.* **53B**, 1–14.

Kerstan, U. (1969). Die beeinflussung des Geschlechterverhältnisses in der Gattung *Heterodera*. II. Minimallebenraum—selektive Absterberate der Geschlechter—Geschlechterverhältnis (*Heterodera schachtii*). *Nematologica* **15**, 210–228.

Ketudat, U. (1969). The effects of some soil-borne fungi on the sex ratio of *Heterodera rostochiensis* in roots. *Nematologica* **15**, 229–233.

Khan, F. A. (1985). Hatching response of *Rotylenchulus reniformis* to root leachates of certain hosts and nonhosts. *Revue Nématol.* **8**, 391–393.

King, K. L. and Hutchinson, K. J. (1976). The effects of sheep stocking intensity on the abundance and distribution of mesofauna in pastures. *J. appl. Ecol.* **13**, 41–55.

King, K. L. and Hutchinson, K. J. (1983). The effects of sheep grazing on invertebrate numbers and biomass in unfertilized natural pastures of the New England Tablelands (NSW). *Aust. J. Ecol.* **8**, 245–255.

Kinloch, R. A. and Hinson, K. (1973). The Florida program for evaluating soybean (*Glycine max*. L. Merr.) genotypes for susceptibility to root-knot nematode disease. *Proc. Soil Crop Sci. Soc. Fla.* **32**, 173–178.

Kinsinger, F. E. and Hopkins, H. H. (1961). Carbohydrate content of underground parts of grasses as affected by clipping. *J. Range Mgmt* **14**, 9–12.

Kiraly, Z., Barna, B. and Ersek, T. (1972). Hypersensitivity as a consequence, not the cause, of plant resistance to infection. *Nature, Lond.* **239**, 456–457.

Kirkendall, L. R. and Stenseth, N. C. (1985). On defining "breeding once". *Am. Nat.* **125**, 189–204.

Kirkpatrick, J. D., van Gundy, S. D. and Mai, W. F. (1964). Interrelationships of plant nutrition, growth, and parasitic nematodes. *Pl. anal. Fert. Problems* **4**, 189–225.

Kort, J. and Jaspers, C. P. (1973). Shift of pathotypes of *Heterodera rostochiensis* under susceptible potato cultivars. *Nematologica* **19**, 538–545.
Kraus-Schmidt, H. and Lewis, S. A. (1979). Seasonal fluctuations of various nematode populations in cotton fields in South Carolina. *Pl. Dis. Reptr.* **63**, 859–863.
Krauss, A. (1980). Influence of nitrogen on tuber initiation of potatoes. *Proc. Colloquium Int. Potash Inst.* **15**, 175–184.
Ladygina, N. M. (1975). [Crossing of potato stem nematode *Ditylenchus destructor* Thorne, 1945 with nematodes *Ditylenchus dipsaci* (Kuhn, 1857) Filipjev, 1936 from parsley and parsnip.] *Biologicheskie Nauki* **18**(10), 118–120. [See *Biol. Abst.* **62**(no. 22204), 1976.]
Larsson, S. (1985). Seasonal changes in the within-crown distribution of the aphid *Cinara pini* on Scots pine. *Oikos* **45**, 217–222.
Law, R. and Lewis, D. H. (1983). Biotic environments and the maintenance of sex—some evidence from mutualistic symbioses. *Biol. J. Linn. Soc.* **20**, 249–276.
Lea, P. J. and Miflin, B. J. (1980). Transport and metabolism of asparagine and other nitrogen compounds within the plant. In *The Biochemistry of Plants* (Ed. by B. J. Miflin), Vol. 5, pp. 569–607.
Lee, J. A., McNeill, S. and Rorison, I. H., Ed. (1983). *Nitrogen as an Ecological Factor*. Oxford: Blackwell, 470 pp.
Lenne, J. M. 1981. Controlling *Meloidogyne javanica* on *Desmodium ovalifolium* with grasses. *Pl. Dis.* **65**, 870–871.
Lewis, W. M. (1985). Nutrient scarcity as an evolutionary cause of haploidy. *Am. Nat.* **125**, 692–701.
Lyda, S. D. (1981). Alleviating pathogen stress. *Am. Soc. agric. Engng Monogr.* **4**, 195–214.
McBrien, H., Harmsen, R. and Crowder, A. (1983). A case of insect grazing affecting plant succession. *Ecology* **64**, 1035–1039.
McKenry, M. V. (1984). Grape root phenology relative to control of parasitic nematodes. *Am. J. Enol. Vitic.* **35**, 206–211.
McKercher, R. B., Tollefson, T. S. and Willard, J. R. (1979). Biomass and phosphorus contents of some soil invertebrates. *Soil Biol. Biochem.* **11**, 387–391.
Maggenti, A. R. (1981). *General Nematology*. New York: Springer-Verlag, 372 pp.
Magnusson, C. (1983). Abundance, distribution and feeding relations of root/fungal feeding nematodes in a Scots pine forest. *Holarctic Ecol.* **6**, 183–193.
Mahmood, I. and Saxena, S. K. (1980). Effect of different doses of ammonium sulphate and urea on the plant growth and resulting biochemical changes in eggplant cv pusa purple long infected with *Rotylenchulus reniformis*. *Acta Bot. Ind.* **8**, 171–174.
Mai, W. F. (1984). The importance of taxonomy to nematode control strategies. *Pl. Dis.* **67**, 716.
Mamiya, Y. (1983). Pathology of the pine wilt disease caused by *Bursaphelenchus xylophilus*. *A. Rev. Phytopath.* **21**, 201–220.
Marshall, L. D. (1982). Male nutrient investment in the Lepidoptera: what nutrients should males invest? *Am. Nat.* **120**, 273–279.
Marumoto, T., Anderson, J. P. E. and Domsch, K. H. (1982). Mineralization of nutrients from soil microbial biomass. *Soil Biol. Biochem.* **14**, 469–475.
Masamune, T., Anetai, M., Takasugi, M. and Katsui, N. (1982). Isolation of a natural hatching stimulus, glycinoclepin A, for the soybean cyst nematode. *Nature, Lond.* **297**, 495–496.
Mateille, T., Cadet, P. and Quénéhervé, P. (1984). Influence du recépage du

bananier Poyo sur le développement des populations de *Radopholus similis* et *d'Helicotylenchus multicinctus*. *Revue Nématol.* **7**, 355–361.
Mayr, E. (1969). *Principles of Systematic Zoology*. New York: McGraw-Hill, 428 pp.
Montenegro, G., Araya, S., Aljaro, M. E. and Avila, G. (1982). Seasonal fluctuations of vegetative growth in roots and shoots of central Chilean shrubs. *Oecologia* **53**, 235–237.
Mugniery, D. and Fayet, G. (1984). Détermination du sexe de *Globodera rostochiensis* Woll. et influence des niveaux d'infestation sur la pénétration, le développement det le sexe de ce nématode. *Revue Nématol.* **7**, 233–238.
Myers, R. F. and Krusberg, L. R. (1965). Organic substances discharged by plant-parasitic nematodes. *Phytopathology* **55**, 429–437.
Nigon, V. and Dougherty, E. C. (1949). Reproductive patterns and attempts at reciprocal crossing of *Rhabditis elegans* (Maupas, 1900), and *Rhabditis briggsae* (Dougherty and Nigon, 1949) (Nematoda: Rhabditidae). *J. exp. Zool.* **112**, 485–503.
Norton, D. C. (1978). *Ecology of Plant-parasitic Nematodes*. New York: Wiley, 268 pp.
Norton, D. C. and Burns, N. (1971). Colonization and sex ratios of *Pratylenchus alleni* in soybean roots under two soil moisture regimes. *J. Nematol.* **3**, 374–377.
Olthof, T. H. A. (1982). Effect of age of alfalfa root on penetration by *Pratylenchus penetrans*. *J. Nematol.* **14**, 100–105.
Oteifa, B. A. (1953). Development of the root-knot nematode, *Meloidogyne incognita*, as affected by potassium nutrition of the host. *Phytopathology* **43**, 171–174.
Oteifa, B. A. and Diab, K. A. (1961). Significance of potassium fertilization in nematode infested cotton fields. *Pl. Dis. Reptr.* **45**, 932.
Owen, D. F. and Wiegert, R. G. (1981). Mutualism between grasses and grazers: an evolutionary hypothesis. *Oikos* **36**, 376–378.
Pant, V., Hakim, S. and Saxena, S. K. (1983). Effect of different levels of NPK on the growth of tomato cv Marglobe and on the morphometrics of the root knot nematode, *Meloidogyne incognita*. *Ind. J. Nematol.* **13**, 110–113.
Papadopoulou, J. and Triantophyllou, A. C. (1982). Sex differentiation in *Meloidogyne incognita* and anatomical evidence of sex reversal. *J. Nematol.* **14**, 549–566.
Parker, L. W., Santos, P. F., Phillips, J. and Whitford, W. G. (1984). Carbon and nitrogen dynamics during the decomposition of litter and roots of a Chihuahuan desert annual, *Lepidium lasiocarpum*. *Ecol. Monogr.* **54**, 339–360.
Parrott, D. M. (1972). Mating of *Heterodera rostochiensis* pathotypes. *Ann. appl. Biol.* **71**, 271–274.
Parry, G. D. (1981). The meanings of *r*- and *K*-selection. *Oecologia* **48**, 260–264.
Pease, J. L., Vowles, R. H. and Keith, L. B. (1979). Interaction of snowshoe hares and woody vegetation. *J. Wildl. Mgmt* **43**, 43–60.
Perry, R. N. and Clarke, A. J. (1981). Hatching mechanisms of nematodes. *Parasitology* **83**, 435–449.
Perry, R. N., Clarke, A. J., Hennessey, J. and Beane, J. (1983). Role of trehalose in the hatching mechanism of *Heterodera goettingiana*. *Nematologica* **29**, 323–334.
Perry, R. N., Plowright, R. A. and Webb, R. M. (1980). Mating between *Pratylenchus penetrans* and *P. fallax* in sterile culture. *Nematologica* **26**, 125–129.
Persson, T. (1983). Influence of soil animals on nitrogen mineralisation in a northern scots pine forest. *Proc. Int. Colloquium Soil Zool.* **8**, 117–126.
Petelle, M. (1982). More mutualisms between consumers and plants. *Oikos* **38**, 125–127.

Pianka, E. R. (1970). On *r*- and *K*-selection. *Am. Nat.* **104**, 592–597.
Pierce, N. M. (1985). Lycaenid butterflies and ants: selection for nitrogen-fixing and other protein-rich food plants. *Am. Nat.* **125**, 888–895.
Pillai, J. K. and Taylor, D. P. (1967). Influence of fungi on host preference, host suitability, and morphometrics of five mycophagous nematodes. *Nematologica* **13**, 529–540.
Pitcher, R. S. (1967). The host-parasite relations and ecology of *Trichodorus viruliferus* on apple roots, as observed from an underground laboratory. *Nematologica* **13**, 547–557.
Pitcher, R. S. and McNamara, D. G. (1970). The effect of nutrition and season of year on the reproduction of *Trichodorus viruliferus*. *Nematologica* **16**, 99–106.
Poinar, G. O. and Hansen, E. (1983). Sex and reproductive modifications in nematodes. *Helminth. Abst.* **52B**, 145–163.
Ponin, I. Y., Voinilo, V. A., Gladkaya, R. M. and Timofeev, N. (1977). [Nematode-resistant potato varieties and ways of using them.] *Pl. Breed. Abst.* **47**, 7545.
Pradhan, G. B. and Dash, M. C. (1984). Rhizosphere effect of *Andropogon pumilus* Roxb. on soil nematodes, soil organic matter and nitrogen. *Proc. Ind. Acad. Sci.* (*Anim. Sci.*) **93**, 77–82.
Prasad, K. S. K. and Rao, Y. S. (1984). Effect of foliar application of systemic pesticides on the development of *Meloidogyne graminicola* in rice. *Ind. J. Nematol.* **14**, 125–127.
Prestidge, R. A., Lauren, D. R., van der Zijp, S. G. and di Menna, M. E. (1985). Isolation of feeding deterrents to Argentine stem weevil in cultures of endophytes of perennial ryegrass and tall fescue. *NZ J. agric. Res.* **28**, 87–92.
Price, M., Caviness, C. E. and Riggs, R. D. (1978). Hybridization of races of *Heterodera glycines*. *J. Nematol.* **10**, 114–118.
Price, N. S. and Sanderson, J. (1984). The translocation of calcium from oat roots infected by the cereal cyst nematode *Heterodera avenae* (Woll.). *Revue Nématol.* **7**, 239–243.
Prot, J. C. (1980). Migration of plant parasitic nematodes towards plant roots. *Revue Nématol.* **3**, 305–318.
Quénéhervé, P. and Cadet, P. (1985). Localisation des nématodes dans les rhizomes du bananier cv. Poyo. *Revue Nématol.* **8**, 3–8.
Raupp, M. J. (1985). Effects of leaf toughness on mandibular wear of the leaf beetle *Plagiodera versicolora*. *Ecol. Ent.* **10**, 73–79.
Raven, A. J. (1983). Phytophages of xylem and phloem: a comparison of animal and plant sap-feeders. *Adv. ecol. Res.* **13**, 135–234.
Rich, J. R., Keen, N. T. and Thomason, I. J. (1977). Association of coumestans with the hypersensitivity of lima bean roots to *Pratylenchus scribneri*. *Physiol. Pl. Path.* **10**, 105–116.
Riggs, R. D., Hamblen, M. L. and Rakes, L. (1977). Development of *Heterodera glycines* pathotypes as affected by soybean cultivars. *J. Nematol.* **9**, 312–318.
Rodriguez-Kabana, R. and Collins, R. J. (1979). Relation of fertilizer treatments and cropping sequence to populations of two plant parasitic nematode species. *Nematropica* **9**, 151–166.
Rogers, W. P. (1962). *The Nature of Parasitism*. New York: Academic Press, 287 pp.
Rohde, R. A. and Jenkins, W. R. (1958). Basis for resistance of *Asparagus officinalis* var. *altilis* L. to the stubby-root nematode *Trichodorus christiei* (Allen 1957). *Bull. Md. agric. exptl Stn* **A97**, 1–19.
Room, P. M. and Thomas, P. A. (1985). Nitrogen and establishment of a beetle for biological control of the floating weed *Salvinia* in Papua New Guinea. *J. appl. Ecol.* **22**, 139–156.

Ross, G. J. S. and Trudgill, D. L. (1969). The effect of population density on the sex ratio of *Heterodera rostochiensis*: a two dimensional model. *Nematologica* **15**, 601–607.

Rossner, J. (1972). Vertikalverteilung wandernder Wurzelnematoden im Boden in abhangigkeit von Wassergehalt and Durchwurzelung. *Nematologica* **18**, 360–372.

Rovira, A. D., Foster, R. C. and Martin, J. K. (1979). Origin, nature and nomenclature of the organic materials in the rhizosphere. In *The Soil–Root Interface* (Ed. by J. L. Harley and R. S. Russell), pp. 1–4. London: Academic Press.

Ruehle, J. E. (1972). Nematodes of forest trees. In *Economic Nematology* (Ed. by J. M. Webster), pp. 312–334. London: Academic Press.

Russell, G. B., Sutherland, O. R. W., Hutchins, R. F. N. and Christmas, P. E. (1978). Vestitol: a phytoalexin with insect feeding-deterrent activity. *J. chem. Ecol.* **4**, 571–579.

Saiki, H. and Yoneda, K. (1981). Possible dual roles of an allelopathic compound, *cis*-dehydromatricaria ester. *J. chem. Ecol.* **8**, 185–194.

Sawhney, R. and Webster, J. M. (1975). The role of plant growth hormones in determining the resistance of tomato plants to the root-knot nematode *Meloidogyne incognita. Nematologica* **21**, 95–103.

Schiemer, F. (1982). Food dependence and energetics of freeliving nematodes. II. Life history parameters of *Caenorhabditis briggsae* (Nematoda) at different levels of food supply. *Oecologia* **54**, 122–128.

Schiemer, F., Duncan, A. and Klekowski, R. A. (1980). A bioenergetic study of a benthic nematode, *Plectus palustris* de Man 1880, throughout its life cycle. II. Growth, fecundity and energy budgets at different densities of bacterial food and general ecological considerations. *Oecologia* **44**, 205–212.

Setty, K. G. H. and Wheeler, A. W. (1968). Growth substances in roots of tomato (*Lycopersicon esculentum* Mill.) infected with root-knot nematodes (*Meloidogyne* spp.). *Ann. appl. Biol.* **61**, 495–501.

Shepherd, A. M. (1962). The emergence of larvae from cysts in the genus *Heterodera. Commonw. Bur. Helminth. Tech. Communs* **32**, 1–90.

Shepherd, A. M. (1970). Extraction and estimation of *Heterodera*. In *Laboratory Methods for Work with Plant and Soil Nematodes* (Ed. by J. F. Southey), pp. 23–33. London: HMSO.

Shirazi, M. A., Lighthart, B. and Gillett, J. (1984). A method for scaling biological response of soil microcosms. *Ecol. Model.* **23**, 203–226.

Shorrocks, B. and Rosewell, J. (1986). Guild size in drosophilids: a simulation model. *J. Anim. Ecol.* **55**, 527–541.

Sidhu, G. S. and Webster, J. M. (1981). The genetics of plant-nematode parasitic systems. *Bot. Rev.* **47**, 387–419.

Simon, A. and Rovira, A. D. (1985). The influence of phosphate fertilizer on the growth and yield of wheat in soil infested with cereal cyst nematode (*Heterodera avenae* Woll.). *Aust. J. exp. Agric.* **25**, 191–197.

Singh, B. and Choudhury, B. (1973). The chemical characteristics of tomato cultivars resistant to root-knot nematodes (*Meloidogyne* spp.). *Nematologica* **19**, 443–448.

Singh, V., Singh, S. P., Yadav, R. and Saxena, S. K. (1985). Effect of different plants on the morphometrics of females of root-knot nematode, *Meloidogyne incognita. Nematol. Medit.* **13**, 81–85.

Slobodchikoff, C. N. (1982). Why asexual reproduction?: Variation in populations of the parthenogenetic wasp, *Venturia canescens* (Hymenoptera: Ichneumonidae). *Ann. ent. Soc. Am.* **76**, 23–29.

Smolik, J. D. and Dodd, J. L. (1983). Effect of water and nitrogen, and grazing on nematodes in a shortgrass prairie. *J. Range Mgmt* **36**, 744–748.
Smolik, J. D. and Rogers, L. E. (1976). Effects of cattle grazing and wildfire on soil-dwelling nematodes of the shrub–steppe ecosystem. *J. Range Mgmt* **29**, 304–306.
Solov'eva, G. I. (1975). *Parasitic Nematodes of Woody and Herbaceous Plants: A Review of the Genus* Paratylenchus *Micoletzky, 1922*. New Delhi. 134 pp.
Southey, J. F., Ed. (1978). *Plant Nematology*. London: HMSO, 440 pp.
Sparling, G. P., Whale, K. N. and Ramsay, A. J. (1985). Quantifying the contribution from the soil microbial biomass to the extractable P levels of fresh and air-dried soils. *Aust. J. Soil Res.* **23**, 613–621.
Spaull, A. M. and Murphy, B. (1983). Effect of *Paratrichodorus anemones* and other plant-parasitic nematodes on the growth of spring wheat. *Nematologica* **29**, 435–442.
Spaull, V. W. (1980). Preliminary observations on the rooting habit of some sugarcane varieties in South Africa. *Proc. S. Afr. Sug. Technol. Assoc.* **54**, 177–180.
Spiegel, Y., Cohn, E. and Kafkafi, U. (1982). The influence of ammonium and nitrate nutrition of tomato plants on parasitism by the root-knot nematode. *Phytoparasitica* **10**, 33–40.
St John, T. V., Coleman, D. C. and Reid, C. P. P. (1983). Growth and spatial distribution of nutrient-absorbing organs: selective exploitation of soil heterogeneity. *Pl. Soil* **71**, 487–493.
Stanton, N. L. (1983). The effect of clipping and phytophagous nematodes on net primary production of blue grama, *Bouteloua gracilis*. *Oikos* **40**, 249–257.
Steen, E. (1985). Root and rhizome dynamics in a perennial grass crop during an annual growth cycle. *Swed. J. agric. Res.* **15**, 25–30.
Stinson, C. S. A. and Brown, V. K. (1983). Seasonal changes in the architecture of natural plant communities and its relevance to insect herbivores. *Oecologia* **56**, 67–69.
Stone, A. R. (1979). Co-evolution of nematodes and plants. *Symb. Bot. Upsal.* **22**, 46–61.
Stone, A. R. (1983). Three approaches to the status of a species complex, with a revision of some species of *Globodera* (Nematoda: Heteroderidae). In *Concepts in Nematode Systematics* (Ed. by A. R. Stone, H. M. Platt and L. F. Khalil) pp. 221–233. London: Academic Press.
Storey, G. W. (1982). The relationship between potato root growth and reproduction of *Globodera rostochiensis* (Woll.). *Nematologica* **28**, 210–218.
Strong, D. R., Lawton, J. H. and Southwood, T. R. E. (1984). *Insects on Plants: Community Patterns and Mechanisms*. Oxford: Blackwell Scientific, 313 pp.
Sturhan, D. (1969). Das Rassenproblem bei *Ditylenchus dipsaci*. *Mitt-biol. Bund-Anst. Ld-u Forstw.* **136**, 87–98.
Suatmadji, R. W. (1968). *Studies on the Effects of* Tagetes *Species on Plant Parasitic Nematodes*. Wageningen: Veenman & Zonen. 134 pp.
Suatmadji, R. W. (1984). A technique for subtracting nematodes from soil by mechanical destruction, with special reference to studies on the relation between *Paratrichodorus* sp. and wheat. *Aust. Pl. Path.* **13**, 17–18.
Sykes, G. B. (1979). Yield losses in barley, wheat and potatoes associated with field populations of 'large form' *Longidorus leptocephalus*. *Ann. appl. Biol.* **91**, 237–241.
Taylor, A. L. and Sasser, J. N. (1978). *Biology, Identification and Control of*

Root-knot Nematodes (Meloidogyne *species*). Raleigh: North Carolina State University. 111 pp.

Tefft, P. M. and Bone, L. W. (1984). Zinc-mediated hatching of eggs of soybean cyst nematode, *Heterodera glycines*. *J. chem. Ecol.* **10**, 361–372.

Tempel, A. S. (1981). Field studies of the relationship between herbivore damage and tannin concentration in bracken (*Pteridium aquilinum* Kuhn). *Oecologia* **51**, 97–106.

Thomas, P. R. (1981). Migration of *Longidorus elongatus, Xiphenema diversicaudatum* and *Ditylenchus dipsaci* in soil. *Nematol. Medit.* **9**, 75–81.

Thomas, S. H. (1978). Population densities of nematodes under seven tillage regimes. *J. Nematol.* **10**, 24–27.

Thompson, J. N. (1983). Selection pressures on phytophagous insects feeding on small host plants. *Oikos* **40**, 438–444.

Thompson, K. and Uttley, M. G. (1982). Do grasses benefit from grazing? *Oikos* **39**, 113–115.

Townshend, J. L. and Potter, J. W. (1973). Nematode numbers under cultivars of forage legumes and grasses. *Can. Pl. Dis. Surv.* **53**, 194–195.

Triantophyllou, A. C. (1960). Sex determination in *Meloidogyne incognita* (Chitwood, 1949) and intersexuality in *M. javanica* (Treub, 1885; Chitwood, 1949.) *Ann. Inst. Phytopath. Benaki, N.S.* **3**, 12–31.

Triantophyllou, A. C. (1973). Environmental sex differentiation in relation to pest management. *Ann. Rev. Phytopathol.* **11**, 441–462.

Triantophyllou, A. C. and Hirschmann, H. (1980). Cytogenetics and morphology in relation to evolution and speciation of plant-parasitic nematodes. *Ann. Rev. Phytopathol.* **18**, 333–359.

Trudgill, D. L. (1972). Influence of feeding duration on moulting and sex determination of *Meloidogyne incognita*. *Nematologica* **18**, 476–481.

Trudgill, D. L., Evans, K. and Parrott, D. M. (1975). Effects of potato cyst-nematodes on potato plants II. Effects on haulm size, concentration of nutrients in haulm tissue and tuber yield of a nematode resistant and a nematode susceptible potato variety. *Nematologica* **21**, 183–191.

Tuomi, J., Hakala, T. and Haukioja, E. (1983). Alternative concepts of reproductive effort, costs of reproduction, and selection in life-history evolution. *Am. Zool.* **23**, 25–34.

Turner, S. J., Stone, A. R. and Perry, J. N. (1983). Selection of potato cyst-nematodes on resistant *Solanum vernei* hybrids. *Euphytica* **32**, 911–917.

Veech, J. A. (1978). An apparent relationship between methoxy-substituted terpenoid aldehydes and the resistance of cotton to *Meloidogyne incognita*. *Nematologica* **24**, 81–87.

Veech, J. A. (1981). Plant resistance to nematodes. In *Plant Parasitic Nematodes* (Ed. by B. M. Zuckerman and R. A. Rohde), Vol. 3, pp. 377–403. New York: Academic Press.

Veresoglou, D. S. and Fitter, A. H. (1984). Spatial and temporal patterns of growth and nutrient uptake of five co-existing grasses. *J. Ecol.* **72**, 259–272.

Viglierchio, D. R. and Croll, N. A. (1968). Host resistance reflected in differential nematode population structures. *Science, N.Y.* **161**, 271–272.

Wallace, H. R. (1966). Factors influencing the infectivity of plant parasitic nematodes. *Proc. R. Soc. B* **164**, 592–614.

Wallace, H. R. (1973). *Nematode Ecology and Plant Disease*. London: Edward Arnold, 228 pp.

Wallace, H. R. (1983). Interactions between nematodes and other factors on plants. *J. Nematol.* **15**, 221–227.

Weaver, D. B., Rodriguez-Kabana, R. and Robertson, D. G. (1985). Performance

of selected soybean cultivars in a field infested with mixtures of root-knot, soybean cyst, and other phytonematodes. *Agron. J.* **77**, 249–253.
White, T. C. R. (1978). The importance of a relative shortage of food in animal ecology. *Oecologia* **33**, 71–86.
Wilbur, H. M. (1980). Complex life cycles. *Ann. Rev. ecol. Syst.* **11**, 67–93.
Williams, C. B., Elkins, C. B., Haaland, R. L., Hoveland, C. S. and Rodriguez-Kabana, R. (1983). Effects of root diameter, nematodes, and soil compaction. *Proc. Int. Grassld Congr.* **14**, 121–124.
Williams, T. D. (1978). Cyst nematodes: biology of *Heterodera* and *Globodera*. In *Plant Nematology* (Ed. by J. F. Southey), pp. 156–171. London: HMSO.
Woodward, R. A., Harper, K. T. and Tiedemann, A. R. (1984). An ecological consideration of the significance of cation-exchange capacity of roots of some Utah range plants. *Pl. Soil* **79**, 169–180.
Wouts, W. M. and Sturhan, D. (1978). The identity of *Heterodera trifolii* (Goffart, 1932), and the description of *H. daverti* n.sp. (Nematoda: Tylenchida). *Nematologica* **24**, 121–128.
Yeates, G. W. (1970). Failure of *Heterodera carotae, H. cruciferae* and *H. goettingiana* to interbreed *in vitro*. *Nematologica* **16**, 153–154.
Yeates, G. W. (1971). Feeding types and feeding groups in plant and soil nematodes. *Pedobiologia* **11**, 173–179.
Yeates, G. W. (1972). Population studies on *Ditylenchus dipsaci* (Nematoda: Tylenchida) in a Danish beech forest. *Nematologica* **18**, 125–130.
Yeates, G. W. (1973). Annual cycle of root nematodes on white clover in pasture. 1. *Heterodera trifolii* in a yellow-grey earth. *NZ J. agric. Res.* **16**, 569–574.
Yeates, G. W. (1974a). Studies on a climosequence of soils in tussock grasslands. 2. Nematodes. *NZ J. Zool.* **1**, 171–177.
Yeates, G. W. (1974b). Effects of *Heterodera trifolii* on the growth of clover in a yellow-grey earth under greenhouse conditions. *NZ J. agric. Res.* **17**, 379–385.
Yeates, G. W. (1976). Effect of fertiliser treatment and stocking rate on pasture nematode populations on a yellow-grey earth. *NZ J. agric. Res.* **19**, 405–408.
Yeates, G. W. (1978). Populations of nematode genera in soils under pasture. I. Seasonal dynamics in dryland and irrigated pasture on a southern yellow-grey earth. *NZ J. agric. Res.* **21**, 321–330.
Yeates, G. W. (1982). *Pungentus maorium* Clark, 1963 (Nematoda: Dorylaimida) population changes under pasture during thirty-six months. *Pedobiologia* **24**, 81–89.
Yeates, G. W. (1984a). *Helicotylenchus pseudorobustus* (Nematoda: Tylenchida) population changes under pasture during thirty-six months. *Pedobiologia* **27**, 221–228.
Yeates, G. W. (1984b). Variation in soil nematode diversity under pasture with soil and year. *Soil Biol. Biochem.* **16**, 95–102.
Yeates, G. W. (1986). Stylet and body lengths as niche dimensions in plant–parasitic nematodes. *Zool. Anz.* **216**, 327–337.
Yeates, G. W., Stannard, R. E. and Barker, G. M. (1983). Vertical distribution of nematode populations in Horotiu soils. *NZ Soil Bureau Sci. Rep. 60*, 14 pp.
Yeates, G. W., Watson, R. N. and Steele, K. W. (1985). Complementary distribution of *Meloidogyne, Heterodera* and *Pratylenchus* (Nematoda: Tylenchida) in roots of white clover. *Proc. 4th Aust. Conf. Grassld Invert. Ecol.*, 71–79.
Yoshikawa, M. (1978). Diverse modes of action of biotic and abiotic phytoalexin elicitors. *Nature, Lond.* **275**, 546–547.
Zucker, W. V. (1982). How aphids choose leaves: the role of phenolics in host selection by a galling aphid. *Ecology* **63**, 972–981.

Ecological Studies at Lough Hyne

J. A. KITCHING

I. SUMMARY

Lough Hyne, County Cork, is a very sheltered sea lough which receives very little fresh water from streams and has a considerable tidal turnover. It is nearly 50 m deep in the Western Trough, which connects the North Basin and South Basin (Fig. 2). Lough Hyne opens southwards through the Rapids into Barloge Creek, which is long and narrow, very sheltered at its north end but opening at its south end into the Atlantic (Fig. 3). At the mouth of Barloge Creek stands the steep fully wave-exposed rocky promontory of Carrigathorna.

ADVANCES IN ECOLOGICAL RESEARCH Vol. 17
ISBN 0-12-013917-0

Fig. 1. Lough Hyne and the Rapids.

Carrigathorna supports a very wide supralittoral zone of lichens, an extensive mid-littoral mussel bed, and a sublittoral *Laminaria* forest. *Laminaria hyperborea* extends down to 17 m below MLWS, and is limited by insufficient irradiance. The flora and fauna of tide pools at Carrigathorna are described in relation to conditions at a range of depths in these pools. The distribution of plants and animals in the Bullock Island Cave is interpreted in relation to illumination and to protection from the unfavourable effects of exposure to air.

The Rapids control the tide in the lough, reducing and rendering asymmetrical the rise and fall. The lough is very sheltered. The narrow fucoid zones, in the North Basin, suffer damage from the sea urchin *Paracentrotus lividus*, which may wander upwards intertidally during daytime high water. The small rise and fall of the tide has made possible the investigation of limpet (*Patella vulgata*) feeding behaviour and a study of the zoning behaviour of the top shell *Gibbula umbilicalis*.

The limited amount of wave action experienced within the lough maintains a relatively clean band of rock and boulders down to 1–2 m below the level of low water. This "shallow sublittoral" supports a very rich flora and fauna—bushy algae with associated animals in much of the South Basin, and the sea urchin *Paracentrotus lividus*, with a very different associated flora and fauna, widely distributed over the North Basin and in patches in the South Basin. Associated with *Paracentrotus* are crustose coralline algae and

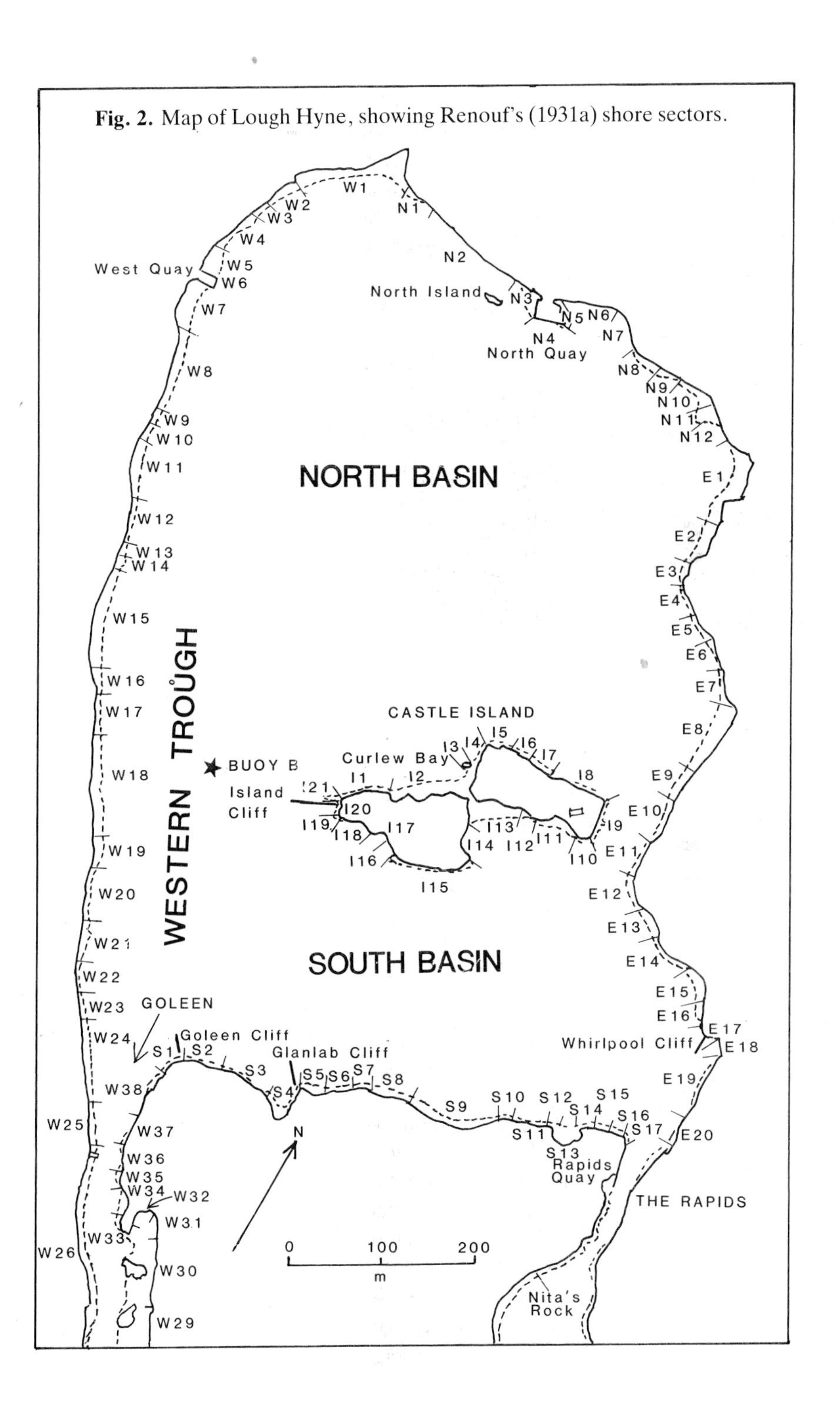

Fig. 2. Map of Lough Hyne, showing Renouf's (1931a) shore sectors.

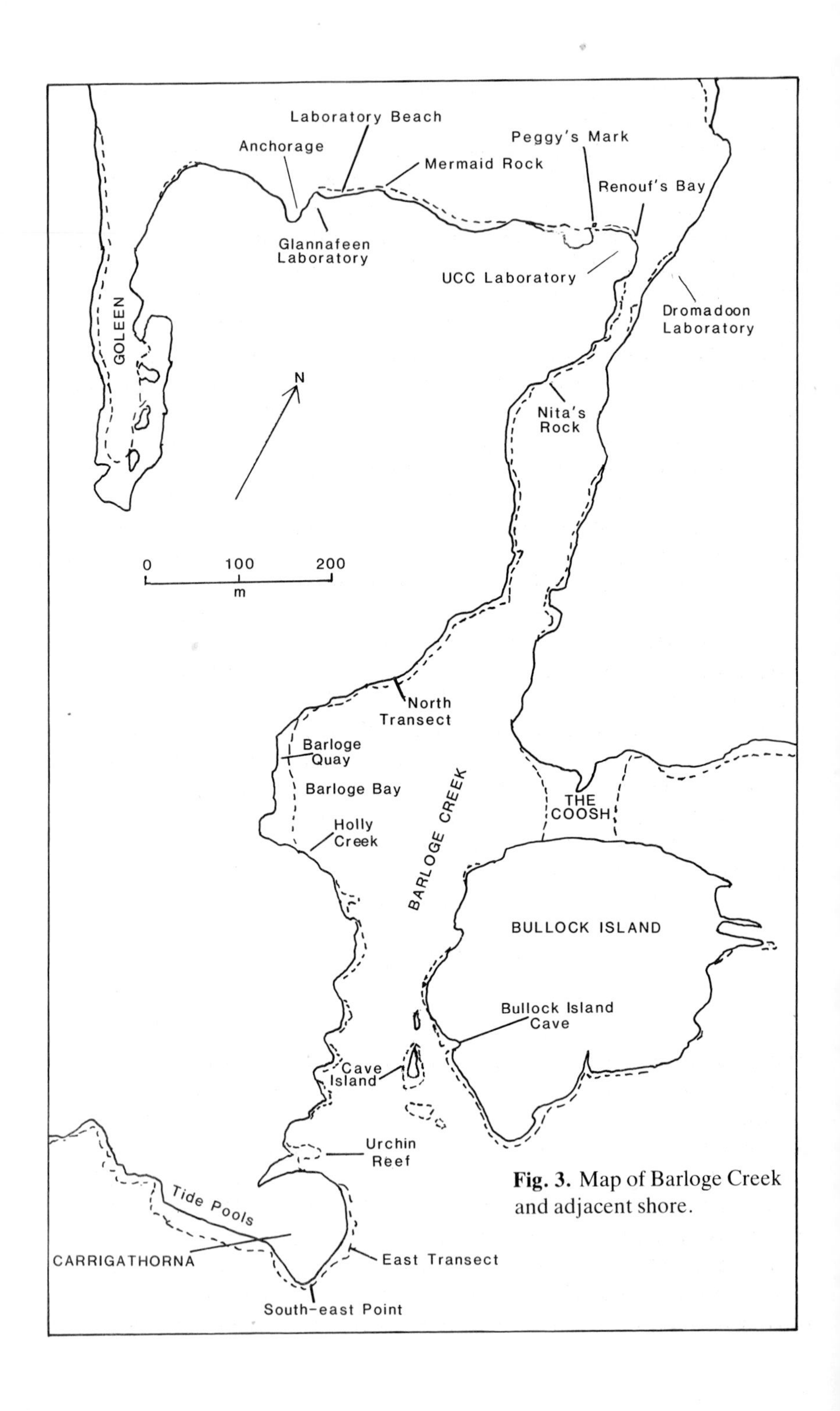

Fig. 3. Map of Barloge Creek and adjacent shore.

the worms which burrow in these, and animals with hard shells, capable of resisting grazing but subject to predation by starfish. *Paracentrotus* also appears to benefit the green alga *Codium fragile* ssp. *tomentosoides* by destroying its slower growing but ultimately more successful competitors.

With increasing depth the boulder scree and rock which line the shores of the lough become buried in mud, although near the "inflow area" adjoining the Rapids the sediment is coarser. Mud covers the bottom of the lough down to the bottom of the Western Trough. However, rock is exposed on four sublittoral cliffs which have received some preliminary study.

Lough Hyne is closely surrounded by low hills, so that its catchment area is small; its tidal exchange is considerable. As a result the salinity of the open water, except sometimes in the surface film, remains close to that of the sea water outside. This unique feature exercises a special influence on the pattern of the hydrography, with regular summer stratification, followed by anoxia of the deep water of the Western Trough, and all its consequences for the fauna, and with regular winter recirculation and reoxygenation. Related to the hydrography, the benthos of the Western Trough can be subdivided into depth zones—the faunistically rich *Audouinella floridula* zone, the mud burrow zone with the decapod Crustacea *Nephrops norvegicus* and *Calocaris macandreae*, and (below the thermocline) the spionid zone. The fauna of the spionid zone is destroyed every summer by anoxia. The vertical distribution of zooplankton is also controlled by lack of oxygen.

The special features of the flora and fauna of the lough, with its many rarities, are related in part to its somewhat Mediterranean features and in part to the intensive study which it has received. Finally, the geological history of the lough will be outlined as far as is possible. A study of diatoms in cores of bottom mud has shown that the lough originated as a freshwater lake, which was ultimately invaded by the sea.

II. HISTORICAL NOTE

Lough Hyne (or Lough Ine, as it is called locally) is a sea lough on the south coast of Ireland, near its western end. It has been intensively studied ever since the marine station of University College, Cork, was set up there by Professor L. P. W. Renouf over the period 1926–1930 (Renouf, 1931a). Renouf's lectures about Lough Hyne attracted many marine biological visitors. Work carried out there by parties from the University of Bristol was reviewed by Kitching and Ebling (1967) in this journal. The present review surveys work carried out there by various investigators since 1965. Over the period 1952–1957 J. A. Kitching set up the Dromadoon Laboratory to provide facilities for experimental research, and the Glannafeen Laboratory to provide alternative facilities for parties of students in view of the impending collapse of the wooden U.C.C. buildings. On the afternoon of 7 March 1962 that collapse was brought about by a high tidal surge. The Glannafeen Laboratory accommodated a long series of visits by students from the University of East Anglia in association with senior members from there and elsewhere. A strong interest in Lough Hyne was also shown by marine biologists from University College, Cork, from University College, Galway, from the Irish National Museum, Dublin, from the Irish Department of Fisheries and Forestry, and by An Taisce (The Irish National Trust). In June 1981, in response to a proposal from An Taisce, the Ministry of Fisheries and Forestry declared the lough a Marine Nature Reserve. A new research laboratory was established at Lough Hyne by University College, Cork, in 1986.

The chart shown in the previous review, based on soundings reported by Renouf (1931a), is now outdated (as to detail) by an echo-sounding survey undertaken by the Irish Department of Fisheries and Forestry. Maps of the lough and of Barloge Creek, showing sites referred to in the present review, are given in Figs 2 and 3.

III. THE ROCKY SEA COAST

A. Supralittoral and Littoral

The vertical and topographical distribution of some common maritime lichens, Algae and invertebrates on the open coast and in Barloge Creek are in general in accordance with expectations, based on descriptions for many parts of the British Isles (Lewis, 1964). Zonation on the open coast at Carrigathorna and in the shelter of the north side of Barloge Bay are illustrated in Figs 4 and 5, being determined from vertical distribution within 1 m on either side of the transect line.

At Carrigathorna the common maritime lichens *Xanthoria parietina, Lecanora atra, Caloplaca marina, Ramalina siliquosa* and *Verrucaria maura* (and many others not investigated) occupied wide zones only contained because the coastal strip where the transect is situated does not reach any higher. These lichens (and others) are also found on the summit—unmeasured in height—of the Carrigathorna headland. *Littorina neritoides* and *Littorina* of the shell form *L. patula* also reach high above MHWS. Fucoids are entirely missing from the fully unsheltered open coast, but plants of *Fucus vesiculosus evesiculosus* form a zone in a gully extending eastwards from Long Pool. The lower littoral is extensively covered by mussels with tufts of *Ceramium*; barnacles (*Chthamalus montagui, C. stellatus, Semibalanus balanoides*) (all badly corroded), limpets (*Patella vulgata, P. aspera*) and dogwhelks (*Nucella lapillus*, mainly in crevices) abound. *Alaria esculenta* and *Laminaria digitata* form a sublittoral fringe. It is difficult to

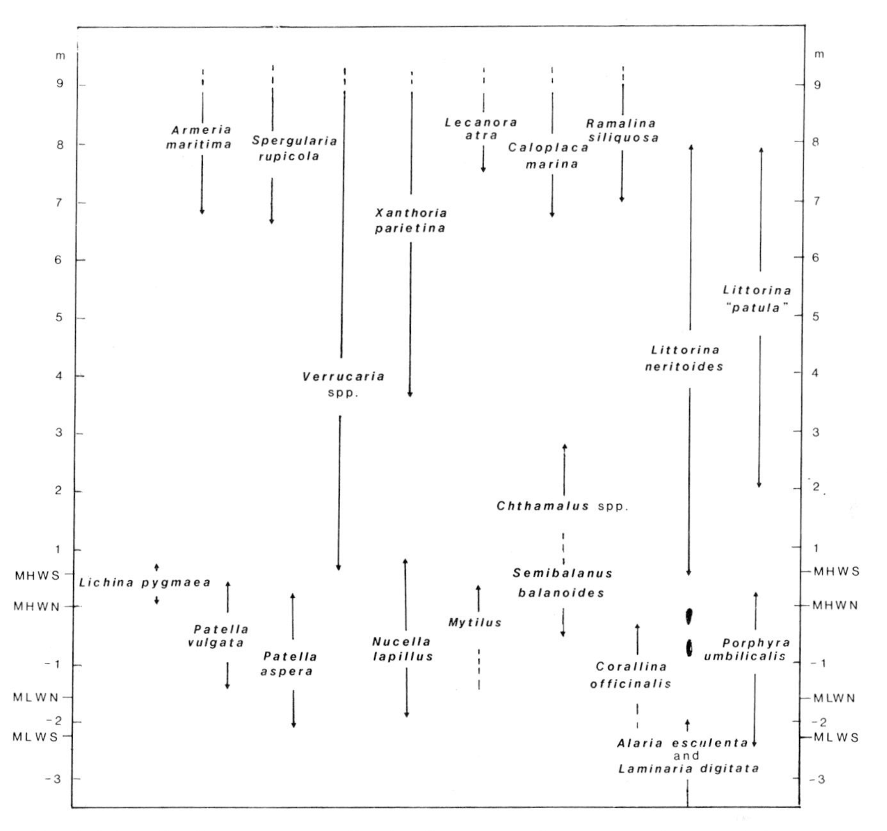

Fig. 4. Zonation in a transect at Carrigathorna, passing between East and West Twin Pools. Reference level is at MHWN, between these pools.

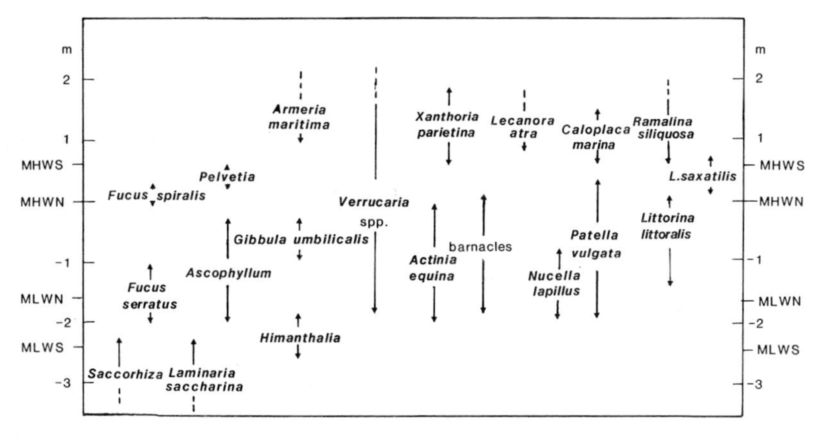

Fig. 5. Zonation in a transect at Barloge Bay North. Reference level is at MHWN.

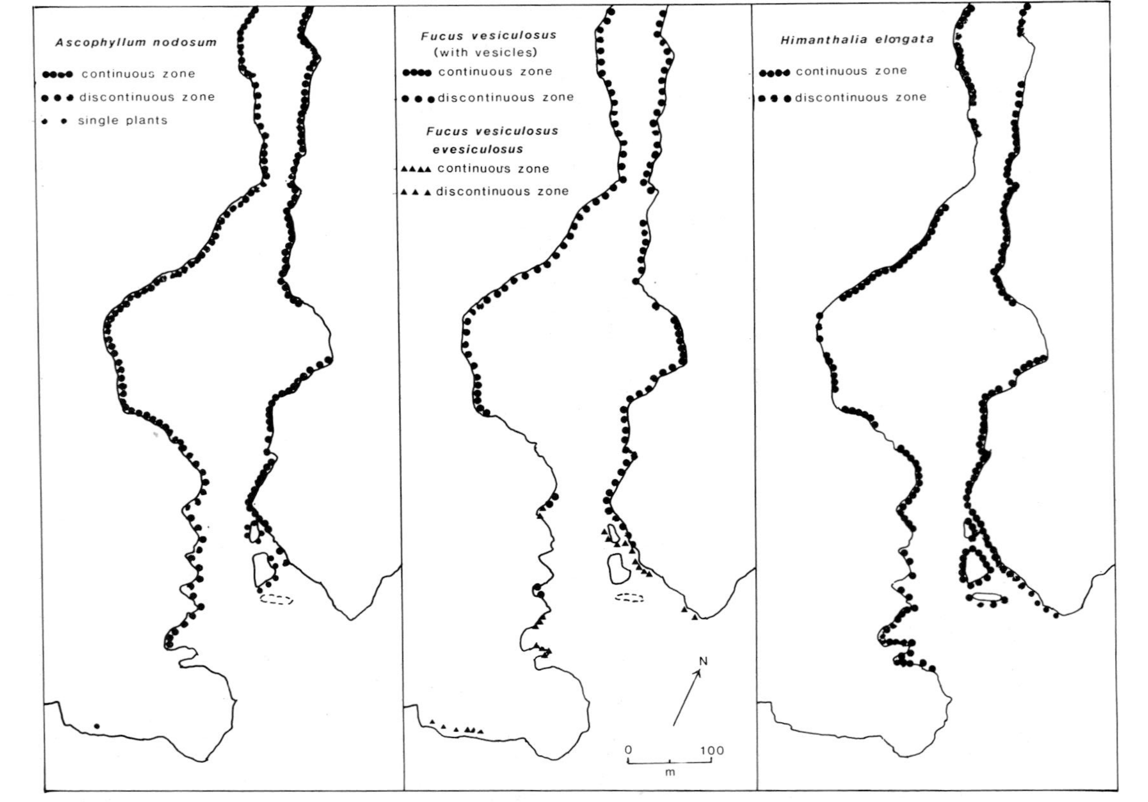

Fig. 6. Distribution of *Ascophyllum nodosum*, *Fucus vesiculosus* and *Himanthalia elongata* in Barloge Creek and on the adjoining open coast, recorded in April 1983 and (for *F. vesiculosus evesiculosus*) April 1986.

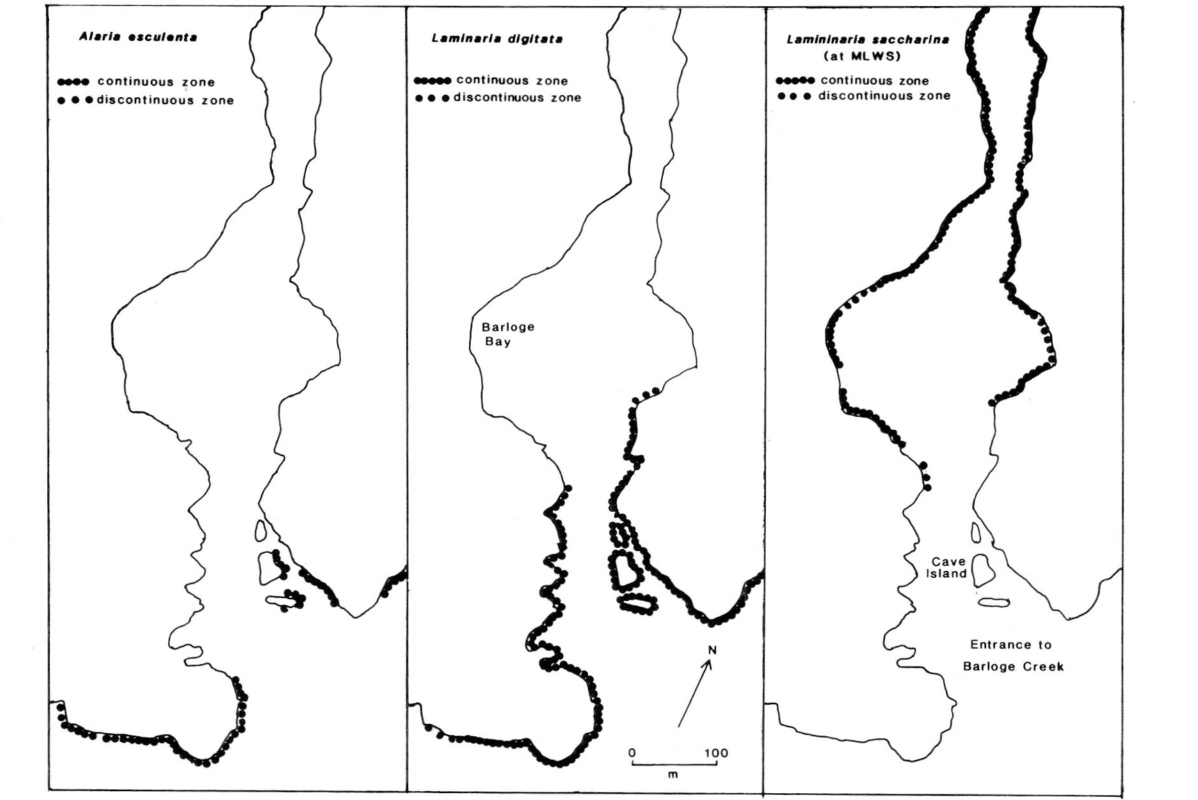

Fig. 7. Distribution of *Alaria esculenta*, *Laminaria digitata* and (at the littoral/sublittoral junction) of *Laminaria saccharina* in Barloge Creek and on the adjoining open coast, recorded in April 1983.

place Carrigathorna precisely on Ballantine's (1961) scale of wave-exposure. It corresponds with grade (2), "very exposed", rather than grade (1), "extremely exposed", in that *L. digitata* contributes to the sublittoral fringe, but it appears to exceed grade (1) in wave-exposure in that *Fucus vesiculosus evesiculosus* dies out with increasing wave-exposure at the mouth of Barloge Creek (Fig. 6) and is absent from the fully open coast of Carrigathorna (although present in a gully). Moreover, *Himanthalia elongata* is found at Carrigathorna only in the partial shelter of tide pools and not on the fully open low littoral. It is found extensively in shelter, but dies out at the mouth of Barloge Creek. It is difficult to explain these discrepancies without an understanding of how wave-exposure acts on the various indicator species and—as pointed out by Ballantine—there are geographical differences in its operation.

Alaria esculenta, used by Ballantine as an indicator of severe wave-exposure in west Wales, is plentiful on the open coast at Carrigathorna and penetrates only into the mouth of Barloge Creek (Fig. 7). It is also associated with wave-exposure in the Sound of Jura (Kitching, 1935; his Fig. 7), but endures more shelter there than in west Wales or Barloge Creek. This is probably due to the flushing of the Sound of Jura with cold water in summer. Sundene (1962) found that plants transferred in autumn from the west coast of Norway to the Oslofjord flourished over the winter but died in the following late spring or summer. He concluded that the distribution of *Alaria esculenta* on European coasts is limited by high temperature. No doubt this is true for the Lough Hyne area. The outflow temperature in the Lough Hyne Rapids can rise in summer well above the critical 16°C; there are no records for Barloge Creek, but it is likely to be affected.

Laminaria digitata, presumably more tolerant of high temperature than is *Alaria*, penetrates further into Barloge Creek, but in still greater shelter it gives way to *Laminaria saccharina* at sublittoral fringe level. *L. digitata* reappears on the Rapids Sill, where no doubt it benefits from the very strong current over a steep shallow bottom during outflow.

The zonation of lichens in Barloge Bay is much less extended than at Carrigathorna (Fig. 5), although their upper limits are ill-defined. *Pelvetia, Fucus spiralis, Ascophyllum nodosum* and *Fucus serratus* form dense zones. *Chthamalus montagui, C. stellatus, Semibalanus balanoides* and (at the lowest level) *Verruca stroemia* are present. *Littorina neritoides, P. patula, Patella aspera* and *Lichina pygmaea* are absent, but *Littorina saxatilis* and *L. littorea* are common or abundant. This shore conforms in general with Ballantine's grade (7), "very sheltered".

B. The Rocky Sublittoral

On the open coast at Carrigathorna *Laminaria hyperborea* extends down densely from the sublittoral fringe to a depth of about 15 m (below MLWS);

below that depth it thins out, and it disappears at a depth of about 17–18 m (Norton, Hiscock and Kitching, 1977). There are occasional plants of *Saccorhiza polyschides* among it.

The standing crop of Algae forming this *Laminaria* forest was estimated by Norton *et al.* (1977) from one 1 m × 1 m quadrat at each of a series of depths on the very wave-exposed south-east point, and two quadrats at each depth on the possibly slightly less wave-exposed east transect (Table 1). The number of quadrats which could be collected was limited by the diving time available, and is sufficient to give an approximate estimate of standing crop at various levels but not enough to define variation at particular depths. However, the number of individual *L. hyperborea* plants collected on the east transect was substantial, and the variation for estimates of their structural characteristics is defined.

Estimates of standing crop of Algae reached around 12 kg m^{-2} at depths of 6 m and 10 m on the south-east point and at 3 m on the east transect. For greater depths, all estimates of standing crops diminish sharply. All the *L. hyperborea* plants collected on the east transect had their ages estimated by Kain's (1963) method, from the growth lines in the holdfast, for which there may be an error of a year. Very few plants exceeded seven growth lines, and there were many with 0–1 lines in shallow water. The length of stipe, length of blade and blade thickness were all considerably less, age for age, for plants collected from the lower limit (17·5 m) than for plants from lesser depths. This lower growth rate implies that the plants were limited by some physical or chemical condition, rather than that they were being destroyed, at their lower limit, by a herbivore such as the sea urchin *Echinus esculentus*, as was found on submerged sections of the Port Erin breakwater (Jones and Kain, 1967). *E. esculentus* is present at Carrigathorna but thinly distributed; it could help to limit *L. hyperborea* because the growth of new plants at that depth is slow, but is probably not an important factor. Turbidity of coastal water, with the stirring up of sediment by wave action, reduces the transmission of light, and light is probably the most important factor limiting the depth to which *L. hyperborea* penetrates. Depths at which the irradiance is reduced to 1% of that at the surface are plotted comparatively by Hiscock (1985; his Fig. 21.3), and are regarded as similar (15–20 m) for sites in west Scotland, south-west Ireland and the Isles of Scilly, and at all these sites the *Laminaria* forest reaches a similar depth.

Blades of *Laminaria hyperborea* carried small quantities only of *Obelia geniculata* (Hydrozoa) and *Membranipora membranacea* (Bryozoa) and *Patina pellucida* (Prosobranchia), all of which are plentiful in parts of the Rapids (as discussed in the earlier review). Many stipes carried *Electra pilosa* and other Bryozoa, and a few carried a hydrozoan *Amphisbetia operculata* (formerly *Sertularia*), abundant in swift current in the Rapids and intolerant of sedimentation (Round *et al.*, 1961). Holdfasts of *L. hyperborea* having one or more growth lines carried a wide variety of sedentary or

Table 1
Standing crops (g m^{-2} fresh weight) of the more abundant seaweeds at Carrigathorna.

	South-east Point, July 1970								East Transect, July 1972				
Depth (m) below chart datum:	3	6	10	12	15	18	21	23	1	3	6	13·5	17·5
Epilithic													
Acrosorium uncinatum	9	–	–	–	–	–	–	–					
Bonnemaisonia asparagoides	–	–	–	1	–	24	5	5	Not collected				
Delesseria sanguinea	12	2	2	2	320	108	13	130	Not collected				
Dictyopteris membranacea	–	–	–	1	1	22	5	1	Not collected				
Dictyota dichotoma	–	–	1	5	3	66	2	–	Not collected				
Dilsea carnosa	14	–	–	–	–	–	–	–	Not collected				
Heterosiphonia plumosa	–	–	1	1	29	5	32	–	Not collected				
Laminaria hyperborea	5 553	11 904	11 820	2 705	1 487	–	–	–	4 152[c]	12 510	4 370	1 397	161
Lithothamnia[a]	60[b]	85[b]	64[b]	76[b]	46[b]	?	?	?	86[b]	53[b]	81[b]	73[b]	?
Plocamium cartilagineum	–	–	8	–	–	<1	–	–	Not collected				
Polyneura hilliae	–	<1	1	<1	3	10	26	10	Not collected				
Phyllophora crispa	–	–	–	–	–	<1	–	6	Not collected				
Saccorhiza polyschides	1 600	–	–	–	–	–	–	–	Not collected				
Rhodymenia pseudopalmata	<1	–	–	2	–	14	13	–	Not collected				
Epiphytic													
Brongniartella byssoides	–	–	–	–	–	–	–	–	?	1	6	–	–
Cryptopleura ramosa	3	70	121	26	10	10	5	10	?	1	44	12	1
Membranoptera alata	52	3	7	1	1	–	–	–	?	41	1	1	–
Nitophyllum punctatum	–	2	11	3	3	5	–	–	–	–	3	1	–
Palmaria palmata	460	70	–	–	–	–	–	–	?	207	–	–	–
Phycodrys rubens	120	130	147	7	4	–	5	–	?	39	87	19	1
Polysiphonia broadiaei	75	2	1	1	–	–	–	5	–	–	–	–	–
P. urceolata	–	–	–	–	–	–	–	–	?	155	1	–	–

[a] Includes all polystromatic, encrusting, calcareous Corallinaceae.
[b] Percentage cover.
[c] Sample taken in 1971.

attached invertebrates (Norton *et al.*, 1977; their Table 5), which are widely distributed and found in the Rapids area.

Conditions on the rock surface are complicated by changing angle of slope as well as by decrease in water movement and increase in siltation which are associated with increasing depth, and it is not yet possible to assess the separate effects of so many changing environmental conditions.

Measurements of light intensity at a range of depths outside the forest and within the forest were made in energy units, with photocells sensitive only to a very narrow waveband at 550 nm (Fig. 8). Transmission was about 80% per metre, and the unshaded irradiance at the lower limit of the forest amounted to about 5% of that measured in a bowl of water on the shore. Within the forest it ranged from 3% to 11% of its value at the same depth in open water, as compared with 1% in the very dense *Laminaria* forest in the Sound of Jura (Kitching, 1941).

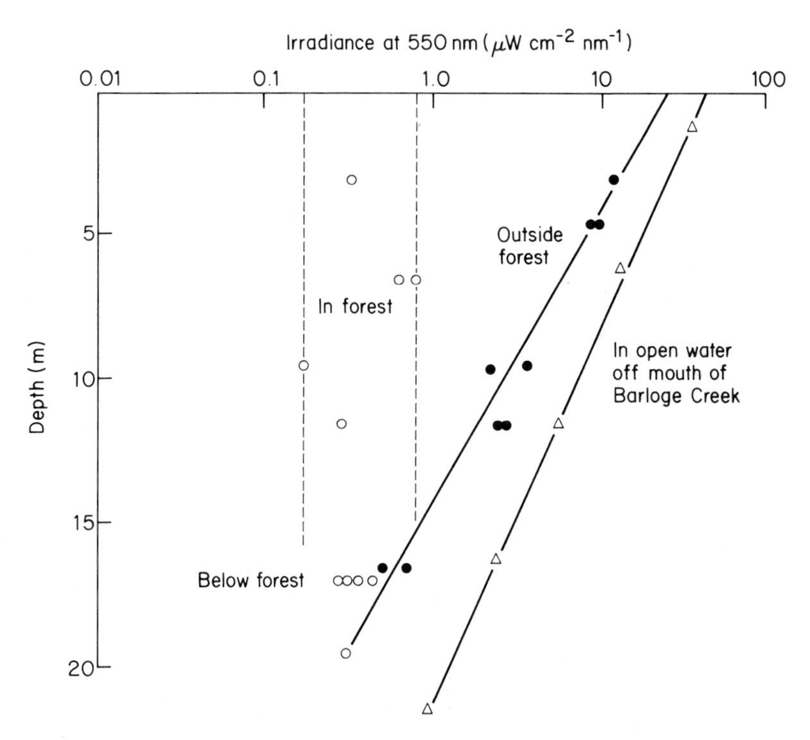

Fig. 8. Irradiance at various depths within the *Laminaria* forest (○) and outside it (●) from readings taken by a diver off Carrigathorna on 24 August 1971, and from readings taken from a boat in open water off the mouth of Barloge Creek (△) on 22 July 1968. (Reproduced from Fig. 2 in Norton, T. A., Hiscock, K. and Kitching, J. A. (1977). The ecology of Lough Ine XX. The *Laminaria* forest at Carrigathorna. *J. Ecol.* **65**, 919–941.)

In very shallow water at Carrigathorna there was an abundance of *Palmaria palmata, Membranoptera alata* and *Phycodrys rubens* growing on the stipes of *Laminaria hyperborea* (Table 1). With increasing depth *Palmaria* disappeared and *Crypyopleura ramosa* became abundant; but all epiphytes became scarce towards the lower limit of the forest. Decreasing illumination was probably responsible for these changes. Undergrowth Algae were scarce within the forest, but below the lower limit of the forest there was a plentiful cover of low bushy Algae; *Delesseria sanguinea, Dictyopteris membranacea, Dictyota dichotoma* and others, all of which must tolerate low irradiance.

C. Tide Pools

Tide pools are a special habitat, neither fully littoral nor fully sublittoral. Their flora and fauna does not have to endure exposure to air, but is subjected to a range of other environmental fluctuations. The pattern of these fluctuations depends on the level of the pools in question with respect to the rise and fall of the tide, the topography, the action of waves, the climate, and the weather experienced from day to day. The environmental conditions under consideration include water temperature, salinity, pH and quantity of dissolved oxygen. A full account of all these over a range of depths would require records on a scale not yet attempted.

Information about conditions in tide pools is available from the work of de Virville (1934, 1935) on the Atlantic and Channel coasts of France and around the Channel Islands, and from the survey by Pyefinch (1943) of tide pools on Bardsey Island, North Wales. Swedish rock pools have been surveyed by Ganning (1971). Daniel and Boyden (1975) have described the diurnal fluctuations in environmental conditions in some pools at St Bride's Haven, Pembrokeshire. Fluctuations in oxygen and carbon dioxide concentrations in the water of rock pools by day and by night are described by Truchot and Duhamel-Jouve (1980). Conditions in tide pools in the Quoddy region of New Brunswick have been surveyed by Thomas (1983). Early literature is summarized in an important discussion by Lewis (1964; Ch. 10).

Much more information is needed about the flora and fauna associated with these various environmental conditions. In the account by Goss-Custard, Jones and Kitching (1979) of the very fine collection of tide pools on the fully wave-exposed coast at Carrigathorna and on Urchin Reef in the mouth of Barloge Creek (shown in Fig. 3), consideration is given to the control of dominant species by physicochemical and biotic conditions and to the relation of tide-pool communities with these conditions. For purposes of description, tide pools are grouped as supratidal (above MHWS but reached by splash), high-tidal (between MHWS and MHWN), mid-tidal (between MHWN and MLWN), low-tidal (between MLWN and MLWS) and subtidal (below MLWS). Mid-tidal pools could with advantage be subdivided into upper mid-tidal and lower mid-tidal, above and below mean tide level, as used by Thomas (1983) (his pool types 1C and 1B) for tide pools in New Brunswick. In this brief account we shall work upwards. A map of the Carrigathorna tide pools is given in Fig. 9, and diagrammatic sections are shown in Fig. 10.

Springtide Pool, over 5 m deep and connected with the sea by a narrow channel even at the lowest tides, carries a sublittoral fringe of *Laminaria digitata* and *Alaria* around its margin, and below this a forest of *L. hyperborea* and *Saccorhiza*, with *Corallina officinalis* and other Algae as undergrowth. Thus far, it resembles the sublittoral region of the open coast, of which it is an enclosure. However, the *Laminaria* forest on the steep sides of this pool rapidly thins out, and disappears at a depth of 2 m. Below this level extend *Dictyota dichotoma* (notably tolerant of a low level of irradiance) and other Algae found as undergrowth in the *Laminaria* forest of the open coast. The rock barrier which encloses the pool no doubt gives some protection against wave action, but other environmental conditions are not likely to deviate significantly from those of the open coast. The lack of laminarian Algae below 2 m is probably due to overshadowing by the *Laminaria* forest above. The abundance of the small limpet *Patina pellucida* may possibly be attributed to better conditions for the settlement of larvae provided by periods of quiet water. Very large numbers of *P. pellucida* were also found on *Palmaria palmata* growing on the stipes of *L. hyperborea*. Vahl (1971) has inferred, on indirect evidence, that larvae of *P. pellucida* first settle on other Algae, perhaps less tough to graze, but later migrate to *Laminaria*. A tide pool might favour such a migration.

Important and sometimes severe fluctuations may occur in mid-tidal pools, disconnected from the sea (and reconnected) with every tide. Shallow pools reach greater extremes than deep pools, and in deep pools extreme conditions are only experienced in the shallow water. Conditions remain nearly steady at depths below 1 m (Figs 11 and 12). These deep pools support in their deep water a forest of *Laminaria hyperborea* with up to seven presumably annual lines (as described by Kain, 1963), and the usual epiphytes (notably *Palmaria palmata*) on the stipes of all but the youngest plants. Although the necessarily small number of samples does not permit a close comparison with the sublittoral of the open coast, the standing crop of *Laminaria* spp. in Main Pool was substantial (around 6–7 kg m^{-2}, as compared with a maximum of around 12 kg m^{-2} on the open coast). The blades of *L. hyperborea* were as long and heavy in Main Pool, age for age, as at any depth on the open coast, but the stipes grew little after the age of five growth lines. Although many of the same epibionts were associated with *Laminaria* spp. as in the *Laminaria* forest of the open coast, some significant differences were recorded: in Main Pool the blades of *L. hyperborea* were

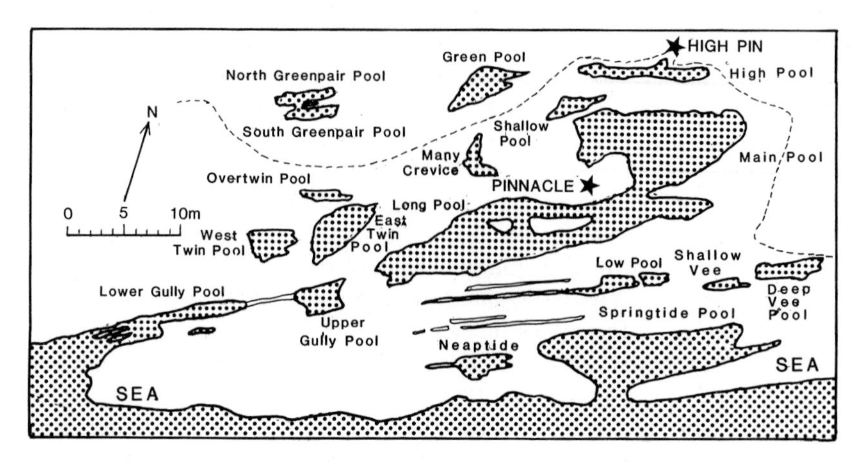

Fig. 9. Map of Carrigathorna tide pools. (Reproduced, with minor modification, from Fig. 2 of Goss-Custard, S., Jones, J. and Kitching, J. A. (1979). Tide pools of Carrigathorna and Barloge Creek. *Phil. Trans. R. Soc. B* **287**, 1–44.)

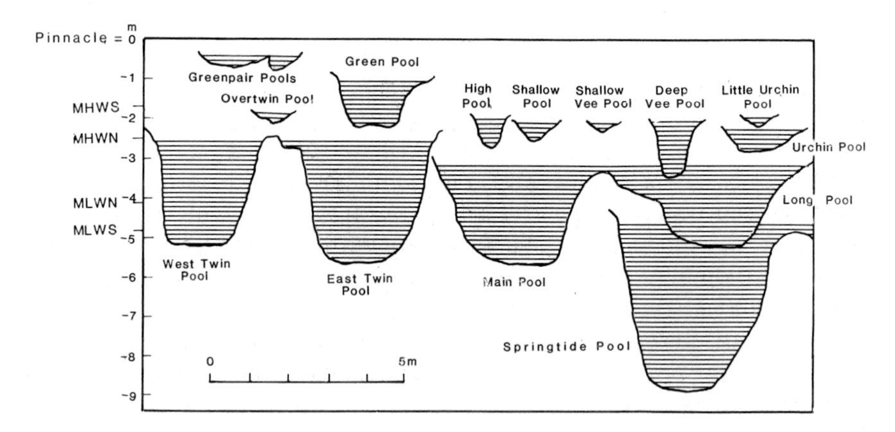

Fig. 10. Diagrammatic sections of Carrigathorna and Urchin Reef tide pools, in relation to their levels on the shore below Pinnacle. (Reproduced, with minor modifications, from Fig. 7 of Goss-Custard *et al.* (1979): for full acknowledgement see Fig. 9.)

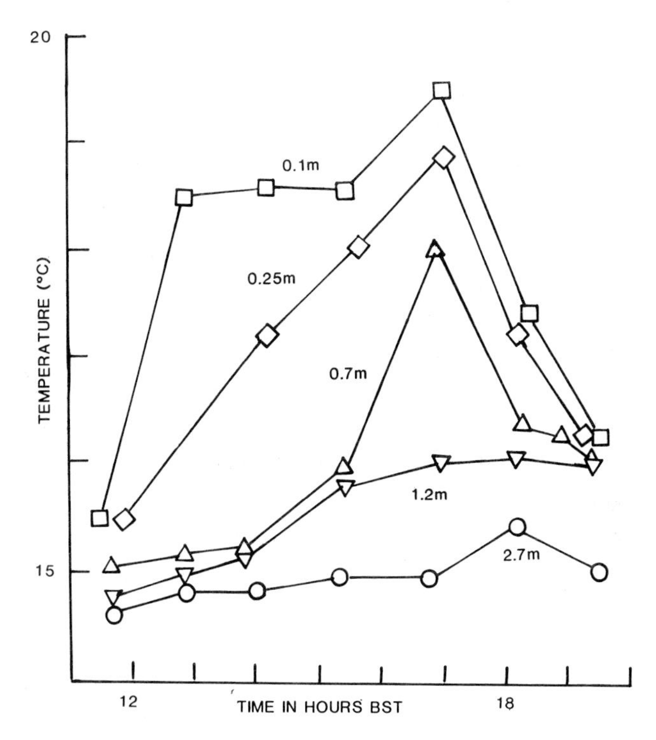

Fig. 11. Temperature at various depths in Main Pool on a fine summer afternoon (29 July 1976) over the period of low water, taken over the deepest part of the pool at the depths stated. (Reproduced from part of Fig. 8 of Goss-Custard *et al.* (1979): for full acknowledgement see Fig. 9.)

clean of *Membranipora membranacea* and *Obelia geniculata* but carried many more *Patina pellucida* than on the open coast. The *Laminaria* forest in Main Pool is not exposed to the high temperatures reached in the shallow water of this pool on fine summer days at low tide. A temperature of around 20°C continuously is probably near the upper limit for *L. hyperborea* sporophytes. These reach their southern limit of distribution half-way down the coast of Portugal, where the water reaches 19–20°C, and they are scarce or absent at sites in the inner Biscay area, where the summer sea temperatures are higher (Kain, 1971). We do not know what would be the response to intermittent high temperatures, but it seems very possible that they might discourage the growth of *L. hyperborea* in shallow water in tide pools.

The shallow regions of mid-tidal pools carry a thick cover of *Corallina officinalis*, also found in the undergrowth of the shallow sublittoral region of the open coast. Many of the same small invertebrates inhabit the under-

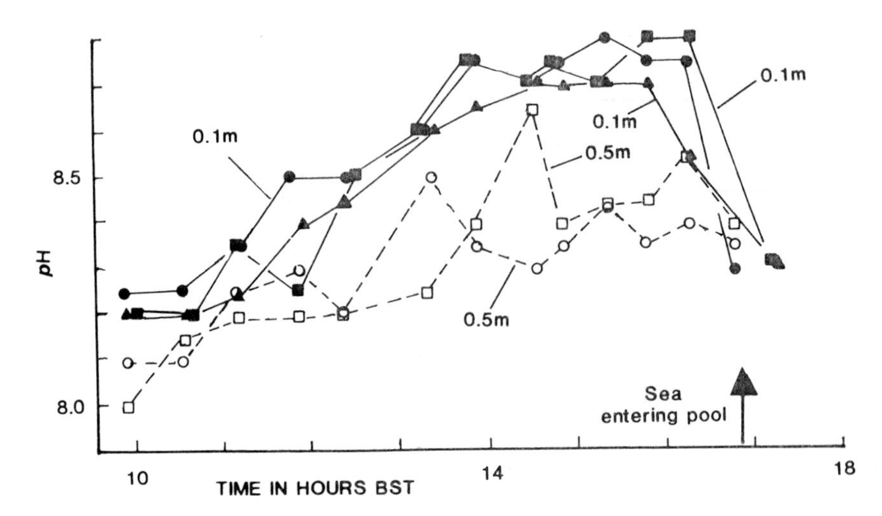

Fig. 12. The pH at various positions in Deep Vee Pool on 5 April 1977. Continuous lines denote sites at the bottom near the edge of the pool, at the depths stated. Broken lines denote sites over the deepest part of the pool, at the depths stated. (Reproduced from part of Fig. 9 of Goss-Custard *et al.* (1979): for full acknowledgement see Fig. 9.)

growth in both these places. However, there are many *Asterina* in the mid-tidal pools belonging to the recently described *Asterina phylactica* (Emson and Crump, 1979, 1984), which is generally confined to tide pools.

Over the narrow strip of shore between high water of neap and of spring tides, there is a very severe increase in adverse conditions for a small increase in height on the shore (Lewis, 1964; his Fig. 11). For tide pools this means an increase in the duration of continuous disconnection from the sea. Subject to a variable correction for splash, this disconnection will make possible a persistent extreme of salinity in especially dry or especially wet weather. Extremes of pH and oxygen content attained through photosynthesis by day depend on quantity of Algae in the pool, and are reversed at night (Truchot and Duhamel-Jouve, 1980), and temperature extremes also tend to reverse by night. All this is illustrated by Pyefinch (1943), and the records obtained by Goss-Custard *et al.* (1979) on Carrigathorna and Urchin Reef conform with this scheme. Extremes attained in shallow water in the Carrigathorna tide pools are summarized in Table 2.

In shallow high-tidal pools at Carrigathorna, the rock is coated with clean crustose coralline Algae scoured by many limpets (*Patella aspera* and *P. vulgata*), both species carrying tufts of *Enteromorpha* on their shells where it is safe from grazing. There were occasional tufts of *Corallina officinalis* in these pools; those in High Pool were regularly pink and healthy in April, but

Table 2
Summary of extreme conditions reached in shallow water (0·1–0·25 m) in Carrigathorna tide pools in July (data from Goss-Custard *et al.*, 1979).

Level	Pool	Maximum depth (m)	Temperature (°C)	Oxygen saturation (%)	pH	Salinity (‰)
Supratidal	Overtwin	0·25	27·9	–	10·0	41·2
	Green	1·4	25	175	9·8	37·0
High-tidal	High	1·1	22	175	–	35·5
	Shallow	0·7	25	194	–	38·2
	Shallow Vee	0·3	25	–	10·0	34·9
	Deep Vee	1·6	22	–	9·0	35·0
Mid-tidal	Main	2·8	20	142	8·7	35·05
	East Twin	2·8	19·5	166	9·0	34·9
	Long	2·0	23	163	9·1	–
Sea	–	–	12	99	8·15–8·25	34·87

white and apparently dead in July, probably as a result of high temperature. *Corallina* in mid-tidal pools carried *Spirorbis corallinae*, three species of Bryozoa, and two crustose corallines, all growing upon it; in high-tidal pools the crustose corallines diminished in quantity, the Bryozoa were missing, and *S. corallinae* was supplemented by another spirorbid, *Janua pagenstecheri*. Many small motile invertebrate species were still present in high-tidal pools, including predators. Little Urchin Pool (high-tidal) and Urchin Pool (on the high-tidal/mid-tidal boundary) on Urchin Reef contain patches of *Corallina officinalis* and areas of limpet-scoured crustose coralline Algae. They also contain tufts of *Cystoseira nodicaulis* (Phaeophyceae) which (especially in summer) accommodates very large numbers of small invertebrates (Goss-Custard *et al.*, 1979; their Table 10). Shelter from wave action probably contributed.

In supratidal pools, above the level of high spring tides in calm weather but subject to splash, environmental conditions deviate most strongly from those experienced in the open sea, and also vary greatly from pool to pool. Much more information is needed about these pools. A few extremes are summarized in Table 2, including some high salinities. The two close-together and very similar Greenpair pools were completely covered with *Enteromorpha intestinalis*. There were no limpets and no *Corallina*, but some underlying crustose corallines. Six quadrats collected at the same time in September yielded only 7 species of Algae and 14 of invertebrates. Several species were present in large numbers, including especially the copepod *Tigriopus fulvus* and larvae of *Halocladius fucicola* (Diptera). These two species and *Enteromorpha intestinalis* tolerate a wide range of salinities.

Thus, these supratidal pools are occupied by a small number of species which are tolerant of very adverse conditions, too severe for most of the general tide-pool flora and fauna. A similar low species diversity is reported by Ganning (1971) for Swedish tide pools where conditions are severe.

The copepod *Trigriopus fulvus*, taken in large numbers in supratidal pools at Port St Mary, Isle of Man (Fraser, 1936a), has been shown to be able to resist high temperatures and salinities but not complete desiccation (Fraser, 1936b; Ranade, 1957). In some very interesting transfer experiments (Dethier, 1980) it has been shown that the related *Tigriopus californicus*, found in high-level pools on the Pacific coast of North America, falls an easy victim to predatory tide-pool fish moved up into these high-level pools. Normally it is protected from predation by the extreme conditions of temperature and salinity which prevent encroachment by these fish.

The slow growing but competitive *Corallina officinalis* forms a dense undergrowth in the very low littoral region of the open coast, and extends down a few metres in the shallow sublittoral. Upwards, it reappears as the dominant undergrowth alga in shallow water, in tide pools up to about the level of MHWN. Above this level limpets take over and clear the rock surface in most high-tidal pools. Still higher, in supratidal pools, limpets fail and *Enteromorpha* grows densely and dominates. Thus the factors which limit the distribution of limpets are of critical importance. Limpets are probably excluded from dense *Corallina* undergrowth, which acts as a barrier both to settlement and to invasion. When a small area in Deep Vee Pool, at the upper margin of dense *Corallina* undergrowth, was scraped clean, and subsequently rescraped of rapidly growing *Enteromorpha*, it became overgrown by other soft Algae and was finally occupied by *Patella* spp., some of which were recent settlers and some older invaders. In another experiment, areas of shallow high-tidal pools (Overtwin Pool and Shallow Pool) were completely cleared of limpets. The cleared areas rapidly became overgrown with *Enteromorpha*. Thus adverse conditions, especially high temperature, limit *Corallina* upwards, making way for *Patella* spp., and *Patella* is limited upwards by still more adverse conditions, making way for *Enteromorpha* in supratidal pools.

There are no doubt deviations from this simple representation, notably the high-tidal Many Crevice Pool. This leaks and partly dries out during neap tides, producing a zone covered with the blue-green alga *Lyngbia confervoides* and inhabited by many larvae of *Halocladius fucicola* (Diptera), both of which evidently tolerate very extreme conditions.

D. The Bullock Island Cave

There are many sea caves along the sea cliffs of County Cork. The Bullock Island Cave is exceptionally rich in plant and animal growth, probably because it is protected from the open sea by the entrance to Barloge Creek.

This cave extends for 97 m into Bullock Island through Devonian slate which dips nearly vertically. At its mouth on Barloge Creek it measures 5–6 m in width and 9 m in height, of which 3 m remains under water at low water of spring tides. The Bullock Island Cave is accessible by boat in calm weather, but when there is a swell setting into the mouth of Barloge Creek the surge piles up within the cave. This cave has been investigated in considerable detail by Norton, Ebling and Kitching (1971), because it acts as a natural—although complicated—experiment on the effects of certain important environmental factors by providing these in unusual combinations.

The habitats available in the cave are the supralittoral and littoral regions, the shallow sublittoral, and some small tide pools. The environmental conditions under special consideration are illumination, which decreases as you go further into the cave, and the protection afforded by the cave against the damaging effects of exposure to air. The assessment of irradiance is complicated. Light is differentially transmitted through sea water and different Algae make use of different wavebands. In the cave the amount of light reaching various distances from the mouth rises greatly when the sun shines directly in, which happens in the afternoon, and is also greater when the site in question is exposed to air. More light will be lost from the water surface as the angle of incidence decreases, and light travelling up the cave under water will suffer much absorption. As the water level increases, the unobstructed aperture of the cave is reduced and, if the site is under water, light striking the surface will undergo some refraction. Extended periods of observation will be required to obtain any useful information.

Light readings within the cave were made, along with readings from "control" cells in the open, every 15 minutes throughout the hours of daylight, and at several stages in the fortnightly tidal cycle. The photocells were sensitive over narrow sharply defined wavebands. They were mounted at sites in the cave so as to face the source of light, but there would be some loss when under water owing to a change in the angle of incidence by refraction.

A study was first carried out of the conditions controlling the penetration of light into the cave. Measurements were made of the light intensity at 550 nm at five sites at stated distances both into the cave and above low water of ordinary spring tides. These are shown in Figs 13 and 14, taken during days of spring and neap tides (note that the scales differ in various parts of these two figures). In each case the peak illumination was during the period when the photocell was exposed to air, and when this happened in the afternoon, with the sky clear and the sun shining into the cave from the west, the readings were highest. Thus the high peak of irradiance for site A in Fig. 13 was cut off sharply by the rise of the tide above the photocell; site B, low down in the littoral region, achieved (Fig. 14) a sharp peak in the late afternoon, when the tide had fallen sufficiently to uncover it. Far into the cave (42 m from the entrance) the irradiance was very low.

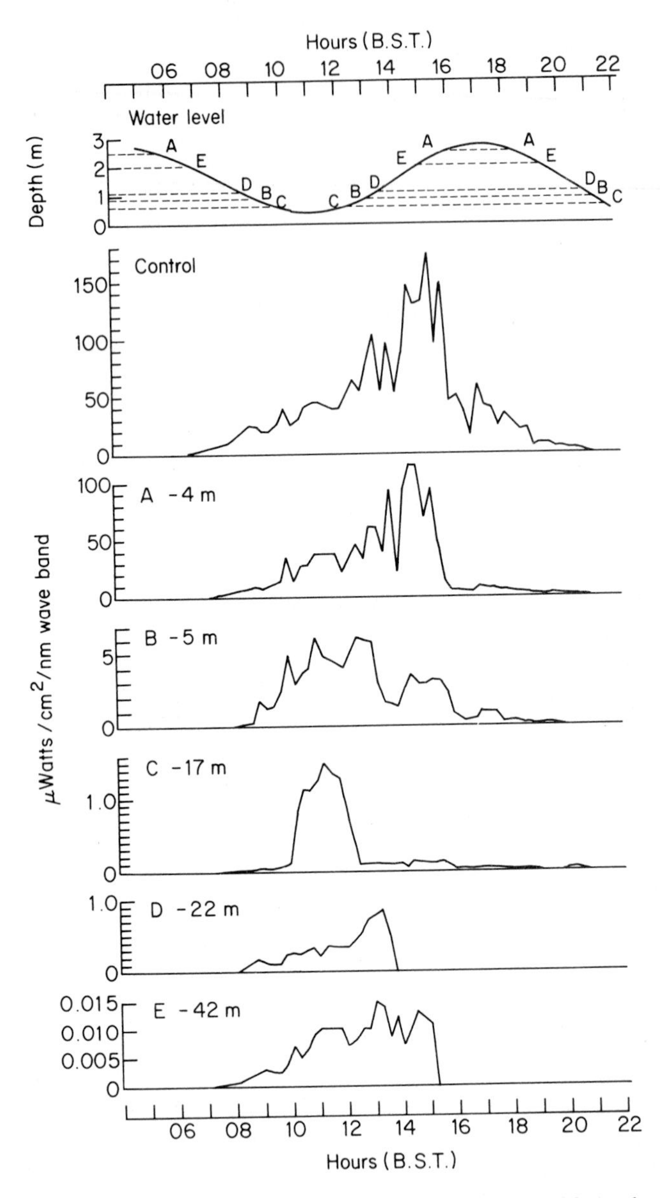

Fig. 13. Illumination at 550 nm, at various positions measured from the cave mouth, in the Bullock Island Cave. Readings were taken every 15 minutes on 6 July 1967. The levels of the photocells and of the water are shown in the top panel. Note the changes in scale made to accommodate the very low irradiance far into the cave. (Reproduced from Fig. 2 of Norton, T. A., Ebling, F. J. and Kitching, J. A. (1971). Light and the distribution of organisms in a sea cave. In *The Fourth European Marine Biology Symposium* (Ed. by D. J. Crisp), pp. 409–432. Cambridge University Press.)

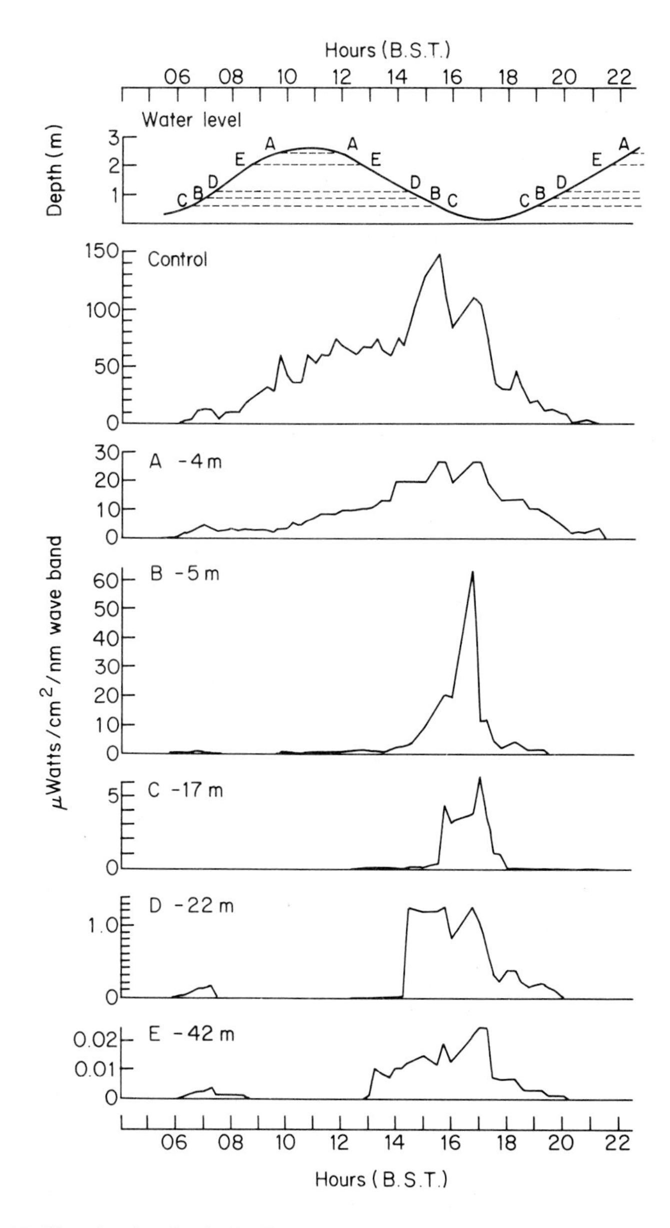

Fig. 14. Illumination in the Bullock Island Cave as for Fig. 13 but on 14 July 1967.

Measurements were made with photocells narrowly sensitive to 515, 550 and 615 nm, stationed just below the level of low water of ordinary spring tides and at the inner limit of *Laminaria hyperborea* and *L. digitata* (7 m from the cave mouth). When the sun was shining into the mouth of the cave and the water level was low, the irradiance amounted to 39–51% of that in the open (and differences between cells were probably not significant). These peak levels lasted only a short time. The total whole-day irradiance, measured at 550 nm at the inner limit of *Laminaria hyperborea* and *L. digitata* on five separate days covering the whole semilunar tidal cycle, amounted to 5% of that in the open. The *Laminaria* forest on the sublittoral rock slopes of Carrigathorna becomes very thin below a depth of 15 m, and stops at 17·5 m (Norton *et al.*, 1977). From a limited number of light readings at 550 nm, a reduction of 3% of the surface value would be expected at a depth of 15 m. This estimate is subject to considerable error when stirring up of bottom sediment and increased reflection at the surface in rough weather are considered. However, it does strongly support the conclusion that irradiance limits both the downward extension of *L. hyperborea* at Carrigathorna and its inward extension into the cave.

The penetration of some of the more abundant Algae and animals into the cave is summarized in Fig. 15. All those listed extend in from outside the cave. More complete data are given in Figs 5, 6 and 7 of Norton *et al.* (1971). The fucoids and laminarians stop only a short distance into the cave. Beyond the inner limit of fucoids the red Algae *Plumaria elegans, Ptilothamnion pluma* and *Lomentaria articulata*, which on the open coast live in crevices or as undergrowth, dominate the unsheltered intertidal walls of the cave and extend along the cave together with some other Algae normally found in protected situations on the open coast. These include *Cladophora* spp., which would be well equipped photosynthetically for the quality of light in the cave so long as there was sufficient irradiance. Various sublittoral Algae found in the undergrowth of the *Laminaria* forest outside the cave penetrate a short distance along the open sublittoral of the cave and remain sublittoral (*Callophyllis laciniata, Delesseria sanguinea, Dilsea carnosa* and others). Other subtidal Algae, *Phyllophora crispa* and *Cryptopleura ramosa*, characteristic of the *Laminaria* forest undergrowth outside the cave, spread upwards within the cave some considerable vertical distance into the littoral region and doubtless were enabled to do so because of the constant high humidity of the atmosphere and the protection from rain and from extremes of temperature. The furthest penetration by plants into the cave was at 26 m from the mouth, but most stopped at 7–15 m into the cave.

Animals penetrated much further with two carnivores, *Asterias rubens* and *Actinia equina*, and two filter feeders, *Pomatoceros* sp. and *Chthamalus* sp., reaching their limits at 80 m from the cave mouth. The filter feeders would collect small items of plankton and the anemone larger swimming

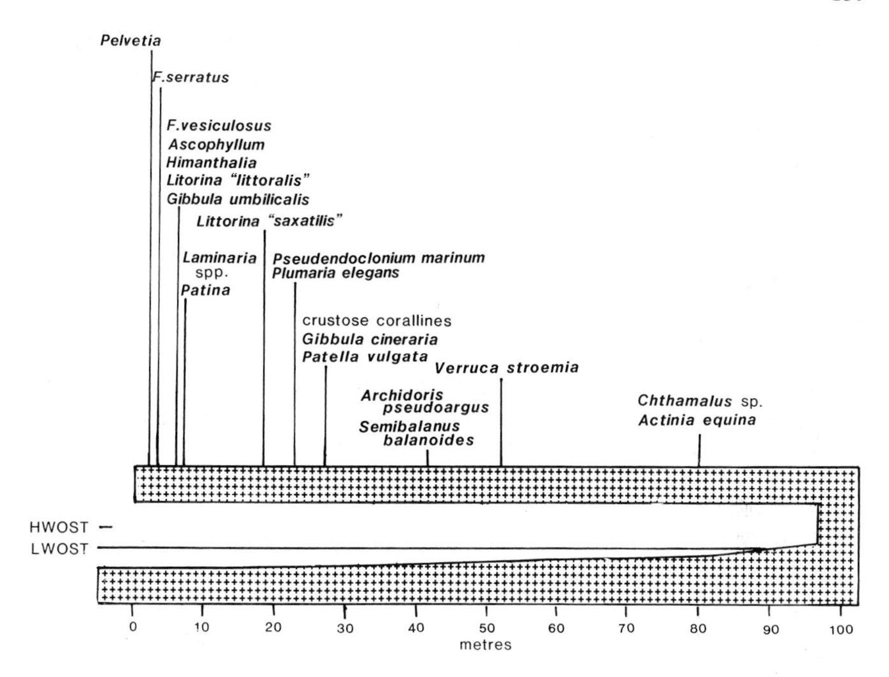

Fig. 15. Limits of penetration of some Algae and invertebrates into the Bullock Island Cave.

prey. The bottom of the cave rises to low water level at 10 m from the end, and surge disturbs the sand and probably makes the walls of the cave uninhabitable. The limitation of many herbivores (*Littorina* spp., *Gibbula umbilicalis, G. cineraria, Patina pellucida, Patella vulgata*) is related to that of various Algae. Various animal species normally sublittoral or low littoral on the open coast extend higher within the cave (*Verruca stroemia, Callopora lineata, Microporella ciliata* and *Archidoris pseudoargus*). Thus the inner limit of invertebrate species is determined by the availability of food and their upper limit by relief from conditions associated with exposure to air.

There is one further example of special conditions brought about by the cave. The anthozoan *Corynactis viridis* grows in sheets on overhanging sublittoral cliffs within the cave, as it does in similar situations on sublittoral cliffs on the open coast and in the lough. In the cave it also forms sheets in small open tide pools. In open tide pools at Carrigathorna it was inconspicuous, isolated, and scattered among the undergrowth; it only forms sheets at Carrigathorna in narrow slit pools deeply shaded from the light. Its occurrence in the Rapids, under the dense canopy of *Saccorhiza* blades, which shade it, will be discussed later (p. 157). It appears to be adversely affected by excessive illumination, as concluded by Muntz *et al.* (1972) from a variety of experiments. It flourishes in the darkness of the cave probably by reason of protection from direct insolation, lack of competition for space by Algae, and freedom from sedimentation.

IV. THE ROCKY SHORE OF THE LOUGH

A. Littoral

1. Designation of Sites

The intertidal shore of Lough Hyne was subdivided by Renouf (1931a) into sectors, each of which is reasonably homogeneous in substrate characteristics: N1, N2, . . . N12 along the north shore, E1 . . . E20, S1 . . . S17, W1 . . . W38, and (for Castle Island) I1 . . . I21 (Fig. 2). He also subdivided Barloge Creek into sectors. Bassindale *et al.* (1948) designated points at 10 m intervals (mark 1 to mark 10) along the Rapids Quay, with corresponding points (D1 to D10, defined by normals from the quay) on the east (Dromadoon) bank.

2. Water Levels and Tides

Water levels in the lough are referred (in this review) to "Standard Level", which was defined by Bassindale *et al.* (1948) as 1·57 m below the edge of the Rapids Quay at mark 3, where there was an iron ring, now disintegrated. The site is still easily recognizable by the notch within which the ring was attached. Depths in the lough are expressed above (+) or below (−) standard level (SL), unless another reference point is specified. Thus SL = Rapids Quay at mark 3 − 1·57 m = Glannafeen Laboratory Quay − 1·29 m = Curlew Rock (in I3) − 1·49 m. Standard level is at about the average level of low tide in the lough. The lowest tides in the lough occur during neap tides (see Figs 16 and 17), but the difference in the level of low water of spring and of neap tides amounts only to about 15 cm.

The account of the effect of the Rapids on the tidal cycle in the lough, already outlined in the previous review, has been fully confirmed by much more comprehensive observations over a spring and over a neap tide in September 1983 (Figs 16 and 17). Much more water came into the lough during inflow of the spring than of the neap tide, and after the turn of the spring tide it took the first five hours of outflow to fall to SL + 0·40 m, whereas at the neap tide the water stood at about this level at high slack water. As a consequence, at spring tides the water had much less time to flow out below this level before the tide turned again to flow in. The volume rate of outflow is low after the water has passed below +0·40 m, mainly because of the small cross-sectional area of the outflow stream. In addition, the rate

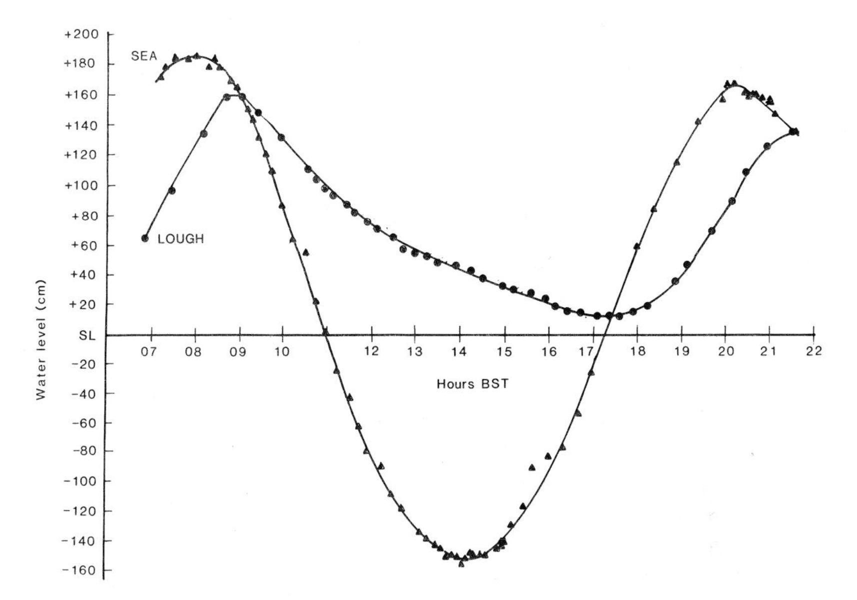

Fig. 16. Spring tide at Lough Hyne, recorded on 10 September 1983. ▲, Water level in the sea, taken at Nita's Rock; ●, water level in the lough, taken at Peggy's Mark.

of outflow is by then independent of the level in the sea; the much lower level reached in the sea at the spring tide does not accelerate the flow. This is attributed to the turbulence of the outflow over and below the sill, due to a combination of fast current and irregular bottom. Thus the tide does not fall as low in the lough at spring as at neap tides. Actually, outflow lasts slightly longer at spring tides because the tide in the sea has to rise higher before it turns in the Rapids.

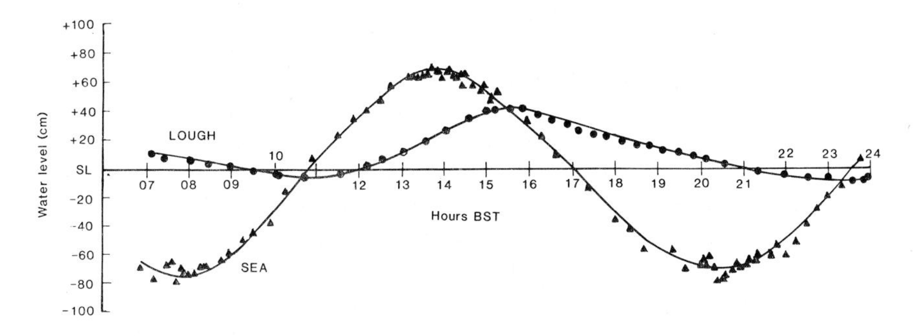

Fig. 17. Neap tide at Lough Hyne, recorded on 16 September 1983. ▲, Water level in the sea, taken at Nita's Rock; ●, water level in the lough, taken at Peggy's Mark.

3. Plant Zones

Nearly the whole of the lough shore is steep and rocky. The tidal rise and fall amounts to about 1·5 m at spring tides and about 0·5 m at neap tides. The zonation is very compressed. It is illustrated in Fig. 18 for a small promontory on the south shore at S7 ("Mermaid Rock"), to a much larger scale than that used for Barloge Bay North and for Carrigathorna. The species present are what one would expect on a very sheltered shore, with narrow but dense

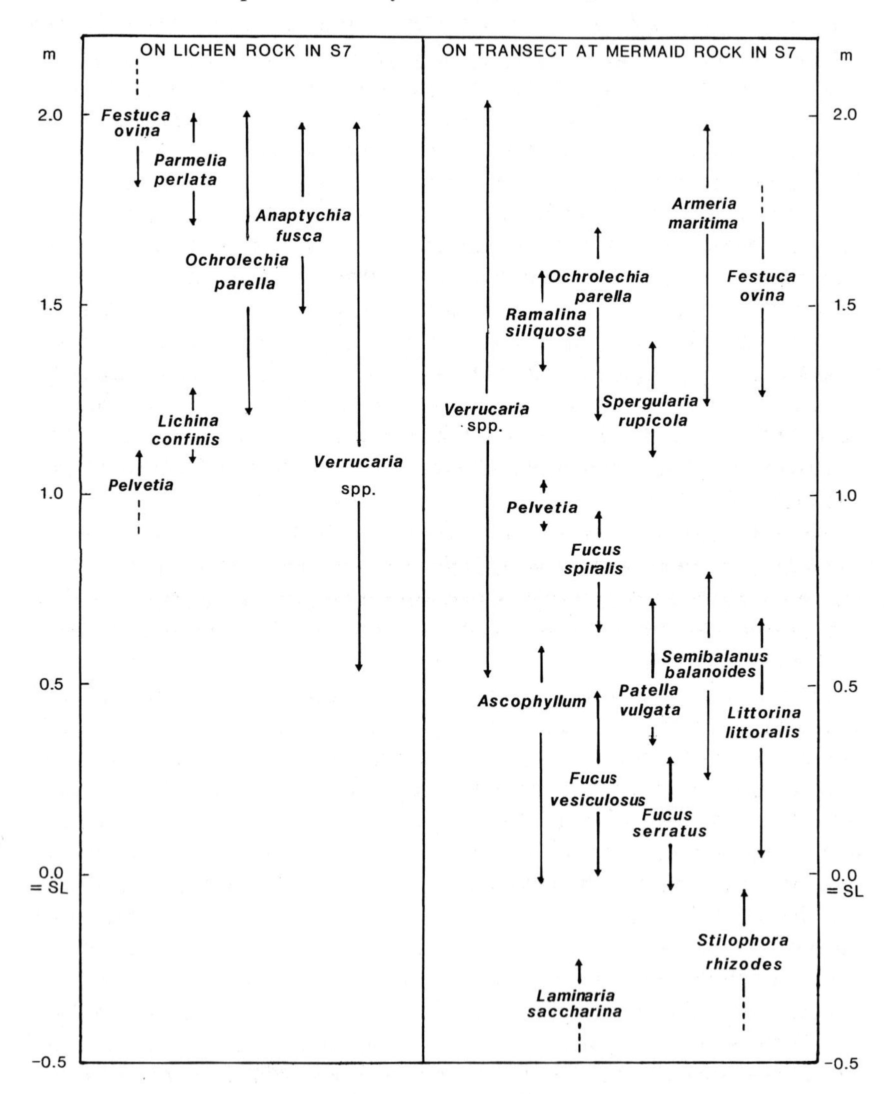

Fig. 18. Zonation in transect at Mermaid Rock (S7) and on adjoining Lichen Rock.

zones of fuccoids. Lichens extend upwards supralittorally, particularly on a steep rock slab adjoining the transect (and also summarized in Fig. 18), on which they only appear to be restricted upwards where soil with flowering plants covers the top of the slab.

Two features of the rocky littoral regions are of particular note: *Fucus serratus* is very scarce along all of the west shore and almost throughout the North Basin, although it is found quite plentifully along the east and south shores of the South Basin. Similarly, *Himanthalia elongata* is found quite plentifully along the east shore of the South Basin and around the Rapids mouth, where the inflow current strikes the shore, but is almost entirely absent from the west shore and from the North Basin. The areas devoid of *Fucus serratus* and of *Himanthalia* are close to shallow sublittoral graze patches cleared by the sea urchin *Paracentrotus lividus*, as illustrated in Fig. 19. When rocks carrying *Fucus serratus* were transferred to the same level in the low littoral of the North Basin (at N6/7 and E5), these plants became very tattered, and the urchins were seen using pieces of plant debris as hats. One *Paracentrotus* was seen holding a piece of *Fucus serratus* in its mouth (Kitching and Thain, 1983). In a similar experiment, rocks with many *Himanthalia* plants in various growth stages were transferred from the Rapids area to the north side of Castle Island (at I5) at the appropriate intertidal level, two of them protected by mesh plastic cages and three of them in the open. All the plants in the open disappeared completely, while the protected plants grew well and appeared healthy over the period 5 July to 5 September 1985.

Paracentrotus lividus was seen by Renouf (1931a) to move upwards into the littoral region of the lough during high water. In two sets of observations, carried out in July 1980 at N4 and in April 1981 at N7, a watch was kept over the movements of populations of *Paracentrotus* by day and by night (Kitching and Thain, 1983). At the two sites chosen for these observations the shore takes the form of a steep but not vertical wall, so that a short excursion can bring the sea urchins high up into the littoral region. Fucoids below the *Pelvetia* zone were sparse and tattered. On both occasions some *Paracentrotus* moved up into the littoral region during the day with the rising tide, and moved down again as the falling water lapped them. There was no upward movement with the rising tide by night. The results of the second set of observations are summarized in Fig. 20, in which the height of the lowest member of the top 5, top 10, and top 25 is shown. There was a fresh wind during the final daytime high water, causing a lapping of the water surface, which probably explains why the rise of the urchins was less extensive. During the daytime rise urchins were observed eating pieces of *Ascophyllum* and *Fucus vesiculosus*, as well as *Enteromorpha* and *Codium*. There is no evidence of any general move in the upward direction with the rising tide by day; it appeared rather that the *Paracentrotus* was moving around randomly

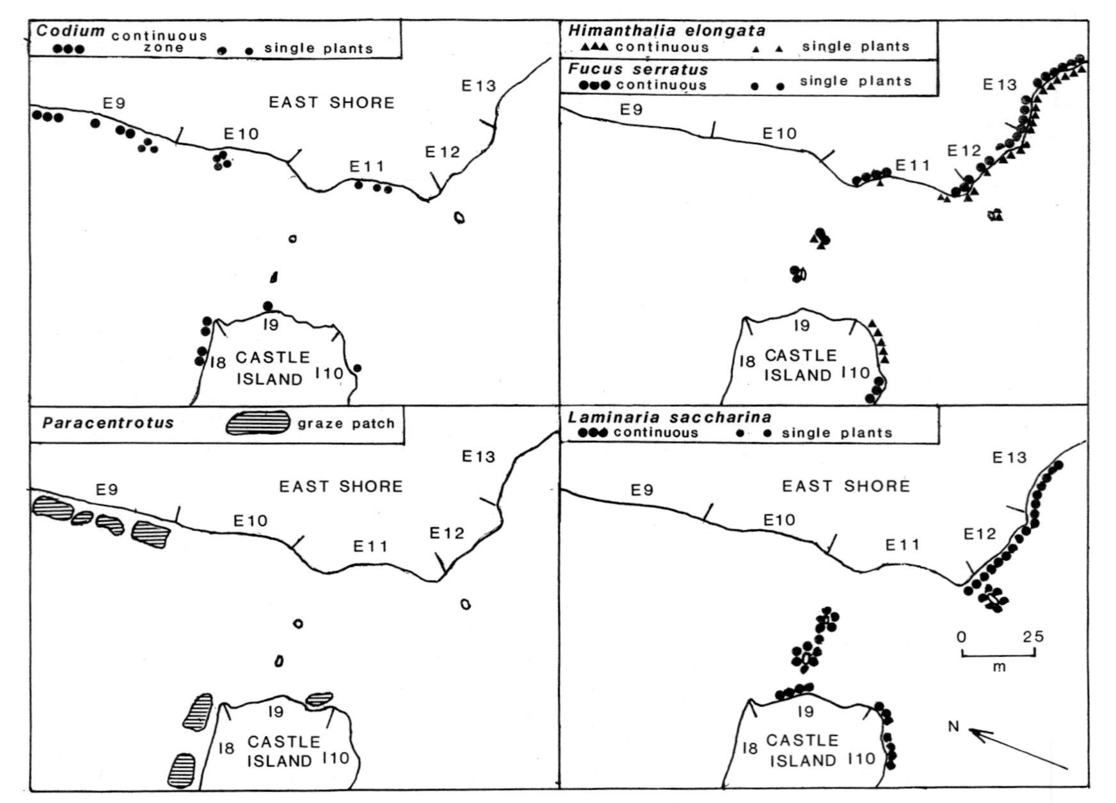

Fig. 19. Distribution of low littoral and of shallow sublittoral Algae, and of graze patches, in the Castle Narrows, in April 1986.

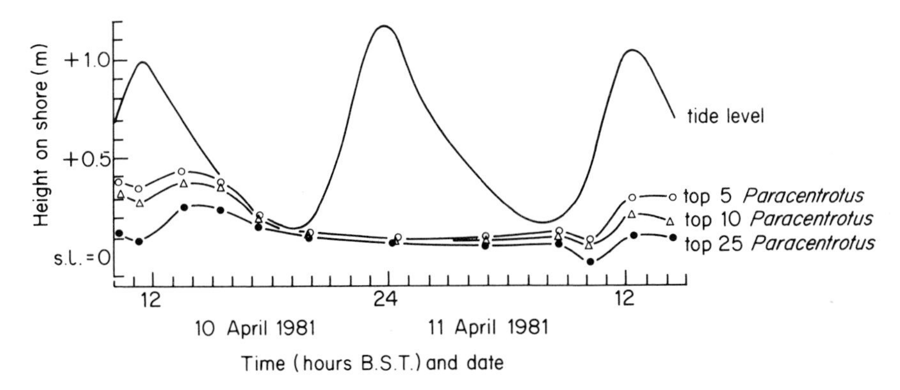

Fig. 20. Vertical migrations of *Paracentrotus lividus* on the North Wall of Lough Hyne at N7 on 10–11 April 1981. (Reproduced from Fig. 8 of Kitching, J. A. and Thain, V. M. (1983). The ecological impact of the sea urchin *Paracentrotus lividus* (Larmack) in Lough Ine, Ireland. *Phil. Trans. R. Soc. B* **300**, 513–552.)

and that those which moved up were able to do so because the water surface had already moved up. The maps of the Castle Narrows (Fig. 19) illustrate the inverse correlation of *Fucus serratus* and of *Himanthalia* with *Paracentrotus* graze patches.

There is a further reason why the low littoral Algae may be prevented from colonizing the west shore of the lough: their potential zones are very narrow, and they may be overshadowed, where the shore is at all steep, by the dense growth of *Himanthalia* (Kitching and Thain, 1983). In a detailed study of sector W19, carried out in April 1986, it was found that there was one single *Fucus serratus* plant, immediately below a large smooth rock populated (and no doubt well grazed) by many large *Patella vulgata* but devoid of *Ascophyllum*.

4. *Feeding Behaviour of Limpets*

A study was carried out by Little and Stirling (1985) of the behaviour of the limpet *Patella vulgata* on a steep rock face on the south shore of Lough Hyne. Because of the small rise and fall of the tide in the lough it was possible to observe the limpets directly, using a view box if necessary and a red light at night. A limpet was regarded as at home when on its home site and in its resting orientation, indicated by a paint mark, but active if either differently oriented or away. Activity was studied in relation to night and day and to condition of tide, and results were grouped in accordance with level of the home site on the shore; low-tidal limpets at +0·00 m to +0·20 m, mid-tidal limpets at +0·25 m to +0·45 m, and high-tidal limpets at +0·50 m to +0·80 m.

For low-tidal limpets there was a major peak of activity, with many of the limpets active, during daytime low water, and a minor peak, involving many of the same limpets, near to the time of low water at night. Mid-tidal limpets had a major peak of activity during night-time low water and a minor peak near to the time of daytime low water. High-tidal limpets had only one peak of activity, during low water by night.

This pattern of behaviour is interpreted tentatively by Little and Stirling in relation to danger from predation and from desiccation. The risk of falling victim to predators such as crabs increases lower down on the shore and at night; the risk of desiccation increases higher up and by day. The time of maximum activity of the limpet population is such as to minimize these risks. The activity of shallow sublittoral herbivores, such as *Gibbula cineraria* and *Paracentrotus lividus*, is adjusted so as to enable them to avoid predators, and is under diurnal control (see previous review). In the littoral region the situation is complicated by the additional risk of desiccation, under tidal control. These risks are differently balanced at different levels on the shore, which may account for variations in limpet behaviour described by other investigators, and reviewed by Branch (1981).

5. *Zoning Behaviour of* Gibbula umbilicalis

An investigation was carried out (Thain, Thain and Kitching, 1985) of the zoning behaviour of the littoral prosobranch *Gibbula umbilicalis*, in which advantage was taken of the freedom from wave action and from human interference, and of the excellent visibility over surfaces cleaned by *Paracentrotus lividus*, on the shores of Lough Hyne. This gastropod had previously been shown by Dr Vivien Thain to find its way back to the littoral region after having been moved to the shallow sublittoral.

Gibbula umbilicalis, collected from a south-facing shore (at N7) and deposited in the shallow sublittoral at a north-facing site (I7) on Castle Island, crawled in the right direction for the littoral region of this new shore, and many of them reached it within the ten-day period of observation. No snails moved in the wrong direction. On the other hand *Gibbula cineraria*, a closely related species which normally inhabits the shallow sublittoral region, scattered around the point of release and showed no tendency to move preferentially towards the littoral.

In further experiments *Gibbula umbilicalis* were released between the low tide mark and an artificial wall-like erection set up in the shallow sublittoral, parallel with the shore-line. In one experiment *G. umbilicalis* were released 1·5 m on the shoreward side of an approximately vertical wall of rocks, topped at the appropriate level with rocks bearing *Ascophyllum nodosum* and *Fucus vesiculosus*; some of the snails were attracted towards the rock wall, and some of them climbed up it. In other experiments a black plastic wall was used, without fucoids; *G. umbilicalis* released 1·5 m from the black plastic wall were attracted to it, while those released 2·5 m away (again on the shoreward side) moved towards the littoral (Fig. 21).

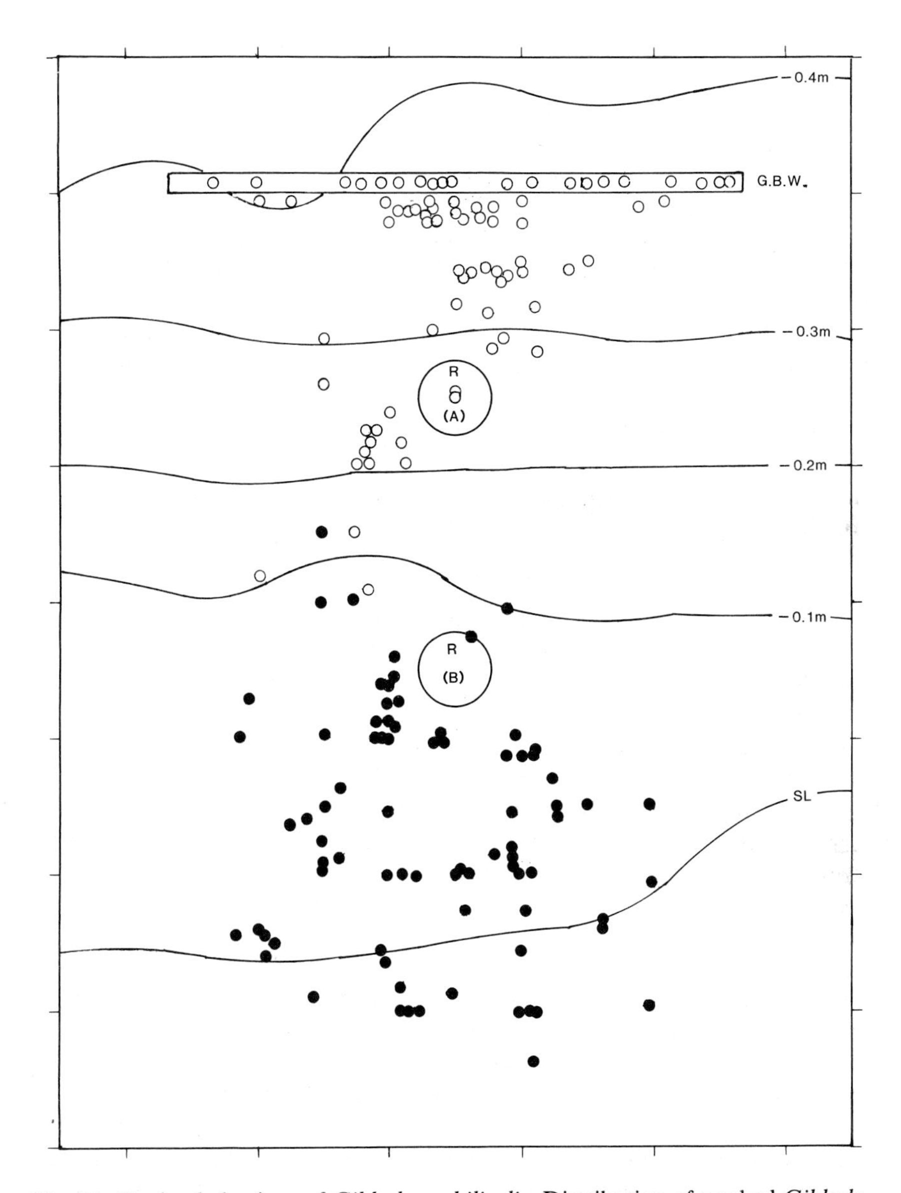

Fig. 21. Zoning behaviour of *Gibbula umbilicalis*. Distribution of marked *Gibbula umbilicalis* in experiment 6 of Thain *et al*. (1985) one day after release. ○, Snails released at position R (A); ●, snails released at position R (B); G.B.W. = Great Black Wall. The intertidal zone is at the bottom of the figure. Continuous lines represent contours of the sea bed at the levels stated (with respect to standard level). The frame of the figure is marked at 1 m intervals.

These experiments exclude any explanation involving sun navigation, compass orientation, memory of topography, or chemo-attraction towards fucoids. It is suggested by Thain *et al.* that the snails are able to perceive an upward extension of the shore visually, perhaps simply as a dark area in their visual field, and that they are sensitive also to gravity, so that they can determine the vertical orientation of this image. The structure of their eyes appears to be appropriate for this operation. A sense of gravity would also play an essential part in enabling the snails to climb out of dark crevices and to continue towards the littoral region. Other senses may also help the snails to remain in their proper zone. Further relevant literature is summarized in Underwood's (1979) review.

B. Shallow Sublittoral

1. General

Around the shores of the lough the surface of boulders and rock masses is kept relatively clean by wind-driven disturbance of the water down to a depth of 1–2 m below low tide level. At greater depths fine sediment increasingly envelops the boulder slopes, and only steep or vertical cliff faces remain clean. This narrow band of clean rock, the "shallow sublittoral", is sheltered from all wave movement generated outside the lough. At the south-east corner of the lough it receives the inflow from the Rapids. A strong current extends from the Rapids mouth across the inflow area to the Whirlpool Cliff and, much less powerfully, up to the Castle Narrows (Figs 2 and 19). The shallow sublittoral of the lough is completely different from corresponding levels in Barloge Creek or on the open coast.

In the transition from the fast current of the Rapids towards areas of gentler water flow, *Saccorhiza polyschides* gives way as the dominant alga to *Laminaria saccharina*. The boundary between the two is sharp and they probably compete. The blade of *Saccorhiza* grows abnormally wide as it stands upright in the sheltered water of Renouf's Bay (Ebling and Kitching, 1950), and this may reduce its power to compete. *Saccorhiza* requires to be turned over from time to time if it is to develop normally (Norton, 1969). *L. saccharina* has a life span of 2–3 years (Parke, 1948). It devotes a considerable proportion of its substance to blade area, and it is probably a powerful and successful competitor in the absence of serious mechanical stress.

Laminaria saccharina extends (with some variation from year to year) up the east shore of the lough as far as the Castle Narrows, but only a short way along the south shore apart from a few isolated plants or patches (Fig. 22). Further into the South Basin it gives way to mixed algal bush in which *Gelidium* spp., *Laurencia platycephala*, and in summer *Stilophora rhizodes* are important. Thus the main distribution of *L. saccharina* in the lough coincides with that of the inflow current, which may discourage sedimentation and which in summer is cooler than the adjacent shallow water of the lough.

Mixed algal bush in turn gives way to graze patches of the sea urchin *Paracentrotus lividus*. As shown in Fig. 22, these cover much of the shallow sublittoral of the North Basin and certain shallow sublittoral patches of the South Basin. The transition from algal bush to graze patches is notably sharp in the Castle Narrows (Fig. 19). Once again, summer temperature seems to offer a possible explanation. Continuous records of temperature were taken at seven stations along the length of the Castle Narrows in late August 1979, in rather indifferent summer weather. The lowest temperatures were recorded from the south end (Kitching and Thain, 1983; their Fig. 2), where the temperature fell with each inflowing tide. Further north through the Narrows the temperature fluctuated diurnally, rising each day and falling each night, and the general level was distinctly higher. Warmer water could be encouraging the *Paracentrotus* or could be discouraging some predator or competitor.

These dominant species exercise an overwhelming influence on the rest of the flora and fauna associated with them (Kitching and Thain, 1983). Mixed algal bush provides accommodation for many attached species, especially epiphytic Algae and Bryozoa, and for many motile species as well, some (such as various prosobranchs) feeding on Algae and others carnivorous. Grazed areas are free from bushy Algae except around their margins. They lack the accommodation and soft algal food provided by algal bush: instead the rock surface is coated with crustose coralline algae, especially *Lithophyllum incrustans*. Saddle oysters (*Anomia ephippium*) and limpets (*Patella aspera*) are widely distributed over the crustose corallines. Filter-feeding *Anomia* is plentiful on the lower surfaces of boulders. Limpets are mainly on the tops, where they presumably graze on minute algal growth. In this habitat they are living below the lowest tide; elsewhere they are not normally found sublittorally, presumably because they cannot penetrate a dense growth of Algae. The small scallop *Chlamys varia* and the polycheate *Pomatoceros triqueter* are also plentiful. All these have hard calcareous shells or tubes, and can no doubt resist the grazing of *Paracentrotus*. The crustose coralline Algae, especially *Lithophyllum incrustans*, are permeated with the tubes of small polycheates, especially *Polydora giardii, Dodecaceria concharum* and *Fabricia sabella*, and there are also many highly motile *Syllis gracilis*. Thus the sedentary fauna of the graze patches consists of species which are either armoured against damage by sea urchins or are using borrowed mechanical protection. The surface of the rocks is also cleaner of fine sediment than in ungrazed areas, probably because there are no bushy Algae to trap it and any which falls is resuspended by the *Paracentrotus*.

Soft upstanding algal bush may fringe the margins of graze patches, and here it accommodates much the same species as are found in completely ungrazed areas of algal bush. One additional algal species, *Codium fragile*

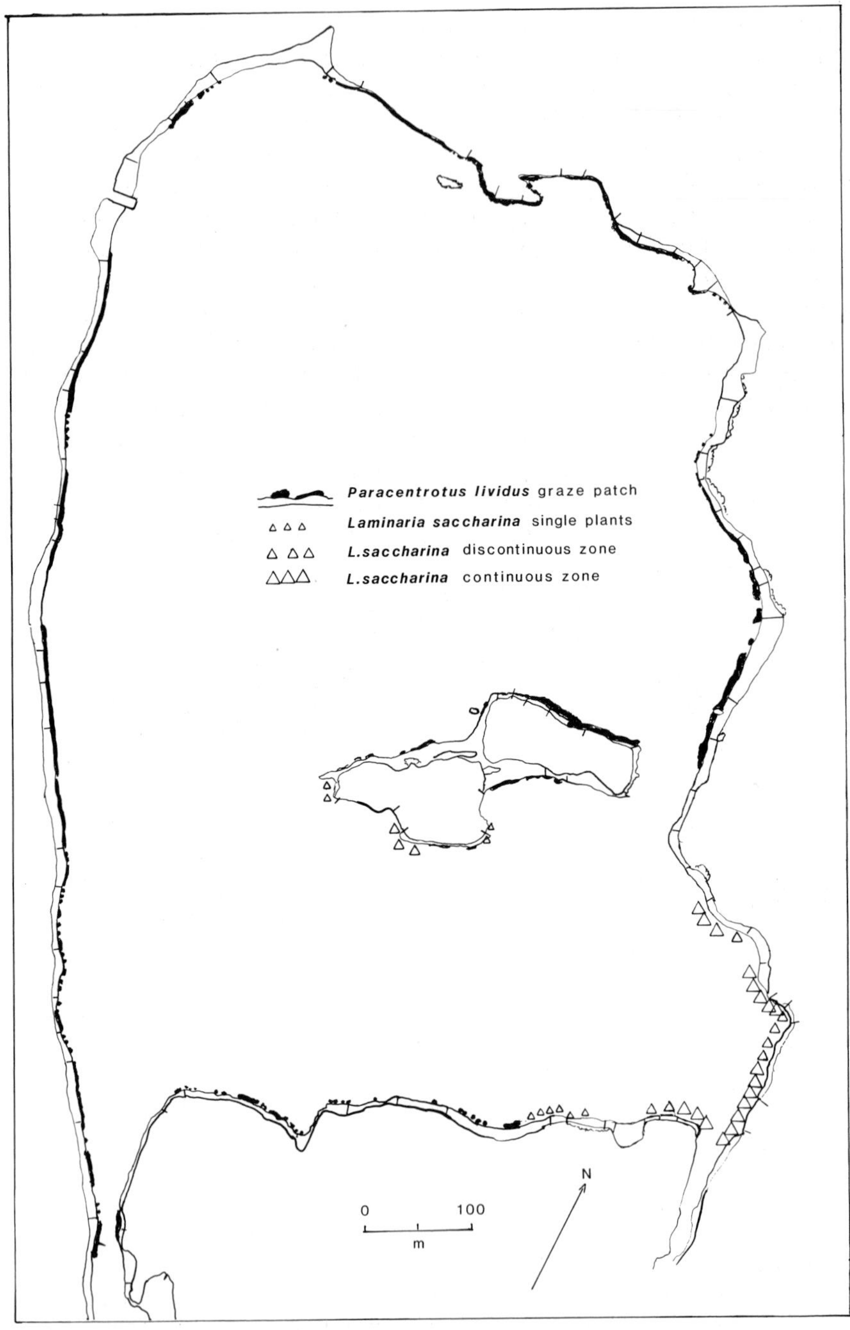

Fig. 22. Distribution of *Laminaria saccharina* and of *Paracentrotus* graze patches in Lough Hyne in July 1979 and July 1980 respectively. Data from Kitching and Thain (1983, Figs 1 and 4).

ssp. *tomentosoides*, provides much of the peripheral algal bush. The association of this species of *Codium* with graze patches along the south shore is illustrated in Fig. 23. When *Codium* plants, with the rocks to which they were attached, were transferred to heavily grazed areas, the *Paracentrotus* ate the outer branches and in some cases the whole of the *Codium* plants. A graze patch which was cleared of all *Paracentrotus* became overgrown a year later with *Codium fragile* ssp. *tomentosoides* and *Enteromorpha clathrata*. Scrapings of the surfaces of graze patches examined by Dr E. M. Burrows revealed unorganized *Codium* filaments having the utricles of *C. fragile* ssp. *tomentosoides*, as well as small tufts of *Enteromorpha clathrata, Cladophora* spp. and other filamentous Algae (Kitching and Thain, 1983). Thus, although *Paracentrotus* eats *Codium* and *Enteromorpha*, on balance it benefits them by clearing away slow growing but ultimately more successful competitors. *Codium* is well populated by most of the species characteristic of ungrazed areas, and in addition it supports a population of the opisthobranch *Elysia viridis*, with which it has a very special relationship (Trench, Boyle and Smith, 1973; Gallop, Bartrop and Smith, 1980).

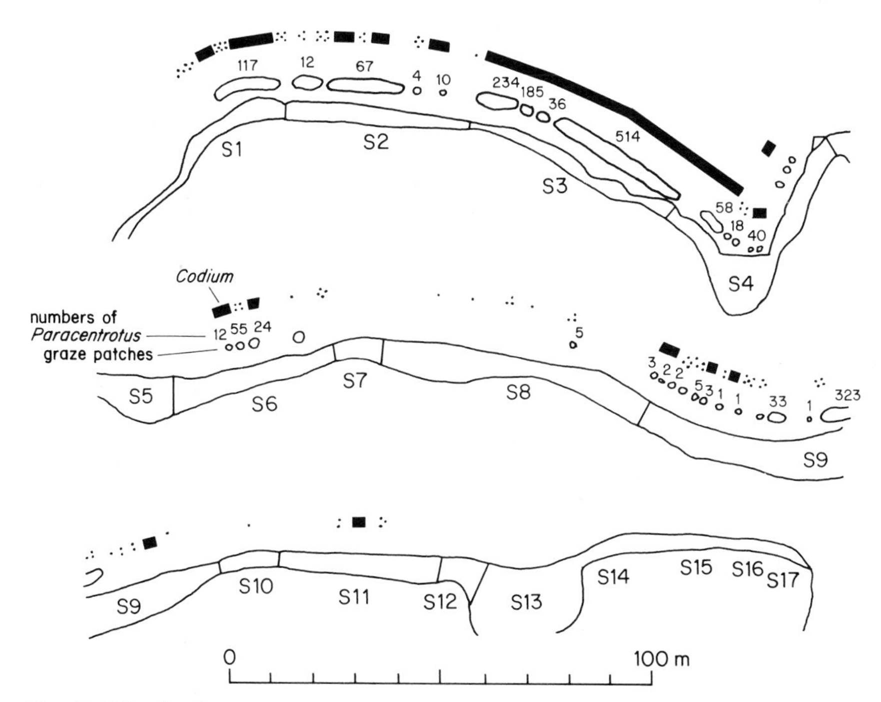

Fig. 23. Distribution of *Codium*, numbers of visible *Paracentrotus*, and graze patches along the south shore of Lough Hyne in July 1977. The map indicates distribution parallel with the shore; the distances of symbols from the low water line have no significance. Dots indicate single plants of *Codium*, and a solid black band indicates more than five *Codium* plants per metre of shore. (Reproduced from Fig. 3 of Kitching and Thain (1983): for full acknowledgement see Fig. 20.)

Urchin-resistant species with hard calcareous shells are vulnerable to starfish. Small *Marthasterias glacialis* are found in shallow water close to standard level around most of the South Basin but become rarer further north, while sublittoral *Patella* and *Anomia* diminish in numbers southwards. When rocks with *Anomia* attached to them were transferred to sites in the shallow sublittoral of the South Basin, the *Anomia* were attacked and eaten by *Marthasterias*, while almost all the *Anomia* protected in cages survived. Large *Marthasterias* were found to move down onto a soft bottom in slightly deeper water, where they were eating buried bivalves (*Venerupis rhomboides* and *Venus verrucosa*) as well as *Anomia* on isolated stones, where available. In July 1985 many of these large *Marthasterias* had come up along the north shore of the lough to less than −2 m, at the lower margin of the graze band. It is not known what restricts small *Marthasterias* from shallow water at the north end of the lough, but water temperature might be implicated. The feeding habits of large *Marthasterias* also need further study. Clam digging by starfish has been widely reported (for literature, see Allen, 1983).

2. Graze Patches

The extent of graze patches in the South Basin varies from year to year, and there is clearly a delicate balance between grazed areas and algal bush. Grazed areas, and the number of *Paracentrotus* visible, depend on the intensity of grazing and the growth rate of Algae; thus the numbers of urchins visible is a complex characteristic that is difficult to interpret. A few quadrats taken in 1971 suggested that the real number of *Paracentrotus*, including those under the boulders, could be 2–3 times that of those visible.

The numbers of *Paracentrotus* visible had increased greatly by July 1964 as compared with the previous year, and so an annual census was started of *Paracentrotus* visible in the South Basin. Numbers continued to be high in the following years, and the census was abandoned. In July 1971 it was found that numbers had fallen drastically, and that previously grazed areas had been overgrown by *Codium fragile* ssp. *tomentosoides*. Accordingly, censuses of visible *Paracentrotus* and of *Codium* in the South Basin were thenceforward carried out every July (Table 3). *Paracentrotus* reached another peak in 1978 and 1979, and the quantity of *Codium* was substantial. Both urchins and *Codium* then fell to new low levels, and algal bush took over much of the previously grazed area. It is possible to suggest—as no more than a speculation—that the warm summers of 1975 and 1976 led to a high spawning or settlement rate for *Paracentrotus*, and that these grew large enough over the next few years to migrate to the tops of the boulders and make graze patches. Observations of growth in cages in the lough and a study of growth rings in the shell plates support this possibility (Kitching and Thain, 1983; their Figs 9 and 10). The small quantities of *Codium* over the

Table 3
Annual census, taken in the South Basin in July, of numbers of *Paracentrotus* visible from above and of quantities of *Codium*, 1971–1985.

Year	Number of visible *Paracentrotus*	Quantity of *Codium* as area fully covered (m^2)
1971	3 230	587
1972	1 405	515
1973	1 378	781
1974	1 203	277
1975	797	160
1976	6 292	5
1977	5 031	92
1978	10 562	240
1979	15 530	230
1980	4 933	126
1981	2 230	200
1982	2 719	5
1983	1 211	12
1984	1 025	44
1985	535	2

Data for 1971–1981 are reproduced from Table 4 of Kitching, J. A. and Thain, V. M. (1983). The ecological impact of the sea urchin *Paracentrotus lividus* (lamarck) in Lough Ine, Ireland. *Phil. Trans. R. Soc. B* **300**, 513–552.

years 1982–1985 are to be ascribed to the small numbers of *Paracentrotus* and to the resulting small extent of grazing. However, it is likely that the *Paracentrotus* count for 1985 was depressed by the unusually heavy growth of blanket weed (*Enteromorpha clathrata, Ceramium echionotum, Ectocarpus siliculosus, Tribonema* spp.) in that year. In general, there are considerable fluctuations in the quantities of *Codium*, which may depend on the rate of change in numbers of *Paracentrotus* large enough to need to make diurnal migrations; a sudden decline, leaving empty a cleared area, is likely to produce an increase of *Codium*, and a slower decline, with a less cleared area, possibly a decrease. As a final complication, in April 1986, after the disappearance of blanket weed, many dead *Paracentrotus* shells, undamaged and of all sizes, were found lying on the bottom in the shallow sublittoral of parts of the west shore. The summer of 1985 was exceptionally wet; the presence of large quantities of *Tribonema* sp. (Xanthophyceae), determined by Professor Tyge Christensen and Dr Yvonne Chamberlain, seems to imply the presence of a layer of fresh water. Professor Christensen pointed out that from its poor state of preservation and from the separation into H pieces without any sign of swarmer formation, it was probably dead when collected, possibly due to salt water having invaded the layer of fresh water in which it had developed. In still further speculation, a reduction of salinity at the bottom might account for the empty *Paracentrotus* shells.

Certain sites have been occupied by graze patches every year: in S9 on the south shore, W35 to W37 in the Goleen, and a large patch from I12 to I13 on the south side of Castle Island. These regular sites are in small bays or recesses. Additional sites may develop in a favourable year. In the North Basin the balance is tilted further in favour of graze patches, and these are almost continuous along the north shore (Fig. 22). What is the critical condition and how does it operate?

In the Mediterranean the larvae of *Paracentrotus* spend a month in the plankton (Pressoir, 1959): if they spend as long as this in the plankton of Lough Hyne, they must be distributed all over the lough. It follows that larvae must reach sites occupied by algal bush as well as graze-patch sites. Quadrat-sampling along the south shore demonstrated that small and medium-sized *Paracentrotus* are present in small numbers under boulders covered in algal bush, but that they are present in much larger numbers under boulders supporting graze patches (Fig. 24). It is possible that there is a better settlement from the planktonic stage at these favoured sites—a site already grazed might even encourage settlement—or it is possible that *Paracentrotus* is less liable to predation during the course of its further

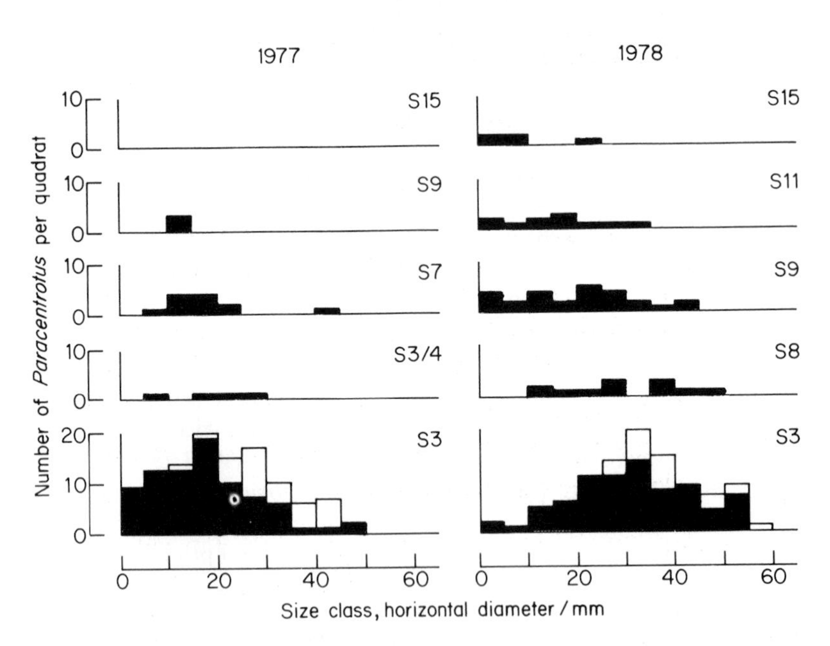

Fig. 24. Size–frequency histograms for *Paracentrotus lividus* in 1 m^2 quadrats on the south shore of Lough Hyne in July 1977 and July 1978. White, above rocks; black, below rocks. (Reproduced from Fig. 7 of Kitching and Thain (1983): for full acknowledgement see Fig. 20.)

development. The possibility of predation by crabs has already been discussed in the earlier review; the urchins might be restricted at an earlier stage and by other unknown predators. Small *Paracentrotus* have been destroyed in cages by small *Marthasterias*. There are no doubt many other possibilities. The interactions of some of the most important members of the shallow sublittoral community, both with the abiotic environment and with each other, are depicted diagrammatically in Fig. 25.

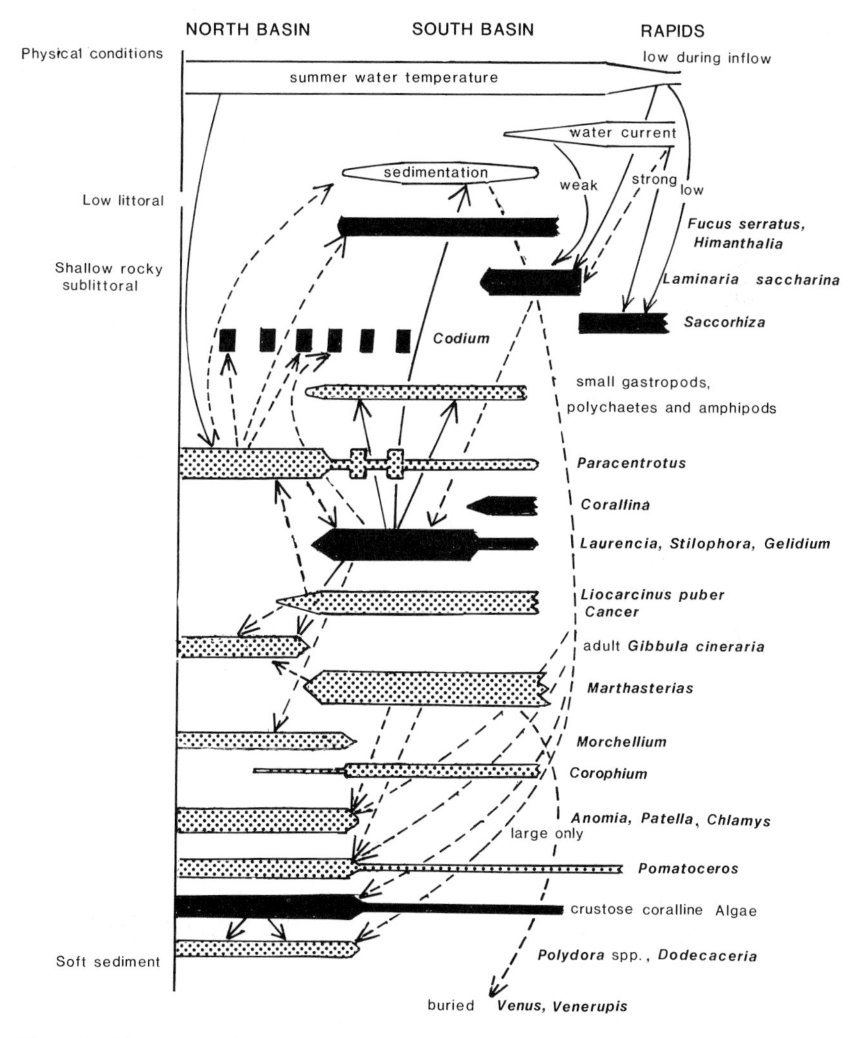

Fig. 25. Diagrammatic representation of some important interactions of physical factors, plants and animals in the shallow sublittoral of Lough Hyne. An arrow with continuous line indicates "encourages"; an arrow with broken line indicates "discourages". (Reproduced from Fig. 11 of Kitching and Thain (1983): for full acknowledgement see Fig. 20.)

3. Diurnal Migrations

Diurnal migrations between the tops of boulders, where food is available for herbivores but where there is greater danger from predators, and with crevices between and under boulders, less well supplied with algal food but safer from predators, have already been described in the earlier review. Crabs and starfish come up by night, when predatory birds are not feeding, and *Paracentrotus* and the prosobranch *Gibbula cineraria* come up by day, when they are unlikely to meet crabs or starfish. The mechanisms controlling these migrations have been investigated by Thain (1971) in *Asterias rubens* in Norfolk and in *Gibbula cineraria* at Lough Hyne.

Asterias rubens on the Norfolk coast (like *Marthasterias glacialis* at Lough Hyne) came out onto the tops of the rocks by night but went underneath by day. In the laboratory it was active and distributed all over the rocks in the aquarium during the hours of darkness but retreated away from light whatever the duration of the imposed light–dark cycle. However, the starfish would stay in the open if feeding, especially if they had been starved. There was no evidence of any endogenous rhythm. For *Gibbula cineraria* the mechanism proved more complicated. This animal behaved abnormally in aquarium tanks, so that the investigations had to be carried out on the sea bed in Curlew Bay, Lough Hyne, Artificial darkness was produced by means of a large black polythene box–lid structure, which excluded all light from the underlying seabed area. The enclosed sea water was continually renewed, and conditions were monitored. A generator was set up on the south side of Castle Island to provide artificial daylight, and power was led across the island to an overhead frame carrying powerful lights erected over the sea. *Gibbula cineraria* came up at dawn whether it was exposed to normal daylight or kept in total darkness. It stayed up if darkened either suddenly or gradually by day, but went down at dusk. Illumination by artificial daylight in the middle of the night caused a few *Gibbula* to come up, but they soon went down again. Artificial daylight shortly before dawn caused the *Gibbula* to come up early. The *Gibbula* were found to be active by day but to stay still at night. It appears that there is an endogenous activity rhythm coupled with an appropriate behaviour mechanism. When *Gibbula cineraria* were kept all day without food in a cage floating in the sea, and were released back into the sea just before dusk, they did not do down under the rocks that night, and sometimes not for two nights, as though starvation had cancelled that phase of their behaviour. However, the mechanism normally keeps time with dawn and dusk, and responds to the appropriate day length, whatever the time of year. Diurnal migration no doubt provides a means of balancing the need to come out and find food—after all that under the rocks has been eaten—against the risk of being attacked. It is interesting that, near Marseille, *Paracentrotus lividus* was found to come out by night instead of by day as in Lough Hyne, and to climb up both hard surfaces and blades of *Posidonia* (Kempf, 1962); and on the coast of Nova Scotia, *Strongylocentrotus droebachiensis* was found to forage only during the night when predatory fish were active by day (Bernstein, Williams and Mann, 1981). This leads to a suspicion that some rhythmic behaviour may be adjusted in relation to local conditions. It would be interesting to investigate the mechanism controlling the migration of *Paracentrotus*.

4. Autecology of Corynactis viridis *(Anthozoa)*

Corynactis viridis is found extensively on overhanging rock surfaces on Whirlpool Cliff and on the south side of Castle Island. It is also abundant on the bottom of the Rapids, under *Saccorhiza* plants, on the tops as well as on the bottoms of boulders. As already described, it is plentiful on dimly illuminated rock surfaces in the Bullock Island Cave and at Carrigathorna. As outlined in the previous review, transfer experiments to new sites and orientations in the lough, and experiments with floating cages, have suggested that it requires freedom from sedimentation and only dim illumination. This view has been strongly supported by observations on the state of expansion at two natural sites. *Corynactis* which covers the narrow recess at the inner angle of Whirlpool Cliff remains open throughout the night, but closes partially or fully by day, especially when the sun shines into its crevice in the afternoon. In laboratory tests with daylight lamps (controlled to an appropriate level of irradiance with neutral filters) there was also a reversible closure on illumination.

Sheets of *Corynactis* extend over the Rapids bottom under the *Saccorhiza* plants, where they received 1–5% of the irradiance at a "control" cell in a bowl of water on the quay. These *Corynactis* were found to be fully open by day. On removal of the overlying *Saccorhiza* from an area of Rapids bottom, the irradiance rose to 80–90% of that at the control cell, and the *Corynactis* shut by day, although they opened by night. A month later, the rocks freed from shading by *Saccorhiza* had become thickly covered by an undergrowth of small Algae and most of the *Corynactis* had disappeared (Muntz *et al.*, 1972). We conclude that *Corynactis* requires protection from strong illumination for physiological reasons, and also cannot compete with the dense undergrowth which accompanies strong illumination.

C. Sublittoral Cliffs

Work on the sublittoral cliffs of Lough Hyne has not progressed sufficiently far to permit of a general ecological treatment, although important preliminary investigations have been carried out. Algae of the cliffs are listed by

Maggs, Freamhainn and Guiry (1983), zonation on the Whirlpool Cliff has been briefly described by Larkum and Norton (1968), and some animal species found on the Island and Whirlpool Cliffs have been listed by Hiscock and Mitchell (1980). J. R. Turner (in preparation) has made a considerable study of the rhythms of expansion and contraction of the anemone *Anemonia viridis* on the Glanlab Cliff. An ecological study of the Whirlpool Cliff is in preparation as a result of visits by the Bristol University Underwater Club; this cliff is interesting because it is exposed to very strong water currents during inflow.

V. THE DEEP WATER OF THE LOUGH

A. Stratification

Small streams enter the lough at the south end of the Goleen, at the north-west corner of the lough (W1/N1), near the north-east corner (E2), and on the east shore of the inflow area (E19). Although the salinity of the lough water is in general between 34‰ and 35‰, values down to 33·5‰ have been found in the surface water of the Goleen, and it is likely that the surface film near any of the streams will fall well below this level. Nevertheless, there is no evidence for any reduction of salinity below 34‰ in the open water of the lough at depths below −1 m. Temperatures can reach almost to 18°C at −1 m in fine summer weather, and it is likely that higher values are attained locally at the surface, although this has not been investigated.

Measurements have been made of temperature and oxygen saturation, with a Mackereth meter, and of salinity from water samples, over a range of depths and on many occasions at Buoy B (Fig. 2) in the Western Trough. Salinities were determined by members of the Ministry of Agriculture, Fisheries and Food Laboratory at Lowestoft, and densities were calculated from Knudsen's tables.

Values for six different occasions in 1975 are shown in Figs 26 and 27, and values for spring, summer and autumn in 1976 and 1979 in Figs 28 and 29 (for 1970, see Kitching *et al.*, 1976, and for 1978 see Fig. 33 and Thain *et al.* 1981). Less complete hydrographical data, but always including water temperature, are available for other years. In March and early April the water was uniform in temperature, salinity and density throughout the water column, and there was only a slight reduction in oxygen saturation with increasing depth. For mid-April 1976 there is evidence of the beginning of stratification, with a fall in oxygen saturation and an increase in density with increasing depth. By July, in all the years when profiles have been taken, the temperature has risen substantially in the upper half of the water column, with a steep thermocline over a depth of 20–30 m; and on occasions when salinity was measured a density gradient was found over the same range of depths. Salinity contributes little to the pycnocline, and may even fall slightly with increasing depth. The thermocline and resulting pycnocline persist throughout August and September. Readings taken in November 1970 (Kitching *et al.*, 1976) and November 1975 (Fig. 26) show that the thermocline and pycnocline have been driven down to a lower level. Throughout the summer the oxygen saturation has fallen drastically. No oxygen was detected in the deep hypolimnion in November 1975, September 1978 and September 1979; on the last two occasions the deep water smelled of sulphide.

Continuous automatic recording for the South Basin and for the Rapids showed that with autumnal cooling of the epilimnion the difference in temperature between shallow water (−4·5 m) and deep water (−40 m) in the Western Trough disappeared by mid-November (Figs 30 and 31). By this time inflow and outflow temperatures had converged to about the same level. Temperature (and oxygen saturation) readings taken in January 1970 were uniform throughout the water column at Buoy B. However, in December 1968 there was still a slightly lower temperature below −30 m, and a sharp oxycline. In late February 1975 there was a slight inverse thermocline below −20 m and a steep oxycline (Fig. 32); this stratification must have been stabilized by a salinity gradient, although no water samples were taken. Although more complete records are needed, it seems that periods of stratification can occur in winter.

We do not know whether the oxygen in the hypolimnion is consumed throughout the hypolimnion by living or dead plankton or suspended organic matter, or whether it is consumed mainly at the surface of the mud. In the latter case there would probably have to be a vertical circulation of water within the hypolimnion.

In autumn the thermocline is driven downwards. The epilimnion grows colder, but the hypolimnion continues to grow slowly warmer (Figs 30 and 31). This erosion of the thermocline takes place over early November. The inflow of increasingly cold water from the sea, and its mixing into the epilimnion, cools the epilimnion down. It seems probable that turbulent mixing in the South Basin, set up by the inflow current, stirs some epilimnetic water into the hypolimnion, and that this is progressively eroded from above downwards. Fluctuations in temperature have been recorded at thermocline level at the south end of the Western Trough. It is likely that there is considerable activity at thermocline level as a result of tidal inflow, but the processes involved have not yet been studied. Speculatively, there might be an internal seiche.

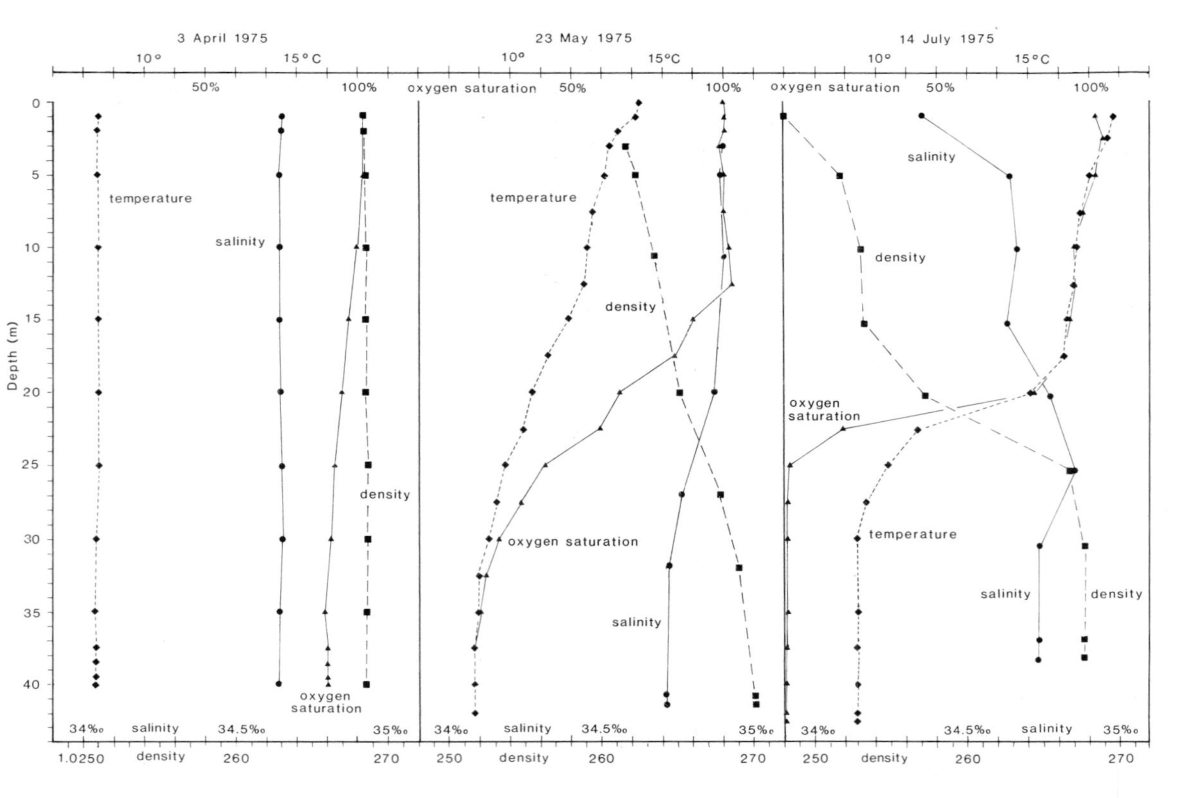

Fig. 26. Temperature, oxygen saturation, salinity and density profiles at Buoy B on 3 April, 23 May and 14 July 1975.

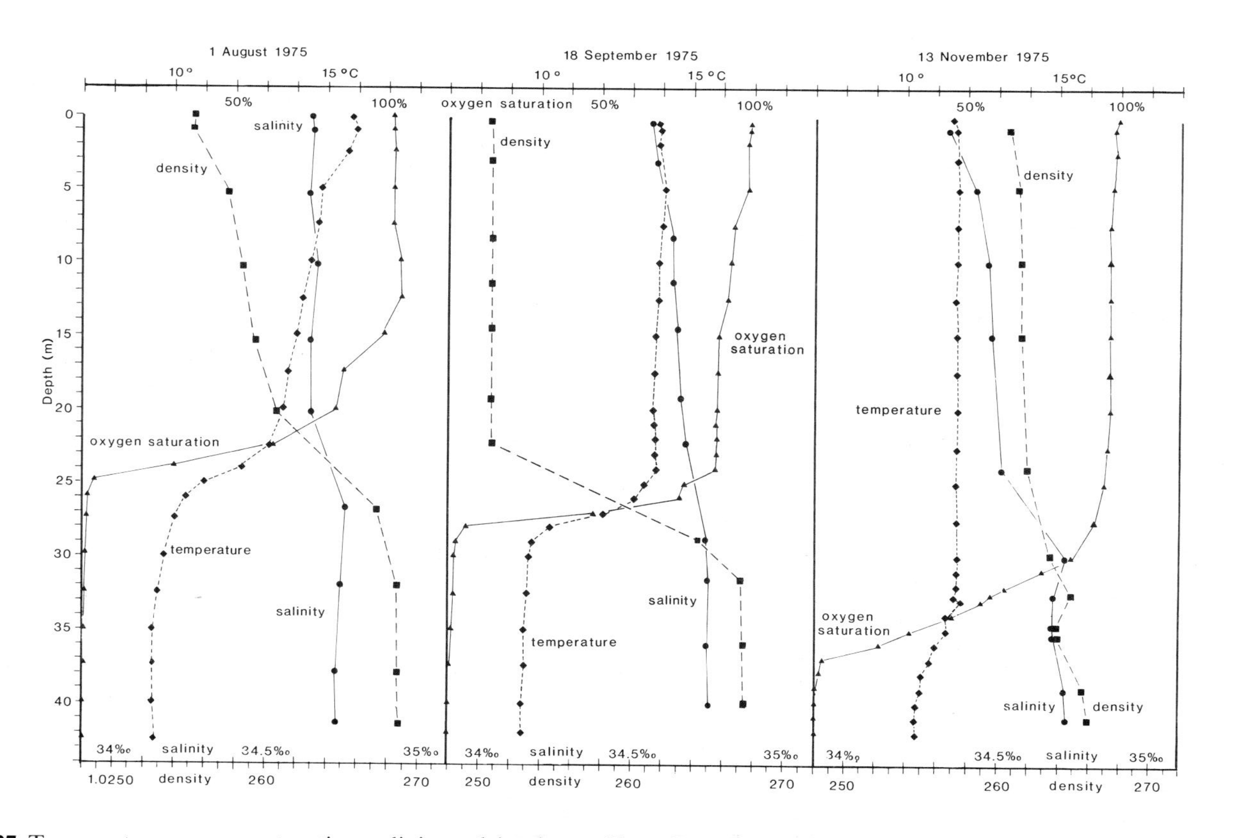

Fig. 27. Temperature, oxygen saturation, salinity and density profiles at Buoy B on 1 August, 18 September and 13 November 1975.

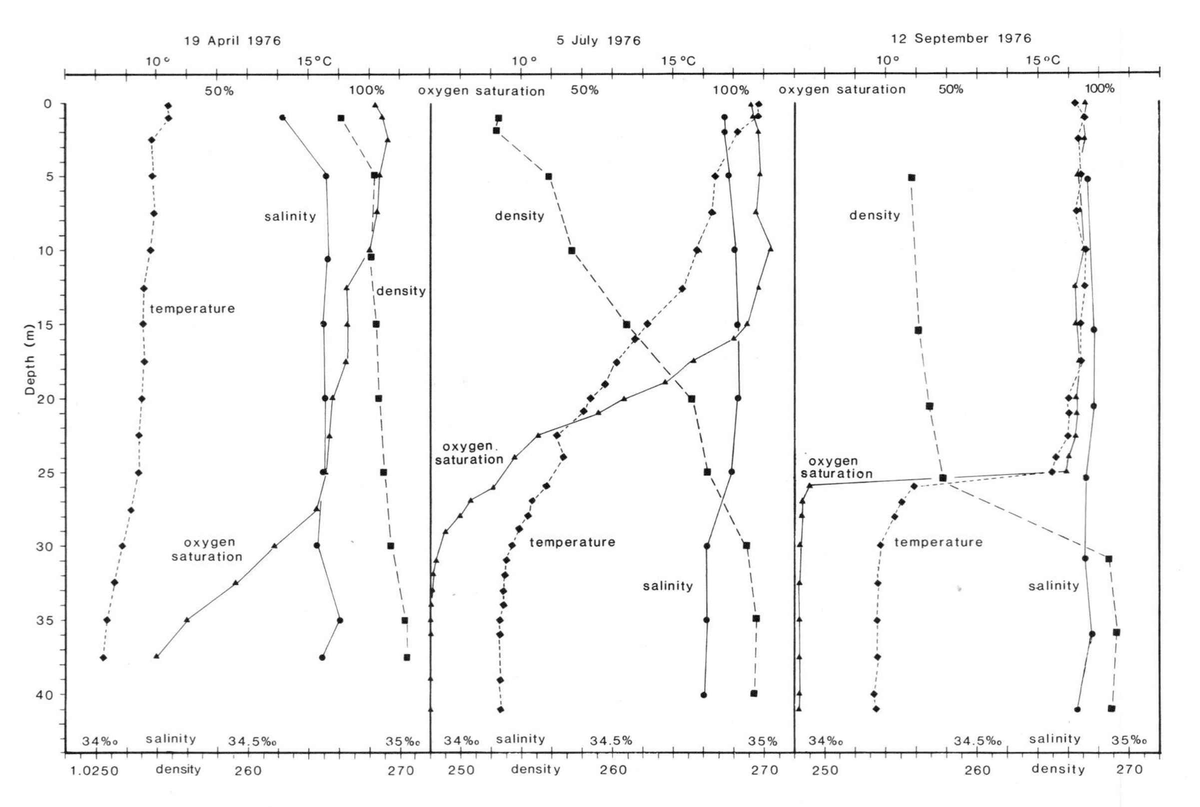

Fig. 28. Temperature, oxygen saturation, salinity and density profiles at Buoy B on 19 April, 5 July and 12 September 1976.

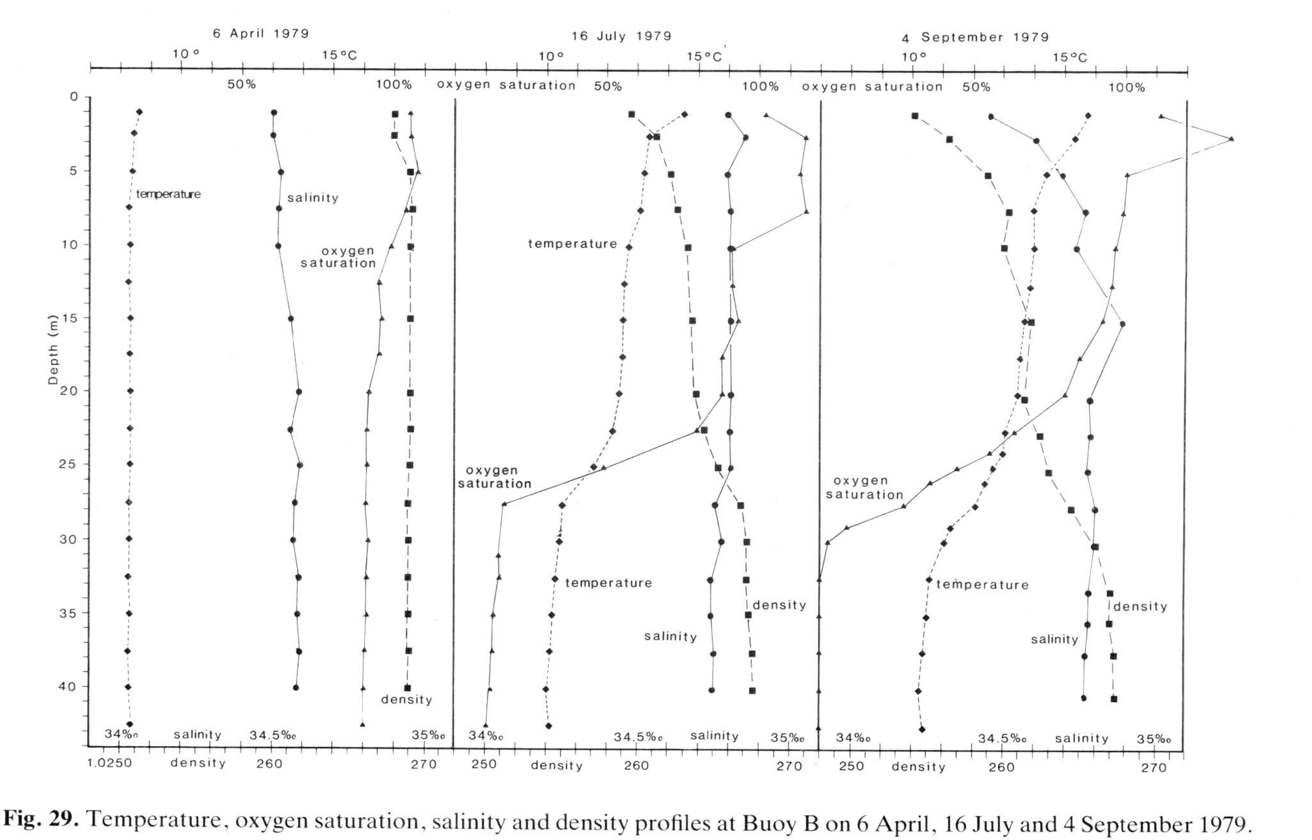

Fig. 29. Temperature, oxygen saturation, salinity and density profiles at Buoy B on 6 April, 16 July and 4 September 1979.

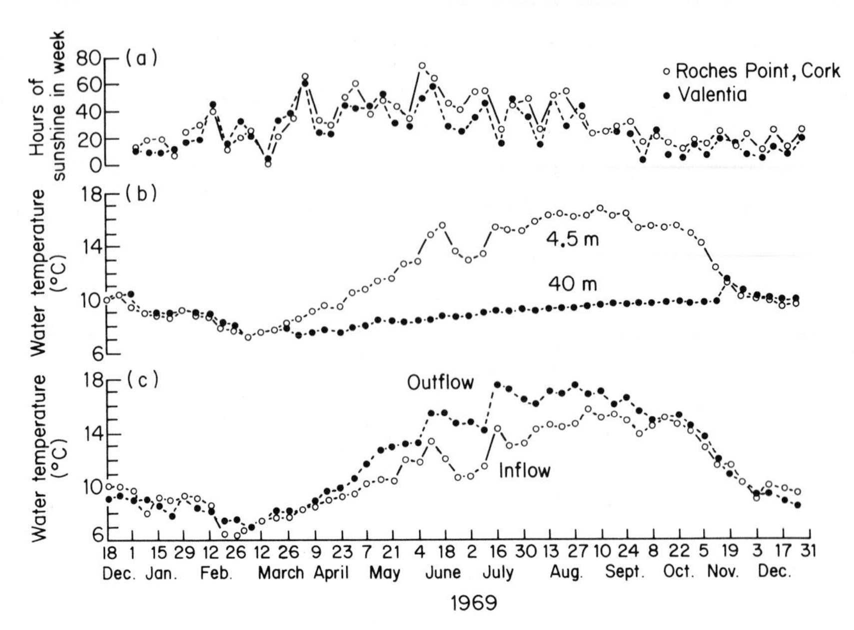

Fig. 30. Weekly hours of sunshine for 1969 from data recorded by Irish Meteorological Service (a). Weekly extracts of water temperature for 1969 for (b) the South Basin of Lough Hyne, and (c) the Rapids of Lough Hyne. (Reproduced from Fig. 2 of Kitching, J. A. *et al.* (1976). The ecology of Lough Ine. XIX. Seasonal changes in the Western Trough. *J. Anim. Ecol.* **45**, 731–758.)

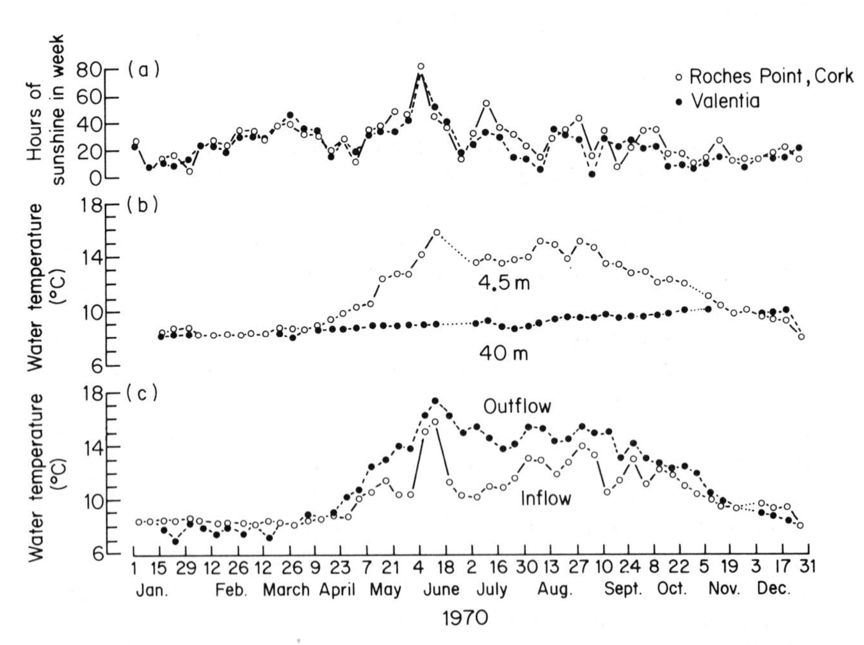

Fig. 31. Data as in Fig. 30, but for 1970. (Reproduced from Fig. 3 of Kitching *et al.* (1976): for full acknowledgement see Fig. 30.)

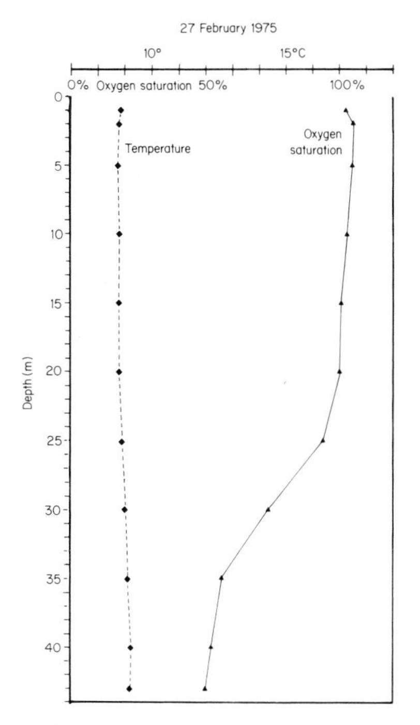

Fig. 32. Temperature and oxygen saturation profiles at Buoy B on 27 Febuary 1975.

B. Effects of Anoxia

In spring and early summer the bottom of the Western Trough below −25 m was densely populated by the polychaete *Pseudopolydora pulchra*, with its tubes standing up out of the mud. The surface of the mud changed from brown to black over the bottom of the Western Trough by mid-August, and below −28 m the *Pseudopolydora* tubes were collapsed and empty. By September *Pseudopolydora* was not found in any grab samples from −20 m downwards, nor any living animals from −30 m or below. Prawns (*Palaemon serratus*) and the shore crab (*Carcinus maenas*) were taken in crab traps at −40 m to −45 m in May, but nothing was captured below −20 m in crab traps set in late August. Two experiments were carried out in which many specimens of living healthy *Carcinus maenas* were suspended for two days in mesh polythene cages, five crabs to a cage, at a series of depths from −2 m to −46 m. Bottom was at −47 m. Out of 35 crabs set up on 18 July 1968, all but three were recovered alive and normally active; three were dead, one each from −40 m, −45 m and −46 m. Of the crabs suspended on 26 August 1968, all crabs in the lowest four cages (−30 m, −40 m, −45 m and −46 m) were dead, except for one which twitched slightly; one from −25 m was lethargic, but all the others from −25 m upwards appeared normal. There can be no doubt that the death of the complete deep-water benthic fauna of the Western Trough, described by Kitching *et al.* (1976),is the result of the reduced oxygen tension of the hypolimnion, probably accentuated by the liberation of some sulphide, which is poisonous (Theede, Ponat and Schlieper, 1969): this happens every summer.

Anoxic conditions are developed in the deeper waters of many land-locked fjords and of several seas with limited outlet. Development of such conditions requires stratification, either of temperature or salinity or both, in order to produce the pycnocline necessary for stability of the water column and the long-term separation of an overlying epilimnion and an underlying hypolimnion. In the Black Sea, over 2000 m deep, stratification is permanent, because of the dilution of the shallow water by inflowing rivers. Below a depth of 200 m there is no oxygen but much H_2S, and there is nothing living other than bacteria (Nikitin, 1931). In land-locked Norwegian fjords, investigated by Strøm (1936), a gradient of salinity maintains the pycnocline in spite of winter cooling of the surface waters, but eventually more saline water spills over the sill and drives the bottom water upwards. Many years may elapse between these ventilations, and much H_2S may accumulate in the bottom waters. Strøm quotes records of a total uplifting of the bottom water by overspill across the sill in Hellefjord in the spring of 1905, which was followed by complete stagnation up to the time of his investigation (1934). The upwelling of H_2S-laden water may completely destroy the fauna of a fjord. Troughs in the Baltic undergo deoxygenation and depopulation for several years on end, and are then reoxgenated by an overspill of dense water which sinks and replaces the bottom water (Fonselius, 1962; Ackerfors, 1966, 1969). Certain parts of Limfjorden in Denmark undergo stagnation for periods of one to several weeks in warm summer weather, and these areas become completely or almost completely anoxic, with accumulation of H_2S. Emergence of buried benthic animals onto the surface of the mud, and their ultimate death if these conditions persist, is graphically described (Jørgensen, 1980). These and other examples are discussed by Rosenberg (1980). In Loch Eil, in western Scotland, the sill is so restricted that overspill is limited to near high water of spring tides (Edwards *et al.*, 1980; Pearson, 1982). Lough Hyne differs from other examples so far described in the complete regularity of the summer deoxygenation and in the lack of overspill. The Sill is very restricted and does not abut on the Western Trough, but is separated from this by the submarine plateau of the eastern South Basin, where mixing takes place. The deep water of the Western Trough reaches its highest temperature in November and December.

C. Benthos of the Soft Sediment

1. Introduction

The mud bottom of the Western Trough has been explored by grab-sampling and by diving (Kitching *et al.*, 1976), and subsequently a detailed study has been made by Thrush and Townsend (1986) of physicochemical conditions and fauna at around −20 m across the South Basin from the Goleen mouth to the Whirlpool. Many more samples are needed for the exploration of the benthos of the lough.

Samples of sediment from the Western Trough (at −29 m and −35 m) contained about 55% silt and clay, and about 95% no larger than "coarse silt" (particle size 0·031 mm). At shallower stations near the north end of Lough Hyne there was an admixture of coarser sediment, but at −20 m in the South Basin (Thrush and Townsend, 1986) there was about 50% clay + fine silt and a further 30% of coarse silt. Near the inflow area the sediment was coarser, contained less organic matter, was more alkaline and had less reductive power. Superimposed on the physical condition of the sediment, the decomposition of organic matter, the metabolic activity of sulphur bacteria, and the changing oxygen content of the overlying water must determine which plants and animals can live on and in the sediment at various depths. The resulting interacting community may show local irregularities or patchiness, brought about by various local events.

2. The Audouinella floridula *Zone*

From about −3 m down to about −17 m the mud bottom carries a carpet of filamentous Algae of which *Audouinella floridula* (formerly *Rhodochorton*) is the most plentiful. It is able to anchor itself on a mud surface without competition from more substantial Algae, for which such a surface would be inadequate, and presumably it profits from the high irradiance experienced in shallow water. *A. floridula* has been reported from a muddy bottom down to a depth of about 10 m in very sheltered Norwegian fjords (Rueness, 1976). It was infertile both in Norway and in Lough Hyne (Norton, pers. comm.). The *A. floridula* carpet in Lough Hyne holds an abundant fauna, not yet investigated. There are large numbers of the prosobranch *Bittium reticulatum*, the polychaete *Capitella*, and the oligochaete *Peloscolex benedeni* among the *A. floridula*, and various more motile animals wander over it, including the starfish *Marthasterias glacialis, Asterias rubens* and *Luidea ciliaris*, and the opisthobranch *Philine aperta*.

It would be interesting to know what gradient of sediment size there is at about −10 m to −15 m between the inflow area and the Goleen mouth, under the South Basin vortex, and what changes in the fauna are associated with this gradient.

3. *The Mud-burrow Zone*

The carpet of *Audouinella floridula* thins out towards its lower limit, and between −17 m and −25 m the bare mud contains the burrows of the decapod Crustacea *Nephrops norvegicus* and *Calocaris macandreae*. The gobiid fish *Lesueurigobius friesii*, well known for its burrowing habits (Rice and Johnstone, 1972) has been reported from the mud-burrow zone by divers, and has been captured in Lough Hyne (Minchin and Molloy, 1978). The mud-burrow zone also contains the buried bivalves *Corbula gibba, Abra alba* and *Abra nitida*, the ophiuroid *Amphiura chiajei*, and many mobile species shared with the *Audouinella floridula* zone. Many species from around −20 m across the South Basin are listed by Thrush and Townsend (1986).

4. *The Spionid Zone*

Below −26 m extends the spionid zone, populated in spring and early summer by large numbers of the spionid polychaete *Pseudopolydora pulchra*. Their tubes stand up out of the mud, and counts of limited areas by divers gave estimates of 3000–18 000 (averaging 8000) per square metre (Kitching *et al.*, 1976; see their Plate 1). With them were many small *Corbula gibba*, with the hydroid *Perigonimus repens* growing on the posterior ends of the shell valves, which stuck out of the mud. Bivalve siphons waved among the worm tubes. Buried in the mud there was also the polychaete *Scalibregma inflatum*, and starfish and crabs moved over the mud surface. All disappeared with the onset of anoxia in August. This fauna must consist of fast-growing opportunists which can repopulate the mud bottom after the collapse of stratification in November. We do not know whether they succeed in producing offspring. Both *Pseudopolydora pulchra* and *Corbula gibba* are found in small numbers above the thermocline, and might restock the deeper water each year. The *P. pulchra* tubes stand up out of the mud by 10–20 mm, and with their pair of long tentacles these worms are well equipped to endure a low oxygen microclimate overlying the anoxic mud. It is even probable that a low oxygen tension would protect them by keeping away various less tolerant predators or competitors. *P. pulchra* is reported from several Scottish lochs (Eleftheriou, 1970) and from the more marine parts of Danish estuaries (Wolff, 1973). It appears to tolerate variations of grain size, oxygen tension and salinity. There is no information as yet about the spacing between individuals in Lough Hyne, as there is for a related species near San Diego (Levin, 1981).

The fauna of the spionid zone at Lough Hyne has some limited resemblance with that of the inner basin of Sullom Voe in the Shetland Islands (Pearson and Eleftheriou, 1981), but has very little in common with that of deep mud stations in Loch Linnhe and Loch Eil (Pearson, 1970). The sediments at all three of these other sites resemble those at the bottom of the Western Trough of Lough Hyne in that they have a high component of silt–clay. However, the oxygen content of the deep water remains high all the year in Loch Linnhe and Loch Eil owing to water renewal, but in Sullom Voe in some years the water becomes completely anoxic and charged with sulphide in the late summer, following the formation of a thermocline (Dooley, 1981; Stanley *et al.*, 1981). The deep fauna at Sullom Voe includes *Scalibregma inflatum, Abra alba* and *Corbula gibba*, all found in the spionid zone at Lough Hyne, although in other important respects the fauna is very different at the two sites.

The upper limit of lethal deoxygenation is not necessarily static. In the Limfjorden (Denmark) there are irregular periods of anoxia and accumulation of H_2S in summer (Jørgensen, 1980). Thrush (in preparation) has reported a temporary upward extension of the anoxic zone in Lough Hyne into parts of the South Basin that are not normally anoxic. He observed starfish (*Asterias rubens*) feeding on bivalves which had emerged from the mud during a limited period of anoxia.

5. *Patchy Distribution*

Much remains to be discovered about local variations in the composition of the bottom communities. Thrush (1986a) has studied the effects of "wracks" of seaweed torn off by storms on the open coast and washed into the lough on the inflowing tide. These wracks are scattered over the mud bottom of the South Basin. The sediment under artificial wracks became more acid and more reductive, but the population densities of the species inhabiting the sediment were very irregular. Many small pits were observed by Thrush in the gravelly sediment of the inflow area, and some of these were occupied by the crab *Cancer pagurus*, which was also seen digging pits. (Pit digging by *Cancer pagurus* is also reported by Minchin (1985) for Lough Hyne and other sites, and broken lamellibranch shells were present.) Changes in the fauna of the sediment, after the digging of artificial experimental pits, proved irregular and difficult to describe (Thrush, 1986b).

D. Plankton

1. *Introduction*

There are two main problems concerning plankton in Lough Hyne which are of special interest. First, what effect has the deficiency of oxygen, with the possible presence of hydrogen sulphide, in the hypolimnion in late summer on the vertical distribution and diurnal migration of zooplankton? The study carried out by Thain, Jones and Kitching (1981) at Buoy B in the Western Trough describes the vertical distribution of zooplankton in April, July and

September in the middle of the day or afternoon when migrating plankton should be at its lowest level, but it does not cover vertical migration, which would require a much more extensive investigation. Secondly, what is the effect of inflow and outflow through the Rapids on the interchange of plankton between the lough and the sea, and how is the supply of plankton related to the energy budget of the lough and the development of anoxic conditions? There has not yet been any investigation of this second problem.

2. Vertical Distribution of Zooplankton

In view of the small area and limited depth of the lough, a plankton pump, rather than a tow-net, was used to collect samples. Water was sucked through the net, so that plankton did not pass through the pump; thus damage to soft-bodied animals such as medusae was avoided. The pump was

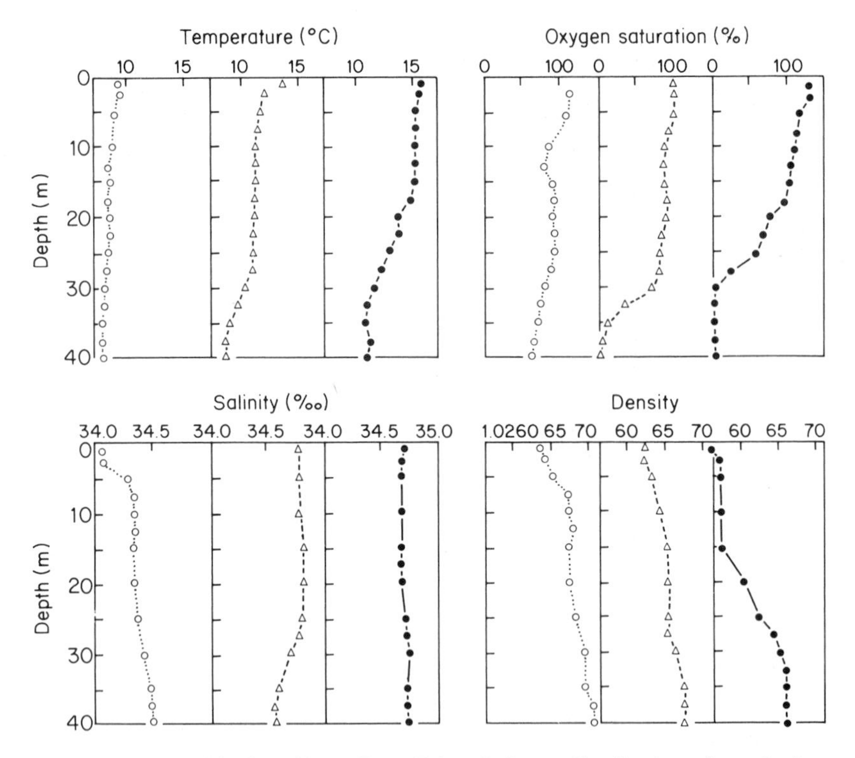

Fig. 33. Hydrographical profiles at Buoy B in relation to distribution of zooplankton, on 3 April (hollow circles and dotted lines), 7 July (hollow triangles and broken lines), and 3 September (fully black circles and continuous lines, 1978. (Reproduced from Fig. 1 of Thain, V. M., Jones, J. and Kitching, J. A. (1981). Distribution of zooplankton in relation to the thermocline and oxycline in Lough Ine, County Cork. *Ir. Nat. J.* **20**, 292–295.)

operated from a rowing boat and was powered by a car battery. The pumping rate was about 10–12 l min^{-1}, and each sample amounted to about 300 l. With operation on such a small scale, it is possible that patchiness of distribution may influence the result, and it is risky to draw conclusions from small numbers of the rarer species (Gibbons and Fraser, 1937).

Hydrographical profiles are shown in Fig. 33 for the occasions of sampling in 1978. In early April there was already some indication of stratification, with temperature and percentage saturation with dissolved oxygen decreasing downwards, but there was plenty of oxygen at all depths. By July there was a sharp decrease in temperature and in percentage oxygen saturation from about −30 m downwards. By early September the temperature and percentage saturation with oxygen fell steeply over the range −20 m to −30 m. No oxygen could be detected at or below −32·5 m and the water from these depths smelled of H_2S.

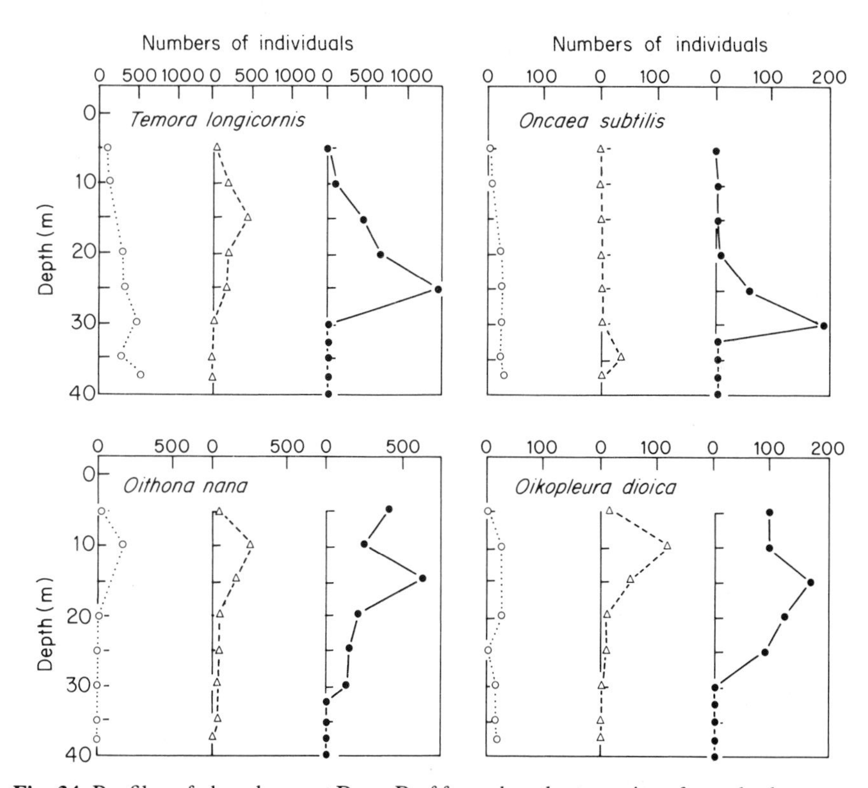

Fig. 34. Profiles of abundance at Buoy B of four abundant species of zooplankton on 2 April (hollow circles and dotted lines), 6 July (hollow triangles and broken lines), and 4 September (fully black circles and continuous lines), 1978. Numbers of individuals are in 300 litres. (Reproduced from Fig. 2 of Thain *et al.* (1981): for full acknowledgement see Fig. 33.)

Of the 30 species of zooplankton reported by Thain *et al.* (1981) in samples taken in April, July and September, 8 were Hydromedusae and 17 were Copepoda. Several were only abundant in April or July, with maxima around −10 m to −20 m, but were not sufficiently plentiful in September to allow conclusions to be drawn about the response to a lack of oxygen. Three Copepoda and one pelagic tunicate were plentiful on all three sampling occasions (Fig. 34). None of these was taken below −35 m in September. Only a few polychaete and lamellibranch larvae were found below −35 m in September, and these are organisms which in due course require to reach the bottom in order to complete their life-cycles. Empty Copepoda were plentiful at all levels and were most numerous in September.

Gradients of temperature or salinity or oxygen concentration (possibly with H_2S) or pH can form a barrier to the penetration or passage of plankton, as concluded in many publications. Petipa, Sazhina and Delalo (1961) discuss extreme conditions in the Black Sea. The lowest concentration of dissolved oxygen permitting continued movement of various planktonic species under experimental observation was found by Nikitine and Malm (1934) to correspond with their vertical distribution in the Black Sea. Some species endure very low concentrations of oxygen. The lower limits of distribution of *Temora longicornis, Oncaea subtilis, Oithona nana, Pseudocalanus elongatus, Acartia clausi* and *Oikopleura dioica* in Lough Hyne in September correspond with the sharp cut-off in oxygen at −30 m to −32·5 m. The gradient of temperature is much more evenly spread out, and the gradient of salinity is negligible. Taken all together, the completely anoxic condition of the water below −32·5 m (with presence of H_2S) was almost certainly responsible for the cut-off in plankton below this level.

VI. SPECIAL FEATURES OF THE FLORA AND FAUNA

The flora and fauna of Lough Hyne is rich in number of species thanks to the variety of its habitats, to shelter from wave action, and to a steady high salinity. The higher summer temperature in the shallow sublittoral, as compared with the water temperature experienced on the less enclosed adjoining coast, may encourage colonization by Mediterranean or other south-westerly species. The principal depth zones of the lough all have some unusual features, as already discussed.

The shallow sublittoral fauna includes many species found also in tide pools. Five species were listed by Emson (1985) as found only in rock pools. All these are found in the Carrigathorna tide pools (Goss-Custard *et al.*, 1979), and all five are reported from Lough Hyne: *Palaemon elegans* (Decapoda) (Holmes, 1980); *Rissoella diaphana, Omalogyra atomus* and *Skeneopsis planorbis* (Prosobranchia) (Kitching and Thain, 1983); and

Asterina phylactica (Emson and Crump, 1984). Species listed by Emson (1985) as "characteristically pool-dwelling", and found plentifully in Lough Hyne, include *Anemonia viridis* (Anthozoa), found in large numbers on Glanlab Cliff by J. R. Turner (pers. comm.), *Fabricia sabella* and *Janua pagenstecheri* (Polychaeta), *Apherusa jurinei* and *Stenothoe monoculoides* (Amphipoda), *Patella aspera* and *Rissoa parva* (Prosobranchia), and *Amphipholis squamata* (Ophiuroidea) (Kitching and Thain, 1983). The dense population of the sea urchin *Paracentrotus lividus* around many parts of the lough is strongly reminiscent of the tide pools all along the west coast of Ireland. Temperature in the very shallow sublittoral of the Castle Narrows (at SL −0·25 m) rose by day in August from 13·5°C to 16°C (Kitching and Thain, 1983; see their Fig. 2), and oscillations of temperature and salinity were recorded near the water surface in the Goleen in April. It is to be expected that substantial diurnal changes in oxygen and carbon dioxide tension will occur among the dense algal bush along the south shore, so that conditions in the shallow sublittoral may approach those of a low or deep mid-tidal pool.

A number of species found in the lough have a south-westerly distribution, ranging from the Mediterranean up the west European Atlantic coast to south-western parts of the British Isles, and in some cases up to the west of Scotland and even to the Orkneys and Shetlands, but not to the east coast of Britain. This distribution follows winter sea-surface temperatures, which are higher to the west, but high summer temperatures might also be necessary. An outstanding example of southwesterly distribution is the sea urchin *Paracentrotus lividus*, found from the Mediterranean to Brittany and up the west coast of Ireland (Southward and Crisp, 1954; see their Fig. 10). The red alga *Scinaia turgida*, found sublittorally in Lough Hyne on the Whirlpool Cliff (Maggs, Freamhainn and Guiry, 1983), has an extensive but entirely westerly distribution, from the Isle of Wight around the west coast of Britain to the Shetland Islands (Maggs and Guiry, 1982; see their Fig. 13). The starfish *Marthasterias glacialis*, common in Lough Hyne, extends to the western English Channel, around Ireland, and up the west coast of Britain to the Orkneys (Earll and Farnham, 1983; their Fig. 7.3). The mussel *Mytilus galloprovinialis* extends from the Mediterranean up the west European coast to Brittany, Cornwall, South Wales, and the south and west coasts of Ireland (Seed, 1978; his Fig. 7). Hiscock (1985, p. 299) records *Dictyopteris membranacea* (Phaeophyceae), found on Whirlpool Cliff, *Anthopleura ballii* (Anthozoa), very common in Lough Hyne, and *Maia sqinado* (Decapoda), found in Lough Hyne, as from south-west coasts only; and *Corynactis viridis* (Anthozoa), plentiful in Lough Hyne and on the adjacent open coast, *Antedon bifida* (Crinoidea), recorded by Renouf (1931a) from the south shore of Lough Hyne, and *Holothuria forskali* (Holothuroidea), as from south-west and western coasts only. *Pare-*

rythropodium coralloides (Pallas) (see Manuel, 1981; synonymous with *Parerythropodium hibernicum* Renouf (1931b), from Lough Hyne) ranges from the Mediterranean to the west coast of Scotland (Hartnoll, 1977; Manuel, 1981).

A possible explanation for the occurrence of rarities in Lough Hyne is that they have a south-westerly distribution but reach their northern limit around the south-west of Ireland. Another explanation is that they have been overlooked elsewhere. A concern with quantitative samples may lead to more thorough searching, but systematically different groups, such as the crustose corallines, tend to be avoided.

Species with few records for Ireland but found in the Lough Hyne area include the following.

Rhodophyta

Atractophora hypnoides Crouan frat. Springtide Pool, Carrigathorna (Goss-Custard *et al.*, 1979); also recorded from Galway (M. D. Guiry, pers. comm.) and Lundy (Hiscock and Maggs, 1984).

Cruoriella armorica Crouan frat. Systematically difficult. Known only from Galway Bay, Lough Hyne and Brittany (M. D. Guiry, pers. comm.; Maggs *et al.*, 1983).

Gracilaria bursa-pastoris (S. G. Gmel.) Silva (Maggs *et al.*, 1983). Of southerly distribution.

Gymnogrongus devoniensis (Grev.) Schotter (Maggs *et al.*, 1983). Insufficiently recorded.

Hymenoclonium serpens (Crouan frat.), a phase of *Bonnemaisonia asparagoides*, very difficult to distinguish (Maggs *et al.*, 1983).

Laurencia platycephala Kutzing. Only recently restored as distinct (Magne, 1980). Plentiful in Lough Hyne.

Pterosiphonia pennata (C. Ag.) Falkenb. Lough Hyne, Galway Bay, and south-west coasts of Britain (Maggs *et al.* 1983).

Porifera

Not discussed because the identifications by M. Burton of the British Museum (Natural History) published by Lilly *et al.* (1953) have been severely criticized by van Soest and Weinberg (1980).

Nemertini

Paradrepanophorus crassus (Quatr.). Found on south shore of Lough Hyne by Renouf (1931a) and identified by Sheppard (1935). Otherwise only known with certainty from the Mediterranean. In Lough Hyne it is found commensal with the polychaete *Staurocephalus rubrovittatus* (reported by Sheppard, 1935).

Polychaeta

Oriopsis hynensis Phyllis Knight-Jones (1983): E10, E11, S5; newly described from Lough Hyne, from material submitted by Kitching and Thain (1983).

Bryozoa

Caberia boryi Prenant and Bobin. From Codium Bay (Humphries, 1953) and Poleen (Lilly *et al.*, 1953). Mediterranean and south-west (Ryland and Hayward, 1977).

Membraniporella nitida (Johnston): wide distribution; identification by Lilly *et al.* (1953) confirmed.

Reptadeonella violacea (Johnston): from 18 m deep on open coast at Carrigathorna (Norton *et al.*, 1977). A warm temperate species (Hayward and Ryland, 1979).

Schizomavella umbonata (Busk): misidentified; delete this record from Norton *et al.* (1977).

Schizoporella magnifica Ryland. From 18 m deep on open coast at Carrigathorna (Norton *et al.*, 1977). Widespread throughout the Mediterranean, but in British waters it reaches only the south-western shores (Hayward and Ryland, 1979).

Umbonula ovicellata (Hastings); reported by Lilly *et al.* (1953) from boulders in the Rapids area. "A southern species reaching its northern limit in the southwest of the British Isles" (Hayward and Ryland, 1979).

Copepoda

Stephos rustadi Strömgren, previously described from water at a very low oxygen tension in a Norwegian fjord (Strömgren, 1969); found in September at −30 m in the Western Trough of Lough Hyne, also at a very low oxygen tension (Thain *et al.*, 1981).

Anchistrotus lucipetus Holmes (1985a); taken in light trap in Lough Hyne, and not known elsewhere. Associated with several Gobiidae (Holmes, pers. comm.).

Tanaidacea

Leptochelia savignyi (Krøyer). Widespread on shallow sublittoral rocks in Lough Hyne (Kitching and Thain, 1983). Of widespread south and westerly distribution (Holdich and Jones, 1983, see their Fig. 6).

Isopoda

Munna armoricana Carton (1961), described from collections made near Roscoff. Reported from Lough Hyne by Kitching and Thain (1983).

Amphipoda

Aora spinicornis Afonso; found in Lough Hyne by Myers and Costello (1984) and off Whirlpool Cliff by Holmes (1985b). Known distribution

Azores, Canaries and Mediterranean; only recently recognized systematically (Myers and Costello, 1984).

Gammarus insensibilis Stock. Two males from S5 in Lough Hyne confirmed by Professor J. H. Stock (Kitching and Thain, 1983); found in Mediterranean and on west European coasts (Lincoln, 1979) and in low salinity creeks or lagoons in southern England (Sheader and Sheader, 1985).

Microdeutopus stationis Della Valle. One specimen taken on Whirlpool Cliff (Holmes, 1985b); otherwise known only from Mediterranean, Atlantic coast of France and Guernsey (Myers, 1968; Lincoln, 1979).

Gammaropsis lobata (Chevreux). Found by Holmes (1983) and again in 1985 (pers. comm.) in shell gravel in Barloge Creek. Reported by Myers and McGrath (1982) from Galway Bay, Isle of Man, Plymouth and Atlantic coast of France.

Stenothoe elachistoides Myers and McGrath (1980), described from Galway; nearest relative is *S. elachista*, which is known only from Mediterranean. Found in 1985 by Holmes (pers. comm.) in Barloge Creek. No other records as yet.

Iphimedia spp. This genus is being reviewed by Myers, McGrath and Costello (in preparation). Two new species are found at Lough Hyne.

Opisthobranchia

Aeolidiella sanguinea (Norman), on west coasts of Ireland and of France (Wilson and Picton, 1983).

Runcina ferruginea Kress, newly found wild in Lough Hyne; more records needed (Wilson and Picton, 1983).

Eubranchus doriae (Trinchese), also found at several other sites in western Ireland (Wilson and Picton, 1983).

Dicata odhneri Schmekel. At 15 m depth on Island Cliff in Lough Hyne. Described from Mediterranean and possibly of south-westerly distribution, but insufficiently recorded as yet (Picton and Brown, 1981).

Facelina dubia Pruvot-Fol. Reported by Picton and Brown (1981) from Island Cliff, Lough Hyne; previously known only from Arcachon and from Mediterranean.

Cuthona genovae (O'Donoghue), a Mediterranean species now reported from Lough Hyne and from Clifden, Co. Galway (Picton and Wilson, 1984).

The pulmonate *Onchidella celtica* was reported incorrectly from Carrigathorna by Norton, Hiscock and Kitching (1977). Delete this record. Our thanks are due to Dr N. J. Evans for determining this material as *Onchidoris muricata* (Müller).

Pisces

Gobius couchi Miller and El-Tawil (1974). This goby was decribed from Helford, Cornwall. Recorded at sites along south and east shores of

Lough Hyne (Kitching and Thain, 1983). Not known from any localities besides Helford and Lough Hyne. Closest relative in morphological terms is *G. auratus*, a Mediterranean species (Miller and El-Tawil, 1974).

Gobius cruentatus Gmelin. Single specimens reported from Bantry Bay, Barloge Creek and Lough Hyne (Wheeler, 1970), and since then from S9 in Lough Hyne (Kitching and Thain, 1983). Widespread in Mediterranean and on Atlantic coast of Spain and Portugal.

From the information available a number of these species do appear to be near their northern limit in Lough Hyne, and they are possibly taking advantage of the warm summer water temperature available there. However, more information is needed from other more northerly localities on the west coast of Ireland.

Invertebrate species of special zoological interest found in the lough include *Physalia physalis* (the Portugese man of war), found once in the Whirlpool, the coral *Caryophyllia smithii*, the large turbellarian *Prostheraeus vittatus*, the floating barnacle *Lepas fascicularis*, also found in the Whirlpool, the brachiopod *Crania anomala*, the chaetognath *Spadella cephaloptera*, and finally *Phoronis hippocrepia*. (See especially Renouf, 1931a; Lilly *et al.*, 1953; Kitching and Thain, 1983.)

A few lists or accounts of particular groups of organisms in Lough Hyne are available: Rees (1935) for Algae, Maggs, Freamhainn and Guiry (1983) for Algae on sublittoral cliffs; Holmes (1980, 1983, 1985b) for Crustacea; Wilson and Picton (1983) for Opisthbranchia (70 species); and Minchin (in preparation) for Pisces. Wilson (1984) has also produced a "Bibliography of Lough Hyne (Ine) 1687–1982".

VII. GEOLOGICAL HISTORY

The origin of the excavation in the Devonian slates and sandstone which is occupied by Lough Hyne remains a matter for speculation. It seems likely that it is a result of glacial activity. It was originally suggested by Coe and Selwood, as reported in the previous review, that it arose by ice scouring in a corrie. However, work in progress by Cruickshank and Baldwin (Cruickshank, pers. comm.) on the analysis of aerial photographs has revealed a complicated network of steep-sided valleys, and suggests that Lough Hyne was excavated by subglacial meltwater erosion. In any case the lough would have originated as a freshwater lake. Seismic reflection and echo-sounding profiles (Baldwin, Cruickshank and Evans, in preparation) have revealed four rock basins in the lough bottom, of which the Western Trough basin reaches a maximum depth of 70 m, with 20 m of sediment overlying this.

Fig. 35. Summary diagram of Lough Ine core L15. A, pollen; B, diatoms. (Reproduced from Fig. 12 of Buzer, J. S. (1981). Diatom analyses of sediments from Lough Ine, Co. Cork, southwest Ireland. *New Phyto.* **89**, 511–532.)

Depth (cm): 0, 50, 100, 150, 175, 180, 190, 200, 210, 220, 230, 240, 250, 260, 270, 280, 285, 287·5, 478·5

Stratigraphy	changes/notes	Pollen: Changes		Zone	Remarks
Grey-green clay mud with shells (0–175)	marine lough	*Pinus* increasing. *Betula* decreasing. *Ulmus* and *Fagus* recorded. Ericaceae, Cyperaceae, Gramineae and herbs increase. Filicales and *Pteridium* abundant but falling		LIP 6	Some afforestation. Fall in oak possibly 17th century
		Alnus and *Quercus* dominant, Cyperaceae, Gramineae, Pteridophytes and herbs abundant	*Pinus* rising slowly. Drop in *Alnus* and *Quercus*. *Ulmus* rare. Rise in *Pteridium* and Filicales. NAP rising	LIP 5c	3rd clearance phase
Banded sediment grey-green to brown (175–c. 195)	gradual transition		Increase in *Pinus*. Small drop in herbs. *Fraxinus* first noted. Low *Pteridium* peak	LIP 5b	2nd, less clearly defined clearance phase
Dark brown fine-textured organic detritus mud (to 287·5)	freshwater lough		Rise in NAP. *Pinus* drops to a minimum. *Corylus* and *Salix* increase a little. *Isoetes* peaks and then declines with *Equisetum*. Low peak of *Pteridium*. Aquatic pollen noted	LIP 5a	Phase of land clearance, use and forest regeneration
		Alnus increases as *Pinus* declines. *Quercus* abundant. *Hedera* and *Ilex* noted. *Equisetum* rising to a maximum. Herbs and Pteridophytes rare. Aquatic pollen noted		LIP 4	Maximum mixed forest growth
		Pinus dominant. NAP and *Corylus* decreasing. *Isoetes* common	*Pinus* receding from maximum. *Alnus* recorded	LIP 3b	Extension of pine and pine–oak forests
			Pinus rising. *Quercus* falling	LIP 3a	
		NAP high, aquatics noted. Ferns, *Isoetes* and *Equisetum* common	*Corylus* dominant. *Quercus* common	LIP 2	Increase in trees and shrubs
			Betula and *Salix* abundant. *Corylus* rising	LIP 1	Open vegetation
Pale grey clay with bands (287·5–478·5)		No pollen or spores			

	Diatom			Brief summary
	Changes		Remarks	
	)iatoms not common. Marine spp. redominate. *Paralia sulcata* dominant. Vavicula lyra, *Podosira stelliger*,)imerogramma minor var. *nana* amongst hose regularly recorded	LID 4	Polyhalobous and indifferent spp. co-dominant. Mostly benthic forms	
	mall peaks of brackish water spp. everal marine spp. first noted. 'aralia sulcata increasing. Most freshwater pecies die out	LID 3	Mesohalobous spp. common. Benthic spp. increasing	Alkaline, freshwater lake becoming rather more neutral before a gradual marine transgression. The latter resulted in brackish water before the lough became truly marine.
spp. dominant	Maximum of attached spp. *Cyclotella* at a minimum. No real decrease in plankton frequencies	LID 2b	Planktonic, alkaliphilous species dominant. Halophilous species common. Peaks of attached spp.	
	Frequencies of *Cyclotella comta* drop rapidly. Attached species increase Brackish water spp. first regularly recorded	LID 2a		
dominant. ↑ragılarıa spp. common	*Cyclotella comta* dominant. *Fragilaria* spp. common. Other *Cyclotella* spp. less frequent	LID 1b	Slight increase in acidophilous spp.	
	Cyclotella comta and *Fragilaria brevistriata* co-dominant. Attached and epipelic spp. present	LID 1a	Some halophobous forms	
	No diatoms			Possibly glacial outwash

Cores were taken in Lough Hyne and in the small adjacent freshwater lough, Ballyally, with a Mackereth corer by Dr Jenny Buzer, and were thoroughly analysed for pollen (Buzer, 1980) and for diatoms (Buzer, 1981). The diatoms are excellent ecological indicators as they were living in the lough water and have known tolerances for salinity and for acidity or alkalinity, and known style of habitat (planktonic, part planktonic, attached, and benthic (Foged, 1954). The results for diatoms in Lough Hyne are based on a core 564 cm long, taken in the North Basin from a water depth of 36 m. The sediment at the bottom of this core is devoid of diatoms and pollen, and is a proglacial laminated clay. It is followed higher up by sediment containing freshwater diatoms. Still higher there is a progressive change to salt-tolerant and finally to marine diatoms.

Results for pollen are based on two cores from Lough Hyne as well as cores from Ballyally. These are more difficult to interpret as they mainly reflect the state of vegetation on the surrounding land. A decline in the relative importance of arboreal (in comparison with non-arboreal) pollen suggests periods of forest clearance by Neolithic man. According to this interpretation, Neolithic man inhabited the country around Lough Hyne while it was still a freshwater lake, and continued to do so as it was invaded by the sea. However, it was pointed out by Buzer that climatic change or disease might also be involved. As a recently reported example, the discovery of elm bark beetle in a neolithic deposit on Hampstead Heath has been linked with a decline of elm in that deposit (Girling and Greig, 1985). Buzer's (1981) diagrammatic representation of the probable sequence of events in Lough Hyne is reproduced in Fig. 35.

The marine transgression was presumably the result of a rise in sea level associated with the melting of the ice which covered the land, this rise outpacing—at the margin of the land—the isostatic rise of the land mass. On indirect evidence it was suggested by Buzer (1981) that this took place 4000 years before the present time. Carbon and/or palaeomagnetic dating need to be carried out. It would also be interesting to know what other fossils are present in these deposits, as a possible guide to the oxygen content of the water.

ACKNOWLEDGEMENTS

I acknowledge with gratitude the contributions of many people to the work at Lough Hyne: colleagues who are co-authors of many papers with me, students of the University of East Anglia who have taken part in field work, and specialists who have identified difficult material including, most recently, Dr Yvonne Chamberlain for certain Algae, Professor Tryge Christensen for *Tribonema*, Dr E. M. Burrows for many algae, and Dr A.

Fletcher for many specimens of lichens. I have had helpful correspondence from Mr Mark Costello, Dr M. D. Guiry, Dr J. M. C. Holmes, Dr Dan Minchin Professor T. A. Norton, Dr B. E. Picton, and Dr R. Seed. I am grateful for much friendly local help, especially to the Donelans of Skibbereen and the Bohanes of Dromadoon.

REFERENCES

Ackerfors, H. (1966). Plankton and hydrography of the Landsort Deep. *Veroff. Inst. Meeresforsch. Bremerh.* **2**, 381–385.

Ackerfors, H. (1969). Ecological zooplankton investigations in the Baltic proper 1963–1965. *Rep. Inst. Mar. Res. Lysekil* **18**, 1–139.

Allen, P. L. (1983). Feeding behaviour of *Asterias rubens* (L.) on soft bottom bivalves: a study in selective predation. *J. exp. mar. Biol. Ecol.* **70**, 79–90.

Ballantine, W. J. (1961). A biologically defined exposure scale for the comparative description of rocky shores. *Fld Stud.* **1**, 1–19.

Bassindale, R., Ebling, F. J., Kitching, J. A. and Purchon, R. D. (1948). The ecology of the Lough Ine Rapids with special reference to water currents. I. Introduction and hydrography. *J. Ecol.* **36**, 305–322.

Bernstein, B. B., Williams, B. F. and Mann, K. H. (1981). The role of behavioral responses to predators in modifying urchins' (*Strongylocentrotus droebachiensis*) destructive grazing and seasonal foraging patterns. *Mar. Biol.* **63**, 39–49.

Branch, G. M. (1981). The biology of limpets: physical factors, energy flow, and ecological interactions. *Oceanogr. mar. Biol. ann. Rev.* **19**, 235–380.

Buzer, J. S. (1980). Pollen analyses of sediments from Lough Ine and Ballyally Lough, Co. Cork, S. W. Ireland. *New Phytol.* **86**, 93–108.

Buzer, J. S. (1981). Diatom analyses of sediments from Lough Ine, Co. Cork, southwest Ireland. *New Phytol.* **89**, 511–533.

Carton, Y. (1961). Étude des representants du genre *Munna* Kröyer sur les côtes françaises de la Manche. *Bull. Soc. linn. Normandie* **10**, 222–242.

Daniel, M. J. and Boyden, C. R. (1975). Diurnal variations in physico-chemical conditions within intertidal rockpools. *Fld Stud.* **4**, 161–176.

Dethier, M. N. (1980). Tidepools as refuges: predation and the limits of the harpacticoid copepod *Tigriopus californicus* (Baker). *J. exp. mar. Biol. Ecol.* **42**, 99–111.

Dooley, H. D. (1981). Oceanographic observations in Sullom Voe, Shetland, in the period 1974–78. *Proc. R. Soc. Edinb. B* **80**, 55–71.

Earll, R. and Farnham, W. F. (1983). Biogeography. In *Sublittoral Ecology of the Shallow Sublittoral Benthos* (Ed. by R. Earll and D. G. Erwin), pp. 165–200. Oxford: Clarendon Press.

Ebling, F. J. and Kitching, J. A. (1950). Exploration of the Lough Ine Rapids. *Sch. Sci. Rev.* **114**, 222–229.

Edwards, A., Edelsten, D. J., Saunders, M. A. and Stanley, S. O. (1980). Renewal and entrainment in Loch Eil; a periodically ventilated Scottish fjord. In *Fjord Oceanography* (Ed. by H. J. Freeland, D. M. Farmer and C. D. Levings), pp. 523–530. New York: Plenum Press.

Eleftheriou, A. (1970). Note on the polychaete *Pseudopolydora pulchra* (Carazzi) from British waters. *Cah. Biol. mar.* **11**, 459–474.

Emson, R. H. (1985). Life history patterns in rock pool animals. In *The Ecology of Rocky Coasts* (Ed. by P. G. Moore and R. Seed), pp. 220–222. London: Hodder and Stoughton.
Emson, R. H. and Crump, R. G. (1979). Description of a new species of *Asterina* (Asteroidea), with an account of its ecology. *J. mar. biol. Assoc. UK* **59**, 77–94.
Emson, R. H. and Crump, R. G. (1984). Comparative studies on the ecology of *Asterina gibbosa* and *A. phylactica* at Lough Ine. *J. mar. biol. Assoc. UK* **64**, 35–53.
Foged, N. (1954). On the diatom flora of some Funen lakes. *Folia limnol. scand.* **6**, 7–81.
Fonselius, S. H. (1962). Hydrography of the Baltic deep basins. *Rep. Fish. Bd Swed.* **13**, 1–41.
Fraser, J. H. (1936a). The distribution of rockpool Copepoda according to tidal level. *J. Anim. Ecol.* **5**, 23–28.
Fraser, J. H. (1936b). The occurrence, ecology, and life history of *Tigriopus fulvus*. *J. mar. biol. Assoc. UK* **20**, 523–536.
Gallop, A., Bartrop, J. and Smith, D. C. (1980). The biology of chloroplast acquisition by *Elysia viridis*. *Proc. R. Soc. Lond. B* **207**, 335–349.
Ganning, B. (1971). Studies on chemical, physical and biological conditions in Swedish rockpool ecosystems. *Ophelia* **9**, 51–105.
Gibbons, S. G. and Fraser, J. H. (1937). The centrifugal pump and suction hose as a method of collecting plankton samples. *J. Cons. perm. int. Explor. Mer* **12**, 155–170.
Girling, M. A. and Greig, J. (1985). A first fossil record for *Scolytus scolytus* (F.) (Elm Bark Beetle): its occurrence in elm decline deposits from London and the implications for neolithic elm disease. *J. archaeol. Sci.* **12**, 347–357.
Goss-Custard, S., Jones, J. and Kitching, J. A. (1979). Tide pools of Carrigathorna and Barloge Creek. *Phil. Trans. R. Soc. B* **287**, 1–44.
Hartnoll, R. G. (1977). Reproductive strategy in two British species of *Alcyonium*. In *Biology of Benthic Organisms* (Ed. by B. F. Keagan, P. O. Cedigh and P. J. S. Boaden), pp. 321–328. Oxford: Pergamon Press.
Hayward, P. J. and Ryland, J. S. (1979). *British Ascophoran Bryozoans*. London: Academic Press, 312 pp.
Hiscock, K. (1985). Aspects of the ecology of rocky sublittoral areas. In *The Ecology of Rocky coasts* (Ed. by P. G. Moore and R. Seed), pp. 290–328. London: Hodder and Stoughton.
Hiscock, K. and Mitchell, R. (1980). The description and classification of sublittoral epibenthic ecosystems. In *The Shore Environment. Vol. 2. Ecosystems* (Ed. by J. H. Price, D. E. G. Irvine and W. F. Farnham), pp. 323–370. Syst. Ass. spec. Vol. 17a. London: Academic Press.
Hiscock, S. and Maggs, C. A. (1984). Notes on the distribution and ecology of some new and interesting seaweeds from south-west Britain. *Br. phycol. J.* **19**, 73–87.
Holdich, D. M. and Jones, J. A. (1983). The distribution and ecology of British shallow-water tanaid crustaceans (Peracarida, Tanaidacea). *J. nat. Hist.* **17**, 157–183.
Holmes, J. M. C. (1980). Some crustacean records from Lough Ine, Co. Cork. *Bull. Ir. biogeogr. Soc.* **4**, 33–40.
Holmes, J. M. C. (1983). Further crustacean records from Lough Ine, Co. Cork. *Bull. Ir. biogeogr. Soc.* **5**, 19–24 (bulletin for 1981).
Holmes, J. M. C. (1985a). *Anchistrotus lucipetus* sp. nov. (Copepoda, Taeniacanthidae), a parasitic copepod from Lough Ine, South West Ireland. *Crustaceana* **48**, 18–25.

Holmes, J. M. C. (1985b). Crustacean records from Lough Ine, Co. Cork; part III. *Bull. Ir. biogeogr. Soc.* **8**, 19–25 (bulletin for 1983).
Humphries, C. F. (1953). A species of Polyzoa, *Caberea boryi* (Audouin), new to Ireland. *Ir. Nat. J.* **11**, 80.
Jones, N. S. and Kain, J. M. (1967). Subtidal algal colonization following the removal of *Echinus. Helgolander wiss. Meeresunters.* **15**, 460–466.
Jørgensen, B. B. (1980). Seasonal oxygen depletion in the bottom waters of a Danish fjord and its effect on the benthic community. *Oikos* **34**, 68–76.
Kain, J. M. (1963). Aspects of the biology of *Laminaria hyperborea*. II. Age, weight and length. *J. mar. biol. Assoc. UK* **43**, 129–151.
Kain, J. M. (1971). Synopsis of biological data on *Laminaria hyperborea*. F.A.O. Fisheries Synopsis No. 87. Rome: *FAO*.
Kempf, M. (1962). Recherches d'écologie comparée sur *Paracentrotus lividus* (Lmk.) et *Arbacia lixula* (L.). *Récl. Trav. Stn mar. Endoume* **25**, 47–116.
Kitching, J. A. (1935). An introduction to the ecology of intertidal rock surfaces on the coast of Argyll. *Trans. R. Soc. Edinb.* **58**, 351–374.
Kitching, J. A. (1941). Studies in sublittoral ecology. III. *Laminaria* forest on the west coast of Scotland; a study of zonation in relation to wave action and illumination. *Biol. Bull. mar. biol. Lab., Woods Hole* **80**, 324–337.
Kitching, J. A. and Ebling, F. J. (1967). Ecological studies at Lough Ine. *Adv. ecol. Res.* **4**, 197–291.
Kitching, J. A., Ebling, F. J., Gamble, J. C., Hoare, R., McLeod, A. A. Q. R. and Norton, T. A. (1976). The ecology of Lough Ine. XIX. Seasonal changes in the Western Trough. *J. Anim. Ecol.* **45**, 731–758.
Kitching, J. A. and Thain, V. M. (1983). The ecological impact of the sea urchin *Paracentrotus lividus* (Lamarck) in Lough Ine, Ireland. *Phil. Trans. R. Soc. Lond. B* **300**, 513–552.
Knight-Jones, P. (1983). Contributions to the taxonomy of Sabellidae (Polychaeta). *Zool. J. Linn. Soc.* **79**, 245–295.
Larkum, A. W. D. and Norton, T. A. (1968). An investigation of the vertical distribution of seaweeds in the Whirlpool area of Lough Ine. *Br. phycol. Bull.* **3**, 601–602.
Levin, L. A. (1981). Dispersion, feeding behavior and competition in two spionid polychaetes. *J. mar. Res.* **39**, 99–117.
Lewis, J. R. (1964). *The Ecology of Rocky Shores*. London: The English Universities Press, 323 pp.
Lilly, S. J., Sloane, J. F., Bassindale, R., Ebling, F. J. and Kitching, J. A. (1953). The ecology of the Lough Ine Rapids with special reference to water currents. IV. The sedentary fauna of sublittoral boulders. *J. Anim. Ecol.* **22**, 87–122.
Lincoln, R. J. (1979). *British marine Amphipoda: Gammaridea*. London: British Museum (Natural History), 658 pp.
Little, C. and Stirling, P. (1985). Patterns of foraging activity in the limpet *Patella vulgata* L.—a preliminary study. *J. exp. mar. Biol. Ecol.* **89**, 283–296.
Maggs, C. A., Freamhainn, M. T. and Guiry, M. D. (1983). A study of the marine Algae of subtidal cliffs in Lough Hyne (Ine) Co. Cork. *Proc. R. Ir. Acad. B* **83**, 251–266.
Maggs, C. A. and Guiry, M. D. (1982). The taxonomy, morphology and distribution of species of *Scinaia* Biv.-Bern. (Nemaliales, Rhodophyta) in north-western Europe. *Nord. J. Bot.* **2**, 517–523.
Magne, F. (1980). *Laurencia platycephala* Kützing (Rhodophyceae), espèce méconnue des côtes de la Manche. *Cah. Biol. mar.* **21**, 227–237.

Manuel, R. L. (1981). *British Anthozoa*. Linn. Soc., London and Brackish Water Sciences Assn. London: Academic Press, 241 pp.
Miller, P. J. and El-Tawil, M. Y. (1974). A multidisciplinary approach to a new species of *Gobius* (Teleostei: Gobiidae) from southern Cornwall. *J. Zool., Lond.* **174**, 539–574.
Minchin, D. (1985). *Lutraria angustior* Philippi (Mollusca: Lamellibranchia) in Irish waters. *Ir. Nat. J.* **21**, 454–459.
Minchin, D. and Molloy, J. (1978). Notes on some fishes taken in Irish waters in 1977. *Ir. Nat. J.* **19**, 264–267.
Muntz, L., Norton, T. A., Ebling, F. J. and Kitching, J. A. (1972). The ecology of Lough Ine. XVIII. Factors controlling the distribution of *Corynactis viridis* Allman. *J. Anim. Ecol.* **41**, 735–750.
Myers, A. A. (1968). A revision of the genus *Microdeutopus* Costa (Gammaridea: Aoridae). *Bull. Br. Mus. nat. Hist. (Zool.)* **17**, 91–148.
Myers, A. A. and Costello, M. J. (1984). The amphipod genus *Aora* in British and Irish waters. *J. mar. biol. Assoc. UK* **64**, 279–283.
Myers, A. A. and McGrath, D. (1980). A new species of Stenothoe Dana (Amphipoda, Gammaridea) from maerl deposits in Kilkieran Bay. *J. Life Sci. R. Dubl. Soc.* **2**, 15–18.
Myers, A. A. and McGrath, D. (1982). Taxonomic studies on British and Irish Amphipoda: the genus *Gammaropsis*. *J. mar. biol. Assoc. UK* **62**, 93–100.
Nikitin, W. N. (1931). Die untere Planktongrenze und deren Verteilung in Schwarzen Meer. *Int. Revue ges. Hydrobiol. Hydrogr.* **25**, 102–130.
Nikitine, B. N. and Malm, E. (1934). L'influence de l'oxygène, des ions hydrogène et de l'acide carbonique sur la distribution verticale du plankton de la Mer Noire. *Ann. Inst. océanogr., Monaco* **14**, 136–171.
Norton, T. A. (1969). Growth form and environment in *Saccorhiza polyschides*. *J. mar. biol. Assoc. UK* **49**, 1025–1045.
Norton, T. A., Ebling, F. J. and Kitching, J. A. (1971). Light and the distribution of organisms in a sea cave. In *Fourth European Marine Biology Symposium* (Ed. by D. J. Crisp), pp. 409–432. Cambridge: Cambridge University Press.
Norton, T. A., Hiscock, K. and Kitching, J. A. (1977). The ecology of Lough Ine. XX. The *Laminaria* forest at Carrigathorna. *J. Ecol.* **65**, 919–941.
Parke, M. (1948). Studies of British Laminariaceae. I. Growth in *Laminaria saccharina* (L.) Lamour. *J. mar. biol. Assoc. UK* **27**, 651–709.
Pearson, T. H. (1970). The benthic ecology of Loch Linnhe and Loch Eil, a sea loch system on the west coast of Scotland. I. The physical environment and distribution of the macrobenthic fauna. *J. exp. mar. Biol. Ecol.* **5**, 1–34.
Pearson, T. H. (1982). The Loch Eil project: assessment and synthesis with a discussion of certain biological questions arising from a study of the organic pollution of sediments. *J. exp. mar. Biol. Ecol.* **57**, 93–124.
Pearson, T. H. and Eleftheriou, A. (1981). The benthic ecology of Sullom Voe. *Proc.. R. Soc. Edinb. B* **80**, 241–269.
Petipa, T. S., Sazhina, L. S. and Delalo, E. P. (1961). Vertical distribution of zooplankton in the Black Sea in relation to hydrological conditions. *Am. Inst. Biol. Sci. Bull.* **133**, 576–578.
Picton, B. E. and Brown, G. H. (1981). Four nudibranch gastropods new to the fauna of Great Britain and Ireland, with a description of a new species of *Doto* Oken. *Ir. Nat. J.* **20**, 261–268.
Picton, B. E. and Wilson, K. (1984). *Cuthona genovae* (O'Donoghue 1926), an aeolid nudibranch new to the fauna of the British Isles. *J. Conch.* **31**, 349–352.

Pressoir, L. (1959). Contribution a la connaissance des échinopluteus de *Paracentrotus lividus* Lmk. et *Psammechinus microtuberculatus* Blainv. *Bull. Inst. océanogr. Monaco* no. 1142, 1–19.
Pyefinch, K. A. (1943). The intertidal ecology of Bardsey Island, North Wales, with special reference to the recolonization of rock surfaces, and the rock pool environment. *J. Anim. Ecol.* **12**, 82–107.
Ranade, M. R. (1957). Observations on the resistance of *Tigriopus fulvus* (Fisher) to changes in temperature and salinity. *J. mar. biol. Assoc. UK* **36**, 115–119.
Rees, T. K. (1935). The marine Algae of Lough Ine. *J. Ecol.* **23**, 69–133.
Renouf, L. P. W. (1931a). Preliminary work of a new biological station (Lough Ine, Co. Cork, I.F.S.). *J. Ecol.* **19**, 410–438.
Renouf, L. P. W. (1931b). On a new species of alcyonarian *Parerythropodium hibernicum. Acta zool.* **12**, 205–223.
Rice, A. L. and Johnstone, A. D. F. (1972). The burrowing behaviour of the gobiid fish *Lesueurigobius friesii* (Collett). *Z. Tierpsychol.* **30**, 431–438.
Rosenberg, R. (1980). Effect of oxygen deficiency of benthic macrofauna in fjords. In *Fjord Oceanography* (Ed. by H. J. Freeland, D. M. Farmer and C. D. Levings), pp. 499–514. New York: Plenum Press.
Round, F. E., Sloane, J. F., Ebling, F. J. and Kitching, J. A. (1961). The ecology of Lough Ine. X. The hydroid *Sertularia operculata* (L.) and its associated flora and fauna: effects of transference to sheltered water. *J. Ecol.* **49**, 617–629.
Rueness, J. (1976). *Rhodochorton floridulum* (Rhodophyceae) on the Norwegian west coast. *Sarsia* **61**, 71–74.
Ryland, J. S. and Hayward, P. J. (1977). *British Anascan Bryozoans*. London: Academic Press, 188 pp.
Seed, R. (1978). The systematics and evolution of *Mytilus galloprovincialis* Lmk. In *Marine Organisms: Genetics, Ecology and Evolution* (Ed. by B. Battaglia and J. A. Beardmore), pp. 447–468. New York: Plenum Press.
Sheader, M. and Sheader, A. L. (1985). New distribution records for *Gammaropsis insensibilis* Stock, 1966, in Britain. *Crustaceana* **49**, 101–105.
Sheppard, E. M. (1935). On *Paradrepanophorus crassus* (Quatr.), a nemertean worm new to the British fauna. *Ann. Mag. nat. Hist.* **10**, 232–236.
van Soest, R. W. and Weinberg, S. (1980). A note on the sponges and octocorals from Sherkin Island and Lough Ine, Co. Cork. *Ir. Nat. J.* **20**, 1–15.
Southward, A. J. and Crisp, D. J. (1954). The distribution of certain intertidal animals around the Irish coast. *Proc. R. Irish Acad. B* **57**, 1–29.
Stanley, S. O., Grantham, B. E., Leftley, J. and Robertson, N. (1981). Some aspects of the sediment chemistry of Sullom Voe, Shetland. *Proc. R. Soc. Edinb. B* **80**, 91–100.
Strøm, K. M. (1936). Land-locked waters, hydrography and bottom deposits in badly ventilated Norwegian fjords with remarks upon sedimentation under anaerobic conditions. *Skr, norske Vidensk.-Akad. Mat.-naturw.* **7**, 1–85.
Strömgren, T. (1969). A new species of *Stephos* (Copepoda Calanoida) from the Norwegian west coast. *Sarsia* **37**, 1–8.
Sundene, O. (1962). The implications of transplant and culture experiments on the growth and distribution of *Alaria esculenta. Nytt Mag. Bot.* **9**, 155–174.
Thain, V. M. (1971). Diurnal rhythms in snails and starfish. In *Fourth European Marine Biology Symposium* (Ed. by D. J. Crisp), pp. 513–537. Cambridge: Cambridge University Press.
Thain, V. M., Jones, J. and Kitching, J. A. (1981). Distribution of zooplankton in relation to the thermocline and oxycline in Lough Ine, County Cork. *Ir. Nat. J.* **20**, 292–295.

Thain, V. M., Thain, J. F. and Kitching, J. A. (1985). Return of the Prosobranch *Gibbula umbilicalis* (Da Costa) to the littoral region after displacement to the shallow sublittoral. *J. Moll. Stud.* **51**, 205–210.

Theede, H., Ponat, A. and Schlieper, C. (1969). Studies on the resistance of marine bottom invertebrates to oxygen deficiency and hydrogen sulphide. *Mar. Biol.* **2**, 325–337.

Thomas, M. L. H. (1983). Tide pool systems. In *Marine and coastal systems of the Quoddy Region, New Brunswick* (Ed. by M. L. H. Thomas), pp. 95–106. Can. Spec. Publ. Fish. Aq. Sci. **64**, Ottawa.

Thrush, S. F. (1986a). The sublittoral macrobenthic community structure of an Irish sea-lough: effects of decomposing accumulations of seaweed. *J. exp. mar. Biol. Ecol.* **96**, 199–212.

Thrush, S. F. (1986b). Spatial heterogeneity in subtidal gravel generated by the pit digging activities of *Cancer pagurus*. *Mar. Ecol. Prog. Ser.* **30**, 221–227.

Thrush, S. F. and Townsend, C. R. (1986). The sublittoral macrobenthic community composition of Lough Hyne, Ireland. *Estuar. coast. Shelf Sci.* **23**, 551–574.

Trench, R. K., Boyle, J. E. and Smith, D. C. (1973). The association between chloroplasts of *Codium fragile* and the mollusc *Elysia viridis*. *Proc. R. Soc. Lond. B* **184**, 51–61.

Truchot, J.-P. and Duhamel-Jouve, A. (1980). Oxygen and carbon dioxide in the marine intertidal environment: diurnal and tidal changes in rockpools. *Resp. Physiol.* **39**, 241–254.

Underwood, A. J. (1979). The ecology of intertidal gastropods. *Adv. mar. Biol.* **16**, 111–120.

Vahl, O. (1971). Growth and density of *Patina pellucida* (L.) (Gastropoda: Prosobranchia) on *Laminaria hyperborea* (Gunnerus) from western Norway. *Ophelia* **9**, 31–50.

de Virville, D. (1934, 1935). Recherches écologiques sur la flore des flaques du littoral de l'océan atlantique et de la Manche. *Revue gen. Bot.* **46**, 705–721; **47**, 26–43; **47**, 160–177; **47**, 230–243; **47**, 308–323.

Wheeler, A. (1970). *Gobius cruentatus*—a fish new to the northern European fauna. *J. Fish Biol.* **2**, 59–67.

Wilson, K. (1984). A bibliography of Lough Hyne (Ine) 1687–1982. *J. Life Sci. R. Dublin Soc.* **5**, 1–11.

Wilson, K. and Picton, B. E. (1983). A list of the Opisthobranchia: Mollusca of Lough Hyne Nature Reserve, Co. Cork, with notes on distribution and nomenclature. *Ir. Nat. J.* **21**, 69–72.

Wolff, W. J. (1973). *The Estuary as a Habitat. An analysis of data on the soft bottom macrofauna of the estuarine area of the rivers Rhine, Meuse, and Scheldt.* Communication no. 106 of the Delta Institute for Hydrobiological Research. Leiden: E. J. Brill, 242 pp.

Isopods and Their Terrestrial Environment

M. R. WARBURG

ADVANCES IN ECOLOGICAL RESEARCH Vol. 17
ISBN 0-12-013917-0

I. SUMMARY

Terrestrial adaptations of the oniscoid isopods range from structural and behavioural to ecological and physiological traits. Special trends in the structure of the respiratory organs, the pseudotracheae, enable conservation of water. Likewise, the water-conducting system, unique to oniscoid isopods, is modified in the more terrestrial forms to enable the resorption of water after the ammonia has evaporated. A series of receptors enables orientation to the burrow and perception of family members or adversaries in the social desert isopod *Hemilepistus*. Other isopods may possibly be similarly organized into family units. The reproductive strategy of isopods is related to their short lifespan: the semelparous species have usually a 1–2 year lifespan, while the iteroparous species may reach a maximum age of 5 years. The reproductive pattern is generally largely dependent upon environmental factors: light, temperature and moisture conditions in the soil. The role of food in governing population oscillations is generally negligible although some exceptions are known, but much more research is needed on this subject. Moisture appears to be the main single factor of importance for survival of most terrestrial isopod species.

II. INTRODUCTION

Over 30 years have elapsed since the appearance of Edney's (1954) paper reviewing the "woodlice and the land habitat". During this period the number of studies on this interesting arthropod group has, at a conservative estimate, at least tripled. Edney's review has greatly stimulated my own research into this fascinating group, as it has undoubtedly also stimulated others. Oniscoid isopods have apparently emerged on land via the littoral zone and, today, we can still find representatives of the whole series of isopod species as if they were amidst the process of emerging from the seashore onto land (some of them possibly regressing from land back to sea; Verhoeff, 1949). Thus, we find isopods inhabiting littoral zone, beach, grassland, woodland and desert habitats, only a few of which have been studied. All of these species show various degrees of adaptive traits enabling their survival on land in their respective habitats. These adaptations were thought to be largely behavioural but it now appears that there are also well established physiological adaptations, based on anatomical structures. Furthermore, we can still see that many isopods found today demonstrate the different ways by which they have adapted to life on land. Thus, we shall take a close look at our present knowledge of this remarkable group of arthropods.

III. ANATOMICAL ADAPTATIONS OF ISOPODS SETTLING ON LAND

A. The Respiratory Organs

In the isopod crustaceans we can see a wide range of respiratory organs, from the gills of aquatic (Babula, 1979a,b) and marine species (Bubel and Jones, 1974; Babula and Bielawski, 1981a,b; Wägele, 1982) through leaf-like structures in the Tylidae (Ebbe, 1981; Hoese, 1983) to complete lung-like structures in some terrestrial porcellionids (Fig. 1 and Hoese, 1982b). Some of these respiratory organs possess an arrangement enabling them to close their respiratory orifice (*Hemilepistus*, see Fig. 2 and Hoese, 1982b). Earlier studies have already pointed to the variability of these organs in isopods (Stoller, 1899; Bepler, 1909; Verhoeff, 1917b, 1921). The recent work by Hoese (1982b, 1983) has strengthened our previous views on the existence of a series of structural and probably functional adaptations in the respiratory organs. This could well be one of the main key factors for the successful colonization of land.

B. The Integument

The integument of isopods has received much less attention than that of either the insects or the other malacostracan crustaceans. Some of the earlier studies dealt with the presence or absence of lipids in the cuticle (e.g. Mead-Briggs, 1956), or with the general composition of the cuticle (e.g. Lagarrigue, 1969). Whereas this latter interesting aspect was never followed up using modern biochemical techniques, the studies of lipids in the isopod cuticle have received greater attention. It was recently shown that even a mesic inhabiting porcellionid species (*Porcellio laevis*) contains lipids in its cuticle (Hadley and Quinlan, 1984), as do the xeric isopods *Hemilepistus reaumuri, Armadillo officinalis* and *A. albomarginatus* (Hadley and Warburg, 1986; and Fig. 3). The presence of lipids can also be demonstrated by standard histochemical techniques using oil-red-O or Sudan black (Fig. 4).

C. The Digestive Organs

The digestive tracts, as well as the digestive glands or hepatopancreas, have received attention in recent years (Hryniewiecka-Szyfter, 1972, Hryniewiecka-Szyfter and Tyczewska, 1975; Prosi, Storch and Janssen, 1983; Storch, 1982; Wägele, Welsch and Müller, 1981). A number of cells

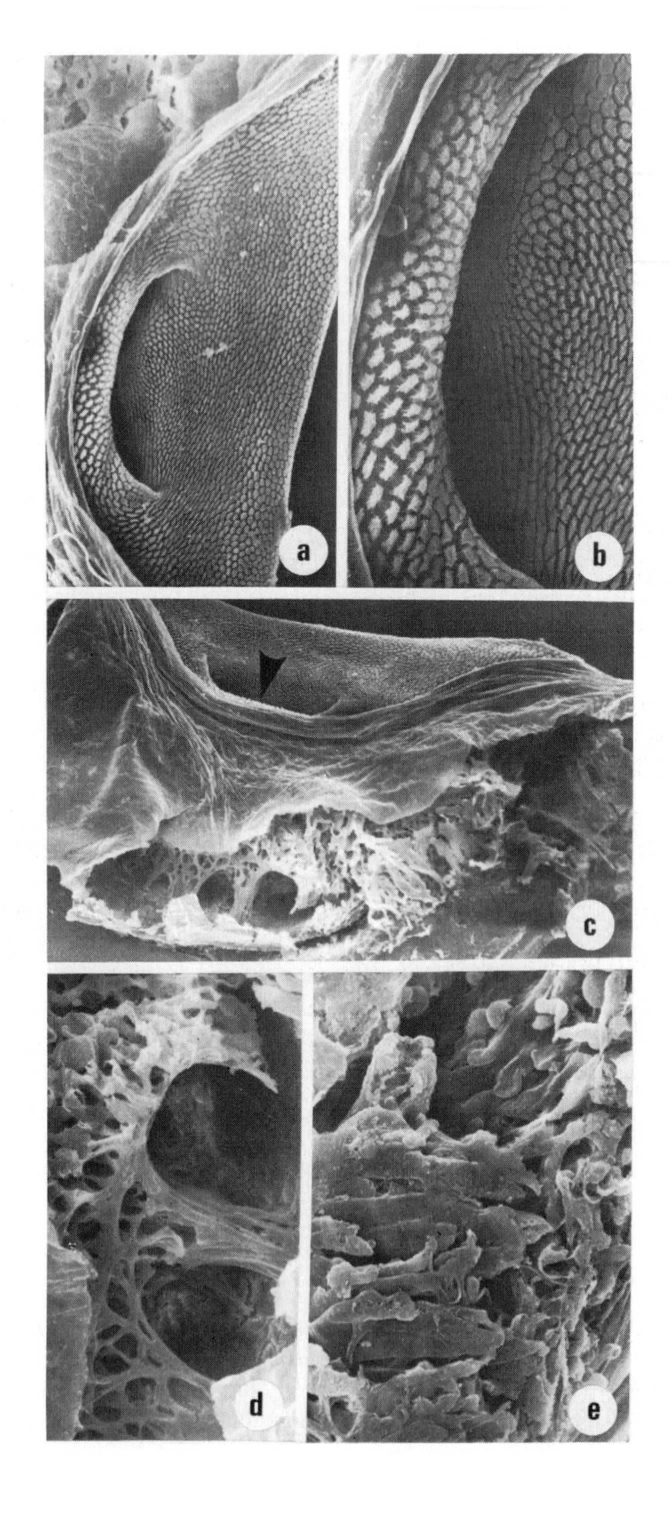

a
b
c
d
e

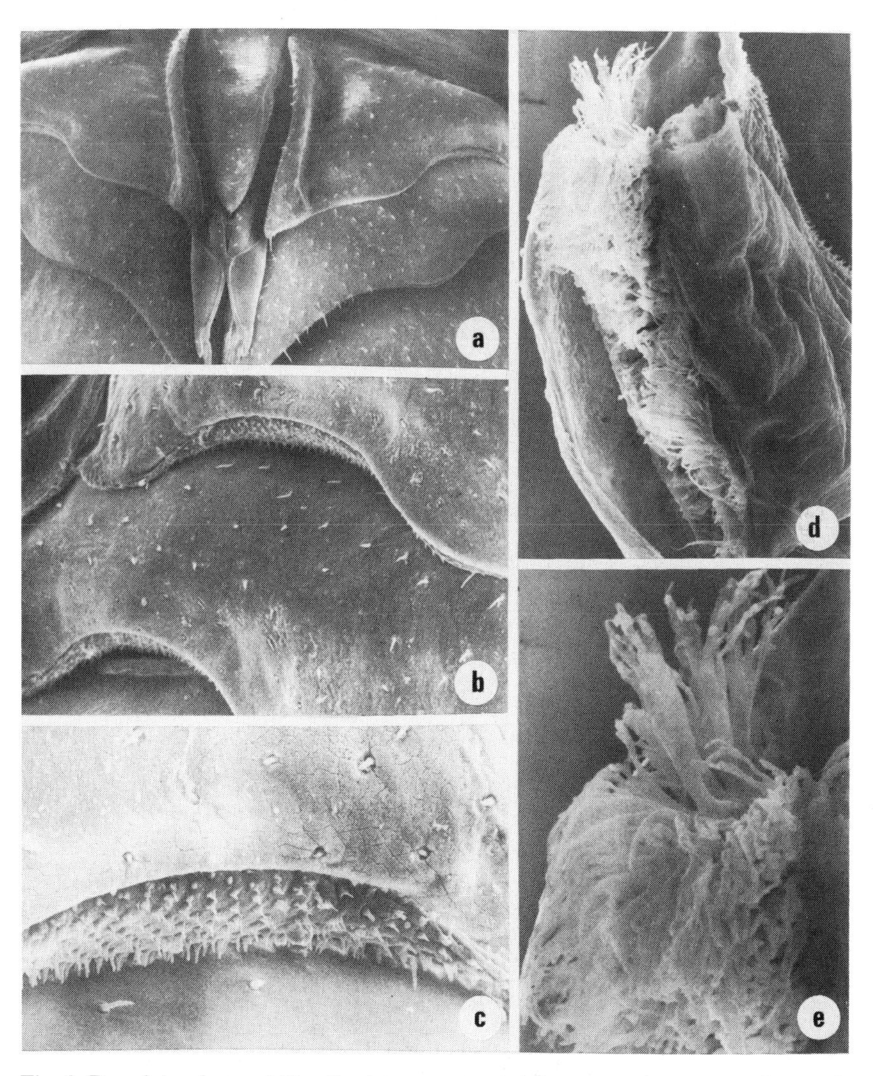

Fig. 2. Pseudotracheae of *Hemilepistus reaumuri*: (a) two openings as seen in a male (× 50); (b) two openings (× 150); (c) opening of pseudotrachea (× 350); (d) pseudotrachea inside (× 150); (e) as in (d) but × 500.

Fig. 1. Pseudotracheae of *Porcellio obsoletus* as seen with the scanning electron microscope: (a) opening to pseudotrachea (× 250); (b) as in (a) but × 500); (c) The pseudotracheae, inside view (× 200)—opening marked by arrow; (d) pseudotrachea inside—note the branching tubes (× 350); (e) as in (d) but × 500.

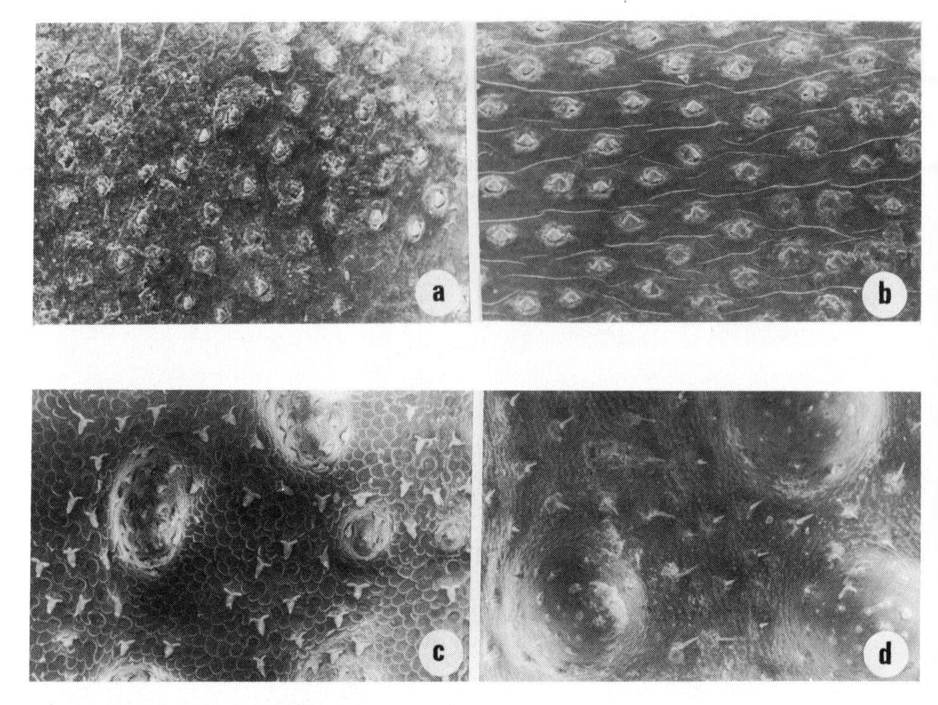

Fig. 3. Cuticle of *Armadillo officinalis* (a,b), and of *Hemilepistus reaumuri* (c,d), as seen by the scanning electron microscope. (a,c) Controls (×200); (b,d) acetone treated (b ×200, d ×250).

have been demonstrated in the glands and in the gut, and different functions have been assigned to some of them (Storch and Lehnert-Moritz, 1980; Hopkin and Martin, 1984; Bettica, Shay, Vernon and Witkus, 1984). Similarly, in the gut, mitochondria-rich cells were suggested to be involved in transport of ions (Coruzzi, Witkus and Vernon, 1982; Palackal, Faso, Zung, Vernon and Witkus, 1984). Recent evidence suggests the presence of ATPase in these cells (Fig. 5, details in preparation).

D. The Excretory Organs

There are no recent studies on excretory organs of isopods and the subject needs attention. Is there any difference between the isopod species inhabiting mesic and xeric habitats?

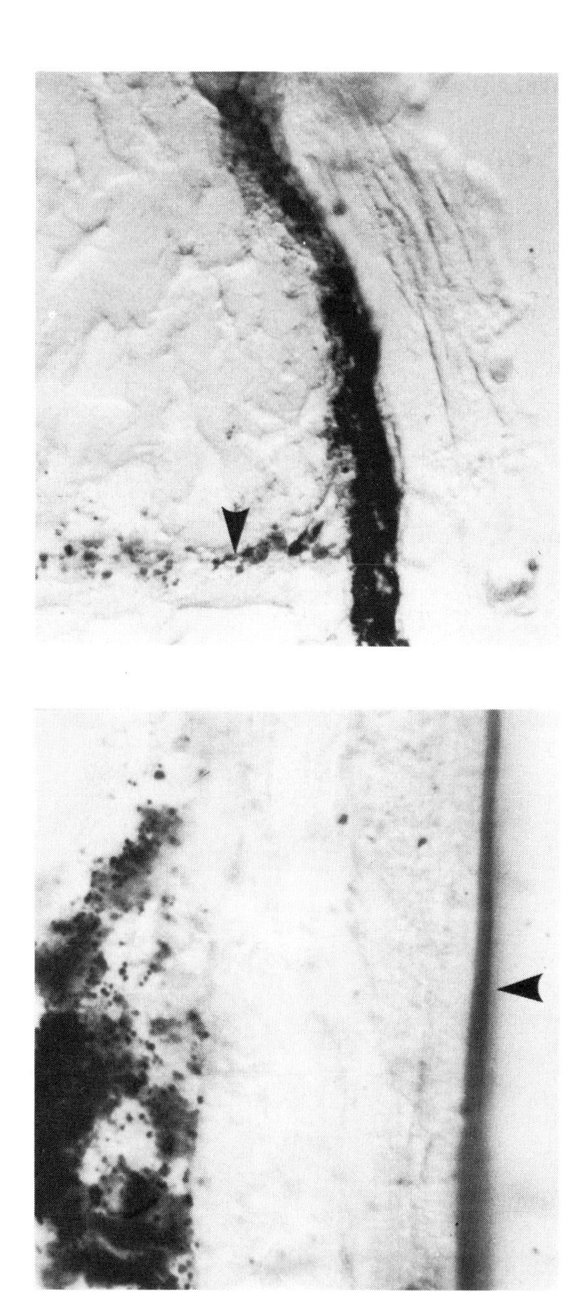

Fig. 4. A lipid seen in the cuticle of *Armadillo officinalis* using oil-red-o: arrows indicate lipid deposits.

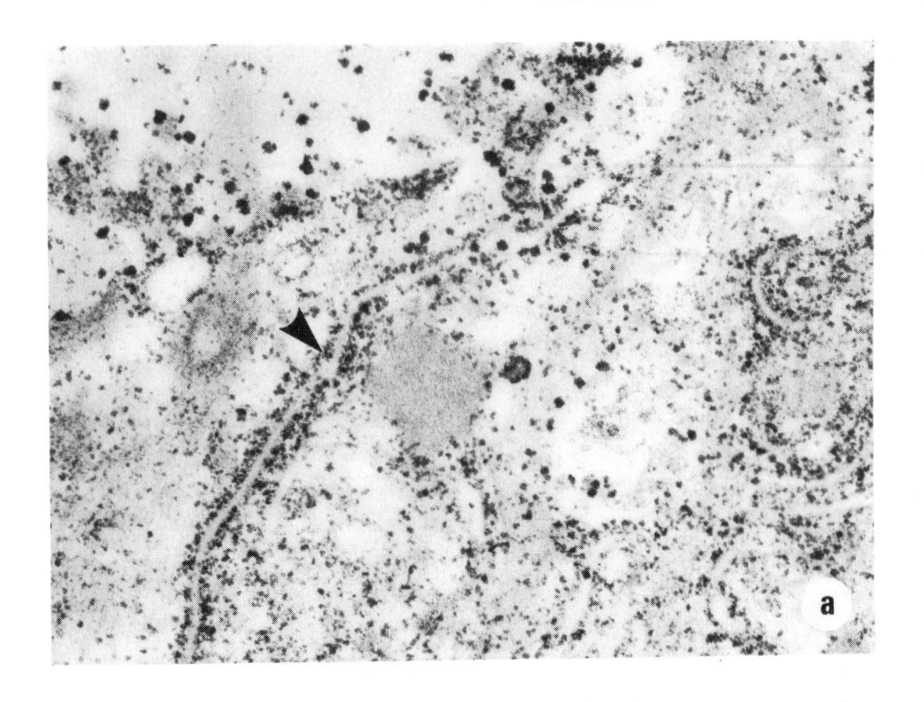

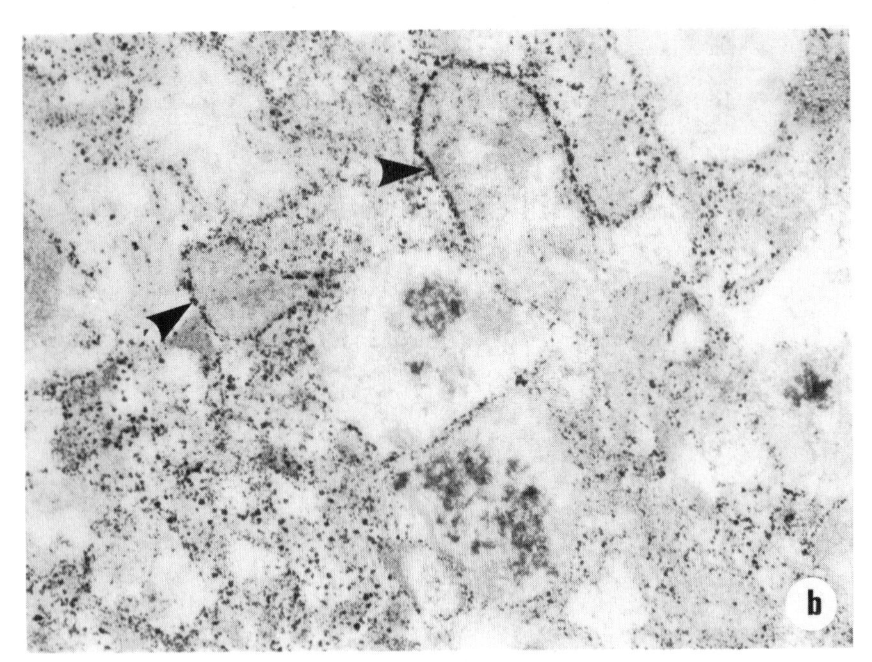

Fig. 5. An ATPase reaction in the hindgut of *Armadillo officinalis*. Note the abundant reaction product (arrows) (a) in the infoldings (× 52 000), and (b) around the mitochondria (× 37 000).

E. The Reproductive System

1. The Marsupium

This structure, the brood pouch, found in many crustacean groups, is undoubtedly one of the main factors that contributed greatly to the isopods' successful emergence on land. Enclosed by the sternal matrix from above and by a membrane of five pairs of oostegites which branch off the sternite from below (Patane, 1940), it is formed of two layers of syncital plasma containing large, vacuolated cells that are possibly endocrine in function. From the sternal matrix certain evaginations ("cotyledons") penetrate into the marsupial cavity (three per somite—the first somite has only one cotyledon; Patane, 1940). The cotyledons degenerate after the second brood (Schöbl, 1880). The marsupial cavity is also filled with an opaline fluid mixed with crystals (Patane, 1940), and a mucus mass surrounding the eggs and embryos (Akahira, 1956). This fluid is apparently of limited nourishment value and egg development can be completed even when the mother remains without food (Verhoeff, 1917a). The crystals found in the marsupial fluid are probably the product of catabolism by the developing embryos (Patane, 1940). This interesting structure merits much more research.

2. The Eggs

The egg contains yolk of two kinds: (1) the mitochondria yolk which is rather diffuse around the nucleus but later disperses to the periphery; (2) the Golgi yolk which is evenly distributed (King, 1926). Bilinski (1979) also distinguishes two kinds of yolk, one arising from autosynthesis and the other from micropinocytosis. The disc-shaped bodies which occur in large numbers in the endoplasmic reticulum are precursors of the intra-oocytic yolk.

During the first stage of vitellogenesis the exogenous yolk penetrates the oocytes and villi are formed in both the follicles and the oocytes (in *Idotea baltica*; see Souty, 1980). During the second stage of vitellogenesis granular material is synthesized which will later form the chorion (Souty, 1980).

The follicle cells are not nutritive but the ovarium epithelium is rich in neutral mucopolysaccharides (Lane, 1980). These follicle cells degenerate shortly before the egg laying (in *Armadillidium vulgare*: Souty, 1980).

Vitellogenin is synthesized by the fat body (in *Porcellio dilatatus*), and stored in special cells of the hepatopancreas (Picaud, 1980). The main vitellogenin proteins found in the ovary, fresh eggs and the haemolymph are two slow-moving glycoproteins (Munuswamy and Subramoniam, 1980). They disappear during the embryonic development. A new protein is then formed which may take part in digesting the complex protein.

The eggs that are extruded into the marsupium have one membrane, the chorion (Stromberg, 1964). In *Hemioniscus* the fertilized egg is enclosed in two envelopes (Goudeau, 1976). The embryo secretes five successive sacs or embryonic envelopes. Therefore, during embryonic development the moulting cycle already exists, and ten embryonic stages have been recognized (Goudeau, 1976).

During embryonic development, eggs of *Ligia oceanica* show a progressive increase in ash content (4·4% to 31·6%) and in non-protein nitrogen (from 1·5% to 2·7%). At the same time, a decrease in protein (from 41·8% to 36·8%), in fat (from 48·8% to 27·4%) and in carbohydrates (from 3·5% to 1·5%) was noted (Pandian, 1972). This is accompanied by a decline in energetic values of the developing embryo from 5 956 cal per g dry wt to 4 175 cal per g dry wt. The water content of the hatched juveniles was 72·2%, whereas that of the egg was 76·1%.

We have recently found that in *Hemilepistus reaumuri* the ash content of the eggs in the marsupium could reach 10% (calculated value). The total energy allocated to egg production reached 34% and for larvae 23% (details in Warburg, 1986c).

IV. BEHAVIOURAL ADAPTATIONS IN ISOPODS

Behavioural adaptations are to our knowledge amongst the most important mechanisms enabling isopods to survive under various terrestrial conditions. However, despite the extensive research on the behavioural responses in isopods (since last reviewed in Warburg, 1968b), we still know surprisingly little about the behavioural patterns of this group and only 38 species have been studied so far (Table 1 here, and Table 7 in Warburg, 1968b). Simple behavioural mechanisms enable the isopods to perceive and respond to

Table 1

Isopod species in which some aspect of behaviour was studied (for earlier work, see Table 7 in Warburg, 1968b).

Species	Family	Reference
(1) *Excirolana chiltoni* Richardson	Cirolanidae	Enright (1972, 1976), Klapow (1972)
(2) *Eurydice pulchra*		Hastings (1981a,b), Hastings and Naylor (1980)
(3) *E. inermis*		Macquart-Moulin (1980)
(4) *Pseudaega punctata* Thomson		Fincham (1973)
(5) *Sphaeroma serratum* (F.)	Sphaeromidae	Elkaim *et al.* (1980)
(6) *Exosphaeroma obtusum* (Dana)		Fincham (1974)

Table 1—*Continued*

Species	*Family*	*Reference*
(7) *Gnorimosphaeroma oregonensis* (Dana)		Standing and Beatty (1978)
(8) *Idotea baltica* (Pallas)	Idotheidae	Hørlyck (1973)
(9) *I. granulosa* Rathke		Hørlyck (1973)
(10) *Tylos punctatus* Holmes & Gay	Tylidae	Holanov and Hendrickson (1980)
(11) *T. granulosus* Miers		Imafuku (1976), Kensley (1972), Ondo (1958, 1959)
(12) *T. latreille* Audouin		Mead and Mead (1972/73), present study
(13) *Ligia exotica* (Roux)	Ligiidae	Farr (1978)
(14) *L. italica* Fabricius		Perttunen (1963), present study
(15) *L. oceanica* (L.)		Alexander (1977)
(16) *Ligidium hypnorum* (Cuvier)		Risler (1978)
(17) *Oniscus asellus* L.	Oniscidae	Kuenen (1959), Schäfer (1982a,b), Warburg (1964)
(18) *Porcellio scaber* Latreille	Porcellionidae	Kuenen and Nooteboom (1963), Lindqvist (1968, 1972), Warburg (1964), Ludwig (1978)
(19) *P. olivieri* (Audouin & Savigny)		Warburg (1968b), present study
(20) *P. obsoletus* B-L.		Present study
(21) *Tracheoniscus rathkei* (Brandt)		Lindqvist (1968, 1972)
(22) *Metoponorthus sexfasciatus* B-L.		Mead *et al.* (1976)
(23) *Hemilepistus reaumuri* (Audouin & Savigny)		Bodenheimer (1935), Warburg (1968b), Seelinger (1977, 1983), Linsenmair (1983)), Hoffmann (1983a,b), present study
(24) *H. aphganicus* Borutzky		Schneider (1971)
(25) *Armadillidium vulgare* Latreille	Armadillidiidae	Kuenen (1959), Warburg (1964), Paris (1965), Lindqvist (1968), Takeda (1980)
(26) *Armadillo officinalis* Dumeril	Armadillidae	Warburg (1968b), Warburg and Berkovitz (1978a,b)
(27) *A. albomarginatus* Dollfus		Warburg (1968b), present study
(28) *A. tuberculatus* Vogl		Present study
(29) *Venezillo arizonicus* Mulaik & Mulaik		Warburg (1964, 1968b)
(30) *Schizidium festai* Verhoeff		Present study

environmental factors such as light, humidity and moisture, chemical stimuli and temperature. More complicated behavioural stimuli were recently described. These include the ability to orient their movements and navigate towards their burrow (Hoffmann, 1984b) as well as maintain a social structure based on the family unit (Linsenmair, 1984).

A. Receptors and Perception

Most of the earlier work was concerned with this aspect of simple responses in isopods: these studies have been reviewed previously in Warburg (1968b).

1. Chemoreception

Work by Kuenen and Nooteboom (1963) has established that *Oniscus asellus, Porcellio scaber* and *Armadillidium vulgare* respond to their own specific smells. Chemoreceptors have recently been located on the antennae. Thus, Alexander (1977) has shown that there are 100 specialized sensory hairs, probably chemoreceptors, on the antennal tip of *Ligia oceanica*, some of which open through a sub-terminal orifice. Other receptors described in Alexander (1977) are probably mechanoreceptors. A similar situation has been described in *Metoponorthus sexfasciatus* (Mead, Gabouriaut and Corbièr-Tichané, 1976), in which the chemoreceptors located on the antennae are protected by cuticular structures that possibly also provide protection against excessive evaporation. The receptors on the antennae of *Ligidium hypnorum* and *Porcellio scaber* have been described by Risler (1977, 1978), who attributed similar functions to them.

Schneider (1973) was the first to describe the olfactory receptors on the antennae of *Hemilepistus aphganicus*. These were studied recently in detail by Seelinger (1977, 1983) in *Hemilepistus reaumuri*. They are mechanical, olfactory and gustatory in nature. Two types of receptors are involved in perceiving air-borne odours: butyric acid cells and amine cells. Four other groups are gustatory receptors: sugar cells, calcium cells, amino-acid cells and special cells responding to the aqueous rinse of other isopod specimens. These receptors probably play a large role in maintaining the family bond in these social isopods.

2. Hygroreaction

There is a report of putative temperature or hygroreceptors in *Porcellio scaber, Oniscus asellus* and *Armadillidium vulgare* (Jans and Ross, 1963), but so far no experimental evidence has been provided. On the other hand, hygroreaction or the response of isopods to the humidity of the air, is a well established phenomenon (Warburg, 1964; see review in Warburg, 1968b). This response is affected by the physiological condition of the animal, as well as other environmental factors, largely temperature and light conditions (Warburg and Berkovitz, 1978a).

In the spheromatid *Gnorimosphaeroma oregonensis*, the humidity response depends on the osmotic condition of the animal (Standing and Beatty, 1978). Thus, if the animal is acclimatized to 100% sea water the intensity of the response increases as compared to acclimatization at a lower salt concentration (25% sea water). This could indicate that the hygroreceptors could in fact be osmoreceptors, playing a role in osmoregulation.

In the sand-beach isopod *Tylos punctatus*, the burrows reach down to where sand moisture is 1%. Lower sand moisture causes increased mortality, probably due to dehydration. The depth of the burrow is primarily controlled by requirements for saturation humidity (Holanov and Hendrickson, 1980).

The activity of these animals (*Tylos granulatus*) is largely affected by temperature (20–30°C), low light intensities and humidity, and the nocturnal behaviour is the result of these requirements (Imafuku, 1976) as well as the high tide (Marsh and Branch, 1979).

In *Ligia italica* the moisture conditions of the animals largely affect their behavioural response (Perttunen, 1963). Thus, desiccated animals reverse their normal light reaction from a photonegative to a photopositive one. Similar responses have been observed in other isopods (Warburg, 1964; Lindqvist, 1968). We have recently studied the phenomenon in more detail in *Armadillo officinalis* (Warburg and Berkovitz, 1978a).

The larvae in the marsupium (of *Porcellio scaber* and *Tracheoniscus rathkei*) already seem to respond to humidity, especially if their mother has been previously desiccated (Lindqvist, 1972).

3. Photoreaction

The response to light is to a large extent affected by temperature. Thus, *Armadillidium vulgare* is positively photokinetic at low and medium temperatures irrespective of humidity (Warburg 1964). A similar pattern was observed in *Armadillo officinalis* (Warburg and Berkovitz, 1978b), and a number of other isopods from the Mediterranean region (Table 2). On the other hand, the desert isopod *Venezillo arizonicus* is mostly photonegative except at high temperatures when it became positive photokinetic (Warburg, 1964). Other desert species (*Porcellio olivieri* and *Armadillo albomarginatus*) respond to light at high temperatures in a similar way by switching from a more negative response to a less negative one (Warburg, 1968b; and Table 3). The normally positive photoreacting isopod *Hemilepistus reaumuri* becomes largely photonegative at high temperatures (Warburg, 1968b; and Table 3). On the other hand, moisture may also affect the photoreaction: this was observed in *Tylos latreille*, in which animals became less negatively phototactic when out of water (Table 4).

Table 2
The response of isopods from mesic habitats to light.

	Intensity index for photoreaction of individual animals	Intensity index for photoreaction of groups of 10 animals	Average number of shifts by individual animals
Porcellio obsoletus			
RH 0–5% $T = 20°C$	−86·2	−6·4	4·8
$T = 30°C$	−62·2	−68·8	11·8
RH 95–100% $T = 20°C$	−39·0	−72·2	17·4
$T = 30°C$	−62·8	−72·8	10·8
Schizidium festai			
RH 0–5% $T = 20°C$	−95·6	−10·7	20·2
$T = 30°C$	−62·2	−68·8	4·8
RH 95–100% $T = 20°C$	−34·2	−49·7	18·8
$T = 30°C$	−62·8	−72·8	21·8
Armadillo tuberculatus			
RH 0·5% $T = 20°C$	+27·2	+3·5	7·6
$T = 25°C$	−32·6	+38·6	11·1
RH 95–100% $T = 20°C$	+10·1	−49·3	15·2
$T = 25°C$	−44·6	−32·6	8·1

Positive and negative values indicate positive and negative photoreactions.

Table 3
Response of some desert isopods to light (further details in Warburg, 1964, 1968b).

	Intensity index for photoreaction of individual animals	Intensity index for photoreaction of groups of 10 animals	Average number of shifts by individual animals
Hemilepistus reaumuri			
RH 0–5% $T = 22–25°C$	+22·7	+24·2	54·0
$T = 37–40°C$	−20·5	−53·5	72·3
RH 95–100% $T = 22–25°C$	+17·8	+12·3	32·6
$T = 37–40°C$	−35·5	−60·6	59·8
Porcellio olivieri			
RH 0–5% $T = 22–25°C$	−62·4	−74·6	51·5
$T = 37–40°C$	−15·4	−13·6	68·3
RH 95–100% $T = 22–25°C$	−87·9	−91·4	33·7
$T = 37–40°C$	+9·5	+12·2	54·5
Armadillo albomarqinatus			
RH 0·5% $T = 22–25°C$	−54·8	−58·9	4·7
$T = 37–40°C$	−36·7	−42·7	17·2
RH 95–100% $T = 22–25°C$	−64·2	−70·7	5·3
$T = 37–40°C$	−28·7	−38·0	8·8

Positive and negative values indicate positive and negative photoreactions.

Table 4
The response of *Tylos latreille* to light.

	Intensity index for photoreaction of individual animals	Intensity index for photoreaction of groups of 10 animals	Average number of shifts by individual animals
Response of animals in sea water	−36·43	+5·0	19·0
Response of animals on wet filter paper	−9·47	−21·6	30·3

Positive and negative values indicate positive and negative photoreactions.

Table 5
Responses to thermal gradients in littoral-zone isopods.

Ligia italica

	Temperature (°C)	I Groups (10)	II Individuals (18)
Position counts	11–16	0·5	0·44
	16–21	1·0	3·66
	21–24	7·0	0·77
	24–27	0·0	16·38
	27–30	0·5	8·72

Tylos latreille

		Temperature (°C)	I Groups (10)	II Individuals (13)
Position counts	First temperature gradient	10–16	4·72	18·92
		16–22	3·12	3·92
		22–25	2·07	7·08
	Second temperature gradient	14–20	5·01	19·92
		20–26	1·21	6·61
		26–30	3·77	3·0
Time measurement		10–15	–	19·06
		15–20	–	10·76
		20–25	–	5·85
		25–28	–	8·18

Groups, counts on minute of groups of 10 animals; individuals, counts on minute of individual animals—numbers of runs in brackets. Time measurement of individual animals.

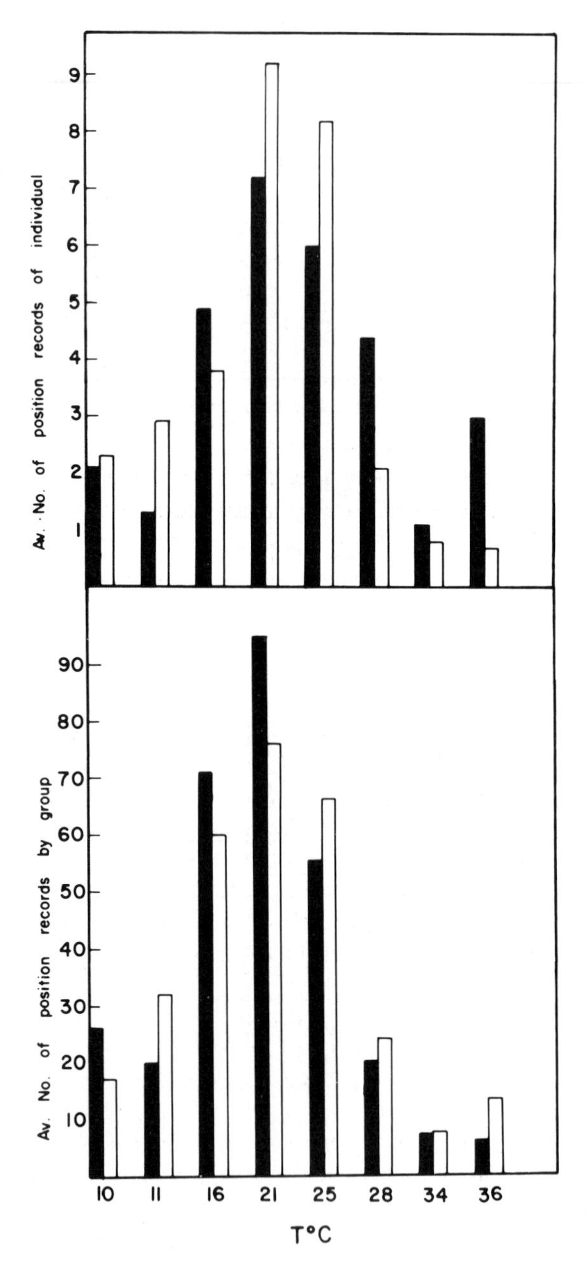

Fig. 6. Thermal responses in *Hemilepistus reaumuri* in a thermopreferendum apparatus as seen in individual animals (top), and groups of 10 animals (bottom). Black bars are females, blanks are males.

4. *Thermoreaction*

Thermoreaction, or the response of isopods to a thermal gradient, is mostly a measure of thermal activity when animals come to rest at a certain "preferred" optimal temperature within a gradient. This range can be different for two isopod species inhabiting the same habitat. Thus, *Ligia italica* and *Tylos latreille*, both littoral zone isopods, differed in their optimal temperatures (Table 5) when studied in similar conditions. In the desert isopod *Hemilepistus reaumuri*, we have again observed a rather different response to temperature. Here the thermal response differs between the male and the female (Fig. 6). Similar observations were reported recently for *Armadillidium vulgare* (Refinetti, 1984).

Increasing temperatures cause increased activity in *Armadillidium vulgare* (Warburg 1964), whereas the desert species *Venezillo arizonicus* responds by lowering its activity at high temperatures. It is possible that acclimatization at different temperatures may affect this behaviour. It has been shown that, when acclimatized for 7–14 days, the lethal temperatures of both *Porcellio laevis* and *Armadillidium vulgare* can also change (Edney, 1964a,b).

B. Orientation

The way in which isopods can orient themselves in their habitat has been studied in recent years for a number of species. These include the beach isopod *Tylos latreille*, which was shown to orientate towards the sea, as it can supposedly perceive both height and relative distances (Mead and Mead, 1972/73). On the other hand, when the waves advance, *Tylos granulatus* turns sideways and moves to the higher part of the shore, thus avoiding the breaking waves, returning afterwards to the surf line (Ondo, 1958). The stimulus for this behaviour is apparently the vibrations caused by the waves. The receptors for this are presumably located on the second basal segment of the antennae (Ondo, 1959).

Another seashore isopod, *Ligia exotica*, tends to aggregate by seeking out conspecific partners. The attraction presumably saves time which would otherwise be spent on seeking out a suitable microhabitat which could help in reducing water loss (Farr, 1978). This phenomenon of aggregation is known in many isopods (reviewed earlier in Warburg, 1968b). An active factor was recently found in the faecal pellets of *Armadillidium vulgare*, and is considered to be responsible for initiating and maintaining the aggregation. The active principle has been shown to occur also in the mid- and hind gut and is presumed to be an aggregation pheromone (Takeda, 1980).

It has been shown that the terrestrial isopods (*Porcellio scaber*) can easily detour small obstacles less than 10 cm large after a brief examination and, approximately, resume their previous route. Larger objects required longer thigmotactic encounter by the isopods, which circle around such objects many times, travelling long distances in vain because of their difficulty in comprehending such obstacles (Ludwig, 1978).

Armadillidium vulgare is known to travel up to 13 m in summer during a day's foraging activity, and shorter distances in winter (Paris, 1965). A more successful navigator appears to be the desert isopod *Hemilepistus reaumuri*, which shows a homing behaviour towards its burrow. It is apparently capable of finding its burrow without reference to landmarks, mostly because of the position of the sun or the polarized light (Hoffmann, 1984b). On the other hand, it locates the exact position of the burrow with the aid of its intrinsic route-searching behavioural pattern (Hoffmann, 1983a). It moves in a spiral-like direction through increasing loops, returning occasionally to the starting point. The average foraging excursion is about 2·6 m, and the average shortest route taken by the isopod to return to its burrow is 1·1 m (Hoffman, 1984a). The longest distance observed was 20 m and the shortest route taken by the isopod was 6 m. The return route stays within 3·7° of the homing direction. There is about a 12% chance that the isopod may overlook the entrance of its burrow. If it misses by as little as 5 cm from the entrance there is about a 17% chance that it will eventually die of water loss, but this remains to be proven.

The capability to negotiate abrupt turns *en route* has been observed in *Oniscus* and *Porcellio* (Schäfer, 1982a,b,). In *Hemilepistus*, chemoreceptors in the terminal segment of the antennae are responsible for the final identification of the burrow, and the isopods must actually touch the edge of the burrow in order to verify it (Hoffmann, 1983b). This is most probably related to a putative "family badge" which would enable each member of the family to identify the burrow (Linsenmair, 1983). A similar phenomenon is also known in another desert isopod, *Hemilepistus aphganicus* (Schneider, 1971). Apparent kin recognition has been demonstrated in *Hemilepistus reaumuri* as well as in another porcellionid from the Canary Islands. Through such a mechanism of kin recognition, the family can be protected against potential intruders (Linsenmair, 1984).

C. Rhythmic Activity

Rhythmic activity has been studied mostly in marine or seashore species of isopods. Thus, the intertidal and beach isopod *Excirolana chiltoni* shows an indigenous tidal rhythm reaching a peak at the tide's crest. Superimposed on this tidal rhythm is an endogenous monthly rhythm of 26–33 days (Enright, 1972; Klapow, 1972). The stimulus which synchronizes the endogenous tidal rhythm is the movement of turbulent waves across the beach (Enright, 1976).

A similar circa-tidal rhythm of spontaneous emergence and swimming activity has also been demonstrated in another intertidal isopod, *Eurydice pulchra* (Hastings and Naylor, 1980; Hastings, 1981a). This rhythm is not affected by the season but it is influenced by temperature changes in the range 10–25°C. It is also affected by a semi-lunar cycle (Hastings, 1981b).

Mechanical agitation mimicking the maximum turbulence occurring at high tide could entrain the endogenous circa-tidal activity rhythm in this species (Hastings, 1981a). Similar phenomena have been described in *Tylos granulatus*, the sand-beach isopod (Kensley, 1972). This isopod emerges at night at low tide only (Marsh and Branch, 1979).

In the southern hemisphere, *Pseudaega punctata*, occupying a similar "niche" as that of *Eurydice*, shows a dual circadian and tidal rhythm which is endogenous (lasting for 10 days under constant laboratory conditions). This swimming rhythm of activity is affected by light as well (Fincham, 1973, 1974).

In the terrestrial oniscoids we have records of rhythmic activity in *Hemilepistus reaumuri* (Bodenheimer, 1935, p. 381; Cloudsley-Thompson, 1956). In the field the activity pattern is related to both light and temperature. Thus in winter (Jan.–Feb.) this isopod is active from 0900–1200 hrs, in spring (March) during most of the day from 0500–1600 hrs. Later during summer it becomes bimodal in its activity from 0400–0800 hrs and again from 1400–1900 hrs until in the late fall (Oct.–Nov.) it is active only between 0400 and 0600 hrs and again between 1500 and 1700 hrs. A similar rhythmic activity correlated with ambient temperature has been described in *Hemilepistus aphganicus* (Schneider, 1971).

It will not be surprising if traces of the rhythmic activity of the seashore isopods will also be observed in various other aspects in more terrestrial isopods.

V. PHYSIOLOGICAL ADAPTATIONS OF ISOPODS

A. The Water Balance

Water is the main factor affecting the survival of isopods on land. Both their distribution and abundance appear to be governed to a large extent by the availability of moisture (Warburg, Linsenmair and Bercovitz, 1984). Thus we find a whole range of isopod species from the littoral zone distributed through mesic habitats and into deserts, showing different degrees of adaptation to water shortage. The subject has been discussed recently in Warburg (1986a,b).

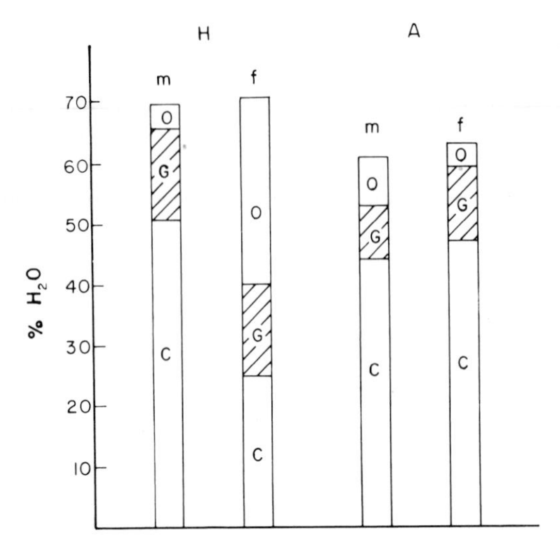

Fig. 7. The distribution of water in body compartments in *Hemilepistus* (H), and *Armadillo* (A), males (m) and females (f). Percentages of water in: C, cuticle and adhering muscle tissue; G, gut, hepatopancreas and gonads; O, other water, mostly haemolymph.

There are comparatively few studies on water uptake in isopods. Some reports indicate that certain species (e.g. *Hemilepistus*), can absorb moisture from the air (Coenen-Stass, 1981). Most isopods inhabiting mesic habitats will take up water from moist surfaces, utilizing their "water channelling" system (Hoese, 1981, 1982a). These cuticular grooves and channels can be best seen in some of the common porcellionid species, but they are also observed in other species.

The water taken up by isopods is distributed throughout the body in such a way that the cuticle normally contains over 50% of the body water, whereas the hepatopancreas, gut and gonads together contain up to 20%. The remaining water is in the haemolymph (Fig. 7). These proportions change throughout the year and differ between the male and the female, as well as between different species (for a more complete discussion see Warburg, 1986b).

Most studies are concerned with ways in which water is lost from the body (see Edney, 1951a, 1957; and reviews, 1967, 1968, 1977). Much of the water evaporates (or transpires) from the body surface. However, the rates at which water is lost varies (Fig. 8) amongst different species. Isopod species inhabiting more terrestrial habitats appear to lose less water than other more mesic species. At the same time, both the air humidity (or the relative amount of moisture in the air) and temperature affect the water loss rate (Fig. 9). Thus both lower humidity and higher temperature enhance the

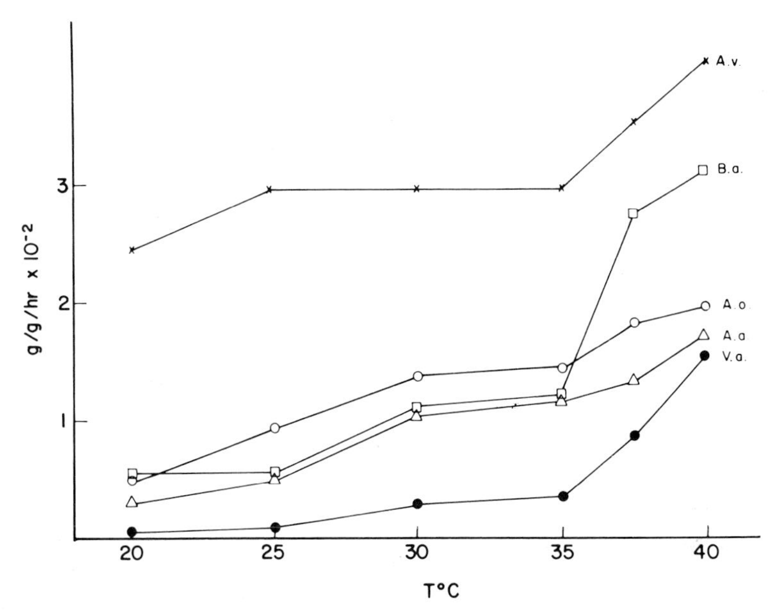

Fig. 8. Water loss rates in some isopods at different temperatures: A.v., *Armadillidium vulgare* (after Warburg 1986a); B.a., *Budellundia albinogrisescens* (after Warburg 1965b); A.o., *Armadillo officinalis* (after Warburg 1986a); A.a., *Armadillo albomarginatus* (after Warburg 1986b); V.a., *Venezillo arizonicus* (after Warburg 1965a).

rates of water loss. The phenomenon is noticeable in most isopod species studied (Warburg, 1965a,b; Edney, 1968) including, to a lesser extent, the desert species (Fig. 9). It seems likely that the water loss rates are governed by the amounts of water either on the surface of the cuticle among the pleopods or in the water-channelling system, as well as through the cuticle from the interior tissues. In the latter event the cuticular structure and the presence or absence of lipids would probably affect the water loss rates (see Fig. 4, and Hadley and Warburg, 1986).

The amount of water in the isopod's body affects the haemolymph's osmolality. The osmotic pressure of the haemolymph is normally comparatively high (Fig. 10). This is apparently a remnant of littoral life, in which most malacostracans show similarly high osmotic values. The ions accounting for this high osmolality are mostly Na^+ (and Cl^-); both haemolymph osmolality and ion (Cl^-) concentrations show some variability throughout the year (Figs 10 and 11), with higher osmotic pressure and ion concentration in the hot and dry summer months and the lower values during the winter (see also Warburg, 1986b). Thus it is not justifiable to consider one single value of water loss rate, but rather a whole range of rates. The same is true for osmolalities. Again, haemolymph osmolalities vary throughout the year,

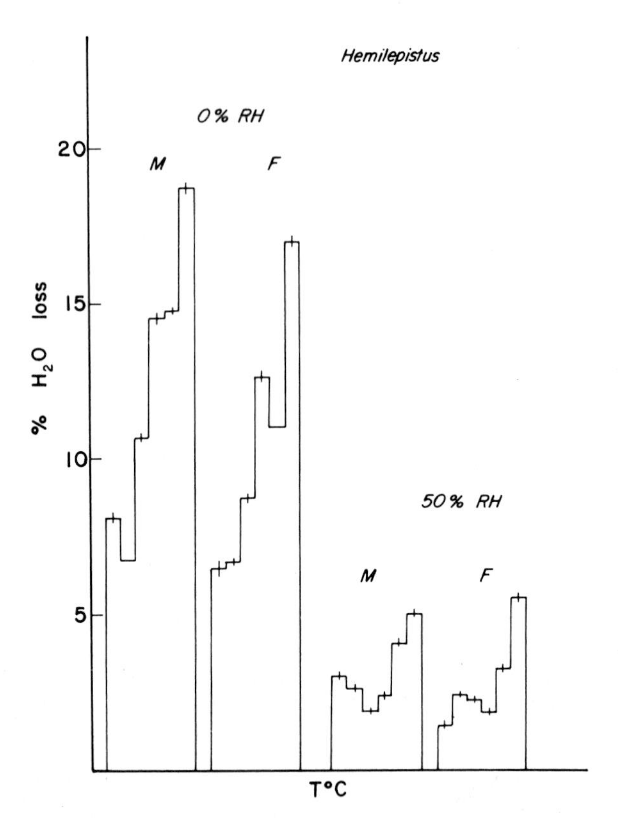

Fig. 9. Percentage water loss of *Hemilepistus reaumuri*: males (M) and females (F), at 0% and 50% RH and different temperatures: 15, 20, 25, 30, 35 and 40°C (each histogram shows a different temperature). Vertical bars are standard errors.

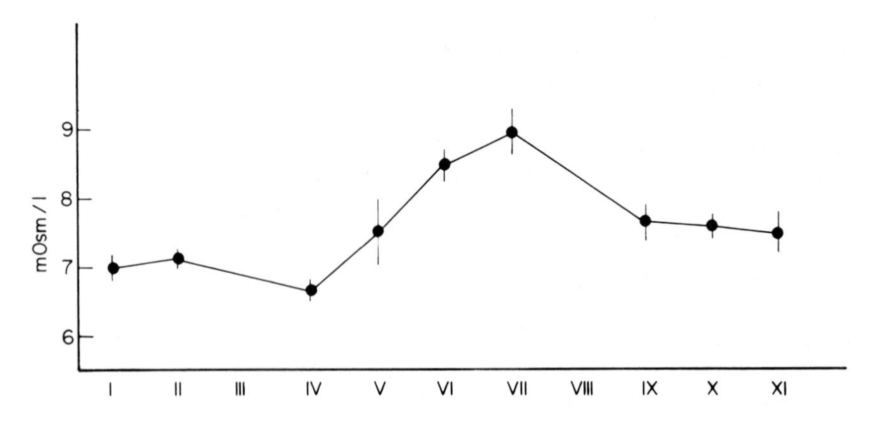

Fig. 10. Haemolymph osmolality of *Armadillo officinalis* throughout the year.

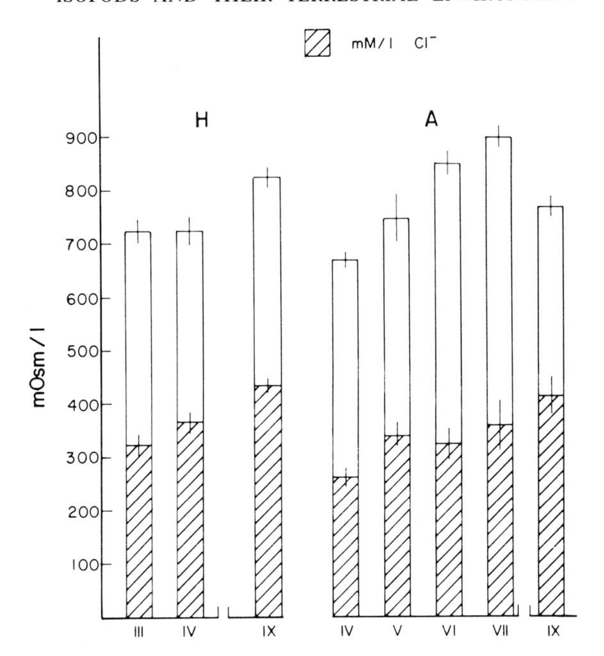

Fig. 11. Haemolymph osmolarity as indicated by ion (Cl^-) concentration, of *Hemilepistus reaumuri* (H), and *Armadillo officinalis* (A) during different seasons.

and the whole range of values should be compared amongst the species. However, some of the mean values are compared in Fig. 12 for different species. Moulting animals and ovigerous females should be excluded from such studies for obvious reasons, especially as we have previously seen here the difference between the sexes in the way water is distributed amongst the various body compartments.

Comparatively little is known about excretion in isopods. Most isopod species studied excrete ammonium (Wieser, 1984), but urea and uric acid have also been demonstrated (Dresel and Moyle, 1950). This mode of ammonotelic excretion is wasteful in water and thus can be utilized only in species with an abundant water supply. In some species ammonium is excreted in gaseous form (Wieser and Schweizer, 1970). There is no comparable study on the more terrestrial species, xeric or arid.

B. The Thermal Balance

Although isopods are generally photonegative they are occasionally found outside in the sun when it is not too hot. Thus it was of interest to see how heat affected their body temperature (Edney, 1951b). The early studies by Edney (1951b, 1953) have already shown that some isopod species (*Ligia*

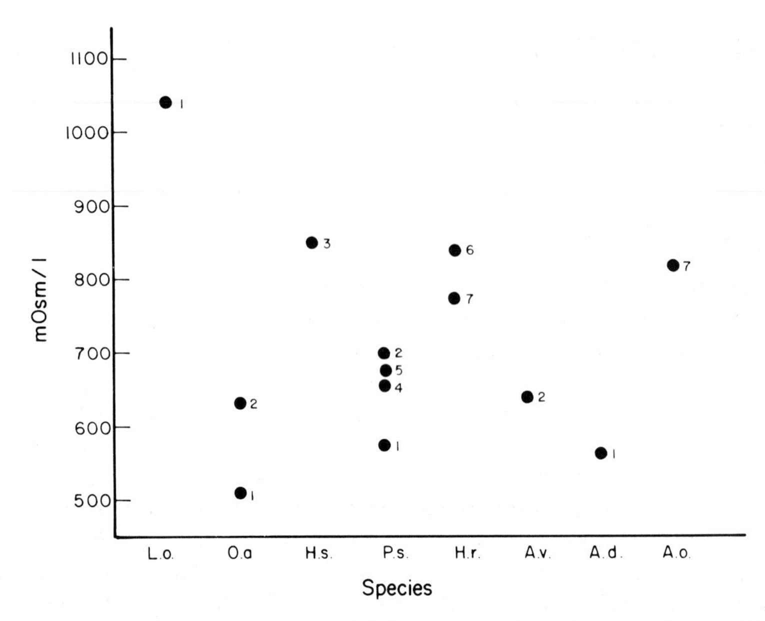

Fig. 12. Haemolymph osmolalities of different isopod species according to different sources.

L.o. *Ligia oceanica*
O.a. *Oniscus asellus*
H.s. *Haloniscus searlei*
P.s. *Porcellio scaber*
H.r. *Hemilepistus reaumuri*
A.v. *Armadillidium vulgare*
A.d. *Armadillidium depressum*
A.o. *Armadillo officinalis*

1 Price and Holdich (1980)
2 Parry (1955)
3 Bayly and Ellis (1969)
4 Lindqvist (1970)
5 Horowitz (1970)
6 Coenen-Stass (1985)
7 Warburg (1986b)

oceanica, Oniscus asellus, Porcellio scaber and *Armadillidium vulgare*) are capable of depressing their body temperature depending on the species and on the ambient temperature (20°C or 37°C). It has also been demonstrated that the body temperature of isopods depends on the amount of moisture in the air and thus the drier the air the easier it becomes to evaporate and cool (Edney, 1951a, 1953, 1956). This was later shown to be true only of some isopod species, whereas others (*Armadillidium vulgare* and *Venezillo arizonicus*) are not capable of depressing their body temperature (Warburg, 1965a). On the other hand, survival of isopods at high temperature is not related to their inability to reduce their body temperature through evaporative cooling. In *Oniscus asellus* and *Porcellio scaber* temperature does not seem to affect the permeability of the cuticle (Mead-Briggs, 1956; Hadley & Quinlan, 1984; respectively).

VI. HOW ISOPODS UTILIZE THEIR FOOD RESOURCES

This aspect of isopodan life has so far been studied in about 20 species (see Table 6).

A. Feeding Habits

Isopods are detritophagous (Gere, 1956, 1962, studying *Protracheoniscus politus*) and omnivorous animals feeding on live and dead leaves, fungi, or live or dead animals (Edney, Allen and McFarlane, 1974). They also feed on their own faecal pellets (Paris, 1963, studies on *Armadillidium vulgare*), and fresh pellets are more attractive than decayed ones (Hassall & Rushton 1982 studying *Porcellio scaber*). Some isopods such as *Hemilepistus reaumuri* are also soil feeders (Shachak, Chapman and Steinberger, 1976). There is evidence that *Armadillidium vulgare* shifts from herbivory during winter and spring when they graze on green leaves, to being scavengers as a result of drought (Paris and Sikora, 1965). Isopods seem to be efficient digesters but somewhat inefficient assimilators. This may be the reason why they also depend to a certain extent on coprophagy, presumably partly in order to extract dissolved nutrients and digestive enzymes as well as minerals such as copper (Wieser, 1965, 1966, 1978; Dallinger and Wieser, 1977). It has recently been shown that some coprophagy is necessary because, when prevented, a decline in growth was noticeable (in *Porcellio scaber*; see Hassall and Rushton, 1982). There is a relationship between the copper content of both litter and soil and that of the isopods (*Trachelipus rathkei* and *Oniscus asellus*) inhabiting the area (Wieser, Busch and Büchel, 1976). The selection of copper in the laboratory is affected by the isopods' own copper deficiency (Dallinger, 1977). Coprophagy may also enable them to increase their fungal infestation, as has been shown by Marikovsky (1969) in *Hemilepistus rhinoceros*. The isopodan gut has been shown to contain spores of fungi naturally (Menon, Tandon and Jolly, 1969).

B. Food Preferences

There is good evidence to show that isopods prefer certain leaves over others (Hassall and Rushton, 1984). Thus, *Armadillidium vulgare* prefers thistle (*Silybum*), tarweed (Rem) and vetch (*Vicia*), both as green or dead leaves (Paris, 1963; Paris and Sikora, 1967), while in Japan it feeds on *Morus* and *Tilia* (Watanabe, 1978). Wieser (1965) found that *Porcellio scaber* feeds on poplar leaves, whereas *Porcellio laevis* feeds on *Morus indica* leaves (Nair, 1976a). Similarly, *Hemilepistus reaumuri* prefers *Hammada* and *Artemisia* in the Negev (Shachak *et al.*, 1976). It has recently been shown that dead

Table 6
Isopod species in which aspects of energetics or population dynamics have been studied.

No.	Species	Family	Energetics	Population dynamics	Reference
(1)	*Idotea phosphorea*	Idotheidae	+		Robertson and Mann (1980)
(2)	*Ligia dilatata*	Ligiidae	+	+	Koop and Field (1980, 1981)
(3)	*L. pallasii*	Ligiidae	+		Carefoot (1973a)
(4)	*L. oceanica*	Ligiidae	+		Pandian (1972)
(5)	*Ligidium hypnorum*	Ligiidae	+	+	Stachurski (1968, 1972, 1974)
(6)	*L. japonicum*	Ligiidae	+	+	Saito (1965)
	L. japonicum	Ligiidae		+	Kato (1976)
(7)	*Trichoniscus pusillus*	Trichoniscidae		+	Sutton (1968, 1970a)
	Trichoniscus pusillus	Trichoniscidae		+	Standen (1970, 1973)
	Trichoniscus pusillus	Trichoniscidae		+	Sunderland *et al.* (1976)
(8)	*Philoscia muscorum*	Onsicidae	+		Hassall (1977, 1983)
	Philoscia muscorum	Onsicidae	+		Hassall and Sutton (1977)
	Philoscia muscorum	Onsicidae		+	Sutton (1968, 1970a)
	Philoscia muscorum	Onsicidae		+	Standen (1970, 1973)
	Philoscia muscorum	Onsicidae		+	Sunderland *et al.* (1976)
	Philoscia muscorum	Onsicidae		+	Davis and Sutton (1977b)
(9)	*P. africana*	Onsicidae		+	Kheirallah (1975)
(10)	*Cylisticus convexus*	Cylisticidae	+		Reichle (1967, 1968)
	Cylisticus convexus	Cylisticidae		+	Hatchett (1947)

cotyledonous leaves enhance growth and fecundity in *Armadillidium vulgare*, whereas moncotyledons cause mortality (Rushton and Hassall, 1983a,b).

Hemilepistus appears to feed on many plant species, both on dry and green parts of the plants. They also feed on lichens or algae. On the other hand, they occasionally feed on dead insects or dead members of their own species as well as on faeces of various animals (goats and other domestic stock, rabbits, rodents and reptiles). In some locations faeces appear to be of some importance for their nourishment. Moreover, they are particular in bringing different food items to their burrows when they feed their offspring (Warburg *et al.*, 1984).

Apparently, the palatability of leaves depends to some extent on their phenolic content. This was shown when four isopod species were tested with 25 species of plants (Neuhauser and Hartenstein, 1978). Preference experiments were also conducted on *Ligia* (Carefoot, 1973a, 1979) feeding on various brown and red algae. Feeding *Oniscus asellus* on agar pellets containing various leaves has shown that growth is better on fresh leaves (Beck and Brestowsky, 1980). It has previously been shown that growth is greatly affected by food quality (Merriam, 1971). Of the artificial food offered, *Armadillidium* thrives best on rabbit food. It has recently been shown that the growth of *Porcellio olivieri* is accelerated, when it is fed on *Hordeum* (Kheirallah and El-Sharkawy, 1981).

C. Food Consumption

Food consumption varies according to the isopod species. Thus it ranges between 1·5% body wt per day in *Armadillidium vulgare* to 3·5% body wt per day in *Metoponorthus pruinosus* (Reichle, 1967, 1968). It also varies according to the diet. *Hemilepistus reaumuri*, when fed on soil, consumes much more (25 mg per individual per day) as compared to 0·8–3·4 mg per individual per day when fed on plants (Shachak *et al.*, 1976).

It has also been shown that the rate of absorption or assimilation varies even among different animals of the same species according to the type of food they consumed (see Table 3 in Crawford, 1981). Thus, *Ligia pallassi* assimilates 78% on a diet of *Ulva* sp., and only 55–76% on a diet of brown algae *Hereocytes* (Carefoot, 1973a). The efficiency of assimilation increases with the concentration of copper ions in the food, particularly in *Porcellio laevis, Porcellio scaber* and *Oniscus asellus* (Dallinger and Wieser, 1977; Debry and Lebrun, 1979). Copper-enriched litter is more easily assimilated and this causes an increased weight in *Oniscus asellus* (Debry and Muyango, 1979). Total nitrogen is not a sole factor in growth or survival; rather, the rate of digestion is important (Rushton and Hassall, 1983a,b).

D. Energy Expenditure

Isopods have also been studied from an energetic approach by several authors (Table 6). It appears that the energy spent on growth as compared with that allocated to reproduction varies with the species. Thus, *Ligia japonicum* used 1·40 kcal for growth and 2·13 kcal for reproduction. The conglobating forms, *Cylisticus convexus* and *Armadillidium vulgare*, differ in the amount of energy allocated to growth and reproduction (Reichle, 1967). Thus, *Cylisticus convexus* spent about 15·4 kcal g^{-1} on growth and 16·4 kcal g^{-1} on reproduction, whereas *Armadillidium vulgare uses 20·4 kcal* g^{-1} for growth and only 7·6 kcal g^{-1} for reproduction. On the other hand, Lawlor (1976a,b) found that reproducing *Armadillidium vulgare* females devote 8·26% more energy to reproduction (and growth) than non-reproductive females. Finally, *Hemilepistus reaumuri* appears to spend similar amounts of energy on growth (52%) and reproduction (see Shachak, 1980).

These different patterns of energy allocation cannot be correlated with reproductive patterns, because both semelparous species (e.g., *Hemilepistus reaumuri*) and iteroparous ones (e.g., *Cylisticus convexus, Armadillidium vulgare* and *Ligia japonicum*) behave in a similar way energetically. We must, however, consider the fact that, for *Hemilepistus reaumuri*, the total energy expenditure on reproduction should also take into account the energy expenditure on brood care, at least during the first few weeks. On the other hand, there are some differences amongst the iteroparous species. Thus, *Ligia japonicum* allocates more energy to reproduction than to growth, whereas for *Armadillidium vulgare* we have seemingly conflicting evidence (Reichle, 1967; Lawlor 1976a,b,).

Finally, there is a report that suggests that isopodan biomass is greater in grassland than in woodland (Davis and Sutton, 1977b). This could imply that isopod numbers are greater due to increased reproduction or that growth is faster, thus allowing for rapid turnover. More research is needed in that direction, preferably of a comparative nature, between grassland and woodland isopods.

VII. ISOPOD HABITAT SELECTION AND DISPERSAL PATTERNS

A. Habitat Selection

Climate and edaphic factors are the most important ones for the isopod's distribution on land (the subject has recently been reviewed extensively in Warburg *et al.* 1984). Thus both temperature and light, as well as the amount of precipitation, are of great significance in determining the activity and dispersal of isopods.

Temperature and moisture conditions seem to be the main factors controlling the "vertical distribution" of isopods. This term describes here the movement of isopods up the trees and down into the ground, which has been observed in various northern hemisphere isopods, mostly European, by several authors (Heeley, 1941; Cole, 1946; Brereton, 1957; Den Boer, 1961; Paris, 1963). Whether temperature affects this pattern of migration or soil moisture appears to depend both upon the edaphic condition and the species. Lower soil water content causes increased aggregation in *Porcellio scaber* Paris, 1963). In the desert isopod *Hemilepistus reaumuri*, soil moisture controls the depth of the burrow (Shachak, 1980). The continuous dehydration of the soil during summer stimulates the isopod to dig deeper burrows. The selection of suitable burrowing sites is to a certain extent dependent upon geological formation and slope aspect (Brown and Steinberger, 1983). In *Ligidium japonicum*, vertical migration is also affected by season (Saito, 1965).

B. The Role of Climatic Factors in the Distribution of Isopods

The early studies by Herold (1925), Verhoeff (1931) and Miller (1938), have already shown the importance of environmental factors on the distribution of the various isopod species within the habitat. It seems that a number of species "prefer" a moister microhabitat, whereas others "prefer" a somewhat drier one. Later studies have shown similar patterns of distribution in a variety of habitats and isopod species (Brereton, 1957; Beyer, 1957/58, 1964; Davis and Sutton, 1977a; Chelazzi and Ferrara, 1978; Warburg, Rankevich and Chasanmus, 1978; Kheirallah, 1980a; Watanabe, 1980). The subject has been extensively reviewed in Warburg *et al.* (1984).

Recently (Warburg *et al.*, 1984), in a detailed analysis of the Mediterranean region in northern Israel, we have found 15 species sharing one habitat (apparently isopods are capable of peaceful coexistence in one place; Schneider and Jakobs, 1977). Some of these species reach their population peak during the winter months (*Philoscia*), whereas others are more abundant during early (*Metoponorthus*) or late (*Armadillo*) spring (Warburg *et al.*, 1984). It turned out that the precipitation pattern had some effect on this pattern of abundance of the various species, probably because of their different soil moisture requirements. This effect was also apparent during the following year.

VIII. REPRODUCTIVE PATTERNS, ENERGETICS AND STRATEGIES

A. Ovarian Oocytes and Marsupial Eggs

Oocytes developing in the ovaries vary in size and number, both in individual females of the same species as well as between species (Table 7; see also Warburg, 1986c). There appears to be a linear relationship between the number of eggs, embryos or larvae in the brood, and the mother's weight (Phillipson and Watson (1965) for *Oniscus asellus*, and Snider and Shaddy (1980) for *Trachelipus rathkei*). The eggs leave the ovaria via the oviduct through a string-like tube into the marsupium (Schöbl, 1880). In the marsupium, some eggs fail to develop and these are found frequently together with the embryos (in *Cylisticus convexus* by Hatchett (1947) and in *Porcellio laevis* by Nair (1978a), as well as our own observation in *Porcellio obsoletus, Hemilepistus reaumuri* and *Armadillo officinalis*—Fig. 13).

The larvae that remain in the brood pouch feed on the eggs that do not hatch (in *Ligia oceanica*; Saudray and Lemercier, 1960). It is not unusual to see large members of the brood feeding on the smaller ones; thereby part of the brood provides food for the rest (in *Cylisticus convexus*; see Hatchett, 1947).

The duration of time the larvae remain in the brood pouch varies from 3 days (in *Philoscia muscorum*) to 9 days (in *Porcellio dilatatus*) (see Heeley, 1941), or 16 days (in *Porcellio scaber*; see Verhoeff, 1917a). Longer periods are given for various species, reaching over 65 days in *Oniscus murarius* (Verhoeff, 1920).

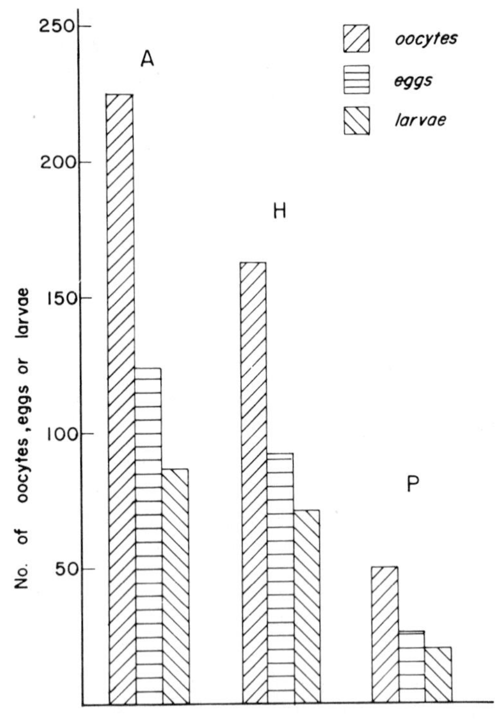

Fig. 13. Number of oocytes, eggs and larvae of *Armadillo officinalis* (A), *Hemilepistus reaumuri* (H), and *Porcellio obsoletus* (P).

B. Gestation and Reproductive Period

Isopods generally breed in spring and early summer (in the northern hemisphere), but breeding is also known in late summer and fall. Thus, *Metoponorthus pruinosus* breeds several times starting in mid-May (mating takes place at the end of April) until mid-November (Shimoizumi, 1952). In *Ligidium hypnorum* the gestation period is between May and August (Tomescu, 1973). In *Porcellio dilatatus* gestation lasts 2·5 months (Mocquard *et al.*, 1976) whereas, in India, *Porcellio laevis* breed between February and October depending on the monsoon rains (Nair, 1976a). In some species (such as *Eluma purpurascens*) breeding starts at the age of 2 years (Juchault *et al.*, 1980a). Later breeding can take place in other isopods.

C. Factors Affecting Breeding

Egg-laying (in *Metoponorthus pruinosus*) is related to moulting and starts 3 hrs after its completion (Shimoizumi, 1952). Steel (1980) found (in *Oniscus asellus, Porcellio spinicornis* and *Cylisticus convexus*) that a trigger initiates moulting and that vitellogenesis starts during the intermoult phase. The second phase of vitellogenesis is controlled by ecdysone. The endocrine factors that control reproduction are synthesized by the nervous tissue and are regulated by both photoperiodicity and temperature (Legrand *et al.*, 1982). In *Armadillidium vulgare* ovarian maturation is accelerated by the presence of males (Jassem, Juchault and Mocquard, 1982).

Several external factors affect reproduction. Thus when *Armadillidium vulgare* females are in constant light conditions they start reproducing; higher temperatures accelerate this process. Long day-length increases the duration of the reproductive period (Mocquard, Pavese and Juchault, 1980; Juchault, Mocquard and Legrand, 1981). Apparently, low light intensity can also initiate reproduction in *Armadillidium vulgare*, which is capable of integrating light intensities up to $2{\cdot}3 \times 10^{-6}$ erg/cm^{-2} (Jassem, Mocquard and Juchault, 1981). The difference between day and night is perceived by the female as a relative difference in light intensities (Jassem, Mocquard and Juchault, 1982).

In *Armadillidium vulgare* there is also a clear connection between reproductive pattern and the latitude at which the original population of isopods was obtained (Juchault, Pavese and Mocquard, 1980b). This may be related to some extent to temperature and light differences which are known to affect breeding in *Oniscus asellus* (McQueen and Steel, 1980). The subject has been reviewed recently (see Warburg *et al.*, 1984).

Table 7
Breeding patterns in isopods.

No.	Species	Author, year	Family	Breeding season[a]	Number of Broods per year	Duration of gravidity (days)	Numbers in marsupium Eggs	Numbers in marsupium Embryos	Number of larvae	Source
(1)	*Idotea emarginata*	Fabricius	Idotheidae	XII–X	1–2					Naylor (1955)
(2)	*Sphaeroma serratum*	Fabricius	Sphaeromoidae		1–2					Raimond and Juchault (1983)
(3)	*Dynamene bidentata*		Sphaeromoidae	V–VII	1					Holdich (1968)
(4)	*Jaera albifrons*	Sars 1859	Jaeridae	III–VIII		30				Holdich (1968)
(5)	*J. forsmani*		Jaeridae	IV–X			20			Jones and Naylor (1971)
(6)	*J. praehirsuta*		Jaeridae	III–XI						Jones and Naylor (1971)
(7)	*J. ischiosetosa*		Jaeridae	IV–VIII						Jones and Naylor (1971)
	J. ischiosetosa		Jaeridae	V–VIII	2		7–30			Steele and Steele (1972)
(8)	*Asellus aquaticus*	Linne 1761	Asellidae	IV–X	1		21–147			Steel (1961)
(9)	*A. intermedius*	Forbes 1876	Asellidae	IV–VI	1		20–250			Ellis (1961)
(10)	*A. meridianus*	Rocovitza	Asellidae	IV–X	1		1–40			Steel (1961)
(11)	*A. tomalensis*	Harfood 1877	Asellidae	III–VI	1		12–58			Ellis (1961)
(12)	*Mancasellus macrourus*	Garman 1890	Asellidae	III–VI			81–248			Markus (1930)
(13)	*Tylos punctatus*	Holmes and Gay 1909	Tylidae	VI–VIII	1	73	100	4–20 (Av. 13·6)		Hamner *et al.* (1969)
(14)	*T. latreille*	Audouin 1825	Tylidae	V–VII			10–38			Vandel (1960)
(15)	*Ligia oceanica*	(Linne 1767)	Ligiidae		–		73			Saudray (1954) Saudray and Lemercier (1960)
	Ligia oceanica	(Linne 1767)	Ligiidae	Spring	1–2	40–100 (Av. 80)				Nicholls (1931)
	Ligia oceanica	(Linne 1767)	Ligiidae	I–X peak VII			17–35 (Av. 27)			Vandel (1960)
	Ligia oceanica	(Linne 1767)	Ligiidae	V–VIII						Meinertz (1950)
	Ligia oceanica			V–VIII	>1					Jöns (1965)
	Ligia oceanica						29–63 (Av. 53)			Pandian (1972)
(16)	*L. dilatata*	Brandt and Stimpson (1897)	Ligiidae		1	35–42				Koop and Field (1980)
(17)	*L. pallasii*	Brandt (1833)	Ligiidae	V–VII			48			Carefoot (1973b)

(18)	*Ligidium hypnorum*	Cuvier (1792)	Ligiidae	V–IX Peak VII			6–21 (Av. 12)	Meinertz (1950, 1951)
	Ligidium hypnorum	Cuvier (1792)	Ligiidae	VI–IX		42	7–19	Beyer (1957/58)
	Ligidium hypnorum	Cuvier (1792)	Ligiidae				7–24 (Av. 14.4)	Herold (1960)
	Ligidium hypnorum	Cuvier (1792)	Ligiidae	V–VIII	2	40–50		Tomescu (1973)
	Ligidium hypnorum	Cuvier (1792)	Ligiidae				10–15	Stachurski (1972)
	Ligidium hypnorum	Cuvier (1792)	Ligiidae	VI–VIII			Av. 9·12	Krumpál (1976)
(19)	*L. japonicum*	Cuvier (1792)	Ligiidae	V–VI			13–39	Saito (1965)
	L. japonicum	Cuvier (1792)	Ligiidae	V–X			13–61	Kato (1976)
(20)	*Trichoniscus pusillus*	Brandt 1833	Trichoniscidae	II–IV	2	37·5	4–7 (Av. 5·5)	Heeley (1941)
	Trichoniscus pusillus	Brandt 1833	Trichoniscidae	V–IX				Meinertz (1951)
	Trichoniscus pusillus	Brandt 1833	Trichoniscidae				4–12	Beyer (1957/58)
	Trichoniscus pusillus	Brandt 1833	Trichoniscidae				2–13 (Av. 6·2)	Herold (1960)
	Trichoniscus pusillus	Brandt 1833	Trichoniscidae	IV–VII				Sutton (1968)
	Trichoniscus pusillus	Brandt 1833	Trichoniscidae	V–IX	2–3	30–35		Tomescu (1973)
	Trichoniscus pusillus	Brandt 1833	Trichoniscidae		1–2		5–20	Standen (1973)
(21)	*Hyloniscus riparius*	(C. L. Koch 1838)	Trichoniscidae	IV–IX	3		9–32	Beyer (1957/58)
	Hyloniscus riparius	(C. L. Koch 1838)	Trichoniscidae				4–35 (Av. 13·2)	Herold (1960)
	Hyloniscus riparius	(C. L. Koch 1838)	Trichoniscidae	IV–IX			Av. 9·1	Krumpál (1976)
(22)	*Haplophthalmus danicus*	B.-L. (1879)	Trichoniscidae	V–IX				Meinertz (1950)
(23)	*Platyarthrus hoffmanseggi*	Brandt 1833	Squamiferidae	VII–IX			2–9 (Av. 5–6)	Vandel (1960) Meinertz (1950)
(24)	*P. schöbli*	B.-L. (1879)	Squamiferidae	IV–VIII			2–4	Vandel (1962)
(25)	*Philoscia muscorum*	Scopoli (1763)	Onsicidae	V–VIII		21·5	13	Heeley (1941)
	Philoscia muscorum	Scopoli (1763)	Onsicidae	VI–VIII				Meinertz (1950)
	Philoscia muscorum	Scopoli (1763)	Onsicidae	IV–IX Peak VII			6–54 (Av. 24)	Vandel (1962)
	Philoscia muscorum	Scopoli (1763)	Onsicidae	VI–VIII			8–69	Sutton (1968)
	Philoscia muscorum	Scopoli (1763)	Onsicidae		1	35		Sunderland *et al.* (1976)
(26)	*Oniscus murarius*	Cuvier 1792	Onsicidae	V		58–65		Verhoeff (1917a)

continued

Table 7 *continued.*

No.	Species	Author, year	Family	Breeding season[a]	Number of Broods per year	Duration of gravidity (days)	Numbers in marsupium Eggs	Numbers in marsupium Embryos	Number of larvae	Source
(27)	*Oniscus asellus*	Linne 1758	Onsicidae	IV–VIII	2	31·5	30			Heeley (1941)
	Oniscus asellus	Linne 1758	Onsicidae	V–IX		41–45	45			Beyer (1957/58)
	Oniscus asellus	Linne 1758	Onsicidae	II–IX			13–17	11	3	Phillipson and Watson (1965)
	Oniscus asellus	Linne 1758	Onsicidae	VI–IX						Meinertz (1950, 1951)
(28)	*Haloniscus searli*	Chilton (1920)	Onsicidae	X–III				42		Ellis and Williams (1970)
(29)	*Cylisticus convexus*	De Geer (1778)	Cylisticidae	V–VIII	2					Verhoeff (1917a)
	Cylisticus convexus	De Geer (1778)	Cylisticidae	VI–VIII						Meinertz (1950)
	Cylisticus convexus	De Geer (1778)	Cylisticidae		2–3	44–62 (Av. 52)	10–70 (Av. 33)	10–40 (Av. 24)		Hatchett (1947)
(30)	*Protracheoniscus politus*	C. L. Koch (1841)	Porcellionidae				6–31 (Av. 16·6)			Herold (1960)
	Protracheoniscus politus	C. L. Koch (1841)	Porcellionidae		1–3					Radu and Tomescu (1971)
(31)	*Trachelipus rathkei*	Brandt (1833)	Porcellionidae		1–2					Verhoeff (1917a)
	Trachelipus rathkei	Brandt (1833)	Porcellionidae		1–3	35–42 (Av. 39)		6–29 (Av. 17)		Hatchett (1947)
	Trachelipus rathkei	Brandt (1833)	Porcellionidae				14–65 (Av. 26·8)			Herold (1960)
	Trachelipus rathkei	Brandt (1833)	Porcellionidae	VI–IX		45–48	12–40 (Av. 35)			Beyer (1957/58)
	Trachelipus rathkei	Brandt (1833)	Porcellionidae	IV–IX			(Av. 10·3)			Krumpál (1976)
	Trachelipus rathkei	Brandt (1933)	Porcellionidae	VI	1–3	17·6–51·4	2–25 (Max. 60)			Snider and Shaddy (1980)
(32)	*T. ratzeburgi*	Brandt (1833)	Porcellionidae				17–55 (Av. 34·4)			Herold (1960)
	T. ratzeburgi	Brandt (1833)	Porcellionidae	VI–VIII						Vandel (1962)
(33)	*T. balticus*	Verhoeff (1907)	Porcellionidae	V–VIII	2					Verhoeff (1917a)
(34)	*Porcellium conspersum*	C. L. Koch (1844)	Porcellionidae	VII			9–32 (Av. 18·9)			Herold (1960)

(35)	*Metoponorthus pruinosus*	Brandt (1833)	Porcellionidae	III–IX			15–60 (Av. 30)			Vandel (1962)
	Metoponorthus pruinosus	Brandt (1833)	Porcellionidae	VI–X						Meinertz (1950)
	Metoponorthus pruinosus	Brandt (1833)	Porcellionidae	III–X	3–6		3–35			Menon *et al.* (1969, 1970)
	Metoponorthus pruinosus	Brandt (1833)	Porcellionidae		4–5	16	105	91		Shereef (1970)
	Metoponorthus pruinosus	Brandt (1833)	Porcellionidae	IV	3			14–26		El-Kifl *et al.* (1970)
(36)	*Lepidonsicus minutus*	C. L. Koch (1838)	Porcellionidae				6–19 (Av. 11·2)			Herold (1960)
(37)	*Leptotrichus naupliensis*	Verhoeff (1901)	Porcellionidae		4–6	18	79	67		Shereef (1970)
	Leptotrichus naupliensis	Verhoeff (1901)	Porcellionidae	IV	3			14–26		El-Kifl *et al.* (1970)
(38)	*L. panzeri*	Audouin (1825)	Porcellionidae	IV–X			13–36			Kheirallah (1980b)
(39)	*Porcellio scaber*	Latreille 1804	Porcellionidae						12–30	Collinge (1946)
	Porcellio scaber	Latreille 1804	Porcellionidae	II–III	3	49–72 (Av. 35) Max. 102				Verhoeff (1917a)
	Porcellio scaber	Latreille 1804	Porcellionidae	III–VII	2	33·5	24			Heeley (1941)
	Porcellio scaber	Latreille 1804	Porcellionidae		1–3	36–37 (Av. 43·5)		24	6–42 (Av. 24)	Hatchett (1947)
	Porcellio scaber	Latreille 1804	Porcellionidae	III–X Peak VI			12–90 (Av. 52)			Meinertz (1950, 1951)
	Porcellio scaber	Latreille 1804	Porcellionidae	V–IX		41–45	Av. 45			Beyer (1957/58)
	Porcellio scaber	Latreille 1804	Porcellionidae				26–96 (Av. 50·4)			Herold (1960)
	Porcellio scaber	Latreille 1804	Porcellionidae	III–VIII						Wieser (1963)
(40)	*P. pictus*	Brandt 1833	Porcellionidae	V		68				Verhoeff (1917a)
	P. pictus	Brandt 1833	Porcellionidae	V–VIII						Meinertz (1950)
(41)	*P. dilatatus*	Brandt 1883	Porcellionidae	II–VIII		46·5	17·5			Heeley (1941)
	P. dilatatus	Brandt 1883	Porcellionidae	VI–VIII						Meinertz (1950)
	P. dilatatus	Brandt 1883	Porcellionidae	III–X		25–60				Tomescu (1973)
(42)	*P. laevis*	Latreille 1804	Porcellionidae	IV–X Peak V			48–112 (Av. 55)			Meinertz (1950, 1951)
	P. laevis	Latreille 1804	Porcellionidae	III–IV	4–6	15	91	81		Mahmoud (1954)
	P. laevis	Latreille 1804	Porcellionidae	III	2			14–26		Shereef (1970), El-Kifl *et al.* (1970)
	P. laevis	Latreille 1804	Porcellionidae	II–X		22–26			50–120	Nair (1976a)

continued

Table 7 *continued.*

No.	Species	Author, year	Family	Breeding season[a]	Number of Broods per year	Duration of gravidity (days)	Numbers in marsupium Eggs	Numbers in marsupium Embryos	Number of larvae	Source
(43)	*P. monticola*	Lereboullet 1853	Porcellionidae				29–81 (Av. 49·3)			Herold (1960)
(44)	*P. olivieri*	Aud. et Sav. 1825–6	Porcellionidae						24–38	Kheirallah and El-Sharkawy (1981)
					1–2				8–32	Kheirallah and Awadallah (1981),
									45	El-Kifl *et al.* (1970)
	P. olivieri	Aud. et Sav. 1825–6	Porcellionidae	III–IV	1	Av. 50	35–108			Warburg *et al.* (1984)
(45)	*P. spinicornis*	Say 1818	Porcellionidae		1–2			18–65		McQueen and Carnio (1974) McQueen (1976a)
(46)	*P. rathkei*	Brandt 1833	Porcellionidae	VI–IX						Meinertz (1950)
(47)	*P. obsoletus*	B.-L.	Porcellionidae	II–VI, X–XII	2		26		20	Present study
(48)	*Hemilepistus reaumuri*	Aud. et Sav. 1825	Porcellionidae	IV–V	1		30–150 (Av. 92)			Shachak (1980)
	Hemilepistus reaumuri	Aud. et Sav. 1825	Porcellionidae	III–V	1	40–50	1·0	95	71	Present study
(49)	*Periscyphis granai*	Arcangeli 1929	Eubelidae	I–III		22				Kheirallah (1979a)
(50)	*Eluma purpurascens*	B.-L. 1879	Armadillidiidae	VI–VII	1					Juchault *et al.* (1980a)
(51)	*Schizidium festai*	Verhoeff 1923	Armadillidiidae	II–V	1			60–228 (Av. 124)	76–128 (Av. 92)	Warburg *et al.* (1984)
(52)	*Armadillidium vulgare*	Latreille 1804	Armadillidiidae						50–150 (Max. 267)	Collinge (1946)
	Armadillidium vulgare	Latreille 1804	Armadillidiidae	IV–IX (Peak VII)					48–156	
	Armadillidium vulgare	Latreille 1804	Armadillidiidae	V–VII		33	113		39–175 (Av. 113)	Heeley (1941)
	Armadillidium vulgare	Latreille 1804	Armadillidiidae			37–51 (Av. 43)		28	5–62 (Av. 28, max. 139)	Hatchett (1947)
	Armadillidium vulgare	Latreille 1804	Armadillidiidae			48	9–75			Beyer (1957/58)
	Armadillidium vulgare	Latreille 1804	Armadillidiidae		2–3		30–40			Warburg (1965c)

	Armadillidium vulgare	Latreille 1804	Armadillidiidae		1–2	32			Lawlor (1976b)
	Armadillidium vulgare	Latreille 1804	Armadillidiidae			40–50	22–60		Al-Dabbagh and Block (1981)
(53)	*A. cinereum*	Zencker	Armadillidiidae	IV–VIII					Meinertz (1950)
(54)	*A. pictum*	Brandt 1833	Armadillidiidae	VI–VII					Meinertz (1950)
(55)	*A. pulchellum*	Zencker	Armadillidiidae	VI–VII					Meinertz (1950)
(56)	*A. opacum*	C. L. Koch (1844)	Armadillidiidae	VI–VII					Meinertz (1950)
	A. opacum	C. L. Koch (1844)	Armadillidiidae		1				Verhoeff (1917a)
(57)	*A. rhenanum*	Verhoeff	Armadillidiidae				7–31 (Av. 11·9)		Herold (1960)
(58)	*Armadillo officinalis*	Dumeril 1816	Armadillidae				90–100		Vandel (1962)
	Armadillo officinalis	Dumeril 1816	Armadillidae	VI–VIII	1	26	50	45	Shereef (1970)
	Armadillo officinalis	Dumeril 1816	Armadillidae	IV–V IX–X	2–3		Av. 35	Av. 33	Warburg *et al.* (1984)
(59)	*A. dorsalis*	Dumeril 1816	Armadillidae	VI–VIII			Av. 9·8		Watanabe (1980)
(60)	*Venezillo arizonicus*	Mulaik & Mulaik 1945	Armadillidae	VII–VIII	1		2–4		Warburg (1965c)
(61)	*V. evergladensis*	Schultz 1963	Armadillidae				4–13		Johnson (1982)
(62)	*Cubaris robusta*	Collinge	Armadillidae		1–2		9–38		Menon *et al.* (1970)

[a] I–XII are the months of the year.

D. Numbers of Oocytes, Eggs and Larvae

The number of oocytes in ovaries is related to their size (weight or length) (in *Armadillidium vulgare*; Sorensen & Burkett, 1977). Furthermore, these numbers vary both with species (Table 7) as well as with the number of broods during the same year. A few eggs are found in a small species such as *Platyarthrus hoffmanseggi*, and a few hundreds in *Schizidium festai*. The size of the juveniles is a function of the mother's feeding condition. Brody and Lawlor (1984) found that larger *Armadillidium vulgare* juveniles were produced when food supply was scarce!

Very little is known about the number of larvae that actually hatch. Only recently were we able to study this aspect carefully in a few species (Table 8, Fig. 13). Apparently, only a certain percentage of marsupial embryos mature into larvae that will hatch. Much more needs to be learned about this problem.

E. Mortality of Eggs and Embryos

The numbers of eggs and embryos varies greatly and it is impossible to generalize (Table 7). It is difficult to estimate the mortality of eggs when the number of larvae hatching from the marsupium is not known. Brood pouch mortality is most probably much greater than is generally assumed—closer to 20% (see Table 8). In *Tylos punctatus* brood pouch mortality was reported to be less than 1% (Hamner, Smyth and Mulford, 1969). In *Ligidium hypnonorum* mortality was 6–10%, in *Hyloniscus riparius* 2·6–3·7%, and in *Tracheoniscus rathkei* 5·7–6·7% (see Krumpál, 1976). In *Porcellio laevis* 3·9% brood pouch mortality was reported by Nair (1978a), as compared with 21% in *Porcellio obsoletus* and 23% in *Hemilepistus reaumuri* found in the present study (Fig. 13). In *Philoscia muscorum* Sutton (1968) found 0·8% mortality, as compared with 3·6% (first brood) and 4·5 (second brood) reported by Sunderland, Hassall and Sutton (1976). Finally, in *Armadillidium vulgare*, Paris and Pitelka (1962) found 8% mortality.

Table 8

Losses and energy spent during oogenesis, egg maturation and larval hatching in *Hemilepistus reaumuri*.

	Numbers (n)[a]	Loss (%)	Energy (% kcal/body wt)
Oocytes	162·96 ± 3·98 (98)	43·8	–
Eggs	91·64 ± 4·69 (31)	22·9	37·58
Larvae	70·67 ± 5·73 (18)	–	22·30

[a] n = number of females examined.

Much more information is needed on representatives of various families in order to be able to draw some general conclusions as to the extent and significance of brood pouch mortality and how it is related to the reproductive period and duration.

F. The Energy Spent on Egg Production

This is probably the first time that some energetic data for isopod reproduction can be presented (Table 8; see also Warburg, 1986c). From these studies on *Hemilepistus reaumuri* we are able to show that about 35% of the energy is spent on marsupial egg production and about 20% on the larvae. These comparatively high values for the energy allocated to reproduction indicates the great effort it costs *Hemilepistus* to reproduce once during its lifetime: they are three times as high as the values given by Lawlor (1976a,b) for *Armadillidium vulgare*.

G. Reproductive Strategies in Isopods

The number of times isopods can reproduce depends on their growth rate, time to maturation, and survival after reproduction. Generally, isopods are short-lived, up to 5 years (see Warburg *et al.*, 1984).

Semelparous organisms reproduce only once in their lifetime, whereas iteroparous ones reproduce more than once (Cole, 1954). It is generally thought that semelparous species are short-lived (Bell, 1980). *Hemilepistus reaumuri* is such an animal, living for about 18 months. In other semelparous species reproduction is probably delayed to a later age (*Schizidium festai*, studies in progress).

When the probability of adult survival is greater than that of the juveniles (in a variable environment) iteroparity is the favoured reproductive strategy (Fritz, Stamp and Halverson, 1982). This is not true in isopods. There are examples of isopods which are semelparous, whereas most are probably iteroparous. As indicated by Giesel (1976) the energy allocated to reproduction requires a large proportion of the energetic resources. An isopod reproducing early in life is unlikely to survive long enough to reproduce again. *Armadillidium vulgare* females, when dehydrated, do not grow but rather reproduce. This may reduce their chance to reproduce again (Brody, Edgar and Lawlor, 1983). Is it possible that a species which is generally iteroparous (such as *Armadillidium vulgare*) becomes semelparous as a result of stressful conditions? The energetic expenditure on reproduction in iteroparous species under different environmental conditions needs to be studied.

As isopods appear to be generally iteroparous (proven semelparous species are apparently rare), there must be some advantage for this reproductive strategy in isopods in spite of their short lifespan. On the other hand, *Hemilepistus reaumuri*—a highly successful desert invertebrate, judging from both its distribution and abundance—is semelparous! This subject needs to be further investigated before any conclusions can be drawn.

IX. POPULATION STRUCTURE AND FLUCTUATIONS

This subject has been reviewed recently (Warburg *et al.*, 1984). All our present knowledge is based on studies of 17 isopod species (Table X in Warburg *et al.*, 1984), which do not even represent all the families. The main feature of an isopod population is the large proportion of younger cohorts comprising the bulk of the population. Juvenile growth rate is at an exponential rate and dependent upon ambient temperature. The same is true for the maturation of the young (Hubbell, 1971). Thus, the population structure changes from a bimodal pattern when the young leave the marsupium to a unimodal one when they grow and mature. This has been observed in a variety of species (*Cylisticus convexus* by Hatchett (1947), in *Armadillidium vulgare* by Paris and Pitelka (1962) and Al-Dabbagh and Block (1981), in *Ligidium japonicum* by Saito (1965), in *Trichoniscus pusillus* by Sutton (1968) and in *Trichoniscus rathkei* by Breymeyer and Brzozowska (1967)). Some species live only about one year (*Porcellio laevis*; Nair, 1976b, 1978a: *Porcellio spinicornis*; McQueen, 1976a: and *Hemilepistus reaumuri*), whereas others live for 4–5 years (*Armadillidium vulgare*; see Paris and Pitelka, 1962). Longevity of isopods is limited to 5 years (see Collinge, 1946; and Table XI in Warburg *et al.*, 1984).

Causes of mortality vary from cannibalism and predation of larvae or adults (Heeley, 1941; Brereton, 1957; Menon, Tandon and Jolly, 1969; Linsenmair, 1972; Sutton, 1970b; Sunderland and Sutton, 1980), to climatic factors (Kheirallah, 1979a; McQueen, 1976c; Al-Dabbagh and Block, 1981).

Most mortality in isopods takes place during their first month of life outside the brood pouch. Only 10% of *Porcellio scaber* survived longer (Brereton, 1957) and about 60% of *Porcellio spinicornis* survived the first 50 days (McQueen and Carnio, 1974). The following 2 months after release from the brood pouch are similarly critical. Thus, survival was 50% for *Porcellio spinicornis* (McQueen and Carnio, 1974), 35% for *Philoscia muscorum*, and 20% for *Trichoniscus pusillus* (Sutton, 1968, 1970a). This high mortality rate was due in part to various predators, whether vertebrates or invertebrates (see Warburg *et al.*, 1984), but mostly to climatic factors: drought, high temperatures and floods (see Davis (1984) and Warburg *et al.* (1984) for discussion).

The population of terrestrial isopods fluctuates in numbers during the first year largely due to mortality of the young recruits when they are released into the population, and later when they decline in numbers after 1–3 months. In long-lived isopods, the next 2–3 years are the main reproductive years, especially in the iteroparous species. Some species fluctuate greatly in numbers: *Armadillidium vulgare* (Davis and Sutton, 1977b; Al-Dabbagh and Block, 1981), *Armadillo officinalis, Metoponorthus* sp. and *Philoscia* sp. (Warburg *et al.*, 1984). Others show only small fluctuation: *Philoscia muscorum* (Sunderland *et al.*, 1976), *Porcellio scaber* (Davis and Sutton, 1977b), *Periscyphys granai* (Kheirallah, 1979b).

Finally, on rare occasions, what appeared to be a population explosion of isopods was observed. This phenomenon, documented for the first time in Lokke (1966) was discussed in detail recently (Warburg *et al.*, 1984). It could be a kind of migration or local dispersal movement of isopods towards a non-specific goal. This could possibly be related to weather conditions (Lokke, 1966). The phenomenon has so far not been studied in detail but is of great interest, especially in order to find out what triggered such outbursts.

X. DISCUSSION AND CONCLUSIONS

Most early studies on the oniscoid isopods were concerned with the anatomical aspects of respiratory and excretory organs (Stoller, 1899; Bepler, 1909; Herold, 1913; Verhoeff, 1917b, 1921) without relating these features to the ecology of the various species. Only later was the ecological outlook stressed, in which each isopod was placed in a well defined "niche" within the habitat (Herold, 1925; Verhoeff, 1931). Some isopod species were found in moister habitats whereas others appear to be better adapted to drier ones. This thesis was then studied experimentally in a more complete way by Miller (1938). In his research area in California he could study the whole range of habitats from seashore to forest with their respective isopod inhabitants. He stressed the role of moisture in this distribution pattern of the various isopod species. A more complete analysis of the various adaptive traits necessary for successful survival of the oniscoid on land can be found in Vandel's (1943) study of the evolution of the Oniscoidea. However, it was Edney's early experimental work on the physiological aspects of these adaptations which demonstrated how some of them were related to water balance (see review in Edney, 1954 and again later in 1968). Other ecological aspects of the adaptation of isopods to life on land were reviewed by Cloudsley-Thompson (1962, 1975).

From the present review, it is apparent that the terrestrial adaptations of the oniscoid isopods include a wide range of traits. Thus behavioural adaptations enable the isopods to maintain an appropriate rhythmic activity and to respond to moisture and light in a way that will keep them in comparatively sheltered microhabitats. (Warburg, 1968b). At the same time, a range of physiological adaptations include a reduced rate of water loss in the more terrestrial species. This may be because of a more impermeable cuticle due to cuticular lipids (Hadley and Warburg, 1986), and appropriately modified respiratory structures (Figs 1 and 2). These range from simple folded pseudotracheae (or lungs) in the Tylidae (Ebbe, 1981; Hoese, 1983) to highly developed tubulous organs in *Hemilepistus* (Hoese, 1982b; and Fig. 2). Furthermore, the water-conducting system unique to the terrestrial isopods plays a part in getting rid of the ammonia, the main execretory product (Hoese, 1981). For completion of their life-cycle,

Table 9
Reproductive strategies in isopods.

Semelparous (1 brood)	Iteroparous (more than 1 brood)
Tylos punctatus Hammer *et al.* (1969)	*Ligidium hypnorum* Tomescu (1973)
Philoscia muscorum Sunderland *et al.* (1976)	*Trichoniscus pusillus* Heeley (1941)
Porcellio olivieri Warburg *et al.* (1984)	*Hyloniscus riparius* Beyer (1957/58)
Hemilepistus reaumuri Warburg (1986c)	*Oniscus asellus* Heeley (1941)
Eluma purpurascens Juchault *et al.* (1980)	*Cylisticus convexus* Verhoeff (1917a)
Schizidium festai Warburg (1986c)	*Protracheoniscus politus* Radu and Tomescu (1972)
Venezillo arizonicus Warburg (1965c)	*Trachelipus riparius* Hatchett (1947)
	Metoponorthus pruinosus Menon *et al.* (1969)
	Leptotrichus naupliensis Shereef (1970)
	Porcellio scaber Verhoeff (1917a)
	P. laevis Mahmoud (1954)
	P. spinicornis McQueen (1976a)
	P. obsoletus Warburg (1986c)
	Armadillidium vulgare Warburg (1965c)
	Armadillo officinalis Warburg (1986c)
	Cubaris robusta Menon *et al.* (1970)

isopods rely heavily on the marsupium or brood-pouch which evolved into a remarkable structure protecting the eggs, embryos and larvae. When the young emerge from the brood-pouch they face a high mortality rate which causes a drastic decline in the population. Much of the energetic effort is spent both on growth and reproduction. The proportions between them are so far not well established. The reproductive strategies, whether semelparous or iteroparous, are both represented in isopods by apparently successful species (Table 9). The energetic efforts of a semelparous species need to be compared with a single generation of an iteroparous one. So far very little work on the reproductive strategy is available.

ACKNOWLEDGEMENTS

It was my first university teacher, Professor Saul Adler, F.R.S., who showed me the beauty of the experimental design and the scientific way of thinking. Later, I was greatly influenced by Professor Eric Edney's physiological approach to the question of terrestrial adaptations in arthropods. My ecological outlook on the problems involved in life on land and, in particular, deserts was first influenced through the teaching of Professor F. S. Bodenheimer and later, to a large extent, by Professors G. E. Hutchinson and H. G. Andrewartha. To all of them I owe much. Last, but not least, several ideas were influenced by many stimulating discussions with my two friends and colleagues, Professors C. S. Crawford and K. E. Linsenmair, to both of whom I am greatly indebted.

REFERENCES

Akahira, Y. (1956). The function of thoracic processes found in females of the common wood-louse, *Porcellio scaber*. *J. Fac. Sci. Hokkaido Univ., Ser. 6, Zoology* **12**, 493–498.

Al-Dabbagh, K. Y. and Block, W. (1981). Population ecology of a terrestrial isopod in two breckland grass heaths. *J. Anim. Ecol.* **50**, 61–77.

Alexander, C. G. (1977). Antennal sense organs in the isopod *Ligia oceanica* (Linn). *Mar. Behav. Physiol.* **5**, 61–77.

Babula, A. (1979a). Structure of the respiratory organs of the fresh-water isopod *Asellus aquaticus* L. (Crustacea). *Bull. Soc. Amis Sci. Lett. Poznan, Ser. D, Sci. Biol.* **19**, 75–82.

Babula, A. (1979b). Ultrastructure of respiratory epithelium in fresh-water isopod, *Asellus aquaticus* L. (Crustacea). *Acta Med. Polon.* **20**, 355–356.

Babula, A. and Bielawski, J. (1981a). Ultramorphological study of gill epithelium in *Mesidotea entomon* (L.) (Isopoda, Crustacea). *Bull. Soc. Amis Sci. Lett Poznan Ser. D, Sci. Biol.* **21**, 51–58.

Babula, A. and Bielawski, J. (1981b). Morphometric and stereological investigations of the gills of *Mesidotea entomon* (L.) (Isopoda, Crustacea). *Bull. Soc. Amis Sci. Lett. Poznan, Ser. D, Sci. Biol.* **21**, 59–71.
Bayly, I. A. E. and Ellis, P. (1969). *Haloniscus searlei*, Chilton: an aquatic "terrestrial" isopod with remarkable powers of osmotic regulation *Comp. Biochem. Physiol.* **31**, 523–528.
Beck, L. and Brestowsky, E. (1980). Auswahl und Verwertung verschiedener Fallaubarten durch *Oniscus asellus* (Isopoda). *Pedobiologia* **20**, 428–441.
Bell, G. (1980). The costs of reproduction and their consequences. *Am. Nat.* **116**, 45–76.
Bepler, H. (1909). Über die Atmung der Oniscoidea. Ph.D. Thesis, Univ. Greifswald, 49 pp.
Bettica, A., Shay, M. T., Vernon, G. and Witkus, R. (1984). An ultrastructural study of cell differentiation and associated acid phosphatase activity in the hepatopancreas of *Porcellio scaber*. *Symp. zool. Soc. Lond.* **53**, 199–215.
Beyer, R. (1957/8). Ökologische und brutbiologische Untersuchungen an Landisopoden der Umgebung von Leipzig. *Wiss. Z. Karl-Marx Universität Math. Nat. Reihe* **7**,, 291–308.
Beyer, R. (1964). Faunistisch-Ökologische Untersuchungen an Landisopoden in Mitteldeutschland, *Zool. Jahrb. Syst.* **91**, 341–402.
Bilinski, S. (1979). Ultrastructural study of yolk formation in *Porcellio scaber* Latr. (Isopoda). *Cytobios* **26**, 123–130.
Bodenheimer, F. S. (1935). *Animal Life in Palestine*. Jerusalem: L. Mayer, 506 pp.
Brereton, J. Le G. (1957). The distribution of woodland isopods. *Oikos* **8**, 85–106.
Breymeyer, A. and Brzozowska, D. (1967). Density, activity and consumption of Isopoda on a *Stellario–Deschampsietum* meadow. In *Methods of Study of Soil Ecology* (Ed. by J. Phillipson), pp. 225–230. UNESCO, Paris.
Brody, M. S. and Lawlor, L. R. (1984). Adaptive variation in offspring size in the terrestrial isopod, *Armadillidium vulgare*. *Oecologia* **61**, 55–59.
Brody, M. S., Edgar, M. H. and Lawlor, L. R. (1983). A cost of reproduction in a terrestrial isopod. *Evolution* **37**, 653–655.
Brown, M. F. and Steinberger, Y. (1983). The importance of geological formations and slope aspect to desert isopod survival. *J. Arid Environs* **6**, 373–384.
Bubel, A. and Jones, M. B. (1974). Fine structure of the gills of *Jaera nordmanni* (Rathke) [Crustacea, Isopoda]. *J. Mar. Biol. Assoc. U.K.* **54**, 737–743.
Carefoot, T. H. (1973a). Feeding, food preference, and the uptake of food energy by the supralittoral isopod *Ligia pallasii*. *Mar. Biol.* **18**, 228–236.
Carefoot, T. H. (1973b). Studies on the growth, reproduction and life cycle of the supralittoral isopod *Ligia pallasii*. *Mar. Biol.* **18**, 302–311.
Carefoot, T. H. (1979). Microhabitat preferences of young *Ligia pallasii* Brandt (Isopoda). *Crustaceana* **36**, 209–214.
Chelazzi, G. and Ferrara, F. (1978). Researches on the coast of Somalia, the shore and the dune of Sar Uanle. 19. Zonation and activity of terrestrial isopods (Oniscoidea). *Monitore Zool. Ital. N.S. Suppl.* 11, No. 8, 189–219.
Cloudsley-Thompson, J. L. (1956). Studies in diurnal rhythms. VI. Bioclimatic observations in Tunisia and their significance in relation to the physiology of the fauna, especially woodlice, centipedes, scorpions and beetles. *Ann. Mag. Nat. Hist. Ser.* **12**(9), 305–329.
Cloudsley-Thompson, J. L. (1962). Microclimates and the distribution of terrestrial arthropods. *Ann. Rev. Entomol.* **7**, 199–222.
Cloudsley-Thompson, J. L. (1975). Adaptations of Arthropoda to arid environments. *Ann. Rev. Entomol.* **20**, 261–283.

Coenan-Stass, D. (1981). Some aspects of the water balance of two desert woodlice, *Hemilepistus aphganicus* and *Hemilepistus reaumuri* (Crustacea, Isopoda, Oniscoidea). *Comp. Biochem. Physiol.* **70A**, 405–419.

Coenen-Stass, D. (1985). Effects of desiccation and hydration on the osmolality and ionic concentration in the blood of the desert woodlouse, *Hemilepistus reaumuri* (Crustacea, Isopoda, Oniscoidea). *Comp. Biochem. Physiol.* **81B**, 717–721.

Cole, L. C. (1946). A study of the cryptozoa of an Illinois woodland. *Ecol. Monogr.* **16**, 49–86.

Cole, L. C. (1954). The population consequences of life history phenomena. *Q. Rev. Biol.* **29**, 103–137.

Collinge, W. E. (1946). The duration of life in terrestrial Isopoda. *Ann. Mag. Nat. Hist.* **13**, 719–720.

Coruzzi, L., Witkus, R. and Vernon, G. M. (1982). Function-related structural characters and their modifications in the hindgut epithelium of two terrestrial isopods, *Armadillidium vulgare* and *Oniscus asellus*. *Exp. Cell Biol.* **50**, 229–240.

Crawford, C. S. (1981). *Biology of Desert Invertebrates*. Springer, Berlin, Germany. 314 pp.

Dallinger, R. (1977). The flow of copper through a terrestrial food chain III. Selection of an optimum copper diet by isopods. *Oecologia* **30**, 273–276.

Dallinger, R. and Wieser, W. (1977). The flow of copper through a terrestrial food chain. I. Copper and nutrition in isopods. *Oecologia* **30**, 253–264.

Davis, R. C. (1984). Effects of weather and habitat structure on the population dynamics of isopods in a dune grassland. *Oikos* **42**, 387–395.

Davis, R. C. and Sutton, S. L. (1977a). Spatial distribution and niche separation of woodlice and millipedes in a dune grassland ecosystem. In *Soil Organisms as Components of Ecosystems*. *Ecol. Bull.* **25**, 45–55.

Davis, R. C. and Sutton, S. L. (1977b). A comparative study of changes in biomass of isopods inhabiting dune grassland. *Sci. Proc. R. Dublin Soc.* **6A**, 223–233.

Debry, J. M. and Lebrun, P. (1979). Effets d'un enrichissement en $CuSO_4$ sur le bilan alimentaire de *Oniscus asellus* (Isopoda). *Rev. d'Ecol. Biol. Sol.* **16**, 113–124.

Debry, J. M. and Muyango, S. (1979). Effets du cuivre sur le bilan alimentaire de *Oniscus asellus* (Isopoda) avec référence particulière au cuivre contenu dans le lisier de porcs. *Pedobiologia* **19**, 129–137.

Den Boer, P. J. (1961). The ecological significance of activity patterns in the woodlouse *Porcellio scaber* Latr. (Isopoda). *Arch. Neerl. Zool.* **14**, 283–409.

Dresel, E. I. B. and Moyle, V. (1950). Nitrogenous excretion of amphipods and isopods. *J. Exp. Biol.* **27**, 210–225.

Ebbe, B. (1981). Beitrag zur Morphologie, Ultrastruktur und Funktion des Respirationsapparates von *Tylos granulatus* Krauss (Isopoda, Oniscoidae). *Zool. Jahrb. Anat.* **105**, 551–570.

Edney, E. B. (1951a). The evaporation of water from woodlice and the millipede *Glomeris*. *J. Exp. Biol.* **28**, 91–115.

Edney, E. B. (1951b). The body temperature of woodlice. *J. Exp. Biol.* **28**, 271–280.

Edney, E. B. (1953). The temperature of woodlice in the sun. *J. Exp. Biol.* **30**, 331–349.

Edney, E. B. (1954). Woodlice and the land habitat. *Biol. Rev.* **29**, 185–219.

Edney, E. B. (1956). A new interpretation of the relation between temperature and transpiration in arthropods. *Proc. 10th Int. Congr. Entomol.* **2**, 329–332.

Edney, E. B. (1957). *The Water Relations of Terrestrial Arthropods*. Cambridge University Press, 109 pp.

Edney, E. B. (1964a). Acclimation to temperature in terrestrial isopods. I. Lethal temperatures. *Physiol. Zool.* **37**, 364–377.
Edney, E. B. (1964b). Acclimation to temperature in terrestrial isopods. II. Heart rate and standard metabolic rate. *Physiol. Zool.* **37**, 378–394.
Edney, E. B. (1967). Water balance in desert arthropods. *Science* **156**, 1059–1066.
Edney, E. B. (1968). Transition from water to land in isopod crustaceans. *Am. Zool.* **8**, 309–326.
Edney, E. B. (1977). *Water Balance in Land Arthropods*. Springer-Verlag, Berlin, Germany 282 pp.
Edney, E. B., Allen, W. and McFarlane, J. (1974). Predation by terrestrial isopods. *Ecology* **55**, 428–433.
Elkaim, B., Hoeslandt, H., Lejuez, R. and Plateaux, L. (1980). Sur le thermo-preferendum de *Sphaeroma serratum* (F.) (Isopode Felabellifère). *Arch. Zool. Exp. Gèn.* **121**, 87–96.
El-Kifl, A. H., Wafa, A. K., Shafiee, M. F. and Shereef, G. M. (1970). Studies on land Isopoda in Giza region. *Bull. Soc. Entomol. Egypte* **54**, 283–317.
Ellis, R. J. (1961). A life history study of *Asellus intermedius* Forbes. *Trans. Am. Micr. Soc.* **80**, 80–102.
Ellis, P. and Williams, W. D. (1970). The biology of *Haloniscus searlei* Chilton, an oniscoid isopod living in Australian salt lakes. *Aust. J. Mar. Freshwat. Res.* **21**, 51–69.
Enright, J. T. (1972). A virtuoso isopod, circa-lunar rhythms and their tidal fine structure. *J. Comp. Physiol.* **77**, 141–162.
Enright, J. T. (1976). Plasticity in an isopod's clockworks: Shaking shapes form and affects phase and frequency. *J. Comp. Physiol.* **107**, 13–37.
Farr, J. A. (1978). Orientation and social behavior in the supralittoral isopod *Ligia exotica* (Crustacea: Oniscoidea). *Bull. Mar. Sci.* **28**, 659–666.
Fincham, A. A. (1973). Rhythmic swimming behaviour of the New Zealand sand beach isopod *Pseudaega punctata* Thomson. *J. Exp. Mar. Biol. Ecol.* **11**, 229–237.
Fincham, A. A. (1974). Rhythmic swimming of the isopod *Exosphaeroma obtusum* (Dana). *N. Z. J. Mar. Freshwat. Res.* **8**, 655–662.
Fritz, R. S., Stamp, N. E. and Halverson, T. G. (1982). Iteroparity and semelparity in insects. *Am. Nat.* **120**, 264–268.
Gere, G. (1956). The examination of the feeding biology and the humificative function of Diplopoda and Isopoda. *Acta Biol.* **6**, 257–271.
Gere, G. (1962). Nahrungsverbrauch der Diplopoden und Isopoden in Freilandsuntersuchungen. *Acta Zool.* **8**, 385–415.
Giesel, J. T. (1976). Reproductive strategies as adaptations to life in temporally heterogeneous environments. *Ann. Rev. Ecol. Syst.* **7**, 57–79.
Goudeau, M. (1976). Secretion of embryonic envelopes and embryonic molting cycles in *Hemioniscus balani* Buchholz, Isopoda Epicaridea. *J. Morph.* **148**, 427–452.
Hadley, N. F. and Quinlan, M. C. (1984). Cuticular transpiration in the isopod *Porcellio laevis*: Chemical and morphological factors involved in its control. *Symp. Zool. Soc., Lond.* **53**, 97–107.
Hadley, N. F. and Warburg, M. R. (1986). Water loss in three species of xeric-adapted isopods: correlation with cuticular lipids. *Comp. Biochem. Physiol.* **85A**, 669–672.
Hamner, W. M., Smyth, M. and Mulford, E. D. (1969). The behavior and life history of a sand-beach isopod, *Tylos punctatus*. *Ecology* **50**, 442–453.
Hassall, M. (1977). Consumption of leaf litter by the terrestrial isopod *Philoscia*

muscorum in relation to food availability in a dune grassland ecosystem. In *Soil Organisms as Components of Ecosystems, Ecol. Bull.* **25**, 550–553.

Hassall, M. (1983). Population metabolism of the terrestrial isopod *Philoscia muscorum* in a dune grassland ecosystem. *Oikos* **41**, 17–26.

Hassall, M. and Rushton, S. P. (1982). The role of cophrophagy in the feeding strategies of terrestrial isopods. *Oecologia* **53**, 374–381.

Hassall, M. and Rushton, S. P. (1984). Feeding behaviour of terrestrial isopods in relation to plant defences and microbial activity, *Symp. Zool. Soc., Lond.* **53**, 487–505.

Hassall, M. and Sutton, S. L. (1977). The role of isopods as decomposers in a dune grassland ecosystem. *Sci. Proc. R. Dublin Soc.* **6A**, 235–245.

Hastings, M. H. (1981a). The entraining effect of turbulence on the circa-tidal activity rhythm and its semi-lunar modulation in *Eurydice pulchra. J. Mar. Biol. Assoc. UK* **61**, 151–160.

Hastings, M. H. (1981b). Semi-lunar variations of endogenous circa-tidal rhythms of activity and respiration in the isopod *Eurydice pulchra. Mar. Ecol. Prog. Ser.* **4**, 85–90.

Hastings, M. H. and Naylor, E. (1980). Ontogeny of an endogenous rhythm in *Eurydice pulchra. J. Exp. Mar. Biol. Ecol.* **46**, 137–145.

Hatchett, S. P. (1947). Biology of the Isopoda of Michigan. *Ecol. Monogr.* **17**, 47–79.

Heeley, W. (1941). Observations on the life-histories of some terrestrial isopods. *Proc. Zool. Soc., Lond. Ser. B* **111**, 79–149.

Herold, W. (1913). Beiträge zur Anatomie und Physiologie einiger Landisopoden. *Zool. Jahrb. Anat. Ontog. Tiere* **35**, 456–526.

Herold, W. (1925). Untersuchungen zur Ökologie und Morphologie einiger Landasseln. *Z. Morphol. Ökol. Tiere* **4**, 337–415.

Herold, W. (1960). Die Vermehrungsgrösse einiger Deutscher Land-Isopoden. *Mitt. Zool. Mus. Berlin* **36**, 101–104.

Hoese, B. (1981). Morphologie und Funktion des Wasserleitungssystems der terrestrischen Isopoden (Crustacea, Isopoda, Oniscoidea). *Zoomorphology* **98**, 135–167.

Hoese, B. (1982a). Der *Ligia*-Typ des Wasserleitungssystems bei terrestrischen Isopoden und seine Entwicklung in der Familie Ligiidae (Crustacea, Isopoda, Oniscoidea). *Zool. Jahrb. Anat.* **108**, 225–261.

Hoese, B. (1982b). Morphologie und Evolution der Lungen bei den terrestrischen Isopoden (Crustacea, Isopoda, Oniscoidea). *Zool. Jahrb. Anat.* **107**, 396–422.

Hoese, B. (1983). Struktur und Entwicklung der Lungen der Tylidae (Crustacea, Isopoda, Oniscoidea). *Zool. Jahr. Anat.* **109**, 487–501.

Hoffmann, G. (1983a). The random elements in the systematic search behavior of the desert isopod *Hemilepistus reaumuri. Behav. Ecol. Sociobiol.* **13**, 81–92.

Hoffmann, G. (1983b). The search behavior of the desert isopod *Hemilepistus reaumuri* as compared with a systematic search. *Behav. Ecol. Sociobiol.* **13**, 93–106.

Hoffmann, G. (1984a). Homing by systematic search. In *Localization and Orientation in Biology and Engineering* (Ed. by Varju and Schnitzler), pp. 192–199. Germany: Springer.

Hoffmann, G. (1984b). Orientation behaviour of the desert woodlouse *Hemilepistus reaumuri*: adaptations to ecological and physiological problems. *Symp. Zool. Soc., Lond.* **53**, 405–422.

Holanov, S. H. and Hendrickson, J. R. (1980). The relationship of sand moisture to burrowing depth of the sand-beach isopod *Tylos punctatus* Holmes and Gay. *J. Exp. Mar. Biol. Ecol.* **46**, 81–88.

Holdich, D. M. (1968). Reproduction, growth and bionomics of *Dynamene bidentata (Crustacea: Isopoda)*. *J. Zool., Lond.* **156**, 137–153.
Hopkin, S. P. and Martin, M. H. (1984). Heavy metals in woodlice. *Symp. Zool. Soc., Lond.* **53**, 143–166.
Hørlyck, V. (1973). Seasonal and diel variation in the rhythmicity of *Idotea baltica* (Pallas) and *Idotea granulosa* Rathke. *Ophelia* **12**, 117–127.
Horowitz, M. (1970). The water balance of the terrestrial isopod *Porcellio Scaber*. *Ent. Exp. Appl.* **13**, 173–178.
Hryniewiecka-Szyfter, Z. (1972). Ultrastructure of heptopancreas of *Porcellio scaber* Latr. in relation to the function of iron and copper accumulation. *Bull. Soc. Amis, Sci. Lett. Poznan, Ser. D* **12/13**, 135–142.
Hryniewiecka-Szyfter, Z. and Tyczewska, J. (1975). Ultrastructure of rectum epithelium in Isopoda (Crustacea). *Ann. Med. Sect. Polish Acad. Sci.* **20**, 83–84.
Hubbell, S. P. (1971). Of sowbugs and systems: the ecological bioenergetics of a terrestrial isopod. In *Systems Analysis and Simulation in Ecology*, Vol. I (Ed. by B. C. Patten), pp. 269–324. London: Academic Press.
Hubbell, S. P., Sikora, A. and Paris, O. H. (1965). Radiotracer, gravimetric and calorimetric studies of ingestion and assimilation rates of an isopod. *Health Phys.* **11**, 1485–1501.
Imafuku, M. (1976). On the nocturnal behavior of *Tylos granulatus* Miers (Crustacea: Isopoda). *Publ. Seto Mar. Biol. Lab.* **23**, 299–340.
Jans, D. E. and Ross, K. F. A. (1963). A histological study of the peripheral receptors in the thorax of land isopods, with special reference to the location of possible hygroreceptors. *Q. J. Micr. Sci.* **104**, 337–350.
Jassem, W., Juchault, P. and Mocquard, J. P. (1982). Déterminisme de la reproduction saisonnière des femelles d'*Armadillidium vulgare* Latr. (Crustacé, Isopode, Oniscoide). *Ann. Sci. Nat. Zool. 13th Ser.* **4**, 195–201.
Jassem, W., Mocquard, J. P. and Juchault, P. (1981). Seuil de l'intensite lumineuse du signal photoperiodique induisant l'entrée en reproduction chez *Armadillidium vulgare*, Latr. (Crustacé, Isopode terrestre). *Bull. Soc. Zool. France* **106**, 451–455.
Jassem, W., Mocquard, J. P. and Juchault, P. (1982). Déterminisme de la reproduction saisonnière des femelles d'*Armadillidium vulgare* Latr. (Crustacé, Isopode, Oniscoide). IV Contribution à la connaissance de la perception du signal photopériodique induisant l'entrée en reproduction: mode de discrimination entre le jour et la nuit longueurs d'onde actives. *Ann. Sci. Nat. Zool. 13th Ser.* **4**, 85–90.
Johnson, C. (1982). Multiple insemination and sperm storage in the isopod, *Venezillo evergladensis* Schultz. 1963. *Crustaceana* **42**, 225–232.
Jones, M. B. and Naylor, E. (1971). Breeding and bionomics of the British member of the *Jaera albifrons* group of species (*Isopoda: Asellota*). *J. Zool., Lond.* **165**, 183–199.
Jöns, D. (1965). Zur Biologie und Ökologie von *Ligia oceanica* (L.) in der westlichen Ostsee. *Kiel Meeres Forsch.* **21**, 203–207.
Juchault, P., Mocquard, J. P. and Legrand, J. J. (1981). Déterminisme de la reproduction saisonnière des femelles d'*Armadillidium vulgare* Latr. (Crustacé, Isopode, Oniscoide) III. Suppression ou prolongation de la période de repos sexuel saisonnier obtenue par application de programmes photopériodiques. *Ann. Sci. Nat. Zool. 13th Ser.* **3**, 141–145.
Juchault, P., Mocquard, J. P., Bougrier, N. and Besse, G. (1980a). Croissance et cycle reproducteur du crustacé isopode oniscoide *Eluma purpurascens* Budde-Lund. Étude dans la nature et au laboratoire, sous différentes conditions de

température et de photopériode, d'une population due centre-ouest de la France. *Vie Milieu* **30**, 149–156.

Juchault, P., Pavese, A. and Mocquard, J. P. (1980b). Déterminisme de la reproduction saisonnière des femelles d'*Armadillidium vulgare* Latr. (Crustacé, Isopode, Oniscoide), II Étude en conditions expérimentales de femelles d'origines géographiques différentes. *Ann. Sci. Nat. Zool. 14th Ser.* **2**, 99–108.

Kato, H. (1976). Life histories and vertical distributions in the soil of *Ligidium japonicum* and *Ligidium* sp. (Isopoda). Preliminary report. *Rev. d'Écol. Biol. Sol.* **13**, 103–116.

Kensley, B. (1972). Behavioural adaptations of the isopod *Tylos granulatus* Krauss. *Zool. Afr.* **7**, 1–4.

Kheirallah, A. M. (1975). Population density and pattern of distribution of the terrestrial isopods *Leptotrichus panzeri* and *Philoscia africana* in different types of Alexandria soils. *Bull. Soc. Entomol. Egypte* **59**, 217–224.

Kheirallah, A. M. (1979a). The ecology of the isopod *Periscyphis granai* (Arcangeli) in the western highlands of Saudi Arabia. *J. Arid Environs* **2**, 51–59.

Kheirallah, A. M. (1979b). The population dynamics of *Periscyphis granai* (Isopoda: Oniscoidea) in the western highlands of Saudi Arabia. *J. Arid Environs* **2**, 329–337.

Kheirallah, A. M. (1980a). Aspects of the distribution and community structure of isopods in the Mediterranean coastal desert of Egypt. *J. Arid Environs* **3**, 69–74.

Kheirallah, A. M. (1980b). The life history and ecology of *Leptotrichus panzerii* (Crustacea: Isopoda) in Egypt. *Rev. d'Écol. Biol. Sol.* **17**, 393–403.

Kheirallah, A. M. and Awadallah, A. (1981). The life history of the isopod *Porcellio olivieri* in the Mediterranean coastal desert of Egypt. *Pedobiologia* **22**, 246–253.

Kheirallah, A. M. and El-Sharkawy, K. (1981). Growth rate and natality of *Porcellio olivieri* (Crustacea: Isopoda) on different foods. *Pedobiologia* **22**, 262–267.

King, S. D. (1926). Oogenesis in *Oniscus asellus*. *Proc. R. Soc. Lond. Ser. B, Biol. Sci.* **100**, 1–10.

Klapow, L. A. (1972). Natural and artificial rephasing of a tidal rhythm. *J. Comp. Physiol.* **79**, 233–258.

Koop, K. and Field, J. G. (1980). The influence of food availability on population dynamics of a supralittoral isopod, *Ligia dilatata* Brandt. *J. Exp. Mar. Biol. Ecol.* **48**, 61–72.

Koop, K. and Field, J. G. (1981). Energy transformation by the supralittoral isopod *Ligia dilatata* Brandt. *J. Exp. Mar. Biol. Ecol.* **53**, 221–233.

Kozlovskaja, L. S. and Striganova, B. R. (1977). Food, digestion and assimilation in desert woodlice and their relations to the soil microflora. In *Soil Organisms as components of Ecosystems. Ecol. Bull.* **25**, 240–245.

Krumpál, M. (1976). Knowledge from biology of isopod reproduction. In Jurský šúr. *Acta Fac. Rer. Nat. Univ. Comenianae Zool. Bratislava* **20**, 63–67.

Kuenen, D. J. (1959). Excretion and waterbalance in some land-isopods. *Entomol. Exp. Appl.* **2**, 287–294.

Kuenen, D. J. and Nooteboom, H. P. (1963). Olfactory orientation in some land-isopods (Oniscoidea, Crustacea). *Entomol. Exp. Appl.* **6**, 133–142.

Lagarrigue, J. G. (1969). Composition ionique de l'hémolymphe des Oniscoides. *Bull. Soc. Zool. France* **94**, 137–146.

Lane, R. L. (1980). Histochemistry of the reproductive systems of *Armadillidium vulgare* (Latreille) and *Porcellionides pruinosus* (Brandt) (Isopoda). *Crustaceana* **38**, 73–81.

Lawlor, L. R. (1976a) Parental investment and offspring fitness in the terrestrial isopod, *Armadillidium vulgare* (Latr.) (Crustacea: Oniscoidea). *Evolution* **30**, 775–785.

Lawlor, L. R. (1976b). Moulting, growth and reproductive strategies in the terrestrial isopod. *Armadillidium vulgare. Ecology* **57**, 1179–1194.

Legrand, J. J., Martin, G., Juchault, P. and Besse, G. (1982). Contrôle neuroendocrine de la reproduction chez les Crustacés. *J. Physiol. Paris* **78**, 543–552.

Lindqvist, O. V. (1968). Water regulation in terrestrial isopods, with comments on their behavior in a stimulus gradient. *Ann. zool. fenn.* **5**, 279–311.

Lindqvist, O. V. (1970). The blood osmotic pressure of the terrestrial isopods *Porcellio scaber* Latr. and *Oniscus assellus* L., with reference to the effect of temperature and body size. *Comp. Biochem. Physiol.* **37**, 503–510.

Lindqvist, O. V. (1972). Humidity reactions of the young of the terrestrial isopods *Porcellio scaber* Latr. and *Tracheoniscus rathkei* (Brandt). *Ann. zool. fenn.* **9**, 10–14.

Linsenmair, K. E. (1972). Die Bedeutung familienspezifischer "Abzeichen" für den Familienzusammenhalt bei der sozialen Wüstenassel *Hemilepistus reaumuri* Audouin et Savigny (Crustacea, Isopoda, Oniscoidea). *Z. Tierpsyochol.* **31**, 131–162.

Linsenmair, K. E. (1983). Individual and family recognition in subsocial arthropods, in particular in the desert isopod *Hemilepistus reaumuri*. In *Experimental Behavioral Ecology*, Vol. 31 (Ed. by B. Hölldobler and M. Lindauer), pp. 411–436. *Fortschrite der Zoologie*. Stuttgart: Fischer.

Linsenmair, K. E. (1984). Comparative studies on the social behaviour of the desert isopod *Hemilepistus reaumuri* and of a *Porcellio* species. *Symp. Zool. Soc., Lond.* **53**, 423–453.

Lokke, D. H. (1966). Mass movements of terrestrial isopods related to atmospheric circulation patterns. *Trans. Kansas Acad. Sci.* **69**, 117–122.

Ludwig, G. (1978). Zur kinästhetischen Verrechnung von Hindernissen bei der Assel *Porcellio scaber. Zool. Jahrb. Physiol.* **82**, 185–199.

Macquart-Moulin, C. (1980). Effects de la temperature sur les rythmes d'emergence des Peracarides Fouisseurs, *Urothoe elegans* (Amphipode) et *Eurydice inermis* (Isopode). *Mar. Behav. Physiol.* **7**, 65–83.

Mahmoud, M. F. (1954). Some notes on the biology of the terrestrial isopod *Porcellio laevis* Latreille. *Bull. Soc. Zool. Egypte* **12**, 33–41.

Marikovsky, P. I. (1969). A contribution to the biology of *Hemilepistus rhinoceros. Zool. Zhur.* **48**, 677–684.

Markus, H. C. (1930). Studies on the morphology and life history of the isopod, *Mancasellus macrourus. Trans. Am. Micr. Soc.* **49**, 220–237.

Marsh, B. A. and Branch, G. M. (1979). Circadian and circatidal rhythms of oxygen consumption in the sandy-beach isopod *Tylos granulatus* Krauss. *J. Exp. Mar. Biol. Ecol.* **37**, 77–89.

McQueen, D. J. (1976a). *Porcellio spinicornis* Say (Isopoda) demography. II. A comparison between field and laboratory data. *Can. J. Zool.* **54**, 825–842.

McQueen, D. J. (1976b). *Porcellio spinicornis* Say (Isopoda) demography. III. A comparison between field data and the results of a simulation model. *Can. J. Zool.* **54**, 2174–2184.

McQueen, D. J. (1976c). The influence of climatic factors on the demography of the terrestrial isopod *Tracheoniscus rathkei* Brandt. *Can. J. Zool.* **54**, 2185–2199.

McQueen, D. J. and Carnio, J. S. (1974). A laboratory study of the effects of some climatic factors on the demography of the terrestrial isopod *Porcellio spinicornis* Say. *Can. J. Zool.* **52**, 599–611.

McQueen, D. J. and Steel, C. G. H. (1980). The role of photoperiod and temperature in the initiation of reproduction in the terrestrial isopod *Oniscus asellus* Linnaeus. *Can. J. Zool.* **58**, 235–240.

Mead, M. and Mead, F. (1972/73). Étude de l'orientation chez l'isopode terrestre *Tylos latreille*, Ssp Sardous. *Vie Milieu* **23**, 81–93.

Mead, F., Gabouriaut, D. and Corbière-Tichané, G. (1976). Structure de l'organe sensoriel apical de l'antenne chez l'isopode terrestre *Metoponorthus sexfasciatus* Budde-Laud (Crustace, Isopoda). *Zoomorphologie* **83**, 253–269.

Mead-Briggs, A. R. (1956). The effect of temperature upon the permeability to water of arthropod cuticles. *J. Exp. Biol.* **33**, 737–749.

Meinertz, T. (1950). Über die Geschlechtsverhältnisse und die Brutzeit der danischen Landisopoden. *Arch. Zool. Soc. Bot. Fenn. "Vanamo".* **4**, 143–150.

Meinertz, T. (1951). Die Vermehrungsintensität bei Land-isopoden. *Zool. Jahrb. Allg. Zool. Physiol.* **63**, 1–24.

Menon, P. K. B., Tandon, K. K. and Jolly, R. (1969). Bionomics of a terrestrial isopod *Porcellionides pruinosus* (Brandt). *Zool. Polon.* **19**, 369–391.

Menon, P. K. B., Tandon, K. K. and Rait, H. K. (1970). Further studies on the bionomics of terrestrial isopods *Porcellionides pruinosus* (Brandt) and *Cubaris robusta* (Collinge). *Zool. Polon.* **20**, 345–372.

Merriam, H. G. (1971). Sensitivity of terrestrial isopod populations *(Armadillidium)* to food quality differences. *Can. J. Zool.* **49**, 667–674.

Miller, M. A. (1938). Comparative ecological studies on the terrestrial isopod Crustacea of the San Francisco Bay region. *Univ. Calif. Publ. Zool.* **43**, 113–142.

Mocquard, J. P., Besse, G., Juchault, P., Legrand, J. J., Maissiat, J., Martin, G. and Picaud, J. L. (1976). Dureé de la période de reproduction chez les femelles de l'oniscoide *Porcellio dilatatus* Brandt suivant les conditions d'élevage: température, photopériode et groupement. *Vie Milieu* **26**, 51–76.

Mocquard, J. P., Pavese, A. and Juchault, P. (1980). Déterminisme de la reproduction saisonnière des femelles d'*Armadillidium vulgare* Latr. (Crustacé, Isopode, Oniscoide). I. Action de la température et de la photopériode. *Ann. Sci. Nat. Zool. 14th Ser.* **2**, 91–97.

Munuswamy, N. and Subramoniam, T. (1980). An electrophoretic investigation on yolk utilisation in an isopod *Ligia exotica* Roux (Crustacea: Isopoda). *Zool. Jahrb. Physiol.* **84**, 417–422.

Nair, G. A. (1976a). Food and reproduction of the soil isopod, *Porcellio laevis. Ind. J. Ecol. Environ. Sci.* **2**, 7–13.

Nair, G. A. (1976b). Life cycle of *Porcellio laevis* (Latreille) (Isopoda, Porcellionidae). *Proc. Ind. Acad. Sci.* **84B**, 165–172.

Nair, G. A. (1978a). Some aspects of the population characteristics of the soil isopod, *Porcellio laevis* (Latreille), in the Delhi region. *Zool. Anz.* **201**, 86–96.

Nair, G. A. (1978b). Sex ratio of the soil isopod, *Porcellio laevis* (Latreille) in Delhi region. *Proc. Ind. Acad. Sci.* **87B**, 151–155.

Naylor, E. (1955). The life cycle of the isopod *Idotea emarginata* (Fabricius). *J. Anim. Ecol.* **24**, 270–281.

Neuhauser, E. F. and Hartenstein, R. (1978). Phenolic content and palatability of leaves and wood to soil isopods and diplopods. *Pedobiologia* **18**, 99–109.

Nicholls, A. G. (1931). Studies on *Ligia oceanica* I. A. Habitat and effect of change of environment. B. Observation on molting and breeding. *J. Mar. Biol. Assoc. U.K.* **17**, 655–673.

Ondo, Y. (1958). Daily rhythmic activity of *Tylos granulatus*. IV. Characteristic movement of the shore sowbug accompanied with the periodic movement of waves (in comparison with *Talorchestia brito*). *Jap. J. Ecol.* **8**, 84–90.

Ondo, Y. (1959). Daily rhythmic activity of *Tylos granulatus* Miers V. Studies on the mechanisms of periodic behavior accompanied with periodic movement of waves. *Jap. J. Ecol.* **9**, 159–167.

Palackal, T., Faso, L., Zung, J. L., Vernon, G. and Witkus, R. (1984). The ultrastructure of the hindgut epithelium of terrestrial isopods and its role in osmoregulation. *Symp. Zool. Soc. Lond.* **53**, 185–198.

Pandian, T. J. (1972). Egg incubation and yolk utilization in the isopod *Ligia oceanica*. *Proc. Ind. Nat. Sci. Acad.* **38**, 430–441.

Paris, O. H. (1963). The ecology of *Armadillidium vulgare* (Isopoda: Oniscoidea) in California grassland: food, enemies and weather. *Ecol. Monogr.* **33**, 1–22.

Paris, O. H. (1965). Vagility of P^{32}-labelled isopods in grassland. *Ecology* **46**, 635–648.

Paris, O. H. and Pitelka, F. A. (1962). Population characteristics of the terrestrial isopod *Armadillidium vulgare* in California grassland. *Ecology* **43**, 229–248.

Paris, O. H. and Sikora, A. (1965). Radiotracer demonstration of isopod herbivory. *Ecology* **46**, 729–734.

Paris, O. H. and Sikora, A. (1967). Radiotracer analysis of the trophic dynamics of natural isopod populations. In *Secondary Productivity of Terrestrial Ecosystems*. (Ed. by K. Petrusewicz), pp. 741–771. Warsaw, Poland.

Parry, G. (1955). Osmotic and ionic regulation in the isopod crustacean *Ligia oceanica*. *J. Exp. Biol.* **30**, 567–574.

Patane, L. (1940). Sulla struttura e la funzioni del marsupio di *Porcellio laevis* Latreille. *Arch. Zool. Italia* **28**, 271–296.

Perttunen, V. (1963). Effects of desiccation on the light reactions of some terrestrial arthropods. *Ergebn. Biol.* **26**, 90–97.

Phillipson, J. and Watson, J. (1965). Respiratory metabolism of the terrestrial isopod *Oniscus asellus* L. *Oikos* **16**, 78–87.

Picaud, J. L. (1980). Vitellogenin synthesis by the fat body of *Porcellio dilatatus* Brandt (Crustacea Isopoda). *Int. J. Invert. Reprod.* **2**, 341–349.

Price, J. B. and Holdich, D. M. (1980). Changes in osmotic pressure and sodium concentration of the haemolymph of woodlice with progressive desiccation. *Comp. Biochem. Physiol.* **66A**, 297–305.

Prosi, F., Storch, V. and Janssen, H. H. (1983). Small cells in the midgut glands of terrestrial Isopoda sites of heavy metal accumulation. *Zoomorphology* **102**, 53–64.

Radu, V. G. and Tomescu, N. (1971). Reproduction and ontogenetic development in *Trachelipus balticus* Verh. 1907. *Rev. Roumaine Biol.* (*Zool.*) **16**, 89–96.

Radu, V. G. and Tomescu, N. (1972). Studiul populatiei de *Protracheoniscus politus* Koch (Crustacee-Izopode) intr-o pădure de foioase. *Stud. Univ. Babes-Bolyai Ser. Biol.* **1**, 75–82.

Radu, V. G., Tomescu, N., Racovită, L. and Imreh, S. (1971). Radioisotope researches concerning the feeding and the assimilation of Ca^{45} in terrestrial isopods. *Pedobiologia* **11**, 296–303.

Raimond, R. and Juchault, P. (1983). Étude du cycle biologique de *Sphaeroma serratum* Fabricius (Crustacé, Isopode, Flabellifere) dans une population du littoral charentais. Comparaison avec le cycle biologique des populations plus meridionales. *Bull. Soc. Zool. France* **108**, 79–93.

Refinetti, R. (1984). Behavioral temperature regulation in the pill bug, *Armadillidium vulgare* (Isopoda). *Crustaceana* **47**, 29–43.

Reichle, D. E. (1967). Radioisotope turnover and energy flow in terrestrial isopod populations. *Ecology* **48**, 351–366.

Reichle, D. E. (1968). Relation of body size to food intake, oxygen consumption, and trace element metabolism in forest floor arthropods. *Ecology* **49**, 538–542.
Risler, H. (1977). Die Sinnesorgane der Anetnnula von *Porcellio scaber* Latr. (Crustacea, Isopoda). *Zool. Jahrb. Anat.* **98**, 29–52.
Risler, H. (1978). Die Sinnesorgane der Antennula von *Ligidium hypnorum* (Cuvier) (Isopoda, Crustacea). *Zool. Jahrb. Anat.* **100**, 514–541.
Robertson, A. I. and Mann, K. H. (1980). The role of isopods and amphipods in the initial fragmentation of eelgrass detritus in Nova Scotia, Canada. *Mar. Biol.* **59**, 63–69.
Rushton, S. P. and Hassall, M. (1983a). The effects of food quality on the life history parameters of the terrestrial isopod (*Armadillidium vulgare*) (Latreille). *Oecologia* **57**, 257–261.
Rushton, S. P. and Hassall, M. (1983b). Food and feeding rates of the terrestrial isopod *Armadillidium vulgare* (Latreille). *Oecologia* **57**, 415–419.
Saito, S. (1965). Structure and energetics of the population of *Ligidium japonica* (Isopoda) in a warm temperate forest ecosystem. *Jap. J. Ecol.* **15**, 47–55.
Saudray, Y. (1954). Utilisation des réserves lipidiques au cours de la ponte et du développement embryonnaire chez deux Crustacés: *Ligia oceanica* Fab. et *Homarus vulgaris* Edw. *C. r. Soc. Biol.* **148**, 814–816.
Saudray, Y. and Lemercier, A. (1960). Observations sur le développement des oeufs de *Ligia oceanica* Fabr. Crustacé Isopode Oniscoide. *Bull. Inst. Ocenogr. Monaco.* No. 1162, 1–11.
Schäfer, M. W. (1982a). Gegendrehung und Winkelsinn in der Orientierung verschiedener Arthropoden. *Zool. Jahrb. Physiol.* **86**, 1–16.
Schäfer, M. W. (1982b). Ein idiothetischer Mechanismus im Gegendrehungsverhalten der Assel *Oniscus asellus* L. *Zool. Jahrb. Physiol.* **86**, 193–208.
Schneider, P. (1971). Lebensweise und soziales Verhalten der Wüstenassel *Hemilepistus aphganicus*. Borutzky 1958. *Z. Tierpsychol.* **29**, 121–133.
Schneider, P. (1973). Über die Geruchsrezeptoren der afghanischen Wüstenassel. *Naturwiss.* **60**, 106–107.
Schneider, P. and Jakobs, B. (1977). Versuche zum intra-und interspezifischen Verhalten terrestrischer Isopoden (Crustacea, Oniscoidea). *Zool. Anz.* **199**, 173–186.
Schöbl, J. (1880). Ueber die Fortpflanzung isopoder Crustaceen. *Arch. Mikr. Anat.* **17**, 125–140.
Seelinger, G. (1977). Der antennenendzapfen der tunesischen Wüstenassel *Hemilepistus reaumuri*, ein komplexes Sinnesorgan (Crustacea, Isopoda). *J. Comp. Physiol.* **113**, 95–103.
Seelinger, G. (1983). Response characteristics and specificity of chemoreceptors in *Hemilepistus reaumuri* (Crustacea, Isopoda), *J. Comp. Physiol.* **152**, 219–229.
Shachak, M. (1980). Energy allocation and life history strategy of the desert isopod *H. reaumuri. Oecologia* **45**, 404–413.
Shachak, M., Chapman, E. A. and Steinberger, Y. (1976). Feeding, energy flow and soil turnover in the desert isopod, *Hemilepistus reaumuri. Oecologia* **24**, 57–69.
Shachak, M., Steinberger, Y. and Orr. Y. (1979). Phenology, activity and regulation of radiation load in the desert isopod, *Hemilepistus reaumuri. Oecologia* **40**, 133–140.
Shereef, G. M. (1970). Biological observations on the woodlice (Isopoda) in Egypt. *Rev. d'Ecol. Biol. Sol.* **7**, 367–379.
Shimoizumi, M. (1952). The breeding habits of *Metoponorthus pruinosus* Brandt. *J. Gakugei Tokushima Univ.* **2**, 31–34.

Snider, R. and Shaddy, J. H. (1980). The ecobiology of *Trachelipus rathkei* (Isopoda). *Pedobiologia* **20**, 394–410.

Sorensen, E. M. B. and Burkett, R. D. (1977). A population study of the isopod, *Armadillidium vulgare*, in northeastern Texas. *Southwest. Nat.* **22**, 375–387.

Souty, C. (1980). Electron microscopic study of follicle cell development during vitellogenesis in the marine crustacean Isopoda, *Idotea balthica basteri. Reprod. Nutr. Dévelop.* **20**, 653–663.

Stachurski, A. (1968). Emigration and mortality rates and the food-shelter conditions of *Ligidium hypnorum* L. (Isopoda). *Ekol. Polsk. Ser. A* **16**, 445–449.

Stachurski, A. (1972). Population density, biomass and maximum natality rate and food conditions in *Ligidium hypnorum* L. (Isopoda). *Ekol. Polsk. Ser. A* **20**, 185–198.

Stachurski, A. (1974). Stabilization mechanisms of energy transfer by *Ligidium hyponorum* (Cuvier) (Isopoda) population in Alder wood (Carici Elongatae-Alnetum). *Ekol. Polsk.* **22**, 3–29.

Standen, V. (1970). The life history of *Trichoniscus pusillus pusillus* (Crustacea: Isopoda). *J. Zool., Lond.* **161**, 461–470.

Standen, V. (1973). The life cycle and annual production of *Trichoniscus pusillus pusillus* (Crustacea: Isopoda) in a Cheshire wood. *Pedobiologia* **13**, 273–291.

Standing, J. D. and Beatty, D. D. (1978). Humidity behaviour and reception in the spheromatid isopod *Gnorimosphaeroma oregonensis* (Dana). *Can. J. Zool.* **56**, 2004–2014.

Steel, C. G. H. (1980). Mechanisms of coordination between moulting and reproduction in terrestrial isopod Crustacea. *Biol. Bull.* **159**, 206–218.

Steel, E. A. (1961). Some observations on the life history of *Asellus aquaticus* (L.) and *Asellus meridianus* Racovitza (Crustacea: Isopoda). *Proc. Zool. Soc., Lond.* **137**, 71–87.

Steele, D. H. and Steele, V. J. (1972). The biology of *Jaera* spp. (Crustacea, Isopoda) in the northwestern Atlantic. 1. *Jaera ischiosetosa. Can. J. Zool.* **50**, 205–211.

Stoller, J. H. (1899). On the organs of respiration of the Oniscoidea. *Zoologica* **10**(25), 1–31.

Storch, V. (1982). Der Einfluss der Ernährung auf die Ultrastruktur der grossen Zellen in den Mittledarmdrüsen terrestrischen Isopoda (*Armadillidium vulgare, Porcellio scaber*) *Zoomorphology* **100**, 131–142.

Storch, V. and Lehnert-Moritz, K. (1980). The effects of starvation on the hepatopancreas of the isopod *Ligia oceanica, Zool. Anz.* **204**, 137–146.

Striganova, B. R. and Kondeva, E. A. (1980). Food requirements and growth of land woodlice (Oniscoidea). *Zool. Zhur.* **59**, 1792–1799.

Striganova, B. R. and Valiachmedow, B. V. (1976). Participation of soil saprophages in the leaf-litter decomposition in pistachio stand. *Pedobiologia* **16**, 219–227.

Stromberg, J. O. (1964). On the embryology of the isopod *Idotea. Ark. Zool.* **17**, 421–473.

Sunderland, K. D. and Sutton, S. L. (1980). A serological study of arthropod predation on woodlice in a dune grassland ecosystem. *J. Anim. Ecol.* **49**, 987–1004.

Sunderland, K. D., Hassall, M. and Sutton, S. L. (1976). The population dynamics of *Philoscia muscorum* (Crustacea, Oniscoidea) in a dune grassland ecosystem. *J. Anim. Ecol.* **45**, 487–506.

Sutton, S. L. (1968). The population dynamics of *Trichoniscus pusillus* and *Philoscia muscorum* (Crustacea, Oniscoidea) in limestone grassland. *J. Anim. Ecol.* **37**, 425–444.

Sutton, S. L. (1970a). Growth patterns in *Trichoniscus pusillus* and *Philoscia muscorum* (Crustacea: Oniscoidea). *Pedobiologia* **10**, 434–441.
Sutton, S. L. (1970b). Predation on woodlice: an investigation using the precipitin test. *Entomol. Exp. Appl.* **13**, 279–285.
Takeda, N. (1980). The aggregation pheromone of some terrestrial isopod crustaceans. *Experientia* **36**, 1296–1297.
Tomescu, N. (1973). Reproduction and postembryonic ontogenetic development in *Ligidium hypnorum* (Cuvier) and *Trichoniscus pusillus* (Brandt 1833) (Crustacea, Isopoda). *Rev. Roumaine Biol. (Zool.)* **18**, 403–413.
Vandel, A. (1943). Essai sur l'origine, l'evolution et la classification des Oniscoidea (Isopodes Terrestres). *Suppl. Bull. Biol. France Belgique* **30**, 1–136.
Vandel, A. (1960). *Isopodes Terrestres*, pp. 1–416. Faune de France No. 64. Paris: Lechevalier.
Vandel, A. (1962). *Isopodes Terrestres*, 2nd pt, pp. 417–927. Faune de France, No. 66. Paris: Lechevalier.
Verhoeff, K. W. (1917a). Über die Larven, das Marsupium und die Bruten der Oniscoidea. *Arch. Naturgesch.* **83**, 1–54.
Verhoeff, K. W. (1917b). Zur Kenntnis der Atmung und der Atmungsorgane der Isopoda-Oniscoidea. *Biol. Zentral.* **37**, 113–127.
Verhoeff, K. W. (1920). Zur Kenntnis der Larven, des Brutsackes und der Bruten der Oniscoidea. *Zool. Anz.* **51**, 169–189.
Verhoeff, K. W. (1921). Über die Atmung der Landasseln. *Z. Wiss. Zool.* **118**, 365–447.
Verhoeff, K. W. (1931). Vergleichende geographisch-ökologische Untersuchungen uber die Isopoda terrestria von Deutschland, den Alpenländern und anschliessenden Mediterrangebieten. *Z. Morphol. Ökol. Tiere* **22**, 231–268.
Verhoeff, K. W. (1949). *Tylos*, eine terrestrisch-maritime Rückwanderer Gattung der Isopoden. *Arch. Hydrobiol.* **42**, 329–340.
Wägele, J. W. (1982). Ultrastructure of the pleopods of the estuarine isopod *Cyathura carinata* (Crustacea: Isopoda: Anthuridea). *Zoomorphology* **101**, 215–226.
Wägele, J. W., Welsch, U. and Müller, W. (1981). Fine structure and function of the digestive tract of *Cyathura carinata* (Kroyer) (Crustacea: Isopoda). *Zoomorphology* **98**, 69–88.
Warburg, M. R. (1964). The response of isopods towards temperature, humidity and light. *Anim. Behav.* **12**, 175–186.
Warburg, M. R. (1965a). Water relation and internal body temperature of isopods from mesic and xeric habitats. *Physiol. Zool.* **38**, 99–109.
Warburg, M. R. (1965b). The evaporative water loss of three isopods from semi-arid habitats in South Australia. *Crustaceana* **9**, 302–308.
Warburg, M. R. (1965c). The evolutionary significance of the ecological niche. *Oikos* **16**, 205–213.
Warburg, M. R. (1968a). Simultaneous measurement of body temperature and water loss in isopods. *Crustaceana* **14**, 39–44.
Warburg, M. R. (1968b). Behavioral adaptations of terrestrial isopods. *Am. Zool.* **8**, 545–559.
Warburg, M. R. (1986a). Haemolymph osmolality, ion concentration and the distribution of water in body compartments of terrestrial isopods under different ambient conditions. *Comp. Biochem. Physiol.* (in press).
Warburg, M. R. (1986b). Water balance in terrestrial isopods (in preparation).
Warburg, M. R. (1986c). Reproductive patterns and strategies in terrestrial isopods (in preparation).

Warburg, M. R. and Berkovitz, K. (1978a). Hygroreaction of normal and desiccated *Armadillo officinalis* isopods. *Entomol. Exp. appl.* **24**, 55–64.

Warburg, M. R. and Berkovitz, K. (1978b). Thermal effects on photoreaction of the oak-woodland pillbug *Armadillo officinalis* (Isopoda; Oniscoidea), at different humidities. *J. Therm. Biol.* **3**, 75–78.

Warburg, M. R., Rankevich, D. and Chasanmus, K. (1978). Isopod species diversity and community structure in mesic and xeric habitats of the Mediterranean region. *J. Arid Environs* **1**, 157–163.

Warburg, M. R., Linsenmair, K. E. and Bercovitz, K. (1984). The effect of climate on the distribution and abundance of isopods. *Symp. Zool. Soc., Lond.* **53**, 339–367.

Watanabe, H. (1978). A food selection experiment on palatability of woodlouse, *Armadillidium vulgare* Latreille. *Edaphologia* **18**, 1–8.

Watanabe, H. (1980). A study of the three species of isopods in an evergreen broad-leaved forest in southwestern Japan. *Rev. d'Écol. Biol. Sol.* **17**, 229–239.

White, J. J. (1968). Bioenergetics of the woodlouse *Tracheoniscus rathkei* Brandt in relation to litter decomposition in a deciduous forest. *Ecology* **49**, 694–704.

Wieser, W. (1963). Die Bedeutung der Tageslänge für das Einsetzen der Fortpflanzungsperiode bei *Porcellio scaber* Latr. (Isopoda). *Z. Naturforsch.* **18**, 1090–1092.

Wieser, W. (1965). Untersuchungen über die Ernährung und den Gesamtstoffwechsel von *Porcellio scaber* (Crustacea: Isopoda). *Pedobiologia* **5**, 304–331.

Wieser, W. (1966). Copper and the role of isopods in degradation of organic matter. *Science* **153**, 67–69.

Wieser, W. (1978). Consumer strategies of terrestrial gastropods and isopods. *Oecologia* **36**, 191–201.

Wieser, W. (1984). Ecophysiological adaptations of terrestrial isopods: a brief review. *Symp. Zool. Soc., Lond.* **53**, 247–265.

Wieser, W. and Schweizer, G. (1970). A re-examination of the excretion of nitrogen by terrestrial isopods. *J. Exp. Biol.* **52**, 267–274.

Wieser, W., Busch, G. and Büchel, L. (1976). Isopods as indicators of the copper content of soil and litter. *Oecalogia* **23**, 107–114.

El Niño Effects on Southern California Kelp Forest Communities

MIA. J. TEGNER AND PAUL. K. DAYTON

I. INTRODUCTION

Coincident with the Christmas season in the Peruvian and Equadorian coastal areas, the normally cold water of the north-flowing Peru Current is displaced by a warm, southward current associated with a decrease in nutrients and a temporary reduction in fishing success. This is the generic "El Niño", a brief seasonal anomaly ending by March or April. Occasionally, however, the warm current strengthens and persists for a year or more, usually with a catastrophic effect on the important anchoveta population and the various populations which depend on it, including several species of higher-order fishes, guano birds, marine mammals and, more recently, human fishermen. The latter, much more severe event has become known as the El Niño (Ramage, 1986), which is how we will use the term. It is important to remember that there is a continuum of severity and, as no two events are the same, there are probably multiple causes of the El Niño phenomena.

ADVANCES IN ECOLOGICAL RESEARCH Vol. 17
ISBN 0-12-013917-0

While El Niños were defined in terms of events off South America, Jacob Bjerknes recognized the link with oceanic and atmospheric processes which extend across the Pacific and Indian Oceans in his study of the El Niño of 1957–1958 (Rasmusson, 1984). As early as the 1920s, Sir Gilbert Walker described the Southern Oscillation, a giant seesaw of atmospheric pressure and rainfall patterns between the Indian Ocean from Africa to Australia and the Pacific Ocean. This periodic disruption of climate is associated with severe droughts in some areas and massive floods in others, a redistribution of hurricane tracks, brush fires, crop failures and losses of livestock and human life. Changes in the Southern Oscillation index, calculated by Walker by subtracting pressure in the western Pacific from pressure in the eastern Pacific, correlate with many atmospheric and oceanic anomalies (see Wallace, 1985), and the relationship between El Niños and the Southern Oscillation is so accepted that it has acquired the acronym ENSO.

More difficult than creating a catchy acronym is understanding the mechanism. Over the years the models have focussed on the southeast trade winds driven by the pressure gradient between the South Pacific high-pressure system and the low-pressure system over Indonesia and Australia (reviewed by Ramage, 1986). A large pressure difference corresponds with a high Southern Oscillation index, strong trade winds, and high sea level in the western Pacific. A precipitous drop in the index and a corresponding collapse of the trade winds in the western Pacific signals the onset of an ENSO event. The water piled up in the western Pacific flows eastward near the Equator in subsurface waves known as Kelvin waves. The Kelvin waves generate anomalous eastward currents of warm water, raise sea level, and depress the thermocline in the eastern tropical Pacific. The deepened thermocline makes upwelling ineffectual and water temperatures soar. This results in the classical El Niño phenomenon. Unfortunately, the El Niños of 1976–1977 and especially 1982–1983 failed to follow this model based on a build-up of the trade winds (Ramage, 1986).

While the physical story of ENSO is far from complete, the effects of the El Niño on the Peruvian anchoveta system are reasonably well understood (Cushing, 1982; Arntz, 1986). The anchoveta was the largest fishery in the world, producing more than 12 million tons in 1970 (Cushing, 1982). During "normal" or non El Niño years, a coastal upwelling system feeds a short but highly productive food chain which, in addition to and largely dependent upon the anchoveta, includes other fishes such as different species of mackerel, hake, and rosefish and, offshore, various larger predators such as tunas. Also dependent on the anchoveta are several guano-producing birds such as cormorants, boobies and pelicans, and the guano they produced represented an important export industry once politically powerful enough to interfere with the development of the anchoveta fishery itself. Thus much of the ecosystem depends upon the anchoveta, and this population depends upon the physical structure of the upwelling system. A major El Niño drastically reduces the productivity and greatly alters the spatial scales of the community. As El Niños are a normal part of the ecosystem, the component species have evolved buffers which allow rapid recovery. However, with the development of human overexploitation of the anchoveta, recoveries have been slower until the strong El Niño of 1972–1973, from which there has been little recovery (Cushing, 1982). The biological consequences of tropical El Niños have been reviewed by Cushing (1982), Barber and Chavez (1983, 1986), *Oceanus* **27**(2) (1984), and Arntz (1986).

The 1982–1983 El Niño was the largest ever measured and reached into high latitudes of the eastern Pacific in both the Southern (Arntz, 1986) and Northern (Wooster and Fluharty, 1985; Mysak, 1986) Hemispheres. Wallace (1985) notes two possible mechanisms for transmitting the ENSO signal from the equatorial belt to higher latitudes. The first is the propagation of coastally trapped waves in the ocean. The second involves local forcing of coastal phenomena by anomalous surface winds associated with planetary-scale teleconnection patterns in the atmosphere. There is good evidence for the role of atmospheric teleconnections, but these are restricted to the winter half of the year for the hemisphere in question (Wallace, 1985). Whatever the mechanism, the most direct effects and the most thoroughly studied aspects are oceanic, especially planktonic food webs (see Chelton *et al.*, 1982; McGowan, 1985). Here we review the effects of this massive oceanic phenomenon on the nearshore kelp communities of the southern California Bight.

II. SOUTHERN CALIFORNIA KELP FOREST COMMUNITIES

Shallow subtidal rocky habitats along most temperate coasts are dominated by large brown algae of the order Laminariales, commonly known as kelps. The kelp forests of southern California are rich and complex assemblages of many species of plants and animals, a number of which are important economically. These communities are organized around the giant kelp, *Macrocystis pyrifera*, and are characterized by large standing stocks and high rates of growth, production, and turnover. While subject to a number of small-scale disturbances, these communities exhibit considerable long-term stability (Dayton *et al.*, 1984). However, in recent years it has become increasingly clear that the health of California kelp forests is linked to events thousands of kilometers away. The ENSO event which began in 1982, well known for its devastating effects in the tropical Pacific Ocean, was also associated with the largest disturbance of southern California kelp forests ever recorded (Dayton and Tegner, 1984a).

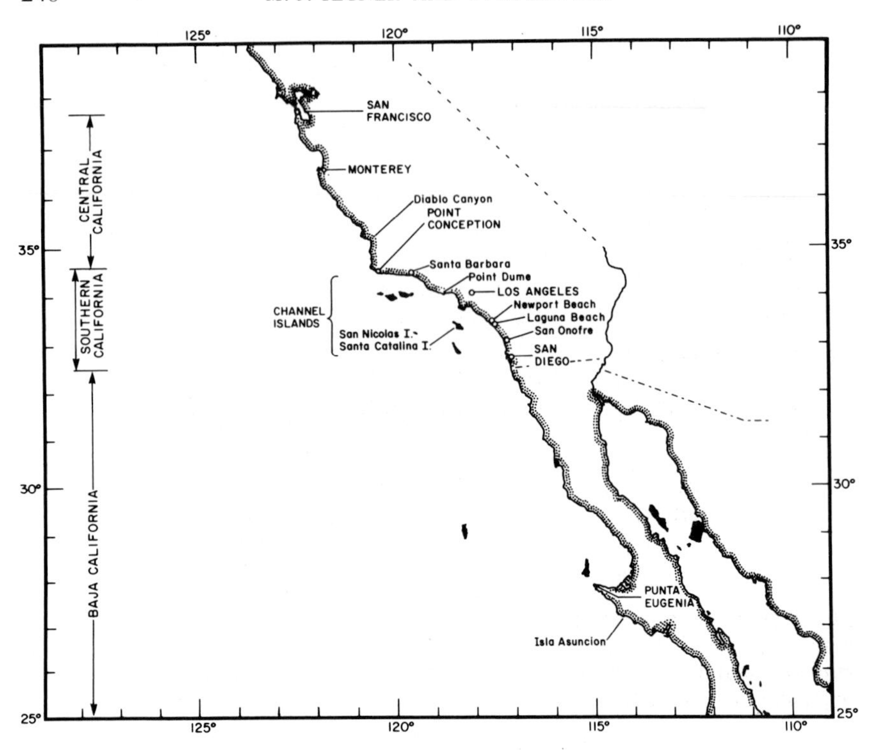

Fig. 1. Map of the coast of the Californias between central Alta California and central Baja California. While its range extends into Alaska (Abbott and Hollenberg, 1976), *Macrocystis pyrifera* forms significant forests only between Isla Asuncion and the coast south of San Francisco. The southern California Bight is the region between Point Conception and the Mexican border including the Channel Islands.

The timing of this event was opportune. The last massive El Niño to affect southern California kelp forests was 1957–1959, a time when biologists were faced with a major anomaly without understanding more typical conditions. Subsequent progress has led to the identification of specific physical parameters and biological processes important to the maintenance of kelp forest community structure. Here we use that knowledge and our 15 years of experience in the Point Loma kelp forest near San Diego to describe the effects of this massive El Niño on an extra-tropical community.

Like most natural communities, kelp forests are composed of distinct patches and are influenced by competition and various physical and biological disturbances (Dayton *et al.*, 1984). The algal patch types or guilds are composed of species of algae that can be categorized into vegetation layers distinguished by morphological adaptations. The layers and some southern California examples are: (1) a canopy supported at or near the surface by

Fig. 2. Sketch of the 15 m site at Point Loma, illustrating patches composed of algae with characteristic canopy guilds, including the floating canopy of *Macrocystis*, the stipitate canopy of *Pterygophora* and *Eisenia*, the prostrate canopy of *Laminaria*, and patches of turf and encrusting coralline algae (from Dayton *et al.*, 1984).

floats (*Macrocystis*); (2) an erect understory in which the fronds are supported well above the substratum by stipes (*Pterygophora californica, Eisenia arborea*); (3) a prostrate canopy in which the fronds lie on or immediately above the substratum (*Laminaria farlowii*); (4) turf composed of many species of coralline, foliose, and filamentous red algae; and (5) a pavement of encrusting coralline algae (Fig. 2). At Point Loma, representatives of these patch types have been observed to persist for many years, which, for some species, is several generations. The mechanisms which enable patch types to resist invasion by other guilds include competition for light, limits to spore dispersal including the physical barrier of extant plants and distance, and spore swamping by patch members following small disturbances. The tallest perennial canopy guild dominates competition for light but is more susceptible to wave stress. Conversely, the lower standing guilds are much more tolerant of wave stress, and dominance hierarchies appear to be reversed in areas exposed to extensive wave stress (Dayton *et al.*, 1984).

Important sources of disturbance for kelp forest communities are storms and grazing (Dayton and Tegner, 1984a; Dayton *et al.*, 1984; Dayton, 1985; Ebeling *et al.*, 1985). The effects of storms vary widely between kelp beds as a function of the frequency and magnitude of storms, degree of exposure to waves, and stability of the substratum. The main source of mortality at Point Loma is storm-dislodged *Macrocystis* which subsequently entangle other plants. The area of the disturbance is a critical component of patch stability. The more common small disturbances are usually colonized by members of the existing patch type, both from preceding sporulations and by spore swamping by surrounding plants. Larger disturbances involve the relative dispersal abilities, physiological thresholds, and reproductive seasonality of the different species. A critical second-order result of large-scale disturbance of the kelps is the effect on the availability of detached, drifting algal material, the major food of sea urchins (Ebeling *et al.*, 1985; Harrold and Reed, 1985).

Numerous animals feed on kelps but sea urchins are the most important herbivores in terms of the frequency and severity of destructive grazing (Leighton, 1971). Sea urchins are occasionally responsible for denuding large areas of algae, but predators such as spiny lobsters and fishes in southern California (Tegner and Dayton, 1981; Cowen, 1983; Tegner and Levin, 1983) seem to generally prevent urchins from devastating entire coastlines, as they have done in Nova Scotia (Mann, 1977; Chapman, 1981), or sea otter-free areas of the North Pacific (Estes and Palmisano, 1974). The scale of sea urchin disturbance varies from the small <1 m ambit from a refuge, to a few square meters in a boulder patch, to an entire kelp forest. A critical factor affecting the scale of disturbance is urchin recruitment, something clearly influenced by large-scale oceanographic processes (Dayton and Tegner, 1984b).

The health of *Macrocystis* forests in southern California is also strongly affected by the relationship between temperature and nutrients. Significant deterioration and occasional disappearance of the surface canopy is associated with warm temperatures during summer and fall; plants appear healthy below the thermocline. *Macrocystis* grows better in colder regimes but laboratory work has shown that the optimal temperature for photosynthesis is between 20 and 25°C, temperatures at which tissue damage has been observed in the field (reviewed by North and Zimmerman, 1984). Nutrient distributions in southern California coastal waters are affected by density stratification and uptake and release by plankton. As temperature is the dominant factor controlling seawater density in this area, there is a clear relationship between temperature and nutrient concentrations, and processes which affect density and temperature distributions, such as shoaling of isotherms near the coast, wind-driven upwelling or downwelling, and internal waves, also affect nutrient distributions (Eppley *et al.*, 1979; Jackson, 1983). A strong negative correlation between temperature and nitrate, the nutrient most likely to limit growth (North *et al.*, 1982), has been established for the southern California Bight; there are negligible amounts of nitrate above 15°C (Jackson 1977, 1983; Gerard, 1982; Zimmerman and Kremer,

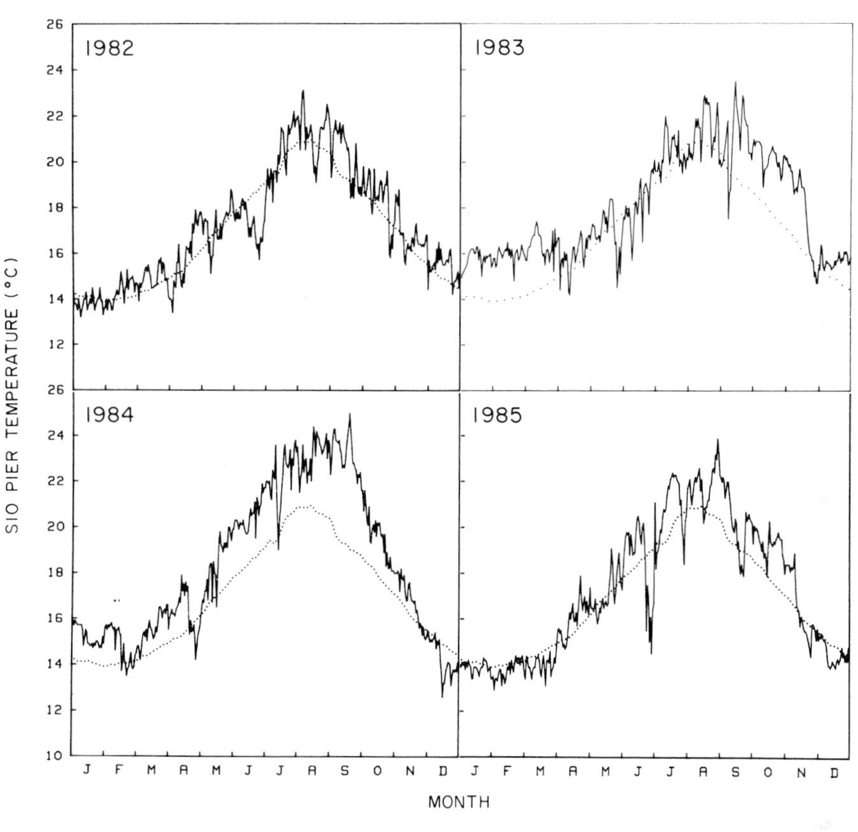

Fig. 3. The daily Scripps Institution of Oceanography pier sea-surface temperature data for 1982–1985 in relation to the 63-year (1920–1982) mean daily temperature (courtesy of E. Stewart, SIO).

1984). The warmer water is generally surface water that has been there for some time; during this time the plankton have depleted it of nutrients. The importance of the correlation between temperature and nutrients for *Macrocystis* is supported by fertilization experiments; nutrient additions decreased summer canopy deterioration relative to controls (North and Zimmerman, 1984). *Macrocystis* growth rates are reduced whenever nitrate concentrations fall below 1 μm, a level reached at about 15·5°C (Zimmerman and Kremer, 1984). Surface temperatures are normally 16°C or warmer from May through November (Fig. 3). *Macrocystis* can build internal nitrogen reserves during periods when external nutrient availability is high, and then use these reserves to maintain relatively rapid growth for at least two weeks to a month in the absence of significant external nutrients (Gerard, 1982; Zimmerman and Kremer, 1986).

III. THE PHYSICAL ENVIRONMENT: RELATIONSHIP OF THE CALIFORNIA CURRENT TO ENSO EVENTS

The lesson of recent years is that kelp forests are not isolated systems responding only to local processes; they are also strongly influenced by much larger scale physical processes which, in this case, may involve all of the southern California Bight and the much larger California Current system. Oceanic and atmospheric processes in the Pacific cause large-scale, low-frequency changes in the California Current leading to highly significant interannual variability in physical and biological parameters (Bernal, 1981; Chelton *et al.*, 1982). These and more local (mesoscale to hundreds of km) oceanic anomalies are probably responsible for the episodic recruitment events which characterize many populations of long-lived animals, including many important kelp forest species.

While El Niños were defined in terms of events off South America, it is now recognized that these changes are connected directly to changes across the entire tropical Pacific and indirectly to changes through much of the world's atmosphere and oceans (Cane, 1983; Rasmusson and Wallace, 1983). Chelton *et al.* (1982) have shown that the strength of the California Current is closely related to ENSO occurrences in the eastern tropical Pacific. The major El Niño events of 1957–1958, 1964, 1969 and 1972, as well as a number of minor events, are reflected in time series data from both areas. El Niños in the tropics are associated with positive California sea-level anomalies which correspond to anomalous poleward flow (Chelton *et al.*, 1982). Negative (or anti) El Niños, which refer to anomalously low sea-surface temperatures and strong coastal upwelling off South America (Quinn *et al.*, 1978), are associated with low sea level and anomalously strong equatorward flow in California (Chelton *et al.*, 1982). For the purpose of this discussion, El Niño will refer to mid-latitude warming events or California "El Niños" (Simpson, 1983) associated with ENSO events in the tropical Pacific Ocean.

An example of a major California El Niño was 1957–1959 when the California Current system was marked by abnormally high water temperatures, salinity, and sea level, increased poleward flow, and decreased zooplankton volume (Chelton *et al.*, 1982). Many species were observed far north of their usual range and some of these spawned successfully off southern California (Radovitch, 1961). Inshore, high temperatures persisted through much of the year. The southern coastal kelp forests from Newport Beach to San Diego apparently suffered to the greatest extent from the warm water. The annual harvest of the Point Loma bed relative to 1952, a year of high yield, was 48% in 1957, 21% in 1958, and 1% in 1959 (Clendenning, 1968). Sea urchins caused extensive grazing damage and effectively prevented kelp recruitment as many once luxuriant kelp beds all but disappeared (North and Pearse, 1970).

IV. THE 1982–1984 EL NIÑO IN CALIFORNIA

Anomalously high sea level, an indication of anomalous poleward flow, became apparent during the late spring of 1982 and was about three standard deviations above normal by fall and early winter (Fig. 4). The January 27, 1983 value was the highest sea level ever recorded off San Diego (Cayan and Flick, 1985). The sea-surface temperature was largely higher than the long-term mean in the latter half of 1982 and was 2°C above normal for the first three months of 1983 (Figs. 3 and 4). Satellite observations provided evidence of mesoscale changes in sea-surface temperatures in the southern California Bight by December 1982 (Fiedler, 1984).

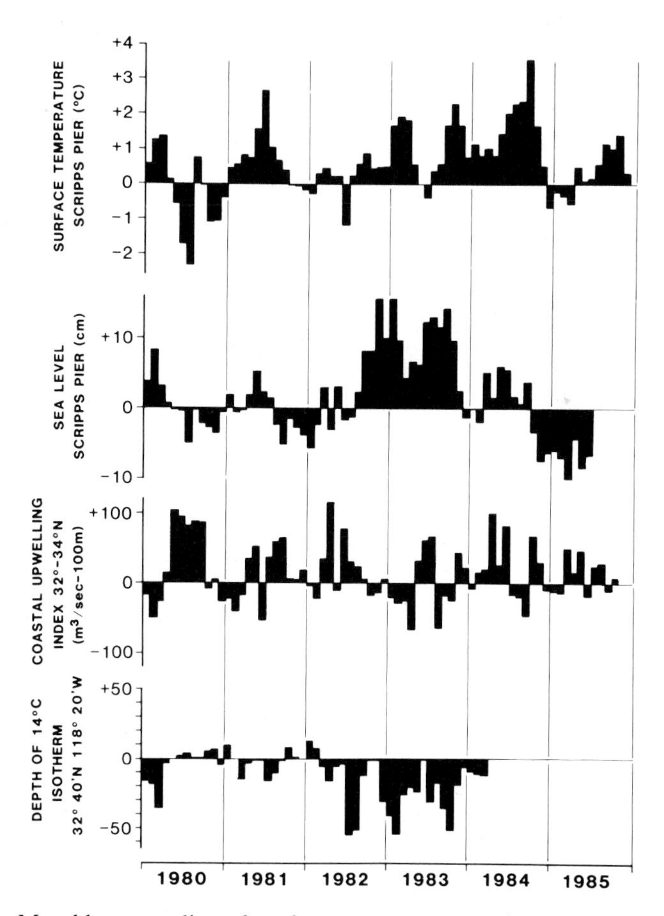

Fig. 4. Monthly anomalies of surface temperature and sea level at the Scripps Institution of Oceanography pier, coastal upwelling index in a 3 × 3 degree square centered at 33°N, 119°W, and depth of the 14°C isotherm at 32·7°N, 118·2°W (adapted from Fiedler *et al.*, 1986).

The winter of 1982–1983 was the most severe storm season in many decades along the west coast of North America (Namias and Cayan, 1984; Seymour *et al.*, 1984), and the extratropical atmospheric circulation was clearly linked to the warming of the east equatorial Pacific Ocean surface (Quiroz, 1983). The unusually deep Aleutian low-pressure center and the intensification of the westerlies resulted in an extraordinary number of severe storms making landfall along the west coast of the United States much further south than normal. There were six events with waves exceeding six meters during the winter of 1982–1983; there were only 18 such events during the period of 1900–1984 (Seymour *et al.*, 1984). A wave monitoring station near Point Loma recorded the highest significant wave height in eight years of measurements (Seymour and Sonu, 1985). Eight storms from January through March had periods of peak energy between 17–22 seconds; only one storm reached 17 seconds in the previous three years. Thus there were more storms, with bigger waves and longer periods, and they devastated *Macrocystis* canopies.

Surface manifestations of the El Niño could not be seen during April and May of 1983; sea-surface temperatures were within the range of normal interannual variability in the California Current system, and coastal upwelling also increased to near normal levels (Fiedler, 1984). However, offshore subsurface temperature anomalies were as great as three standard deviations above normal during this period (Simpson, 1983). By July, large-scale surface warming was again evident with temperatures up to 4°C above normal along the coast (Fiedler, 1984). The temperature, salinity, and dissolved oxygen characteristics of the nearshore waters during this event were consistent with enhanced onshore transport of subarctic water induced by large-scale atmospheric forcing. Onshore transport leads to increases in coastal sea level and large-scale depression of the thermocline (Fig. 4); geostrophic readjustment to these changes causes enhanced poleward flow (Simpson 1983, 1984a). The 14°C isotherm marks the depth of the nutrient-depleted surface waters at this latitude (Fiedler *et al.*, 1986). Depression of the thermocline rendered coastal upwelling ineffective and eliminated internal waves, critical sources of nutrients for kelps, from kelp forest depths.

The tropical ENSO event dissipated by early fall of 1983 (Kerr, 1983). However, the anomalous oceanographic conditions in the California Current system associated with this event persisted through 1984 (Simpson, unpubl. MS). The large-scale subsurface positive temperature and negative salinity anomalies (Simpson, 1984a) were very stable. In the absence of the major atmospheric events required to overturn the water column, these conditions persisted through the mild winter of 1983–1984 and became even warmer in 1984 with normal seasonal heating. Thus the temperature anomaly in 1984 was even greater than in 1983 (Figs. 3 and 4). Slow erosion and storm mixing in the fall of 1984 led to near normal conditions by the end of 1984. The 1940–41 and 1957–59 El Niños also persisted longer off California than in the tropics (Simpson, unpubl. MS).

In the absence of a long-term record of *in situ* bottom temperatures in the kelp forest, the 66 years of sea-surface measurements collected from the Scripps Institution of Oceanography pier constitute the best available data set for evaluating temperature anomalies (North, 1985). List and Koh (1976) used digital filtering and covariance analysis to compare the pier and 10 m temperature records and found no correlation for high-frequency fluctuations (periods of two weeks and less). High-frequency fluctuations largely represent thermocline motion and are local phenomena. There was a very high degree of correlation for intermediate (two weeks to three months) and low-frequency (greater than three months) events, both between the pier and 10 m temperature records and between areas of the southern California Bight (List and Koh, 1976). Because ENSO events are clearly low-frequency, large-scale events, this temperature record is appropriate. The pier sea-surface temperature data for 1982–1985 in relation to the 63–year (1920–1982) mean are shown in Fig. 3.

V. STORM EFFECTS ON KELP FORESTS

The surface manifestations of the storms were dramatic. The *Macrocystis* canopy along the Palos Verdes Peninsula near Los Angeles declined from 196 ha on January 7, 1983 to 18 ha by January 31 (Wilson and Togstad, 1983). The Point Loma canopy was reduced from over 600 ha in fall of 1982 to less than 40 ha. Aerial surveys indicated similar results from Point Conception to Isla Asuncion, Baja California, the worst condition of the kelp resource in the 54-year history of the local kelp harvesting company (R. McPeak, Kelco, pers. comm.).

In addition to canopy loss, there were two major types of storm damage to *Macrocystis*: (1) holdfasts ripped off the substrate by large swells, and (2) entanglement of attached plants with drifting plants and holdfast bundles causing extensive to complete loss of stipes. The recovery of plants damaged but not killed by the storms appears to have been retarded or precluded by the warm, nutrient-poor water associated with the El Niño. The water column averaged 16°C during the stormy period of January through March of 1983, a critical elevation in terms of nitrate availability (Jackson, 1977, 1983; Zimmerman and Kremer, 1984).

Our studies of storm damage were designed to test predictions regarding relative severity of effects on surface canopies versus understory species and damage as a function of depth and location in the Point Loma forest. Diving surveys (see Fig. 5 for the location of our study sites) revealed that the

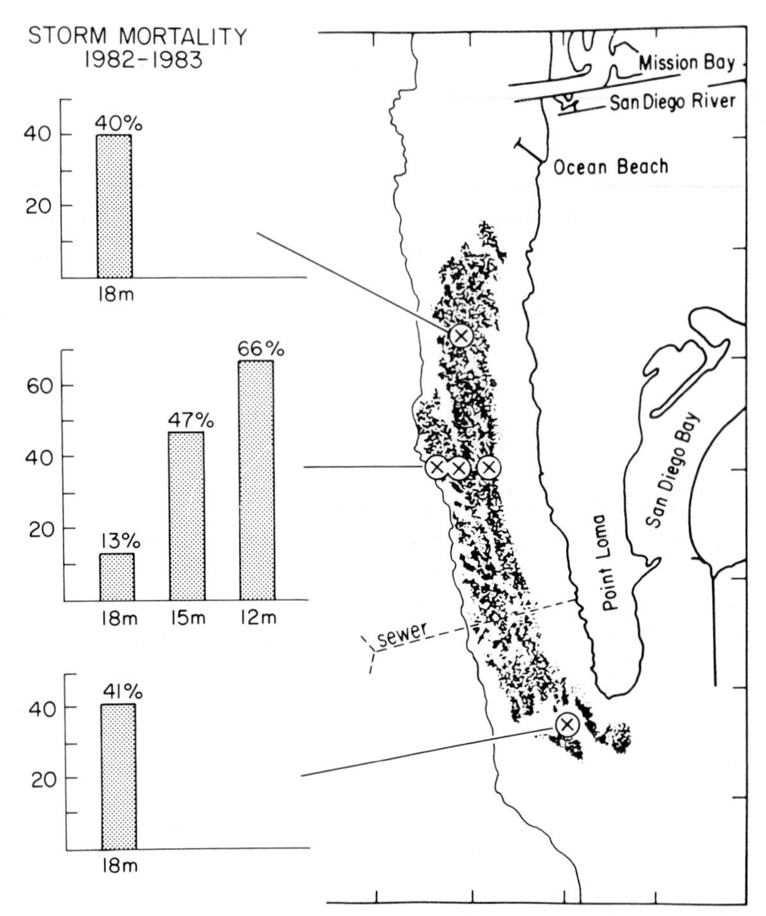

Fig. 5. Mortality of giant kelp, *Macrocystis pyrifera*, attributed to storm damage during the winter of 1982–1983, at five stations in the Point Loma kelp forest (Dayton and Tegner, 1984a). The stippling represents the *Macrocystis* canopy in 1980 and the contour line marks the 18 m (60 ft) depth contour.

damage to *Macrocystis* was not uniform. Mortality, indicated by densities of extant plants and recently killed holdfasts or holdfast scars in areas of 960 to 1 440 m^2, was highest (66%) on the shallow, inner margin of the forest and decreased with depth (Fig. 5). Mortality at the northern and southern ends of the forest was considerably higher than at the central station of the same depth (Dayton and Tegner, 1984a). Two- and three-year-old *Macrocystis* survive better than other ages at Point Loma (Dayton *et al.*, 1984). In the winter of 1975–1976, mortality of two-year-old plants in a 400 m^2 area in the central part of the kelp forest was 6% (N = 84); during the following, more severe winter, mortality of the same cohort of now three-year-old plants was

21% (N = 38). In the winter of 1981–1982, mortality of two-year-old plants in the same area was 7% (N = 56), but the massive storms of 1982–1983 resulted in 44% (N = 48) mortality of the then three-year-olds (Dayton and Tegner, 1984a).

In contrast to the catastrophic effects on *Macrocystis*, storm effects on patches of understory species monitored since 1971 (Dayton *et al.*, 1984) were moderate to nonexistent. Stipitate species were occasionally disturbed by drifting *Macrocystis* holdfast bundles, but there was no apparent damage to prostrate *Laminaria* or turf patches. In each case where an understory patch had bordered *Macrocystis*, the *Macrocystis* was gone or heavily disturbed and the lower standing guild patch had survived (Dayton and Tegner, 1984a).

In marked contrast to Point Loma, storm disturbances caused striking reversals of kelp forest community structure on Naples Reef, a small (2·2 ha), isolated reef near Santa Barbara (Ebeling *et al.*, 1985). A severe storm in 1980 removed the *Macrocystis* canopy but spared most understory kelps. The supply of algal drift, which is largely *Macrocystis*, soon became limiting, causing sea urchins to leave their shelters and consume most living plants. The foraging urchins precluded algal recruitment and weakened the detritus-based food chain, especially the Embiotocid fishes. The storms of 1983 reversed the process because the exposed urchins were subject to the full force of the surge. With grazer populations greatly reduced and good conditions for recruitment and growth of kelps, normal community structure was soon re-established (Ebeling *et al.*, 1985).

VI. EVENTS SUBSEQUENT TO THE STORMS

The spring of 1983 was a brief respite for the kelps at Point Loma. An increase in coastal upwelling to near normal levels (Fig. 4) dropped bottom temperatures in the kelp forest to about 13°C. *Macrocystis* plants which survived the storms formed a sparse surface canopy in late spring. The combination of large amounts of open space cleared by the storms, minimal canopy shading, and the appropriate temperature and light conditions (Dean and Deysher, 1983) led to massive recruitment of many species of algae in May.

There was some giant kelp recruitment at all five sites but we observed hundreds of juvenile *Macrocystis* per square meter at the shallower two sites where storm mortality (Fig. 5) had been highest. Another consequence of the storms was dispersal of algal reproductive material on scales far greater than observed from 1970 to 1982, when we collected dispersal data on the composition of recruits within patches and distance from presumed parent populations (Dayton *et al.*, 1984). *Pterygophora* and *Laminaria* both recruited heavily into areas formerly occupied by *Macrocystis*; however, the reverse was not observed. The most dramatic changes were seen in three species, *Desmarestia ligulata, Dictyopteris undulata*, and *Acrosorium uncinatum*, the first two of which had regularly been seen in low numbers around disturbed areas of our study sites in 1970–1982. A massive *Desmarestia* bloom dominated several hectares around our shallow site, *Dictyopteris* covered large areas of the two deeper central stations, and *Acrosorium* was abundant at the north and south stations. In past years, we had noted the relatively rare occurrences of these species as individual plants. However, McPeak (pers. comm.) reports that blooms of *Desmarestia* are common in barren areas after urchins have been removed. Thus El Niño effects on physical conditions, the structure of surviving *Macrocystis* populations, and scramble competition among competing species of algal recruits all appeared to affect the recovery of community structure at Point Loma.

Another facet of the storm damage at Naples Reef was extensive damage to the reef itself. Massive sections of the shale ridges were broken loose, exposing large areas of virgin rock. Harris *et al.* (1984) found more than ten times greater *Macrocystis* recruitment on these new surfaces than on old rock surfaces, where the presence of coralline algal crusts indicated that they had been in place before the storms. On the old surfaces, most foliage that developed was red algal turf, which apparently regenerated from basal fragments that survived the scouring. The new surfaces were covered with dense stands of filamentous browns. The longer filamentous browns facilitated the survival of *Macrocystis* recruits, apparently by hiding the small sporophytes from herbivorous fishes until the plants attained some refuge in size (Harris *et al.*, 1984).

In contrast to the benign spring, the warm El Niño summer and fall of 1983 had devastating consequences for the Point Loma kelp forest. Major upwelling events, defined as lasting more than six days with a maximum surface temperature reduction of more than 3°C from the long-term mean, usually occur twice a summer (Dorman and Palmer, 1981). Summer upwelling events are apparent in the temperature records for 1982 and 1985, but did not occur in 1983 or 1984 (Fig. 3). Bottom temperatures, measured above 16°C only on rare occasions in the past, were above 16°C from August through October of 1983 and went as high as 21·4°C. (We have had *in situ* recording thermographs at each site since 1983.) The normal thermocline was depressed to depths below the kelp forest (Dayton and Tegner, 1984a) eliminating nutrient input from internal waves. Surface temperatures were even warmer in 1984; with the exception of three brief excursions to the mean, surface temperatures were well above normal until November of 1984 (Fig. 3). Bottom temperatures were again anomalously high.

The *Macrocystis* canopy began to deteriorate during the summer of 1983 and by October, the tops of most plants were 6–8 m below the surface. Since

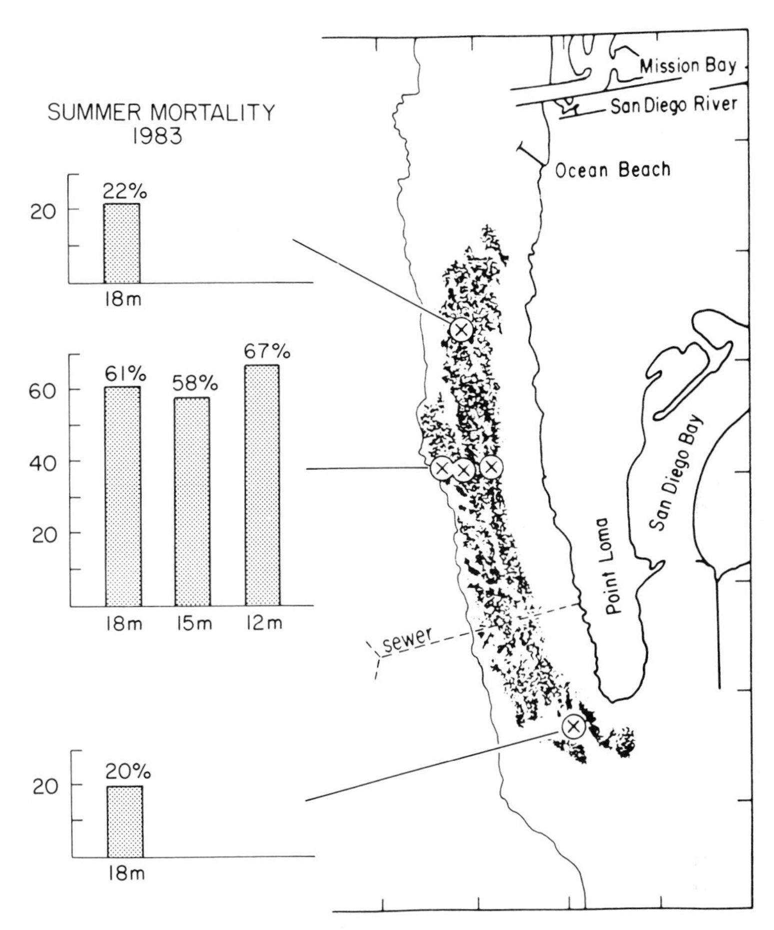

Fig. 6. Mortality of giant kelp, *Macrocystis pyrifera*, plants which survived the storms during the warm summer and fall of 1983. Data are presented from five stations in the Point Loma kelp forest (Dayton and Tegner, 1984a). The stippling represents the *Macrocystis* canopy in 1980 and the contour line marks the 18 m (60 ft) depth contour.

50% or more of the biomass of a healthy *Macrocystis* population is typically found in the upper 1 m of the water column (North *et al.*, 1982), canopy deterioration could have a major impact on drift availability to herbivores. While *Macrocystis* tissue generally appeared healthier near the bottom, there was substantial mortality of adults which survived the storms. Mortality at Point Loma varied with depth and location (Fig. 6). The improved survival at the two end-of-the-forest stations may be a result of their positions; they face into longshore currents where they may be exposed to water not depleted of nutrients by the rest of the forest (Dayton and Tegner,

1984a). In 1975, a non El Niño year, summer mortality of two- and three-year-old *Macrocystis* at the 15 m site in the center of the forest was 14% (N = 128) and 9% (N = 45), respectively. In 1982, pre El Niño summer mortality of the two-year-old plants was 2% (N = 52), and during the 1983 El Niño summer, mortality of the normally robust three-year-old plants was 59% (N = 27) (Dayton and Tegner, 1984a).

Mean nitrogen content (measured as % dry weight), which typically ranges from 1 to 4% in southern California, can be used as an indicator of the nutritional status of *Macrocystis* (Gerard, 1982; North *et al.*, 1982; North and Zimmerman, 1984). Gerard (1982) concluded that the critical level representing no nitrogen reserves for growth was an N content of 1·1% for laminar tissue. During a 20-month study of southern California kelp forests (1979–1980), North *et al.* (1982) found that the N content of canopy blades approached critical levels only once and basal blades not at all. *Macrocystis* at Point Loma had nitrogen reserves after the spring upwelling of 1983; basal blades (five blades averaged per determination) were 2·5 and 2·9% N and canopy blades 1·4 and 1·7% N at two sites in early July. By October 1983, major nitrogen depletion had taken place. Basal blades at the central and southern sites had dropped to 2% N but basal blades at the northern site had dropped to 1·1% N, Gerard's (1982) critical level. "Canopy" (collected as close to the surface as possible) blade averages, which dropped below critical level at all sites, ranged from 0·8 to 1·0% N.

Gerard (1984), working with an experimental population of adult *Macrocystis* at 11 m near Laguna Beach, reported other physiological changes during the summer of 1983. She attributed reductions in chlorophyll content and photosynthetic capacity of canopy blades to nitrogen starvation. Small fronds did not deplete their internal N reserves so their growth, normally dependent on the translocation of photosynthates from canopy fronds, was probably carbon limited. Finally, she suggested that slow growth before N or C content became limiting was due to temperature stress (Gerard, 1984).

There was poor survival of the massive spring *Macrocystis* recruitment at Point Loma, partly because of the heavy understory canopies which are known to interfere with giant kelp recruits (Dayton *et al.*, 1984). The El Niño summer also affected the recruits because those *Macrocystis* which escaped the understory grew slowly, were discolored and often diseased, and by September, the fronds in many areas died 2 to 3 m above the bottom (Dayton and Tegner, 1984a). For example, recruitment at the shallow site was first observed in April but by September the mean plant size was only 137 cm (std. dev. = 61 cm, N = 60), poor growth for a species that can grow as much as 15 cm per day under optimal conditions (Neushul and Haxo, 1963). Some of these recruits survived to form canopy during the spring of 1984.

Growth of juvenile *Macrocystis* in coastal kelp forests of southern California is usually limited by irradiance (Dean and Jacobsen, 1984). Bottom irradiance levels were significantly higher from 1982 through 1984 than in previous years (Dean and Jacobsen, 1986). Dean and Jacobsen (1986) used fertilizer experiments to demonstrate that low juvenile *Macrocystis* growth rates during the El Niño were due to nutrient limitation, not temperature stress. The clear water was apparently related to the southern California Bight-wide reduction in surface phytoplankton pigment concentrations as detected by satellite (Fiedler, 1984) and a deeper chlorophyll maximum layer (McGowan, 1985).

Water temperatures returned briefly to normal during November of 1983 (Fig. 3) and during the winter a *Macrocystis* canopy began to form. However, the surface warmed much faster in 1984 than 1983, and summer temperatures were higher than the year before. Deterioration began as early as June, when the per cent N of surface blades at three sites had dropped below Gerard's critical value (including 0·5% at the shallowest site), and the canopy was largely gone by August. The effects of this El Niño on the *Macrocystis* canopy are apparent in the harvest records; in 1982, Kelco harvested 75% of its long-term average at Point Loma, but nothing was harvested in 1983, and only 9% in 1984, all at the beginning of the year (R. McPeak, pers. comm.).

Kelp forests in the lee of Santa Catalina Is. were protected from the storms but suffered perhaps the most severe effects of the warm summer and fall of 1983 in southern California (Zimmerman and Robertson, 1985). These forests depend upon vertical excursions of the thermocline for nutrient input. The depression of the 15°C isotherm to 50 m, more than twice as deep as 1981, effectively prevented the thermocline from reaching the lower limit of the kelp forest (20 m) on most excursions. Growth rates plummeted, the low rates of frond production were not able to keep up with frond losses at 7 and 10 m, plant sizes declined, and *Macrocystis* was extinct above 10 m by the beginning of November (Zimmerman and Robertson, 1985).

VII. EFFECTS ON HIGHER TROPHIC LEVELS

The decline and delayed recovery of many coastal kelp forests after the El Niño of 1957–1959 were associated with dense populations of sea urchins, especially red (*Strongylocentrotus franciscanus*) and purple (*S. purpuratus*) urchins (North and Pearse, 1970). These species normally feed on drift algae carried by water motion to their protected microhabitats in reefs and rock piles. When drift becomes limiting, hungry urchins leave their habitats to forage on attached plants (Dean *et al.*, 1984; Harrold and Reed, 1985). Moving aggregations or fronts of urchins can denude large areas of kelp forest of all macro-algae, leaving only a pavement of encrusting coralline algae (the "barren areas" of Lawrence, 1975). Enough foraging urchins

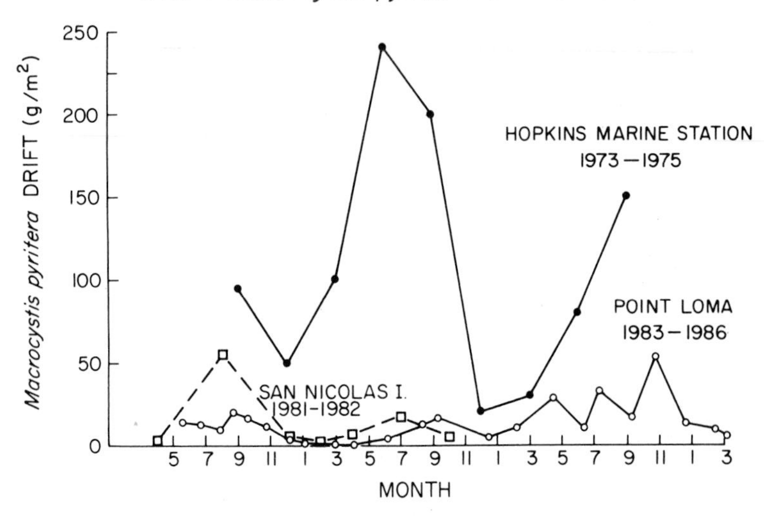

Fig. 7. Drift availability to herbivores. Gerard's (1976) data on the standing crop of *Macrocystis pyrifera* drift at Hopkins Marine Station near Monterey, California in the mid-1970s and Harrold and Reed's (1985) data from a forested area on San Nicolas Is. from 1981–1982 are compared with data collected at Point Loma in the period 1983–1986.

remain to preclude recruitment of kelps so that these barren patches may be stable for months or sometimes years (Dayton and Tegner, 1984b; Harrold and Reed, 1985).

There were three potential impacts of the storms and warm water on sea urchin populations, any of which could lead to further declines in kelp abundance: (1) substantial decrease in algal drift availability; (2) increase in sea urchin recruitment; or (3) decrease in sea urchin predation rates.

Gerard (1976) and Harrold and Reed (1985) demonstrated a seasonal cycle in drift availability; drift is more abundant in summer and fall when senescent fronds accumulate on the forest floor than in winter and spring when higher water movement increases export rates from the kelp forest. It is likely that most of the drift detached by the massive storms was washed out of the forest. We hypothesized that drift would be low during the rest of 1983 and during 1984 due to the reduction in algal standing stocks. Drift availability followed the seasonal cycle (Fig. 7). The quantity was about an order of magnitude less than Gerard (1976) observed at Monterey and similar to what Harrold and Reed (1985) found on San Nicolas Is. during non El Niño periods. No urchin fronts formed at any of our five sites. Much of the drift was decaying, heavily epiphytized *Macrocystis* suggesting that El Niño-induced canopy deterioration was making up for any potential decrease in drift production due to lower standing stock.

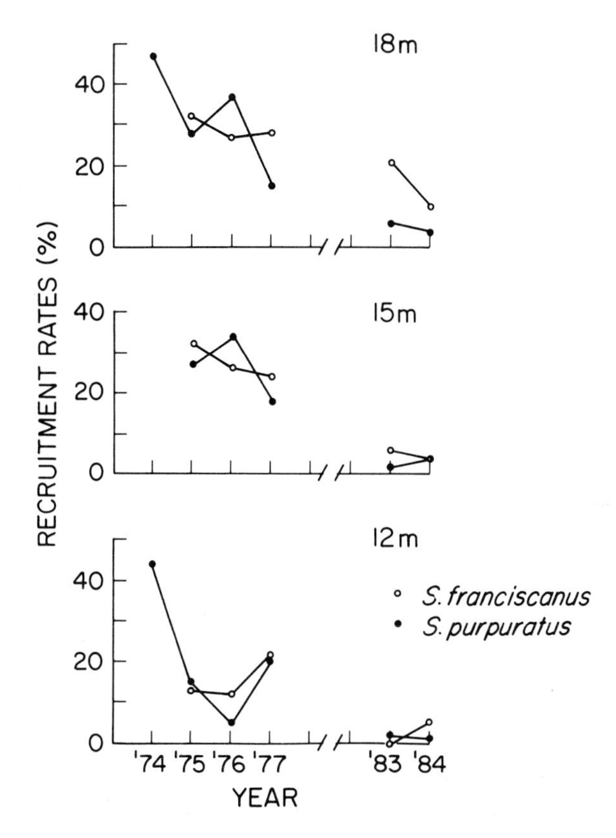

Fig. 8. Sea urchin recruitment rates collected during the mid-1970s (Tegner and Dayton, 1981) and during 1983–1984 at the three central stations in the Point Loma kelp forest. Recruitment rates are expressed as the proportion of the total population represented by young of the year.

A second potential consequence of the storms was increased sea urchin recruitment. In previous years, red urchin recruitment (which we define as the proportion of the population in the zero-year age class) was highest at the outside edge of the *Macrocystis* canopy, or in the center of the forest after localized disruption of the canopy by storms (Tegner and Dayton, 1981). Elimination of the edge effect, after the storms devastated the canopy during the main period when urchins settle, created the potential for strong urchin recruitment throughout the bed. Instead, urchin recruitment was much lower than we have observed previously (Tegner and Dayton, 1981) and zero in some areas (Fig. 8). Complete canopy disappearance during the warm summers of 1983 and 1984 followed by partial recoveries the following winters continued to greatly minimize the edge effect compared to the pre El Niño canopy (Fig. 5) but urchin recruitment remained low in 1984 and only began to recover in 1985. Altered current patterns during these years may

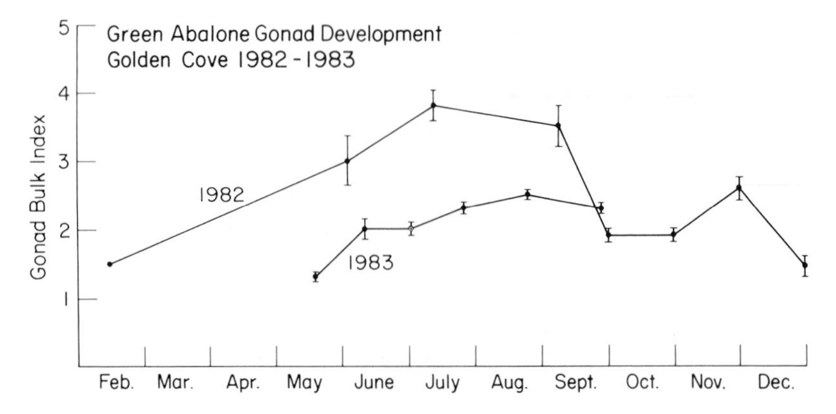

Fig. 9. Green abalone (*Haliotis fulgens*) gonad development during 1982–1983 from a site on the Palos Verdes Peninsula near Los Angeles. The gonad bulk index is a visual index developed by George Lockwood. The scale varies from 1, a spawned-out condition, to 6; spontaneous spawning has been observed at 3·5 or above. For consistency, all grading was done by one person. The curve for 1982 shows the two normal peaks of green abalone spawning. Variability is expressed as ± one standard error (Tegner, in press).

have been unfavorable for advection of these long-lived (30–50 days, Cameron and Schroeter, 1980) larvae. Furthermore, the extremely low nutrient concentrations, surface productivity, and zooplankton biomass in the waters of the southern California Bight during 1983 (McGowan, 1985) suggest that planktotrophic urchin larvae may have starved. With drift always available and no evidence of feeding fronts or a burst in urchin recruitment, it was not surprising that urchin abundances, as censused in 100 m^2 in each of our five sites, did not change significantly from spring 1983 to spring 1984.

The extended stormy period during the winter of 1983 also appears to have interfered with herbivore reproduction. Moderate surge conditions are optimal for the transport of drift algae to herbivores, but in very rough seas, drift may become impossible to catch (Shepherd, 1973). Abalones, like sea urchins, depend primarily on drift algae. We followed gonadal development of green abalones, *Haliotis fulgens*, living in a very shallow (4–6 m) kelp forest on the Palos Verdes Peninsula (Tegner, in press). The storms removed all attached *Macrocystis*; the abalones probably had little if any food from January until algal recruitment in May. Gonadal development in 1983 began later and remained low in comparison to the previous year (Fig. 9). While we did not monitor sea urchin gonadal development during this period, the roe yield of red urchins harvested by the commercial fishery was lower than normal, often to the point of making processing uneconomical, and many processors closed (Kato and Schroeter, 1985). Size–frequency distributions of red (*H. rufescens*) and pink (*H. corrugata*) abalones,

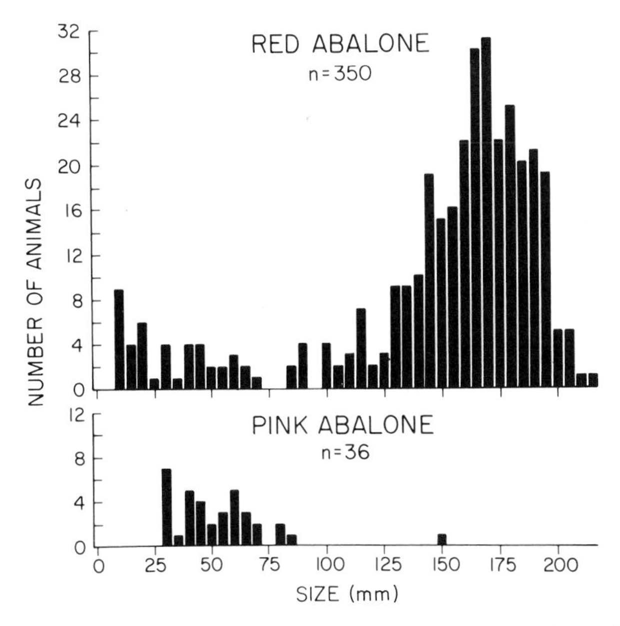

Fig. 10. Size–frequency distributions of red (*Haliotis rufescens*) and pink (*H. corrugata*) abalones collected from Abalone Cove on the Palos Verdes Peninsula in 1986. The nearly inverse relationship of these data suggests that the El Niño affected recruitment of these species in opposing ways.

collected from a mid-depth (6–12 m) location at Palos Verdes in 1986, suggest that El Niño conditions affected the recruitment of these congenors differently (Fig. 10). There was a near absence of sizes corresponding to two- and three-year-old red abalones, animals which would have been spawned during 1983 and 1984; virtually all pink abalones were in these size categories. We speculate that these data reflect altered current patterns during the El Niño.

One of the most dramatic effects of the warm water was on starfish populations, especially the batstar, *Patiria miniata*. Apparent disease led to several episodes of mass mortality of *Patiria* in recent years. Dixon and Schroeter (unpubl. MS) interviewed researchers working at sites from San Diego to Carmel in order to determine the spatial and temporal pattern of the epizootics. The outbreaks began in shallow (<10 m) water in the southern part of the southern California Bight in 1978, a warm water year. (Water temperatures were warmer than normal from 1976 to 1984; Norton *et al.*, 1985.) In 1981, a second outbreak devastated populations in deeper water and as far north as Santa Barbara. The disease reappeared during the El Niño, affecting populations as deep as 18 m and north to Diablo Canyon in central California (Dixon and Schroeter, unpubl. MS). Davis (1985) reports that *Patiria* numbers dropped an order of magnitude in three years from over 1·43 per m^2 in 1982 to 0·86 in 1983 and 0·14 in 1984 in the Channel Islands National Park. Data from 1972 at Point Loma suggest a similar long-term decline of three asteroid species (Table 1). *Patiria* were present in very low numbers at 15 m in early 1983, disappeared by September and have not been re-observed on our transects. (Recruitment began with the onset of cooler temperature in 1985 but juveniles, found only under rocks, are not sampled by our transects.) At two deeper water (18 m) sites, where diseased *Patiria* were noted during the summer of 1983, densities subsequently decreased by 55 and 98%. Adult *Patiria* remain abundant in water deeper than 30 m. *Pisaster giganteus*, predators of molluscs, barnacles, and polychaetes (Morris *et al.*, 1980), declined at Point Loma (Table 1) and on the Channel Is. (Davis, 1985). *Dermasterias imbricata*, predators of purple sea urchins (Rosenthal and Chess, 1972), have also disappeared from mid-depths at Point Loma and are quite rare in deeper water.

Dixon and Schroeter's (unpubl. MS) laboratory experiments strongly suggest a relationship between the disease and temperature. During periods of disease, healthy *Patiria* will sicken and die in a few days at 18°C but animals in 15°C water remain healthy. Sick individuals will heal if placed in cold water. Disease also appeared to be implicated in a sea urchin die-off in shallow water at Santa Catalina Is. and some of the other warm water Channel Is. in the fall of 1984 (Dixon and Schroeter, unpubl. data; Tegner, pers. obs.). Mass mortality of diseased urchins off Nova Scotia, Canada was reported during years of record high sea-surface temperatures (Scheibling and Stephenson, 1984).

Patiria appear to control the distribution of the white sea urchin, *Lytechinus anamesus*, generally restricting this small echinoid to areas outside of kelp forests (Schroeter *et al.*, 1983). Concurrent with the *Patiria* die-offs, the Channel Islands monitoring program found that *Lytechinus* numbers increased from 0·97 per m^2 in 1984 to 8·27 per m^2 in 1985 (Davis, 1985). Destructive grazing by *Lytechinus* in 1985 has been reported from Channel Is. (Davis, pers. comm.), Palos Verdes (Wilson, pers. comm.), and San Onofre (Dixon and Schroeter, unpubl. MS) kelp forests of southern California. Before the effective loss of *Patiria* (densities declined from 1–2 per m^2 in 1981 to near zero in 1983) in the San Onofre kelp forests of

Table 1

Asteroid densities from central Point Loma (15 m), 1971–1985. Data represent number per square meter from a sample of 400 m^2.

	1972	1981	Dec. 1982	Apr. 1983	Sept. 1983	Jan. 1984	Apr. 1984	Aug. 1984	May 1985	July 1985
Patiria miniata	0·23	0·005	0·005	0	0	0	0	0	0	0
Pisaster giganteus	0·22	0·11	0·04	0·05	0·09	0·06	0·04	0·03	0·02	0·07
Dermasterias imbricata	0·08	0·03	0·01	0·02	0·01	0	0	0	0	0

northern San Diego County, the community role of *Lytechinus* was to limit kelps by grazing on small stages, thereby reducing successful recruitment (Dean *et al.*, 1984). After a combination of recruitment and migration increased the *Lytechinus* abundance within the kelp forest, their community role changed. Grazing on adult *Macrocystis* led to giant kelp mortality and created a spatial refuge for competitively inferior understory species. These changes may be long lasting if *Lytechinus* continues to prevent *Macrocystis* recruitment (Dixon and Schroeter, unpubl. MS). To date, *Lytechinus* are rare at our Point Loma study sites but have been observed in high numbers at other areas near the south end of the bed (McPeak, pers. comm.).

The most important predators of red and purple sea urchins in southern California are spiny lobsters, *Panulirus interruptus*, and the labrid fish *Semicossyphus pulcher*, commonly called the California sheephead (Tegner and Dayton, 1981; Tegner and Levin, 1983; Cowen, 1983). Both species have their center of abundance in Baja California, and southern California is near the northern boundary of both distributions. Spiny lobsters recruit into very shallow (0–4 m) waters and enter kelp forests when they get older (Engle, 1979); the general impact of the El Niño on their recruitment is not yet known. Cowen (1985) reports that spiny lobsters showed strong recruitment success at San Nicolas Is. in 1983 after little or no success in the previous four years. Sheephead recruitment reached Monterey in 1983, nearly 250 km north of their normal range, and was very strong in areas below the northern range limit which do not receive regular recruitment. As sheephead live for 20–25 years in these areas (Cowen, 1985), the recruitment effects of this El Niño on kelp forest community structure may be long lived.

The physical events of the 1982–1984 El Niño and their effects on southern California kelp forest communities are summarized in Table 2.

Table 2
Summary of the effects of the 1982–1984 El Niño on Southern California kelp forest communities.

Physical events	Effects on kelp forest communities
Unprecedented series of large storms	Loss of *Macrocystis* canopy, considerable *Macrocystis* mortality, provision of open space
Relative normal spring upwelling	Outstanding recruitment of *Macrocystis* and many other species
Extraordinarily warm summer–fall water temperatures	Nutrient depletion leading to loss of *Macrocystis* canopies, considerable mortality and reduced growth rates
Elevated sea level and depressed thermocline	Nutrient input from coastal upwelling and thermocline motion effectively ceases
Unusually clear water	Normally light limited, *Macrocystis* growth becomes nutrient limited
Altered current patterns	Effects on animal recruitment

VIII. PROSPECTS FOR RECOVERY

Sea-surface temperature finally returned to the mean in November of 1984 and was close to "normal" in 1985 (Fig. 3). *Macrocystis* canopy development followed quickly in most areas of southern California. A notable exception was the stretch of mainland coast west of Santa Barbara, an area protected from most swell activity. These extensive forests were based on a different form of giant kelp, similar to *M. angustifolia*, which produces a very large holdfast, sufficiently large to anchor the plant on sand in some areas. This form reproduces vegetatively as well as sexually (Neushul, 1971). Vegetative reproduction results in large holdfast systems which apparently act as settlement sites for spores. The massive storms of 1982–1983, out of a more westerly direction than usual (Seymour *et al.*, 1984), caused high mortality here as elsewhere. But because these plants grew on unstable sediments, the loss of most holdfasts has left minimal substrate for recruitment. There has subsequently been some recruitment but density is very low (C. Barilotti, D. Glanz, R. McPeak, Kelco, pers. comm.). Recovery here is likely to take many years. W. North (pers. comm.) reports that this variety of *Macrocystis* used to form a continuous forest from Dana Point to San Mateo Rocks in Orange County, a semiprotected stretch of coastline which is largely sand. The forest disappeared during the 1957–1959 El Niño and this form of *Macrocystis* is no longer found in the area.

The massive recruitment of understory algae and two subsequent summer–fall losses of the *Macrocystis* canopy led to the possibility of El Niño-induced changes in kelp community composition. While established stands of understory kelps have considerable resistance to invasion by giant kelp (Dayton *et al.*, 1984), *Macrocystis* rapidly grows above the lower standing species under conditions of scramble competition (Pearse and Hines, 1979; McPeak, 1981; Dayton and Tegner, 1984a; Reed and Foster, 1984). Repeated disturbances of the giant kelp appear to be necessary for understory species such as *Pterygophora* to form a canopy sufficiently dense to inhibit further algal recruitment (Reed and Foster, 1984). Two canopy losses were not adequate disruptions of *Macrocystis* for *Pterygophora* recruits to establish a dense canopy at Point Loma; apparently, the shortened giant kelp plants continued to preempt most of the light. When we compared *Pterygophora* populations under *Macrocystis* with those in a large clearing where the giant kelp was completely removed, it took about a year for the *Pterygophora* in the clearing to establish a canopy dense enough to inhibit further algal recruitment. The *Pterygophora* under *Macrocystis* grew at about one-third the rate of the plants in the clearing, put almost no energy into reproduction and had a significantly higher mortality rate (Tegner *et al.*, in prep.). Similarly, the Channel Is. monitoring program observed a peak in understory percent cover of 79% in 1983, which rapidly declined to 39% in 1984, and 19% in 1985 (Davis, 1985).

While *Macrocystis* retained its competitive dominance during this period, an apparent secondary effect of the El Niño raised havoc with algal populations at Point Loma in 1985 (Tegner *et al.*, in prep.). With minimal storm activity and cool temperatures, the winter of 1985 was very favorable for kelp growth. However, the canopy which formed soon began losing ground to an infestation of grazing amphipods, especially the kelp curler, *Amphithoe humeralis*. Grazing damage occurred throughout the 10 km forest but large areas of the southern end of the forest were completely denuded of algae. Amphipod outbreaks were also observed in coastal forests west of Santa Barbara (R. McPeak, pers. comm.) and near Point Dume (M. Tegner, pers. obs.). The amphipod species were not exotic; all are found in healthy kelp forests. The most likely explanation was a reduction in the populations of fishes which feed on kelp-associated invertebrates, especially the kelp surf perch (*Brachyistius frenatus*). *Brachyistius* are closely associated with the *Macrocystis* canopy and populations may sharply decline or disappear entirely during decreases in kelp density (Coyer, 1979; Ebeling *et al.*, 1980); the canopy at Point Loma was lost three times in a two-year period. The *Macrocystis* canopy declined from 632 ha in January 1985 to 275 ha in July, almost a 60% reduction during the seasons when the canopy should have been increasing in area and density. As a result, Kelco's Point Loma harvest for 1985 was only 31% of an average year (R. McPeak, pers. comm.).

IX. DISCUSSION

The ENSO event of 1982–1983 produced clear physical and biological signals along the west coast of North America as far north as Alaska (see papers in Wooster and Fluharty, 1985; Mysak, 1986). Fiedler (1984) and McGowan (1985) documented the profound effects in the California Current system: pronounced deepening of the thermocline, significant warming of the mixed layer, weakened coastal upwelling, altered surface currents, reduced productivity and a deep chlorophyll maximum layer, and greatly reduced macrozooplankton biomass. Not surprisingly, these changes were reflected in commercial fish landings; catches of chinook salmon, market squid, crab, shrimp, and other species were down by 70% or more in 1983 (McGowan, 1984). However, this depression was partially offset by great increases in the catch of warm-water species such as albacore, yellowtail, yellowfin and skipjack tuna, marlin, dorado, etc. The northern anchovy of the California Current is an ecological analog of the Peruvian anchoveta; both are relatively small, short-lived, pelagic planktivores in highly variable eastern boundary currents (Fiedler *et al.*, 1986). Similar to the anchoveta (see Arntz, 1986), the El Niño affected the growth, spawning range, fecundity, and early larval mortality of the northern anchovy. However, due to strong recruitment of the 1984 year class, stock size of the minimally exploited anchovy returned to pre El Niño levels in 1985 (Fiedler *et al.*, 1986). Evaluation of these physical and biological consequences of El Niños has been possible because of the 37 years of the California Cooperative Fisheries Investigations (see McGowan, 1985) and long-term fisheries data.

In contrast, there are very few examples of long-term data for benthic systems. The effects of this El Niño on intertidal algae in southern California were ambiguous; Gunnill (1985) was unable to find a uniform response in seven species of macro-algae. There was considerable mortality during the storms and warm-water episodes but net recruitment by some species was relatively high. He suggested that elevated sea level and fall cloud cover may have mitigated the effects of the El Niño. Similarly, Paine (1986), in a relatively long-term study, was unable to distinguish any changes in the rocky intertidal on the Washington coast from background variability. We speculate that intertidal species, adapted to a variable, more rigorous environment than the subtidal, are better equipped to deal with El Niños. Interestingly, the Peruvian intertidal was dramatically affected in 1983; apparently the extremely high temperatures (10–11°C above normal) killed the barnacles and invertebrate grazers and released the algae (Arntz, 1986).

The effects of the 1982–1983 El Niño on California kelp forest communities varied both locally and regionally. With the exception of the lee side of Santa Catalina Is. (Zimmerman and Robertson, 1985), the storms caused massive damage to forests from San Diego (Dayton and Tegner, 1984a) in the south, to Palos Verdes (Wilson and Togstad, 1983), Naples Reef (Ebeling *et al.*, 1985), and San Nicolas Is. (Harrold, 1985) in southern California, to Diablo Canyon (Kimura, 1985) and the Monterey Bay region (M. Foster, J. Roughgarden, pers. comm.) in central California. While canopy loss was essentially complete, island sheltering of storm waves may have minimized *Macrocystis* mortality in some areas, especially where the islands are very close to shore (Pawka *et al.*, 1984). Negative impacts of the warm summer temperatures, however, were considerably more limited; the major effects were reported from Santa Catalina Is. (Zimmerman and Robertson, 1985) and along the southern mainland coast from Laguna Beach (Gerard, 1984) to San Diego (Dayton and Tegner, 1984a). In part, the differences can be attributed to temperature decrease with latitude. Comparing Fig. 3 with the 1983 temperatures from Naples Reef (Ebeling *et al.*, 1985, Fig. 2), it is apparent that while values were well above normal for that location, both the magnitude and the duration of the warm event at Santa Barbara were considerably less than at La Jolla, some 300 km to the south. In part, the differences can be attributed to ocean circulation patterns. Santa Catalina Is.'s position in the southern California Bight is within the area affected by an intrusion of warm, southern, offshore water every year during summer, fall, and winter and is consistently warmer than the adjacent mainland. In contrast, San Nicolas Is., which is offshore and slightly to the south of Santa Catalina Is., lies within an area of the California Current which is consistently colder and more eutrophic (Pelaez and McGowan, 1986). At Santa Catalina Is., the El Niño associated depression of the thermocline and the temperature anomaly added to an already warmer than average environment for *Macrocystis,* leading to the worst El Niño summer effects on kelps reported (Zimmerman and Robertson, 1985). There are no temperature data available from San Nicolas Is., but the rapid recovery of *Macrocystis* after the storms suggests that serious nutrient depletion did not take place (Harrold, 1985). Finally, some of the differences may be ascribed to anthropogenic influences. Water temperatures offshore of the Palos Verdes Peninsula were anomalously warm for an extended period of 1983 but *Macrocystis* flourished (Tegner, unpubl. data). B. Jones (pers. comm.) has recently found high concentrations of NH_4 from the nearby Los Angeles County sewage outfall in kelp forest depths during summer thermally stratified conditions.

Storm damage to *Macrocystis* populations in central California was so severe that in many cases, *Nereocystis leutkeana*, an annual surface canopy-former which competes with giant kelp, was able to greatly expand its patch boundaries (Kimura, 1985; M. Foster and G. Van Blaricom, pers. comm.). Kimura (1985) reported that *Nereocystis* populations near Diablo Canyon senesced prematurely during the 1983 El Niño. Diablo Canyon is about the southern range boundary for this species (Abbott and Hollenberg, 1976), so warm-water effects on *Nereocystis* were not surprising. *Macrocystis*, while more susceptible to wave stress than *Nereocystis*, is the competitive dominant, so given adequate sporogenic material and minimal storm activity, is likely to re-establish pre El Niño community structure in the near future (Dayton *et al.*, 1984).

The relevance of El Niños to kelp forest community structure must be related to recurrence rates. Using historical data for evidence of tropical ENSO-type events, Quinn *et al.*, (1978) found an average of 12·3 years between the onsets of strong events and 5·4 years between the onset of strong and moderate El Niños; adding weak events dropped the interval to 3·7 years. While not all tropical El Niños are reflected in the California Current (Chelton *et al.*, 1982), there are mid-latitude warm events which do not occur in concert with tropical El Niños (Simpson, 1984b; Norton *et al.*, 1985). Using the criterion of a minimum time of one month of Scripps pier temperature $\geq 0{\cdot}5°C$ above the 63-year mean (Fig. 3), North (1985) classified one-third of the 1955–1984 period as warm water. There were 33 separate warm-water events, of which seven events lasted six months or more and two

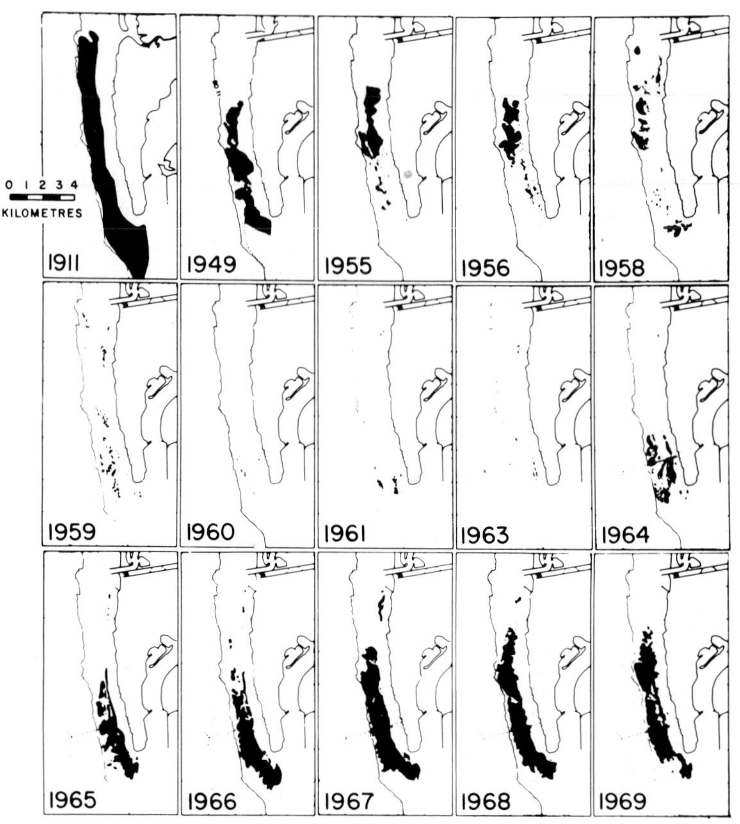

Fig. 11. Historical records of the surface canopy of the Point Loma kelp forest. The contour line represents the 18 m (60 ft) depth contour (courtesy of W. J. North, California Institute of Technology).

lasted 15 months. Thus warmer than normal periods are a regular feature of the California coastal environment. The effects on kelp community structure will depend upon the season of the warm period (e.g., given the normal summer stress on *Macrocystis*, warm-water events then are more serious than a similar temperature elevation in winter), and the duration and magnitude of warming (North, 1985).

Major storms are another factor strongly affecting the results of warm-water periods on kelp community structure. Seymour *et al.* (1984) found a highly significant correlation between strong El Niños and large wave events in southern California. However, not all El Niños result in large wave events. During 1982–1983, the Aleutian low was not only unusually deep (similar to many El Niño winters), but the low was, on average, large enough in areal extent and displaced eastward sufficiently to affect California (Seymour *et al.*, 1984). During other El Niño winters, high pressure prevails along the west coast and storms generated in the North Pacific make landfall

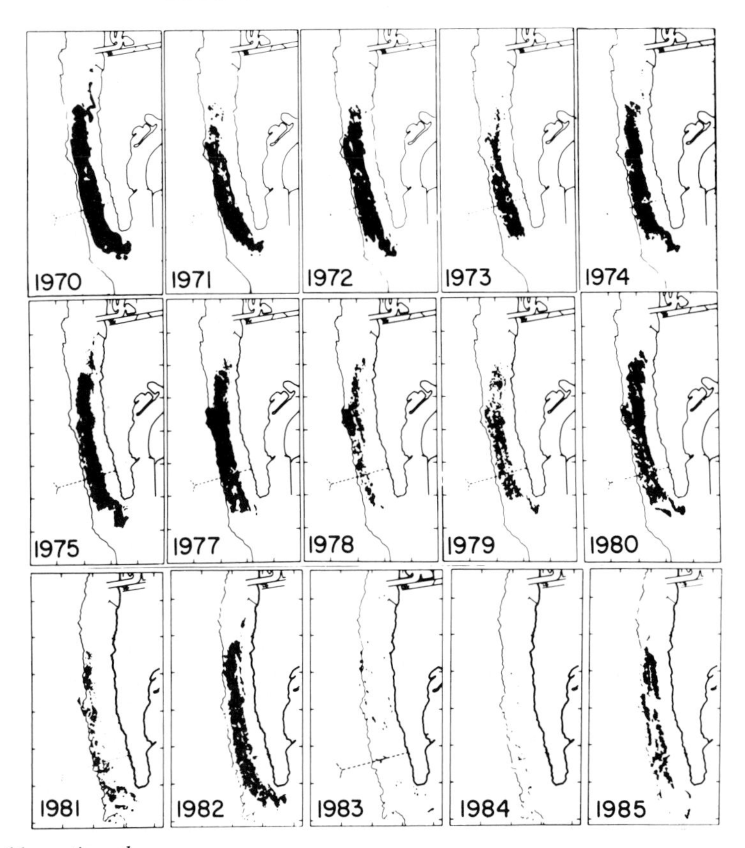

Fig. 11 *continued.*

in Alaska and British Columbia. The El Niños of 1976–1977 and 1982–1984 were both marked by warm water off southern California but the former event was nearly storm-free (Namais and Cayan, 1984; Cayan and Flick, 1985).

Two major El Niños stand out in recent California history, the events of 1957–1959 and 1982–1984. Both were catastrophic-scale disturbances but the two events differed physically and had very different implications for kelp forest community structure. There were four wave episodes exceeding 3 m during 1957–1959; there were ten such episodes during the latter El Niño, an unprecedented record (Seymour *et al.*, 1984). The former warm-water period was longer but the 1982–1984 event included a substantial period of unparalleled high sea-surface temperature anomaly (North, 1985). Because the large temperature anomalies exacerbated the effects of the storms, we believe the recent El Niño to be the larger disturbance, but many coastal kelp forests showed strong recoveries in 1985, something that took several years in the 1960s (Fig. 11). Two critical factors are different. First, spiny lobsters and sheephead, the important predators of red and purple sea urchins in southern California, have been heavily fished for many years. Thus even small reductions in drift availability during 1957–1959 were likely to have been enough to push an already out-of-balance situation towards destructive grazing (Tegner, 1980; Tegner and Dayton, 1981; Tegner and Levin, 1983; Dean *et al.*, 1984; Harrold and Reed, 1985). A red sea urchin fishery developed in the early 1970s (Kato and Schroeter, 1985) and has had a major impact on standing stocks of this species, the largest and most destructive of the California sea urchins. The role of urchin grazing was consequently much less in 1982–1984 than in 1957–1959. Second, waste water disposal practices of the 1940s and 1950s had led to substantial decreases in *Macrocystis* standing stocks in several coastal kelp forests, notably at Palos Verdes and Point Loma, before the El Niño of 1957–1959 (Wilson *et al.*, 1977). The eventual recovery of these forests was facilitated by improvements in waste water treatment, kelp restoration and sea urchin control efforts, and the sea urchin fishery. Thus while the disturbances caused by the El Niño of 1982–1984 were larger, the destabilized kelp forests of the late 1950s were much more severely affected.

While attempting to integrate local ecological factors with the physical parameters of the 1957–1959 and 1982–1984 El Niños, it is important to remember that tropical ENSO events vary considerably in magnitude (Quinn *et al.*, 1978). Furthermore, even the large-scale events may be very different from each other; recall that the recent El Niño was not preceded by an intensification of the trade winds, something formerly believed to be a necessary condition for an ENSO event (Ramage, 1986). Physical differences are also apparent in the California record. Both the 1957–1959 and the 1982–1984 events were characterized by very high temperatures, but the former had anomalously high salinity and the latter had anomalously low salinity water (McGowan, 1985). High-temperature, high-salinity water is consistent with transport from the south, but the only possible low salinity source is Pacific Subarctic water from the west (Simpson, 1984a). This suggests different physical forcing mechanisms. Simpson (1984a,b) argues that anomalous atmospheric circulation, an atmospheric teleconnection to the tropical ENSO event, produced the 1940–1941 and 1982–1984 California El Niños. However, there is biological evidence that planktonic larvae of southern species moved far to the north during 1983 (Cowen, 1985). Other physical oceanographers present evidence for both oceanic and atmospheric forcing of the recent event off the west coast of North America (Huyer and Smith, 1985; Norton *et al.*, 1985; Reinecker and Mooers, 1986). Although there is clearly much to be learned about these large-scale oceanographic phenomena, we know that there are physical differences between events, and that different forcing mechanisms are likely to have very different biological ramifications.

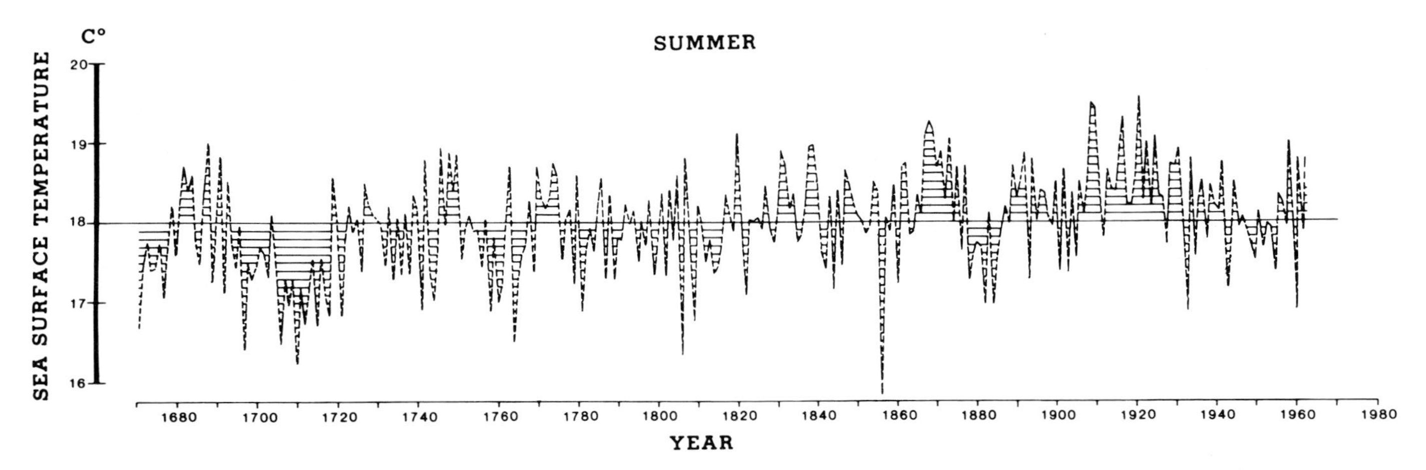

Fig. 12. Summer sea temperatures 5 m below surface at La Jolla, California reconstructed from standardized tree ring indices for *Pseudotsuga macrocarpa* from 1 214 m on the Santa Ana Mountains, California. Data from Douglas (1976, p. 188) were put into graphical form by Gerry Kuhn (Dayton and Tegner, 1984b).

Finally, from the biological perspective, the opposite side of the coin may be even more important to the California Current system. Anti El Niño years, characterized by strong southward transport, low sea level, reduced temperature and salinity, and high zooplankton abundance (Chelton *et al.*, 1982), are years of high biological productivity. Indeed, Barilotti *et al.* (1984) have developed an index based on the depth of the 14°C isotherm which correlates with *Macrocystis* standing crop and two measures of ENSO activity, sea level and the Southern Oscillation index (Fig. 4). Ocean temperatures extrapolated from tree ring data (Fig. 12) suggest that in the decade to century time scales, these anti El Niños could also be very important to the entire coastal system.

ACKNOWLEDGEMENTS

We thank R. Butler, G. Chen, T. Klinger, C. Lennert, P. Parnell, and K. Plummer for countless hours in the field, L. Snider for identifying the amphipods, E. Stewart for Fig. 3, P. Fiedler for Fig. 4, W. North for Fig. 11, and G. Kuhn for Fig. 12. J. Dixon, T. Klinger, C. Lennert, R. McPeak, W. North, P. Parnell, S. Schroeter, J. Simpson, and R. Zimmerman commented on the manuscript. J. Dixon and S. Schroeter kindly shared unpublished data. We gratefully acknowledge continuing discussions with C. Barilotti, G. Jackson, J. McGowan, W. North, and R. McPeak. This work is a result of research sponsored by NSF, the U.C.S.D. Academic Senate, and NOAA, National Sea Grant College Program, Department of Commerce, under grant number NA80AA-D-00120, project number R/NP-1-12F, through the California Sea Grant Program. The U.S. Government is authorized to reproduce and distribute for governmental purposes.

REFERENCES

Abbott, I. A. and Hollenberg, G. J. (1976). *Marine Algae of California*. Stanford, CA: Stanford University Press, 827 pp.

Arntz, W. E. (1986). The two faces of El Niño 1982–1983. *Meeresforsch*. **31**, 1–46.

Barber, R. T. and Chavez, F. P. (1983). Biological consequences of El Niño. *Science* **222**, 1203–1210.

Barber, R. T. and Chavez, F. P. (1986). Ocean variability in relation to living resources during the 1982–1983 El Niño. *Nature* **319**, 279–285.

Barilotti, D. C., McLain, D. R. and Bauer, R. A. (1984). Forecasting standing crops of *Macrocystis* from the depths of the 14°C isotherm. *EOS* **65**, 909.

Bernal, P. A. (1981). A review of the low frequency response of a pelagic ecosystem in the California Current. *California Cooperative Oceanic Fisheries Investigations Reports* **22**, 49–64.

Cameron, R. A. and Schroeter, S. C. (1980). Sea urchin recruitment: effect of substrate selection on juvenile distribution. *Mar. Ecol. Prog. Ser.* **2**, 243–247.

Cane, M. A. (1983). Oceanographic events during El Niño. *Science* **222**, 1189–1195.

Cayan, D. R. and Flick, R. E. (1985). *Extreme Sea Levels in San Diego, California, Winter 1982–1983*. Scripps Institution of Oceanography Reference No. 85-3, 1–58.

Chapman, A. R. O. (1981). Stability of sea urchin-dominated barren grounds following destructive grazing of kelp in St. Margaret's Bay, eastern Canada. *Mar. Biol.* **62**, 307–311.

Chelton, D. B., Bernal, P. A. and McGowan, J. A. (1982). Large-scale interannual physical and biological interactions in the California Current. *J. Mar. Res.* **40**, 1095–1125.

Clendenning, K. A. (1968). A comparison of the annual harvesting yields of certain California kelp beds, 1950–1960. *Calif. Dept. Fish and Game, Fish Bull.* **139**, 213–217.

Cowen, R. K. (1983). The effect of sheephead (*Semicossyphus pulcher*) predation on red sea urchin (*Strongylocentrotus franciscanus*) populations: an experimental analysis. *Oecologia (Berlin)* **58**, 249–255.

Cowen, R. K. (1985). Large scale patterns of recruitment by the labrid, *Semicossyphus pulcher*: causes and implications. *J. Mar. Res.* **43**, 719–742.

Coyer, J. A. (1979). The invertebrate assemblage associated with *Macrocystis pyrifera* and its utilization as a food resource by kelp forest fishes. *Ph.D. Dissertation*, 364 pp. University of Southern California, Los Angeles, California.

Cushing, D. H. (1982). *Climate and Fisheries*. London: Academic Press, 373 pp.

Davis, G. E. (1985). Kelp forest dynamics in Channel Islands National Park, California, 1982–1985. *Channel Islands National Park and National Marine Sanctuary Natural Science Study Reports* CH15-86-001, 1–11.

Dayton, P. K. (1985). Ecology of kelp communities. *Ann. Rev. Ecol. Syst.* **16**, 215–245.

Dayton, P. K., Currie, V., Gerrodette, T., Keller, B., Rosenthal, R. and Ven Tresca, D. (1984). Patch dynamics and stability of some southern California kelp communities. *Ecol. Monogr.* **54**, 253–289.

Dayton, P. K. and Tegner, M. J. (1984a). Catastrophic storms, El Niño, and patch stability in a southern California kelp community. *Science* **224**, 283–285.

Dayton, P. K. and Tegner, M. J. (1984b). The importance of scale in community ecology: a kelp forest example with terrestrial analogs. In: *Novel Approaches to Interactive Systems* (Edited by P. W. Price, C. N. Slobodchikoff and W. S. Gaud). New York: Wiley, pp. 457–481.

Dean, T. A. and Deysher, L. E. (1983). The effects of suspended solids and thermal discharges on kelp. In: *The Effects of Waste Disposal on Kelp Communities* (Edited by W. Bascom). Institute of Marine Resources, University of California, San Diego, pp. 114–135.

Dean, T. A. and Jacobsen, F. R. (1984). Growth of juvenile *Macrocystis pyrifera* (Laminariales) in relationship to environmental factors. *Mar. Biol.* **83**, 301–311.

Dean, T. A. and Jacobsen, F. R. (1986). Nutrient limited growth of juvenile kelp, *Macrocystis pyrifera*, during the 1982–84 "El Niño" in southern California. *Mar. Biol.* **90**, 597–601.

Dean, T. A., Schroeter, S. C. and Dixon, J. (1984). Effects of grazing by two species of sea urchins (*Strongylocentrotus franciscanus* and *Lytechinus anamesus*) on recruitment and survival of two species of kelp (*Macrocystis pyrifera* and *Pterygophora californica*). *Mar. Biol.* **78**, 301–313.

Dorman, C. E. and Palmer, D. P. (1981). Southern California summer coastal upwelling. In: *Coastal Upwelling* (Edited by F. A. Francis). Washington, D.C.: American Geophysical Union, pp. 44–56.

Douglas, A. V. (1986). Past air–sea interactions over the eastern north Pacific Ocean as revealed by tree ring data. *Ph.D. Thesis*, University of Arizona, Tucson, 196 pp.

Ebeling, A. W., Larson, R. J. and Alevizon, W. S. (1980). Habitat groups and island–mainland distribution of kelp-bed fishes off Santa Barbara, California. In: *Multidisciplinary Symposium on the California Islands* (Edited by D. M. Power). Santa Barbara, California: Santa Barbara Museum of Natural History, pp. 403–431.

Ebeling, A. W., Laur, D. R. and Rowley, R. J. (1985). Severe storm disturbances and reversal of community structure in a southern California kelp forest. *Mar. Biol.* **84**, 287–294.

Engle, J. M. (1979). Ecology and growth of juvenile California spiny lobster, *Panulirus interruptus* (Randall). *Ph.D. Dissertation,* 298 pp. University of Southern California, Los Angeles, California.

Eppley, R. W., Renger, E. H. and Harrison, W. G. (1979). Nitrate and phytoplankton production in southern California coastal waters. *Limnol. Oceanogr.* **24**, 483–494.

Estes, J. A. and Palmisano, J. F. (1974). Sea otters: their role in structuring nearshore communities. *Science* **185**, 1058–1060.

Fiedler, P. C. (1984). Satellite observations of the 1982–83 El Niño along the U.S. Pacific coast. *Science* **224**, 1251–1254.

Fiedler, P. C., Methot, R. D. and Hewitt, R. P. (1986). Effects of the California El Niño on the northern anchovy. *J. Mar. Res.* **44**, 317–338.

Gerard, V. A. (1976). Some aspects of material dynamics and energy flow in a kelp forest in Monterey Bay, California. *Ph.D. Dissertation*, 172 pp. University of California, Santa Cruz.

Gerard, V. A. (1982). Growth and utilization of internal nitrogen reserves by the giant kelp *Macrocystis pyrifera* in a low-nitrogen environment. *Mar. Biol.* **66**, 27–35.

Gerard, V. A. (1984). Physiological effects of El Niño on giant kelp in southern California. *Mar. Biol. Lett.* **5**, 317–322.

Gunnill, F. C. (1985). Population fluctuations of seven macroalgae in southern California during 1981–1983 including effects of severe storms and El Niño. *J. Exp. Mar. Biol. Ecol.* **85**, 149–164.

Harris, L. G., Ebeling, A. W., Laur, D. R. and Rowley, R. J. (1984). Community recovery after storm damage: a case of facilitation in primary succession. *Science* **224**, 1336–1338.

Harrold, C. (1985). Impact of storms of 1982–1983 on kelp forest communities of San Nicolas Island, California. *Abstracts of the Western Society of Naturalists' 66th Annual Meeting, Monterey, California,* pp. 38–39.

Harrold, C. and Reed, D. C. (1985). Food availability, sea urchin grazing and kelp forest community structure. *Ecology* **66**, 1160–1169.

Huyer, A. and Smith, R. L. (1985). The signature of El Niño off Oregon, 1982–1983. *J. Geophys. Res.* **90**, 7133–7142.

Jackson, G. A. (1977). Nutrients and production of the giant kelp, *Macrocystis pyrifera*, off southern California. *Limnol. Oceanogr.* **22**, 979–995.

Jackson, G. A. (1983). The physical and chemical environment of a kelp community. In: *The Effects of Waste Disposal on Kelp Communities.* Institute of Marine Resources, University of California, pp. 11–37.

Kato, S. and Schroeter, S. C. (1985). Biology of the red sea urchin, *Strongylocentrotus franciscanus*, and its fishery in California. *Mar. Fish. Rev.* **47**, 1–20.
Kerr, R. A. (1983). Fading El Niño broadening scientists' view. *Science* **221**, 940–941.
Kimura, S. (1985). Differences in long-term patterns of algal change in Diablo Cove, central California subtidal habitats. *Abstracts of the Western Society of Naturalists' 66th Annual Meeting, Monterey, California*, pp. 45–66.
Lawrence, J. M. (1975). On the relationships between marine plants and sea urchins. *Oceangr. Mar. Biol. Ann. Rev.* **13**, 213–286.
Leighton, D. L. (1971). Grazing activities of benthic invertebrates in kelp beds. *Nova Hedwigia* **32**, 421–453.
List, E. J. and Koh, R. C. Y. (1976). Variations in coastal temperatures on the southern and central California coast. *J. Geophys. Res.* **81**, 1971–1979.
Mann, K. H. (1977). Destruction of kelp beds by sea urchins: a cyclical phenomenon or irreversible degradation? *Helgolander wiss. Meeresunters* **30**, 455–467.
McGowan, J. A. (1984). The California El Niño, 1983. *Oceanus* **27**, 48–51.
McGowan, J. A. (1985). El Niño 1983 in the southern California Bight. In: *El Niño North, Niño Effects in the Eastern Subarctic Pacific Ocean* (Edited by W. S. Wooster and D. L. Fluharty). Washington Sea Grant Program, University of Washington, Seattle, pp. 166–184.
McPeak, R. H. (1981). Fruiting in several species of Laminariales. *Proc. Int. Seaweed Symp.* **8**, 404–409.
Morris, R. H., Abbott, D. P. and Haderlie, E. C. (1980). *Intertidal Invertebrates of California*. Stanford, Ca: Stanford University Press, 690 pp.
Mysak, L. A. (1986). El Niño, interannual variability and fisheries in the northeast Pacific Ocean. *Can. J. Fish. Aquat. Sci.* **43**, 464–497.
Namias, J. and Cayan, D. R. (1984). El Niño: implications for forecasting. *Oceanus* **27**, 41–47.
Neushul, M. (1971). The species of *Macrocystis* with particular reference to those of North and South America. *Nova Hedwigia* **32**, 211–222.
Neushul, M. and Haxo, F. T. (1963). Studies on the giant Kelp *Macrocystis*. I. Growth of young plants. *Am. J. Bot.* **50**, 349–353.
North, W. J. (1985). Health of kelp beds. In: *Report on 1984 Data, Marine Environmental Analysis and Interpretation, San Onofre Nuclear Generating Station*. Southern California Edison 85-RD-37 pp. 6-1–6-43.
North, W. J. and Pearse, J. S. (1970). Sea urchin population explosion in southern California waters. *Science* **167**, 209.
North, W. J., Gerard, V. and Kuwabara, J. (1982). Farming *Macrocystis* at coastal and oceanic sites. In: *Synthetic and Degradative Processes in Marine Macrophytes*. Berlin: de Gruyter, pp. 247–262.
North, W. J. and Zimmerman, R. C. (1984). Influences of macronutrients and water temperatures on summertime survival of *Macrocystis* canopies. *Hydrobiologia* **116/117**, 419–424.
Norton, J., McLain, D., Brainhard, R. and Husby, D. (1985). The 1982–83 El Niño event off Baja and Alta California and its ocean climate context. In: *Niño North, Niño Effects in the Eastern Subarctic Pacific Ocean* (Edited by W. S. Wooster and D. L. Fluharty). Washington Sea Grant Program, University of Washington, Seattle, pp. 44–72.
Oceanus, **27**(2), 1984
Paine, R. T. (1986). Benthic community – water column coupling during the 1982–1983 El Niño. Are community changes at high latitudes attributable to cause or coincidence? *Limnol. Oceanogr.* **31**, 351–360.

Pawka, S. A., Inman, D. L. and Guza, R. T. (1984). Island sheltering of surface gravity waves: model and experiment. *Cont. Shelf. Res.* **3**, 35–53.
Pearse, J. S. and Hines, A. H. (1979). Expansion of a central California kelp forest following the mass mortality of sea urchins. *Mar. Biol.* **51**, 83–91.
Pelaez, J. and McGowan, J. A. (1986). Phytoplankton pigment patterns in the California Current as determined by satellite. *Limnol. Oceanogr.* **31**, 927–950.
Quinn, W. H., Zopf, D. O., Short, K. S. and Yang, R. T. W. K. (1978). Historical trends and statistics of the southern oscillation, El Niño and Indonesian droughts. *Fish. Bull., U.S.* **76**, 663–678.
Quiroz, R. S. (1983). The climate of the "El Niño" winter of 1982–83—a season of extraordinary climatic anomalies. *Monthly Weather Rev.* **111**, 1685–1706.
Radovitch, J. (1961). Relationships of some marine organisms of the northeast Pacific to water temperatures particularly during 1957 through 1959. *Calif. Dept. Fish and Game Fish Bull.* **112**, 1–62.
Ramage, C. S. (1986). El Niño. *Scient. Am.* **254**, 76–83.
Rasmusson, E. M. (1984). El Niño: the ocean/ atmosphere connection. *Oceanus* **27**, 4–12.
Rasmusson, E. M. and Wallace, J. M. (1983). Meteorological aspects of the El Niño/southern oscillation. *Science* **222**, 1195–1202.
Reed, D. and Foster, M. (1984). The effects of canopy shading on algal recruitment and growth in giant kelp (*Macrocystis pyrifera*) forest. *Ecology* **65**, 937–948.
Reinecker, M. and Mooers, C. N. K. (1986). The 1982–1983 El Niño signal off northern California. *J. Geophys. Res.* **91**, 6597–6608.
Rosenthal, R. J. and Chess, J. R. (1972). A predator–prey relationship between the leather star (*Dermasterias imbricata*) and the purple urchin (*Strongylocentrotus purpuratus*). *Fish. Bull., U.S.* **70**, 205–216.
Scheibling, R. E. and Stephenson, R. L. (1984). Mass mortality of *Strongylocentrotus droebachiensis* (Echinodermata: Echinoidea) off Nova Scotia, Canada. *Mar. Biol.* **78**, 153–164.
Schroeter, S. C., Dixon, J. and Kastendiek, J. (1983). Effects of the starfish *Patiria miniata* on the distribution of the sea urchin *Lytechinus anamesus* in a southern California kelp forest. *Oecologia (Berlin)* **56**, 141–147.
Seymour, R. J., Strange, R. R. III, Cayan, D. R. and Nathan, R. A. (1984). Influence of El Niños on California's wave climate. In: *Nineteenth Coastal Engineering Conference, Proceedings of the International Conference, Sept. 3–7, 1984, Houston, Texas* (Edited by B. L. Edge). Am. Soc. Civil Engineers, New York, vol. I, pp. 577–592.
Seymour, R. J. and Sonu, C. J. (1985). Storm waves and shoreline damage on the California coast, winter of 1983. *Jap. J. Coast. Engng* **70**, 59–64.
Shepherd, S. A. (1973). Studies on southern Australian abalone (genus *Haliotis*). I. Ecology of five sympatric species. *Aust. J. Mar. Freshwat. Res.* **24**, 217–257.
Simpson, J. J. (1983). Large-scale thermal anomalies in the California Current during the 1982–1983 El Niño. *Geophys. Res. Letts* **10**, 937–940.
Simpson, J. J. (1984a). El Niño-induced onshore transport in the California Current during 1982–1983. *Geophys. Res. Letts* **11**, 241–242.
Simpson, J. J. (1984b). Warm and cold episodes in the California current: a case for large-scale mid-latitude atmospheric forcing. In: *Proceedings of the Ninth Annual Climate Diagnostic Workshop, Oct. 22–26, 1984.* Corvallis, Oregon, Dept. of Atmospheric Sciences, Oregon State University, pp. 173–184.
Tegner, M. J. (1980). Multispecies considerations of resource management in southern California kelp beds. *Can. Tech. Rep. Fish. Aquat. Sci.* **945**, 125–143.

Tegner, M. J. The California abalone fishery: production, ecological interactions, and prospects for the future. In: *Scientific Approaches to Management of Invertebrate Stocks* (Edited by J. F. Caddy). New York: Wiley, in press.

Tegner, M. J. and Dayton, P. K. (1981). Population structure, recruitment and mortality of two sea urchins (*Strongylocentrotus franciscanus* and *S. purpuratus*) in a kelp forest. *Mar. Ecol. Prog. Ser.* **5**, 225–268.

Tegner, M. J. and Levin, L. A. (1983). Spiny lobsters and sea urchins: analysis of a predator–prey interaction. *J. exp. Mar. Biol. Ecol.* **73**, 125–150.

Wallace, J. M. (1985). Atmospheric response to equatorial sea-surface temperature anomalies. In: *El Niño North, Niño Effects in the Eastern Subarctic Pacific Ocean* (Edited by W. S. Wooster and D. L. Fluharty). Washington Sea Grant Program, University of Washington, Seattle, pp. 9–21.

Wilson, K. C., Haaker, P. L. and Hanan, D. A. (1977). Kelp restoration in southern California. In: *The Marine Plant Biomass of the Pacific Northwest Coast* (Edited by R. Krauss). Oregon State University Press, Corvallis, Oregon, pp. 183–202.

Wilson, K. C. and Togstad, H. (1983). Storm caused changes in the Palos Verdes kelp forests. In: *The Effects of Waste Disposal on Kelp Communities.* Institute of Marine Resources, University of California, pp. 301–307.

Wooster, W. S. and Fluharty, D. L. (1985). *El Niño North, Niño Effects in the Eastern Subarctic Pacific Ocean*. Washington Sea Grant Program, University of Washington, Seattle, 312 pp.

Zimmerman, R. C. and Kremer, J. N. (1984). Episodic nutrient supply to a kelp forest ecosystem in southern California. *J. Mar. Res.* **42**, 591–604.

Zimmerman, R. C. and Kremer, J. N. (1986). *In situ* growth and chemical composition of the giant kelp, *Macrocystis pyrifera*: response to temporal change in ambient nutrient availability. *Mar. Ecol. Prog. Ser.* **27**, 277–285.

Zimmerman, R. C. and Robertson, D. L. (1985). Effects of the 1983 El Niño on growth of giant kelp, *Macrocystis pyrifera*, at Santa Catalina Island. *Limnol. Oceanogr.* **30**, 1298–1302.

Communities of Parasitoids Associated with Leafhoppers and Planthoppers in Europe

N. WALOFF AND M. A. JERVIS

ADVANCES IN ECOLOGICAL RESEARCH Vol. 17
ISBN 0-12-013917-0

I. GENERAL INTRODUCTION

Only a few studies, of a preliminary nature, have yet been attempted on the communities of insect parasitoids associated with Auchenorrhyncha (leafhoppers, planthoppers and related insects), partly because identification of Strepsiptera, Dryinidae and Mymaridae (Hymenoptera) and Pipunculidae (Diptera) is difficult. In this respect, we have been extremely fortunate in having available the important monographs of Olmi (1984) on Dryinidae, and Kinzelbach (1971, 1978) on Strepsiptera. However, our knowledge of the ecology and basic biology of these groups of interesting parasitoids is still rudimentary, with even the hosts of many species being unknown and the interactions between the different taxa remaining unexplored. The complexity of parasitoid–host and parasitoid–parasitoid interactions can be immediately surmized by examining the illustrations of community structure in Figs. 1 and 2.

The dramatic increase in populations of Auchenorrhyncha on tropical crops, particularly on rice, which followed the "Green Revolution", has stimulated much basic research both on pest species of planthoppers and leafhoppers and on their natural enemies, the latter with a view to the possibilities of their use in biological control. These investigations are now being actively pursued, and an enormous literature has grown around them. However, far less ecological effort is being directed towards studying the parasitoids of Auchenorrhyncha in temperate regions, and this is why we have attempted to summarize the available literature, and to limit ourselves to the European fauna.

We have divided this review into two parts. The first deals with the biology, systematics and taxonomy of the parasitoids, while the second is concerned with field studies and more general aspects of host specificity and population dynamics (including the role of individual behaviour), and also

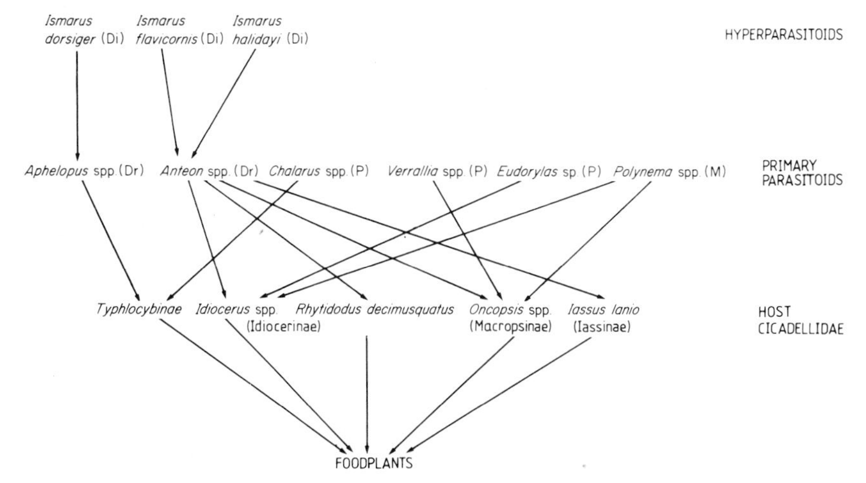

Fig. 1. Assemblages of parasitoids and Auchenorrhyncha (all Cicadellidae) occupying deciduous woodland canopy in Europe. Within these, we have described a parasitoid community associated with Typhlocybine leafhoppers. *Cicadella viridis* and its egg parasitoids are not included in the figure, although Huldén (1984) and Arzone (1974b) found that it occurs in woodland canopy, as well as in open-field habitats (Fig. 2). Arrows point towards the food source. Di, Diapriidae; Dr, Dryinidae; P, Pipunculidae; M, Mymaridae.

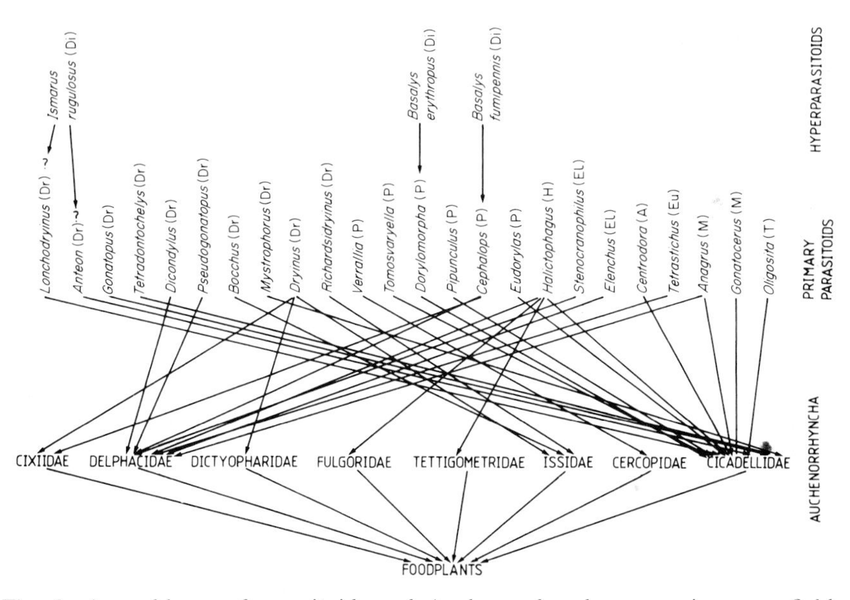

Fig. 2. Assemblages of parasitoids and Auchenorrhyncha occupying open-field habitats, including low shrubs, in Europe. Within these assemblages, we have described two particular parasitoid communities, one associated with non-typhlocybine Cicadellidae, the other with Delphacidae. The parasitoid community associated with Typhlocybine is not included in the figure. Di, Dr, P and M as in Fig. 1; A, Aphelinidae; Eu, Eulophidae; T, Trichogrammatidae; El, Elenchidae; H, Halictophagidae.

current knowledge of community structure. Appendices are also provided, containing some of the original data on which the review is based.

We hope that by emphasizing the enormous gaps in our knowledge, others will be stimulated into investigating these fascinating insects.

Part 1. GENERAL BIOLOGY, SYSTEMATICS AND TAXONOMY OF PARASITOIDS

I. INTRODUCTION

Table 1 lists the known parasitoid groups associated with Auchenorrhyncha worldwide, and indicates, where possible, the host taxa (families) and host

Table 1
The known parasitoid groups that are associated with Auchenorrhyncha, worldwide.

	Parasitoids												
Hosts	HYMENOPTERA Pteromalidae	Encyrtidae	Aphelinidae	Eulophidae	Mymaridae	Trichogrammatidae	Scelionidae	Dryinidae	DIPTERA Pipunculidae	Sarcophagidae	STREPSIPTERA Halictophagidae	Elenchidae	COLEOPTERA Rhipiceridae
Cixiidae								N/A[a]	N/A[a]				
Delphacidae	?			E[a]	E[a]	E	?	N/A[a]	N/A[a]		N/A[a]	N/A[a]	
Dictyopharidae								N/A[a]				N/A	
Fulgoridae			E				E	N/A			N/A[a]	N/A	
Tropiduchidae								N/A					
Tettigometridae		E				E					N/A[a]		
Issidae							?	N/A[a]			N/A[a]		
Flatidae		E		+			E	N/A	N/A			N/A	
Nogodinidae								N/A					
Lophopidae		E		E				N/A					
Acanaloniidae								N/A					
Ricaniidae			E					N/A				N/A	
Eurybrachidae		E									N/A	N/A	
Cicadidae			E[a]			E				A			N
Cercopidae		N	E		E	E			A[a]		N/A		
Membracidae		N[a]	E		E[a]	E		N/A			N/A		
Cicadellidae		N/A	E[a]	E	E[a]	E[a]		N/A[a]	N/A[a]		N/A[a]	N/A	
Eurymelidae								N/A					

From Craighead (1921), Myers (1930), Norris (1970) and Patel (1968); other sources are given elsewhere in this paper.

E, host egg parasitized; N, host nymphs parasitized; A, host adult parasitized.

[a] Recorded in Europe.

Epipyropidae are not included in the table, as it has not been established whether they are in fact parasitoids.

?, questionable.

stages that are parasitized, and whether the parasitoid–host association has been recorded in Europe.

Askew and Shaw (1986) argue that the classification of parasitoids into ectoparasitic and endoparasitic forms is less useful than the one proposed by Häeselbarth (1979) in which parasitoids are divided into "koinophytes" and "idiophytes" (subsequently termed koinobionts and idiobionts by Askew and Shaw—the terminology adopted here). Koinobionts permit the host to grow or metamorphose beyond the stage attacked, and benefit from the continued life of the host, whereas idiobionts consume the host in the location and state in which it is attacked. On the basis of this classification, the principal parasitoid groups (Fig. 3) associated with European Auchenorrhyncha consist of both types: Dryinidae, Pipunculidae and Strepsiptera are koinobionts, while Mymaridae are idiobionts.

II. EGG PARASITOIDS

A. Mymaridae

1. Systematic Position, Taxonomy and General Biology

The family Mymaridae, together with the Trichogrammatidae, includes the smallest known insects (Burks, 1979). Adult mymarids, commonly known as "fairy flies", are highly distinctive in appearance, with slender bodies, exceptionally long antennae, and long narrow wings that are fringed with long setae. All species, so far as is known, develop endoparasitically in the eggs of other insects. Hosts include Odonata, Orthoptera, Hemiptera, Psocoptera, Neuroptera, Coleoptera, Lepidoptera and Diptera (Burks, 1979; Clausen, 1940). Several genera are associated with Auchenorrhyncha (Table 2).

The systematics and morphology of Mymaridae have recently been reviewed by Schauff (1984) and Gibson (1986). Whereas Annecke and Doutt (1961) considered the family to be polyphyletic, Königsmann (1978), Schauff (1984) and Gibson (1986) concluded that it is probably monophyletic. The relationship of the Mymaridae to other Chalcidoidea remains unclear (Gibson, 1986).

Keys to genera are given by Annecke and Doutt (1961), Debauche (1948), Kryger (1950), Peck *et al.* (1964), Schauff (1984) and Subba Rao (1968, 1983), while descriptions and keys to species are contained in Ali (1979), Bakkendorf (1934), Debauche (1948), Kryger (1950), Matthews (1986), Soyka (1946, 1956), Subba Rao (1968, 1983) and Walker (1979).

Confusion surrounds the species-level taxonomy of Mymaridae, especially in the four largest genera, *Anagrus, Anaphes, Gonatocerus* and *Polynema* (Matthews, 1986). This mainly stems from the small size of adult

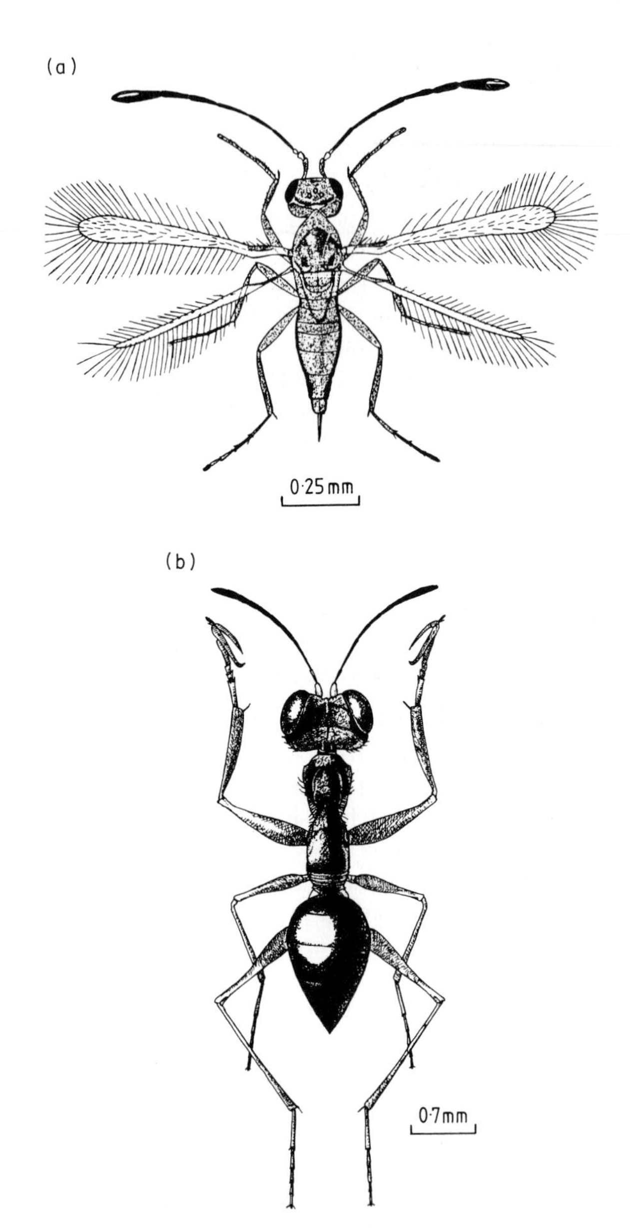

Fig. 3. Representatives of the major parasitoid groups associated with European Auchenorrhyncha: (a) female of *Anagrus flaveolus* (Hymenoptera, Mymaridae); (b) female of *Gonatopus sepsoides* (Hymenoptera, Dryinidae); opposite (c) male of *Chalarus spurius* (Diptera, Pipunculidae); (d) male of *Elenchus tenuicornis* (Strepsiptera, Elenchidae); (e) female of *Ismarus dorsiger* (Hymenoptera, Diapriidae). (a) Reproduced by kind permission of M. F. Claridge; (b) reproduced by kind permission of M. Olmi; (c) after Bankowska (1973); (d) after Kinzelbach (1978); (e) after Hellén (1964).

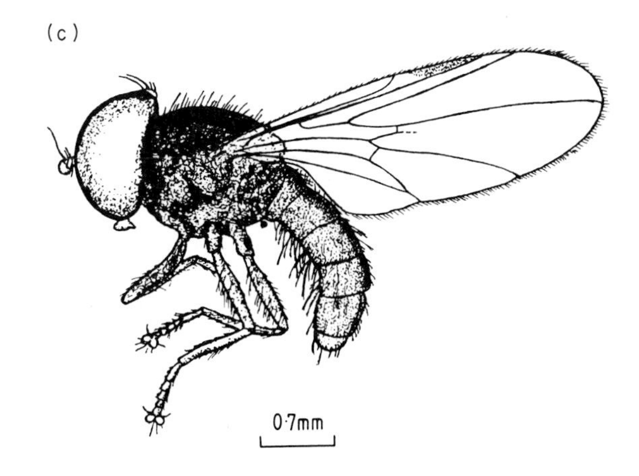
(c)
0·7mm

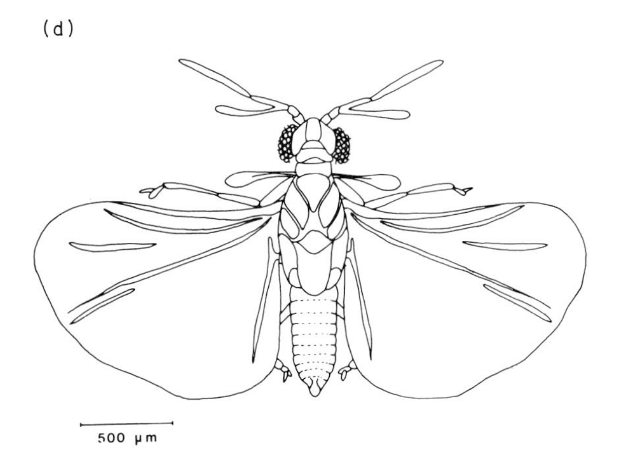
(d)
500 μm

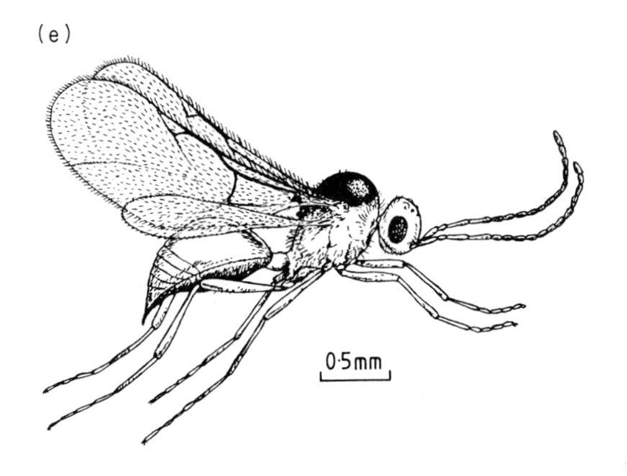
(e)
0·5mm

Table 2
Mymarid genera known to be associated with leafhopper and planthopper hosts.

Parasitoids	Hosts			
	Membracidae	Cercopidae	Cicadellidae	Delphacidae
Gonatocerus	+	−	+	+[a]
Ooctonus	−	+	+	−
Arescon	−	−	+	−
Camptoptera	−	−	+	−
Erythmelus	−	−	+	−
Anaphes	−	−	+	−
Stethynium	+	−	+	−
Anagrus	−	+	+	+
Polynema	+	−	+	+
Acmopolynema	−	+	−	−
Mymar	−	−	+	−
Chaetomymar	−	−	+	+

From Balduf (1928), Chu and Hirashima (1981), Doutt (1961), Dozier (1932), Ewan (1961), Fattig (1955), F. Herard (unpubl.), Miura *et al.* (1979), Pickles (1933), Schauff (1984) and Subba Rao (1983).
[a] Questionable.

mymarids, which makes their collection, preservation and, most importantly, identification difficult. Field collections have, until recently, been made using the least effective techniques (e.g. sweeping; see Kryger, 1950). Also, there have previously been no standardized methods for preserving and mounting adult material (Noyes (1982) has suggested major improvements in techniques, notably the use of the critical-point drying technique for preserving adults). However, much of the confusion has been caused by workers who have either used names, or described new species, without reference to type material.

It is now clear that the most significant taxonomic advances are to be made through biological studies, as has been demonstrated by Walker (1979) and Ali (1979). Both workers have shown that in the genus *Anagrus*, species that are very similar morphologically may differ markedly in aspects of their biology, e.g. oviposition behaviour, courtship and mating behaviour and host range.

2. *Foraging and Oviposition Behaviour*

Female *Anagrus, Polynema* and *Gonatocerus* walk rapidly over plant surfaces when foraging for hosts, drumming the substrate with the clubs of their antennae. They display arrestment, increased antennal drumming and also ovipositor probing in reponse both to host oviposition sites (Ali, 1979; Chandra, 1980c; Perkins, 1905d; Sahad, 1984) and to artificially produced punctures in the plant epidermis (Ali, 1979). Ali (1979) found that *Anagrus* sp. "d" spent significantly less time probing freshly-made artificial punctures than they did older punctures around which a plant wound reaction had developed. The latter most closely resemble host oviposition scars.

The oviposition behaviour of *Anagrus* has been described by Ali (1979), Moratorio (1977) and Witsack (1973), that of *Gonatocerus* by Bakkendorf (1934), and that of *Polynema* by Bakkendorf (1934) and Balduf (1928). Having located a suitable egg site, the female drills with her ovipositor. Some species drill directly into the eggs, via the egg-site opening (Ali, 1979; Bakkendorf, 1934; Balduf, 1928; Perkins, 1905d; Vidano and Meotto, 1968), whereas others drill through the surrounding plant tissue (Ali, 1979; Chandra, 1980c; Rothschild, 1967). *Anagrus mutans* displays both types of behaviour (Moratorio, 1977). Among the *Anagrus* species studied by Ali (1979), including some attacking the same host species, significant interspecific differences in behaviour were recorded with respect both to the time spent performing individual activities, i.e. ovipositor tapping, drilling and withdrawal, and to the total time spent attacking each host egg site. Moratorio (1977) found similar differences between *A. mutans* and *A. silwoodensis*. However, in this case the differences were not constant, but varied both in relation to the host species the parasitoid was reared from, and in relation to the host species attacked.

The genus *Anagrus* includes both solitary and gregarious species. Of the eight investigated by Ali (1979), six are solitary and two are gregarious. *A mutans* and *A. silwoodensis* are facultatively gregarious (Moratorio, 1977). In the laboratory, females deposit several ova in eggs of *Cicadella viridis* (*mutans* 1–3, *silwoodensis* 2–4), while they deposit only one ovum in the smaller eggs of *Dicranotropis hamata*. Both species avoid superparasitism, consistently rejecting already parasitized eggs of *D. hamata* following ovipositor insertion (rejection is indicated by the rapid withdrawal of the ovipositor). They sometimes accept parasitized eggs of *C. viridis* (Moratorio, 1977). Whether females, when ovipositing in parasitized hosts, do so in those containing the smallest number of parasitoid progeny, remains to be determined. The ability of females to distinguish in this way between hosts containing different numbers of eggs, has been demonstrated in a variety of gregarious Chalcidoidea (van Lenteren *et al.*, 1976; van Lenteren, 1981).

According to Arzone (1974a,b), neither *Gonatocerus cicadellae* nor *Polynema woodi* oviposit in host eggs that are already parasitized either by conspecifics or, surprisingly, by other mymarid and trichogrammatid species.

Rearing data obtained by Moratorio (1977) from field-collected eggs of *C. viridis* strongly suggest that females of the arrhenotokous gregarious species *A. mutans* and *A. silwoodensis* deposit fewer male than female progeny per host egg. Their sex allocation behaviour thus appears to resemble that of other egg parasitoids such as *Trichogramma evanescens* (Waage and Ng, 1984). The theory of Local Mate Competition predicts that the evolutionarily stable sex ratio for a parasitoid population should decrease from 0·5 (proportion of males) as mating becomes restricted to the offspring of fewer and fewer female wasps. If very few females colonize each patch of host resource (egg, egg mass) and there is also a high degree of sib-mating, conditions which we suspect are the norm in a number of *Anagrus* species, the optimal sex ratio produced by a female should be low (for reviews see Waage, 1982; Waage and Ng, 1984). This behaviour would account for some of the highly female-biased field and laboratory sex ratios that have been observed generally in both gregarious and solitary arrhenotokous species of *Anagrus* (Ali, 1979; Armstrong, 1936; Bakkendorf, 1934; MacGill, 1934; Maillet, 1960; May, 1971; McKenzie and Beirne, 1972; Meyerdirk and Moratorio, unpubl.; Moratorio, 1977; Ôtake, 1969; Raatikainen, 1967; Sahad, 1984; Tay, 1972; Walker, 1979; Whalley, 1956). Sex ratio is a potentially important component affecting the dynamics of parasitoid–host interactions (Comins and Wellings, 1985; Hassell and Waage, 1984; Waage and Hassell, 1982).

Little detailed information is available on the number of progeny allocated per host egg site by a female. Both *Anagrus* sp. "c" and *A. incarnatus* apparently attack only one egg per *Conomelus anceps* egg site (Ali, 1979; Rothschild, 1967), whereas *A.* sp. "d" deposits a variable number. According to Sahad (1982a, 1984) female *A. optabilis* and *G. cincticipitis* do not leave an egg site until all the eggs have been parasitized. Apparently, *A. mutans* and *A. silwoodensis* do not parasitize host egg batches smaller than 5 (*mutans*) or 6 (*silwoodensis*) (Moratorio, 1977).

Much remains to be discovered about progeny and sex allocation, both per host egg and per host egg site, in parasitoids of leafhopper/planthopper eggs. For a detailed discussion of these aspects of parasitoid reproductive strategies, see Waage and Godfray (1985).

Most *Anagrus* species studied so far are arrhenotokous (Table 3); that is, they have diploid females that develop sexually from fertilized eggs, and haploid males that develop parthenogenetically from unfertilized eggs. A few appear to be either thelytokous or amphitokous. (Thelytokous Hymenoptera have no males, and the diploid females develop parthenogenetically; amphitokous Hymenoptera produce both diploid females and haploid males by parthenogenesis, most eggs yielding females, see Walter (1983) for a review.)

Several, possibly all, *Anagrus* species are pro-ovigenic *sensu* Flanders (1950), i.e. females emerge with their full complement of ripe or nearly ripe ova (Ali, 1979; Meyerdirk and Moratorio, unpubl.; Moratorio, 1977). There is no preoviposition period (MacGill, 1934; May, 1972; Meyerdirk and Moratorio, unpubl.; Witsack, 1973).

There is significant intra- and interspecific variability in the potential

Table 3
Arrhenotokous reproduction in *Anagrus, Gonatocerus and Polynema.*

Species	*References*
A. atomus	Raatikainen (1967), Whalley (1969), Witsack (1973)
A. delicatus[a]	Stiling and Strong (1982a)
A. ensifer	Witsack (1973)
A. flaveolus[a]	Chandra (1980c)
A. sp. nr *flaveolus*[a]	Ôtake (1969)
A. giraulti[a]	Meyerdirk and Moratorio (unpubl.)
A. incarnatus	Whalley (1969), Witsack (1973)
A. mutans	Moratorio (1977), Walker (1979)
A. optabilis[a]	Sahad (1984)
A. silwoodensis	Moratorio (1977), Tay (1972), Walker (1979)
A. stenocrani	May (1971), Walker (1979)
A. sp. "a"	Ali (1979)
A. sp. "d"	Ali (1979)
A. sp. "e"	Ali (1979)
A. sp. "f"	Ali (1979)
A. sp. "g"	Ali (1979)
G. cincticipitis[a]	Sahad (1982a)
P. striaticorne	Balduf (1928)
A. ensifer†	Walker (1979)
A. sp. "b"†	Ali (1979)
A. sp. "c"†	Ali (1979)
A. sp. "c_3"†	Ali (1979)

[a] Non-European species.
† Mode of reproduction uncertain; apparently not arrhenotokous.
The mode of reproduction given in the table was either stated explicitly by the above authors, or was inferred by us from their laboratory and field data.

lifetime fecundity of female *Anagrus* (measured as the number of ovarian eggs in newly emerged adults). Newly emerged *A. silwoodensis* for example, contain between 10 and 105 eggs (Moratorio, 1977). Ali (1979) recorded significant differences in fecundity among several *Anagrus* species.

The potential fecundity of a female is determined by (1) larval rearing temperatures, (2) the species of host in which the female developed, and (3) larval competition (Moratorio, 1977). Although Moratorio did not investigate directly the relationship between fecundity and body size in females, it is clear from this study that (1), (2), and (3) influence fecundity mainly, if not entirely, through their effects on body size. A positive correlation between forewing length and fecundity was previously demonstrated by Orita (1972) in *Gonatocerus* sp.

Data on realized fecundity and oviposition rates of *Anagrus* and *Gonatocerus* are contained in Chandra (1980c), Meyerdirk and Moratorio (unpubl.), Moratorio (1977), Ôtake (1969), Raatikainen (1967), Sahad (1982a, 1984), Tay (1972) and Witsack (1973).

Descriptions of courtship and mating have been given for arrhenotokous species of *Anagrus* (Ali, 1979; Ison, 1959; Maillet, 1960; Moratorio, 1977; Raatikainen, 1967; Rothschild, 1967; Sahad, 1984; Witsack, 1973), *Gonatocerus* (Orita, 1972; Sahad, 1982a) and *Polynema* (Bakkendorf, 1934; Balduf, 1928; Vidano and Meotto, 1968).

There is no premating period (Ali, 1979; Moratorio, 1977; Raatikainen, 1967; Rothschild, 1967; Sahad, 1984; Witsack, 1973). Males tend to emerge earlier than females (Moratorio, 1977; Witsack, 1973) and remain in the vicinity of the emergence site, copulating with females as they emerge (Moratorio, 1977). Aggressive encounters between males have been recorded by Balduf (1928) in *P. Striaticorne* and by Sahad (1982a) in *Gonatocerus cincticipitis*. Territorial males presumably obtain a large proportion of matings, so tending, on average, to reduce outbreeding on egg sites (see above).

Courtship behaviour involves a large olfactory component (Ali, 1979). Ali (1979) showed that virgin females produce a pheromone which releases courtship behaviour in males. Males of *A.* sp. "d" will attempt to copulate with a fine hair brush that has previously been wiped over the bodies of virgin females (Ali, 1979). Production of the sex pheromone apparently ceases shortly after mating has occurred (Ali, 1979; Walker, 1979). There is apparently no interspecific recognition of pheromone (Ali, 1979), i.e. the pheromone is species-specific. The pheromone appears to act as an effective species isolating mechanism, since cross-species matings have never been observed despite attempts to induce them (Ali, 1979; Walker, 1979).

Data on longevity of *Anagrus* and *Gonatocerus* are contained in Chandra (1980c), Meyerdirk and Moratorio (unpubl.), Moratorio (1977), Orita (1972), Ôtake (1969) and Sahad (1982a, 1984). Adult *Anagrus* are relatively short-lived. Female and male *A. giraulti* for example, when given honey and water, live on average for 11·4 and 9·1 days respectively (Meyerdirk and Moratorio, unpubl.)

Adults of both sexes live significantly longer when given honey and water than when supplied with water alone (Meyerdirk and Moratorio, unpubl.; Sahad, 1982a, 1984). While females, being pro-ovigenic, are capable of depositing most of their egg complement during the first 24–48 hours of life when abundantly supplied with hosts (Chandra, 1980c; Meyerdirk and Moratorio, unpubl.; Moratorio, 1977; Ôtake, 1969), they are likely to depend heavily on energy-rich foods for maintenance during periods of host scarcity.

In *G. cincticipitis* and *A. optabilis* longevity decreases significantly with increasing temperature between 15 and 30°C (Sahad, 1982a, 1984). There is likely to be a positive relationship between body size and longevity (e.g. see Waage and Ng, 1984), but this has so far not been examined in mymarids.

According to Clausen (1940) Mymaridae are "true egg parasites in the sense that they normally attack the host eggs before an appreciable development of the embryo has taken place". However, Whalley (1969) showed that female *Anagrus* oviposit readily into eggs containing well-developed embryos. This phenomenon, which Whalley (1969) termed "embryo-parasitism", has been recorded in a number of *Anagrus* species (Ali, 1979; Chandra, 1980c; Ôtake, 1968; Tay, 1972). However, *A. stenocrani* appears to avoid embryo-parasitism altogether (May, 1971).

The eggs of *Anagrus, Polynema* and *Gonatocerus* are of the stalked type (Arzone, 1974a,b; Balduf, 1928; Sahad, 1982b; Witsack, 1973). There are two larval instars in *Anagrus* and some *Gonatocerus* species (Ali, 1979; Bakkendorf, 1934; Meyerdirk and Moratorio, unpubl.; Moratorio, 1977; Sahad, 1982a, 1984), but there appear to be more in *Polynema* and *G. cicadellae* (Arzone, 1974a,b; Balduf, 1928; Vidano, 1967). The first instar larvae of *Anagrus* and *Polynema* are of the sacciform and mymariform types respectively (Clausen, 1940), while the larva of *Gonatocerus* is somewhat intermediate in form between these two (Arzone, 1974a; Bakkendorf, 1934; Sahad, 1982b). The second instar larva of *Anagrus* is considerably different from the first; this "histriobdellid" larva (Ganin, 1869) is segmented and has well-developed mandibles, a pair of cephalic processes, and a pair of caudal processes. It eventually becomes bright yellow, orange or red in colour (as does that of *Gonatocerus*: Chandra, 1980c; Sahad, 1982b), allowing parasitized host eggs to be readily distinguished from unparasitized ones.

Unlike the first instar larva, which lies motionless within the host egg (Moratorio, 1977), the second instar larva of *Anagrus* and *Gonatocerus* is active, its movements having been variously described as churning, writhing, wiggling, rotation, peristalsis and rolling. Whalley (1969) suggested that these movements serve to circulate the contents of the host egg, both for feeding and for facilitating gaseous exchange.

Quantitative data on development are contained in Ali (1979), Bentur *et al.* (1982), MacGill (1934), Meyerdirk and Moratorio (unpubl.), Moratorio (1977), Orita (1972), Raatikainen (1967), Rothschild (1967), Sahad (1982a, 1984), Walker (1979) and Witsack (1973). The temperature threshold for development and the effects of temperature on development rates have been measured in *A. giraulti* (Meyerdirk and Moratorio, unpubl.), *A. optabilis* (Sahad, 1984), *A. silwoodensis* (Moratorio, 1977), and *G. cincticipitis* (Sahad, 1982a). Meyerdirk and Moratorio (unpubl.) and Sahad (1982a, 1984) also measured the thermal constant (i.e. the number of day degrees required for development) in the species they studied.

Males tend to develop faster than females, and in both sexes development is strongly temperature-dependent (Meyerdirk and Moratorio, unpubl.; Moratorio, 1977; Sahad, 1982a, 1984). The rate of development also varies with the species of host attacked. *A. silwoodensis* develops faster in *C. viridis* than in *D. hamata*, whereas *A. mutans* develops faster in *D. hamata* (Moratorio, 1977).

Larvae of *A.* sp. nr *flaveolus* developing in eggs containing fully differentiated host embryos apparently survive as well as those developing in eggs that are at an earlier stage of embryogenesis, but development is somewhat prolonged (Ôtake, 1968). However, Whalley (1969) was unable to detect any difference in development rates in the *Anagrus* he examined.

3. Effects of Larval Competition on Parasitoid Survival and Fitness

Several workers (Ali, 1979; Sahad, 1982; Whalley, 1969; Witsack, 1973) have reported cases of larval mortality resulting from superparasitism, while others (e.g. Tay, 1972) have reported an absence of such mortality. In the solitary species, *G. cincticipitis*, supernumerary larvae do not survive beyond the first instar (Sahad, 1982a). From the various anecdotal accounts available, it may be concluded that in gregarious species, mortality will occur only if very large numbers of larvae are present within the host egg. Unfortunately there is no precise information on this.

The effects of larval competition on parasitoid fitness have been demonstrated by Moratorio (1977). In *A. silwoodensis* there is a marked reduction in the fecundity of progeny even at low larval densities. *A. silwoodensis* females normally deposit a maximum of 4 ova per host egg (Moratorio, 1977), suggesting that they maximize fitness by allocating progeny in such a way as to produce daughters with high overall fecundity. This effect has also been observed in *Trichogramma evanescens* (Waage and Ng, 1984). Sex ratio allocation must also be taken into account here, since depositing male eggs in a host will reduce total progeny fecundity per egg, by reducing female size (Waage and Ng, 1984).

4. Phenological Studies

The most effective means of collecting adult mymarids appears to be the yellow pan trap method (Noyes, 1982). While this technique has been widely used for other insect groups, it seems to have been overlooked by workers on leafhopper/planthopper egg parasitoids, and this may partly account for the lack of detailed information on flight periods.

There are, by contrast, well-proven techniques for rearing adults (Ali, 1979; Bentur *et al.*, 1982; Huldén, 1984; Meyerdirk and Moratorio, unpubl.; Moratorio, 1977; Vidano and Meotto, 1968; Walker, 1979). Walker (1979) succeeded in maintaining continuous cultures of three *Anagrus* species for up to three years.

There is presently little information on the phenologies of individual species. This has undoubtedly been due in large part to problems involved in collecting and identifying both adults and immature stages. The very limited data that are available derive almost exclusively from material reared from field-collected host eggs.

Polynema euchariforme and *P. bakkendorfi* attack the eggs of univoltine *Oncopsis* species. The flight period of *P. euchariforme* coincides closely with the host's summer egg-laying period. On the other hand, adult *P. bakkendorfi* emerge during the host's spring hatching period, suggesting that either the adults spend the summer in diapause, or they attack another host species (Claridge and Reynolds, 1972).

Polynema woodi and *Gonatocerus cicadellae* are bivoltine (Arzone, 1974b), while *Polynema striaticorne*, which attacks the univoltine membracid, *Stictocephala bisonia*, has three generations per year both in North America and Europe (Balduf, 1928; Vidano, 1968; Vidano and Meotto, 1968).

Several *Anagrus* species have been recorded as having up to three generations per year (May, 1971; Rothschild, 1962, 1967; Whalley, 1969; Witsack, 1973), while Raatikainen (1967) and Bakkendorf (1925) recorded up to four and five generations in the *Anagrus* species they studied in Finland and Denmark, respectively. Whalley (1969) noted that *Anagrus* attacking univoltine hosts are able, through embryo-parasitism, to produce several generations.

Anagrus and *Polynema* species overwinter as either prepupae or pupae (Moratorio, 1977; Tay, 1972; Vidano, 1968; Witsack, 1973). An exception to this is *P. woodi*, which overwinters as a first instar larva in eggs of *Cicadella viridis* (Arzone, 1974b). *Gonatocerus cicadellae* also overwinters in this way (Arzone, 1974a). There is disagreement as to whether overwintering stages undergo diapause (Walker, 1979) or are in a state of "thermal quiescence" (Tay, 1972; Witsack, 1973). Witsack (1973) also found that larvae enter a brief diapause-like state during the summer.

Walker (1979) recorded significant interspecific differences among *Anagrus* species in the duration of the "postdiapause development" period.

Raatikainen (1967) and Ôtake (1969, 1970a, 1976) obtained evidence to suggest that adult *Anagrus* are capable of dispersing over relatively long distances. Raatikainen (1967), for example, recorded parasitized eggs of *Javesella pellucida* in spring cereals that were situated as far as 100 m from the nearest overwintering sites of *A. atomus*, Ôtake (1969) suggested that adults disperse on air currents.

5. *Host Specificity*

Mymaridae have so far been reared from over 30 species of European Auchenorrhyncha (Appendix I). Unfortunately, because of the taxonomic confusion within the genera *Anagrus, Gonatocerus* and *Polynema*, we cannot be certain about the identity of most of the species listed in Appendix I. It is therefore difficult at this stage to draw any useful conclusions regarding host range. Nevertheless, we suspect in the light of the studies by Ali (1979) and Walker (1979), that most mymarids attacking Auchenorrhyncha have, as the data suggest, very narrow host ranges, i.e. most are either monophagous or oligophagous.

It has been suggested (Claridge and Reynolds, 1972; Ali, 1979) that in some *Polynema* and *Anagrus* species female ovipositor length is an important limiting factor determining host range. *P. euchariforme* and *P. bakkendorfi*, for example, have mutually exclusive host ranges, despite the fact that the eggs of their hosts occur in similar positions in the buds of birches (*Betula* spp.) during more or less the same period of the year. Claridge and Reynolds (1972) suggested that because the eggs of *O. flavicollis* and *O. subangulata* are laid closer to the bases of the leaf petioles than those of *O. tristis*, and are thereby better concealed, they may be generally unavailable to *P. bakkendorfi*, as females of the latter species have shorter ovipositors than those of *P. euchariforme*. Ali (1979) likewise found that *Anagrus* spp. with long ovipositors attack only hosts whose eggs are laid well below the surface of the host plant, and that species with short ovipositors attack only hosts whose eggs are laid close to the surface.

Ovipositor length is, however, unlikely to be of general importance as a *proximate* factor in host range delimitation, since it fails to explain why mymarids with long ovipositors do not attack both shallow-lying and deep-lying eggs. Explanations must therefore be sought elsewhere for the apparently very narrow host ranges of most mymarids.

Preliminary investigations by Ali (1979) suggest that discriminative searching and oviposition behaviour are important determinants of host specificity in *Anagrus*. Most of the eight species studied by Ali rejected egg sites of unnatural host species following their examination (unnatural host species are those from which the parasitoid species has not been reared in the field). Females of *A*. sp. "a" also rejected the eggs of such hosts when these were exposed in petri dishes. However, females of both *A*. sp. "a" and *A*. sp. "d" accepted eggs of unnatural host species if these were placed among eggs of the natural host, suggesting that olfactory stimuli may be used by females in host identification. Both Ali (1979) and Walker (1979) showed that *Anagrus* species may develop successfully in the eggs of unnatural host species. However, Walker (1979) found that in her *Anagrus* cultures "breeding success", as measured by the number of viable adult offspring produced per parent female, was greatest in the "normal" (natural) hosts. This suggests that host specificity may evolve through the differential effects of different host species on the fitness of parasitoid offspring. Evidence in support of this comes from Moratorio's (1977) studies, where it was shown that female *A. mutans* and *A. silwoodensis* reared from the natural host *C. viridis* were larger and contained more ovarian eggs than females reared from the supposed unnatural host *D. hamata* (see page 291). However, the observation that *A. mutans*, unlike *A. silwoodensis*, develops more rapidly in *D. hamata* than in *C. viridis* (see page 293), suggests that not all components of fitness are affected in the same way.

Comparison of the phenologies of hosts and parasitoids have pointed to the existence of one or more alternative host species that are exploited by the parasitoid during periods when the host under investigation is not vulnerable to attack (i.e. its eggs are not present in the field) (Claridge and Reynolds, 1972; Huldén, 1984; Raatikainen, 1967). The role of such hosts is far better understood in the case of the grape leafhoppers (*Erythroneura* spp.), which are pests in vineyards on the western seaboard of North America. Control of *Erythroneura elegantula* by *Anagrus epos* is significantly better in vineyards situated close to blackberry (*Rubus* spp.) hedges, since the latter harbour populations of *Dikrella cruentata* which serve as an overwintering refuge for the parasitoid (Doutt and Nakata, 1965) (see also Doutt *et al.* (1966) and McKenzie and Beirne (1972)).

Polynema woodi appears to be very habitat-specific in its searching behaviour. Arzone (1974b) found that it parasitizes only *C. viridis* eggs that are situated 250–800 mm above ground level in the shoots and twigs of woody plants, and not those that occur either below 250 mm or in herbaceous plants such as *Juncus*.

B. Other Egg Parasitoids

In addition to Mymaridae, other kinds of Hymenoptera parasitize the eggs of Auchenorrhyncha (Table 4).

In Europe, Trichogrammatidae do not appear to play as important a role in the population dynamics of leafhoppers and planthoppers as they do in some parts of Asia (see Greathead (1983), Miura *et al.* (1979) and Sasaba and Kiritani (1972) for information on trichogrammatid parasitoids of rice-feeding Auchenorrhyncha). Estimates of percentage parasitism are not available for European species. However, the frequency with which individual wasps were reared by Becker (1975) suggests that the contribution by *Oligosita engelharti* to total egg parasitism in *Macrosteles sexnotatus* was small.

Three species of trichogrammatid have been reared from European Auchenorrhyncha: *Oligosita krygeri* Girault from *Cicadella viridis* in Denmark (Bakkendorf, 1934, 1943) and *Macrosteles sexnotatus* in Sweden (Ahlberg, 1925), *O. engelharti* Kryger from *M. sexnotatus* in Britain (Becker, 1975), and *Epoligosita nudipennis* (Kryger) reared from an unidentified typhlocybid in Denmark (Bakkendorf, 1934). Bakkendorf (1934) describes mating and oviposition behaviour and also larval development in *O. krygeri*, while Vungsilabutr (1978) gives a detailed account of the biology of *Paracentrobia andoi* (Ishii).

Table 4

Hymenopterous parasitoids, other than Mymaridae, known to attack the egg stage of Auchenorrhyncha.

	Hosts											
Parasitoids	Delphacidae	Fulgoridae	Tettigometridae	Issidae	Flatidae	Lophopidae	Ricaniidae	Eurybrachidae	Cicadidae	Cercopidae	Membracidae	Cicadellidae
TRICHOGRAMMATIDAE												
Abelloides	–	–	–	–	–	–	–	–	–	–	+	–
Aphelinoidea	–	–	–	–	–	–	–	–	–	–	–	+
Chaetostricha	–	–	–	–	–	–	–	–	–	–	–	+
Giraultiola	–	–	–	–	–	–	–	–	+	–	–	–
Lathromeris	–	–	–	–	–	–	–	–	–	–	+	–
Neolathromera	–	–	–	–	–	–	–	–	+	–	–	–
Oligosita	+	–	–	–	–	–	–	–	–	+	–	+
Paracentrobia	+	–	–	–	–	–	–	–	–	?	+	+
Pseudobrachysticha	–	–	+	–	–	–	–	–	–	–	–	–
Ufens	–	–	–	–	–	–	–	–	–	–	–	+
Uscanoidea	–	–	–	–	–	–	–	–	–	+	–	–
Epoligosita	–	–	–	–	–	–	–	–	–	–	–	+
Uscanopsis	–	–	–	–	–	–	–	–	–	–	+	–
Zaga	–	–	–	–	–	–	–	–	–	–	+	–
EULOPHIDAE												
Testrastichus	+	–	–	–	–	+	–	–	–	–	+	+
APHELINIDAE												
Centrodora	–	+	–	–	–	–	+	–	+	+	+	+
Tumidiscapus	–	–	–	–	–	–	–	–	–	+	–	–
ENCYRTIDAE												
Ectopiognatha	–	–	–	–	+	–	–	+	–	–	–	–
Fulgoridicida	–	–	–	–	–	–	–	+	–	–	–	–
Ooencyrtus	–	–	–	–	–	+	–	–	–	–	–	–
Proleurocerus	–	–	–	–	–	+	–	+	–	–	–	–
Psyllechthrus	–	–	+	–	–	–	–	–	–	–	–	–
SCELIONIDAE												
Aphanomerus	–	+	–	–	+	–	–	–	–	–	–	–
Telenomus	–	–	–	?	–	–	–	–	–	–	–	–

Records from de Azevedo Marques (1925), Arzone (1977), Burks (1979), Domenichini (1966), Doutt (1961), Doutt and Viggiani (1968), Girault (1916), Hayat (1974), Kershaw (1913), Manjunath *et al.* (1978), Miura *et al.* (1979), Myers (1922), Perkins (1905a,b,c,d, 1906b,c), Pickles (1932), Pruthi (1937), Subba Rao (1983) and Tachikawa (1974).

?, questionable.

Two species of *Centrodora* (Aphelinidae) have been reared from eggs of European Auchenorrhyncha. Becker (1975) reared *Centrodora livens* from eggs of *M. sexnotatus* in Britain, while Silvestri (1918, 1920, 1921) reared *C. cicadae* Silvestri from eggs of two cicadas, *Cicada orni* L. and *Lyristes plebeius* (Scopoli).

Domenichini (1966) lists two Palaearctic species of *Tetrastichus* (Eulophidae) which parasitize Auchenorrhyncha. Only one, *T. mandanis* (Walker), has been reared from European hosts (Arzone, 1977; Bakkendorf, 1934; Rothschild, 1967; Whalley, 1956).

T. mandanis is a parasitoid-predator, attacking the eggs of *Conomelus anceps* and *C. dehneli* (Delphacidae) which occur in the stems of rushes (*Juncus* spp.). The female wasps generally oviposit in one egg per host egg mass. The first instar larva is completely endoparasitic in the host egg. However, the second instar larva devours the entire contents of the egg, and begins to feed on the remaining eggs in the mass. Larvae generally attack further egg sites, reaching them by tunneling through the *Juncus* pith. Individual larvae attack, on average, a total of eight eggs during their lifetimes. Eggs of the non-host species, *Cicadella viridis*, are sometimes fed upon by the tunneling larvae, which are also cannibalistic (Rothschild, 1967).

Although a single generation per year was recorded by both Bakkendorf (1934) and Whalley (1956), Rothschild (1967) found evidence to suggest that *T. mandanis* may be bivoltine, the first larval generation developing in some unidentified host species. *T. mandanis* overwinters as either an early (parasitic) or late (predatory) larva.

Up to 32% of eggs may be destroyed in individual *Juncus* stems (Rothschild, 1967). Further information, including descriptions of the larval stages of *T. mandanis*, is contained in Arzone (1977), Bakkendorf (1934) and Rothschild (1967).

Although they are best considered as predators, not parasitoids, four other Chalcidoidea which attack the eggs of European Auchenorrhyncha deserve mention: *Mesopolobus aequus* (Walker), *Panstenon oxylus* (Walker) (Pteromalidae), *Eupelmus cicadae* (Girault) (Eupelmidae) and *Archirileya inopinata* Silvestri (Eurytomidae). For details of their biology, see Raatikainen (1967, 1970) and Silvestri (1920, 1921).

III. PARASITOIDS OF NYMPHS AND ADULTS

A. Dryinidae

1. Systematics, Taxonomy and General Biology

As far as is known, all Dryinidae are parasitic in nymphs and adults of Auchenorrhyncha. The true affinity of the family is considered to be ambiguous by a number of authorities (e.g. O. W. Richards, pers. comm.; Perkins, 1976). Much of what we know about dryinid biology derives from the work of R. C. L. Perkins (1905a, 1906a) and Fenton (1918), and the most recent taxonomic revision of the group is that of Olmi (1984). The information in this section, particularly host records and taxonomic nomenclature, is based on Olmi's two-volume monograph.

Olmi has described many species, and has synonymized many others. He has divided the family into 10 subfamilies and 46 genera containing a total of around 800 species (Table 5); although he considers the true number to be nearer 1000. Sixteen to seventeen per cent of described species occur in the Palaearctic region, and 60% of these (i.e. 10% of the total) are found in Europe (Table 6). Table 5 lists both the geographical distribution of each dryinid subfamily and the known host groups. For clarity, hosts of Palaearctic genera are summarized in Table 7, and lists of known hosts to species level are given in Appendix II. The latter includes not only the records given by Olmi (1984) and Freytag (1985), but also unpublished records.

Thirteen fossil species of Dryinidae, belonging to four subfamilies, have been recorded in Amber deposits (Table 8), and among these *Dryinus antiquus* (Ponomarenko) is the earliest known dryinid. *D. antiquus* was

Table 5

Distribution of subfamilies of Dryinidae and their hosts to family and subfamily levels (after Olmi, 1984).

		Distribution and numbers of known species						Known hosts																
									Cicadellidae															
Dryinidae Subfamilies	No. of genera	Palaearctic	Ethiopian	Oriental	Nearctic	Neotropic	Australian	Membracidae	Idiocerinae	Macropsinae	Typhlocybinae	Other	Dictyopharidae	Cixiidae	Tropiduchidae	Lophopidae	Flatidae	Acanoloniidae	Issidae	Ricaniidao	Nogodinidae	Delphacidae	Fulgoridae	Unknown
(1) Anteoninae	5	25	46	24	22	79	44		•	•		•												
(2) Aphelopinae	2	7	3	8	6	14	5	•			•													
(3) Apodryininae	1					1																		•
(4) Biaphelopinae	1	1																						•
(5) Bocchinae	6	18	4	4	8		5					•							•					
(6) Conganteoninae	3	2	2	1																				•
(7) Dryininae	8	15	26	23	15	48	25						•	•	•	•	•	•	•	•		•	•	
(8) Gonatopodinae	18	76	67	51	59	57	43					•			•	•	•	•		•	•	•		
(9) Thaumato-dryininae	1		5	3	1	6	2										•							
(10) Transdryininae	1						1																	•

Table 6
Numbers of dryinid species in the European part of the Palaearctic region: totals include fossil species (F).

	Number of species	
Dryinid subfamily	In Palaearctic region	In its European part
Anteoninae	25 (1F)	16 (1F)
Aphelopinae	7	5
Biaphelopinae	1	0
Bocchinae	18	11
Conganteoninae	2	0
Dryininae	15 (3F)	8 (3F)
Gonatopodinae	76	46
Thaumatodryininae	— (3F)	— (3F)
	Σ = 144	Σ = 86

found in Siberian Taimyr amber, and according to Ponomarenko (in Olmi, 1984), it is around 100 million years old.

As a group, Dryinidae present a number of unusual morphological and biological features. One of these is the semi-external position of the larva, which lies in a prominent sac formed from moulted exuviae. Females of all subfamilies, except Aphelopinae, are characterized by chelate forelegs, with which they seize and hold hosts during either oviposition or host-feeding, many species being both parasitic and predatory. The chelae have developed from a single tarsal segment and its claw, which is enlarged and opposable. Anteoninae have the simplest chelae, whereas those of Gonatopodinae and Dryininae are larger and more complex. Some species have extremely long chelae, and in *Megadryinus magnificus* they are almost as long as the whole body of the wasp (Richards, 1953). Female dryinids are often brachypterous or apterous, but the males are always fully winged. Wing reduction is typically associated with parasitism of hosts in grassland and other herbage, presumably because wings hinder movement through such vegetation. Apparently, even fully winged females are not particularly active flyers, spending most of their time running around on plants.

Richards (1939) points out that three evolutionary processes have occurred simultaneously within the family Dryinidae: (1) reduction of wings, with associated reduction of the meso- and metathorax; (2) reduction of mouthparts, principally in the number of mandibular teeth and palp segments; and (3) the development of more mobile, larger, more complex and probably more efficient chelae.

Superficially, the wingless females resemble ants, and Donisthorpe (1927) suggested that this is an example of mimicry, whereas Richards (1939) pointed out that the gait of the wasps is quite distinct from that of ants.

Table 7
Palaearctic subfamilies and genera of Dryinidae and their known hosts (after Olmi 1984).

	Known hosts									
	CICADELLIDAE									
DRYINIDAE	Idiocerinae	Iassinae	Macropsinae	Aphrodinae	Deltocephalinae	Typhlocybinae	Dictyopharidae	Cixiidae	Delphacidae	Issidae
ANTEONINAE										
Anteon	•	•	•		•					
Lonchodryinus					•					
APHELOPINAE										
Aphelopus						•				
BOCCHINAE										
Bocchus										•
Mystrophorus					•					
DRYININAE										
Dryinus							•	•		•
Richardsidryinus										•
GONATOPODINAE										
Agonatopus									•	
Dicondylus									•	
Donisthorpina									•	
Echthrodelphax									•	
Gonatopus				•	•					
Haplogonatopus									•	
Pseudogonatopus									•	
Tetrogonatopus					•					

However, the exotic species *Anteon myrmecophilum* appears to be truly myrmecophilous, parasitizing leafhoppers that are frequently attended by ants. Perkins (1905a) states that the resemblance of this wasp to ants is enhanced by its behaviour—females may stand face to face on their hind legs and lick each other's mouths, soliciting food. However, in most species, the apparent association with ants can be explained by the overlap of their habitats.

Detailed observations on courtship and copulation in *Aphelopus melaleucus* are given in Jervis (1979b). The male contacts the female from

Table 8
Distribution of fossil Dryinidae preserved in amber: numbers denote species (after Olmi 1984).

	Regions			
	Palaearctic		Nearctic: Canadian	Neotropic:
Subfamilies	Baltic amber 40–45 000 000 years old	Siberian Taimyr amber 80–100 000 000 years old	Medicine Hat amber 70–75 000 000 years old	Dominica amber 27–30 000 000 years old
Anteoninae	1	—	—	—
Dryininae	4	2	1	1
Gonatopodinae	1	—	—	—
Thaumato-dryininae	3	—	—	—

the rear with his antennae, and seizes her with his first two pairs of legs and mounts her. During copulation, the male reclines backwards and is dragged along the substrate by the female. After a few seconds, the pair disengages. Similar behaviour has been seen in *Anteon brachycerum* in the field (Jervis, 1979b), and it is essentially the same in *A. ephippiger* (Becker, 1975) and *Dicondylus bicolor* (Waloff, 1974). In *Pseudogonatopus distinctus* the pair criss-cross their rapidly vibrating antennae before mating (Waloff, 1974). Swarming behaviour has been observed in male *Aphelopus melaleucus* by Jervis (1979b), who collected 39 males and 9 females within a few seconds of beating a hornbeam (*Carpinus betulus*) tree.

Unlike those of the vast majority of aculeate Hymenoptera, female dryinids deposit their eggs within, instead of upon, the host. The first of the five larval instars lies within the host's haemocoel. Subsequent instars are, however, semi-external: the parasitoid's head and part of its tail region lying inside the host, with the remainder of its body contained within a sac which is comprised of moulted exuviae that are retained and added to with successive larval moults. The position of the sac, which is either black or dark brown, tends to be characteristic of a dryinid species. Ponomarenko (1971) states that larvae of Gonatopodinae feed with the aid of two oral vesicles in the pharynx; these extend directly into the gut of the host, and by their pumping action, transmit material from the host's intestine into that of the parasitoid.

The last instar larva eventually splits the sac, and with its head still inside the host, consumes the contents of the latter. Fifth instar larvae of *Gonatopus sepsoides* empty the contents of their previously active host within 50 min, and become twice their original size (Waloff, 1974). Having

Table 9

Development (in days) of some dryinid species in their non-diapausing generations. Jervis's (1980) data at 18 ± 2°C, Waloff's (1974, 1975) in outdoor insectary at 20–23°C. Limits in parentheses may depend on date of capture of parasitized hosts. In Great Britain unless otherwise stated.

Species	Days from oviposition to appearance of larval sac on host	Duration of larval development (oviposition to emergence of mature larva from host)	Time spent in cocoon (i.e. to adult emergence)	Days from oviposition to adult emergence	Reference
Aphelopus atratus	32·8 (32–34) $n = 16$	46·4 (45–48) $n = 6$	26·9 (25–29) $n = 11$	74·3 (73–76) $n = 3$	Jervis (1980b)
A. melaleucus	—	—	21·3 (17–24) $n = 4$	—	Jervis (1980b)
Anteon ephippiger	9·0 $n = 7$	14·0 $n = 6$	(24–31) $n = 8$	(44–54)	Becker (1975)
A. jurineanum in USSR	—	21·0	—	—	Ponomarenko (1968)
A. pubicorne	—	14·0 $n = 1$	28·0 $n = 2$	42 (40·6–48·2)	Waloff (1974) Khafagi (1986)
Lonchodryinus ruficornis	—	—	47·7 (42–52) $n = 3$	—	Waloff (1974)

Echthrodelphax hortuensis in S. France	—	23·0 $n = 1$	—	—	Abdul-Nour (1971)
Pseudogonatopus albosignatus in S. France	—	30–32 $n = 2$	—	—	Abdul-Nour (1971)
Dicondylus bicolor	—	—	♀ 28·7 (17–35) $n = 44$ ♂ 27·6 (22–35) $n = 19$	—	Waloff (1974, 1975)
Haplogonatopus oratorius in Italy	—	—	♀ (21–53) ♂ (18–32)	—	Currado and Olmi (1972)
Tetrodontochelys pedestris	—	—	(14–24)	—	Currado and Olmi (1972)
Gonatopus distinguendus in S. France	—	32·0 (30–36) $n = 4$	—	—	Abdul-Nour (1971)
G. lunatus in S. France	—	29·0 $n = 1$	—	—	Abdul-Nour (1971)
G. sepsoides	4–5, $n = 10$	21·1 (17–24) $n = 10$	30·3 (23–43) $n = 24$	51·4 (40–67)	Waloff (1974, 1975)
in S. France	—	(15–20)	33·5 (19–61) $n = 22$	—	Abdul-Nour (1971)
in Italy	—	—	(28–47)	—	Currado and Olmi (1972)
in USSR	—	23·0	—	approx. 40	Ponomarenko (1968b)
Bocchus europaea in S. France	—	43·0	—	—	Abdul-Nour (1971)

consumed most or all of the host's soft tissues, the larva moves away by means of strong peristaltic contractions, and eventually spins a double-layered cocoon either in the ground or leaflitter (e.g. *Aphelopus*) or on the surface of the host's foodplant (e.g. most Gonatopodinae and many Anteoninae).

Larval development in *Aphelopus* is somewhat different. In the solitary, sac-producing species, the first instar larva lies freely within the host's haemocoel, enveloped in a mass of cells which are frequently referred to as a "trophamnion", but whose origin and function have yet to be established (for discussion see Jervis, 1978b).

According to Ponomarenko (1975), the mode of larval feeding in Anteoninae is intermediate between that of Gonatopodinae and Aphelopinae.

Larval development times of some European Dryinidae are given in Table 9. Recently, Khafagi (1986) established cultures of *Anteon pubicorne* on *Macrosteles viridigriseus*, and found that development time from oviposition to emergence of the adult parasitoid decreased in the older stages of the host, and that individuals from larger hosts emerged sooner than from smaller ones (Table 9).

Crovettia theliae, a North American species, is highly unusual in that it is polyembryonic and completely endoparasitic within its membracid host (Kornhauser, 1919).

Data on adult longevity are summarized in Table 10. As in other host-feeding parasitoids (Jervis and Kidd, 1986), longevity is increased by host-feeding. Khafagi (1986) demonstrated that longevity of the autogenous *Anteon pubicorne* increased markedly on diets rich in carbohydrates (yeast, sugar or honey) and that in both sexes it was significantly shorter on water or honeydew. Simultaneously, carbohydrate foods had no effect on fecundity, while protein-rich diets of pollen grains, peptone or honeydew significantly decreased it, in comparison with a control diet of water.

Data on host range have been obtained for 49 European species of Dryinidae (Appendix II). Only two species, *Dryinus collaris* and *Anteon pubicorne*, have been reared from more than one subfamily of hosts, while *D. collaris* attacks hosts belonging to different families (Cixiidae and Issidae). *Anteon brachycerum* is almost certainly narrowly polyphagous, while *A. infectum* is probably truly monophagous.

2. Voltinism and Overwintering Stages

Voltinism in dryinids varies with geographical location, particularly latitude, and depends also on the voltinism of the host (e.g. see Jervis, 1980b). With the exception of British *Aphelopus* studied by Jervis (1978b, 1980b,c), the phenologies of European Dryinidae are poorly known. Some data on voltinism and overwintering stages are summarized in Table 11. In Britain,

Table 10
Longevity in days of some dryinid species in Great Britain (unless otherwise stated) at outdoor insectary temperatures: fed on soaked raisins, and female Gonatopodinae also on host nymphs. Limits in parentheses.

Species	Unfed	Fed	Reference
Aphelopus atratus	—	5♀—10·00 (3–14) 2♂—6·50 (4–9)	Jervis (1980b)
A. melaleucus	—	17♀—7·53 (1–16) 16♂—7·88 (3–14)	Jervis (1980b)
A. nigriceps	—	1♂—15·00	Jervis (1980b)
A. serratus	—	3♀—9·66 (6–15) 6♂—6·33 (1–14)	Jervis (1980b)
Diconclylus bicolor	18♀—3·72 (3–7) 9♂—3·89 (1–7)	16♀—6·00 (3–10) 15♂—6·93 (3–12)	Waloff (1974, 1975)
in Finland	♀♀—2–4	≥ 42	Raatikainen (1967)
Pseudogonatopus distinctus	3♀—5·00 (1–7) 2♂—8·00 (7–9)	5♀—21·20 (14–28)	Waloff (1974, 1975)
Gonatopus sepsoides	♀♀ ≤ 10·00	8♀—18·5 (12–27)	Waloff (1974, 1975)
in S. France		♀♀—*c*. 15	Abdul-Nour (1971)
Anteon pubicorne	8♀—9·8 ± 3·7 8♂—5·0 ± 0·6	Carbohydrate-rich diets 24♀—(27·9–34.8) 24♂—(21·4–29·9) Protein-rich diets 24♀—(11·6–20·8) 24♂—(11·6–23·4)	Khafagi (1986)

most *Aphelopus*, together with the two Anteoninae listed in Table 11 are either uni- or bivoltine; one of the Gonatopodinae is univoltine, and the other two are either bi- or trivoltine, depending on weather conditions during a season. *Dicondylus bicolor* is bi- or trivoltine in Britain, but univoltine in Finland, while *Gonatopus sepsoides* is either bi- or trivoltine both in Britain and in southern France.

3. Parthenogenesis and Sex Ratio

Most dryinids appear to be arrhenotokous. Although *Gonatopus sepsoides* is the most common dryinid parasitoid of grassland leafhoppers in southern England, males were never recorded there (Richards, 1939, 1948; Waloff, 1974, 1975). Males have similarly never been recorded in either Italy or the USSR (Currado and Olmi, 1972; Ponomarenko, 1968). Richards suggested that this is a thelytokous species, but Abdul-Nour (1971) reared males in the

Table 11

Voltinism and overwintering stages of some dryinid species in Great Britain (unless otherwise stated).

Subfamily and species	Voltinism	Overwintering stage	(a) Hosts (b) Usual position of larval sac	Reference
ANTEONINAE				
Anteon pubicorne	Univoltine/ Bivoltine	In cocoon spun by mature larva	(a) Cicadellidae–Deltocephalinae (b) Thorax of host	Waloff (1974, 1975) Khafagi (1986)
Lonchodryinus ruficornis	Usually univoltine can be bivoltine	In cocoon spun by mature larva	(a) Cicadellidae–Deltocephalinae (b) Thorax or between head and thorax	Olmi (1984) Waloff (1974, 1975)
APHELOPINAE				
Aphelopus atratus and *A. melaleucus*	Univoltine/ bivoltine	Prepupa in cocoon	(a) Cicadellidae–Typhlocybinae (b) Abdomen of adult host	Jervis (1980b,c)
A. nigriceps	Univoltine/ ?bivoltine	Prepupa in cocoon	(a) Cicadellidae–Typhlocybinae (b) Abdomen of adult host	Jervis (1980b,c)
A. serratus	Univoltine/ bivoltine	Prepupa in cocoon and 1st instar larva	(a) Cicadellidae–Typhlocybinae (b) Abdomen of adult host	Jervis (1980b,c)
GONATOPODINAE				
Dicondylus bicolor	Bivoltine/ trivoltine	1st instar larva	(a) Delphacidae (b) Abdomen of nymph or adult	Waloff (1974, 1975)
in Finland	Univoltine			Raatikainen (1967) as *D. lindbergi*
Gonatopus sepsoides	Bivoltine/ trivoltine	In cocoon spun by mature larva	(a) Cicadellidae–Deltocephalinae (b) Abdomen of adult host	Waloff (1974, 1975)
in S. France	Bi/trivoltine			Abdul-Nour (1971)
Pseudogonatopus distinctus	Univoltine	In cocoon spun by mature larva	(a) Delphacidae (b) Abdomen of nymph or adult	Waloff (1974, 1975)

ratio of 1♂:30♀ in southern France, and Sander (1980) reared males in Germany. *G. sepsoides* is possibly an amphitokous species, males being produced only under certain conditions, such as at high temperature (see Walter's (1983) review). On the other hand, this species may be comprised of different biological races which display different types of parthenogenesis.

Other examples of sex ratios in the literature include those of Behring (in Strübing, 1956), who recorded a male/female ratio of 1:1 in *Dicondylus bicolor*, and Raatikainen (1961), who recorded the same in *D. dichromus*. On the other hand, Raatikainen (1967) reared only 6 males out of a total of 451 *D. bicolor*, i.e. a sex ratio of 1:74, while Ponomarenko (1968) reared *Anteon jurineanum* in the ratio of 1:3. These are female-biased ratios, but in recent work by Khafagi (1986) on *Anteon pubicorne* parasitizing the leafhopper *Macrosteles viridigriseus*, males outnumbered females. Over two consecutive years, the sex ratio of this *arrhenotokous* species in the field samples was 1♂:0·3♀.

Further, in laboratory experiments males always outnumbered females, but the proportions of the sexes of this autogenous parasitoid varied on different diets. The percentages of female progeny of *A. pubicorne* increased on diets rich in carbohydrates, such as honey, sugar or pollen grains, and decreased on those rich in proteins, such as peptone. Thus, on sugar, the proportion of males to females was 146:84, or 1♂:0·58♀, while on peptone it was 136:35, or 1♂:0·26♀.

Also, on third to fifth instar nymphs of the host, Khafagi (1986) reared 87♂ and 27♀ (1♂:0·31♀) dryinids, whereas on adult hosts progeny reared from male *M. viridigriseus* were all male, but those from female hosts were a mixture of male and female wasps in the ratio of 32♂:12♀ or 1:0·38.

4. Host-feeding and Oviposition Behaviour

The host-feeding habit of female Dryinidae, which is associated with the possession of chelate forelegs (and is therefore absent in Aphelopinae), has been noted by several authors, e.g. Becker (1975), Chandra (1980b,c) Chua and Dyck (1982), Chua *et al.* (1984), Heikinheimo (1957), Lindberg (1950), Raatikainen (1961, 1967), Subba Rao (1957), Strübing (1956) and Waloff (1974). Waloff (1974) examined host-feeding and oviposition behaviour in three Gonatopodinae and two Anteoninae, and found that the degree of predation on hosts, which either preceded or accompanied oviposition, varied considerably between species.

Waloff (1974) examined the behaviour of *Gonatopus sepsoides* in the greatest detail, making altogether 52 observations on 14 reared females. Each morning, a female was presented with six cicadellid nymphs or adults at a time. The usual hosts were *Adarrus ocellaris, Psammotettix confinis* and *Arthaldeus pascuellus*. Any host handled by the wasp was removed and replaced. *G. sepsoides* usually fed on, but did not oviposit in the first host encountered during a day. Thereafter, hosts were used either for feeding and oviposition, or for oviposition only. All the individuals which were fed on, including those which were both fed on and oviposited in, died as a result of host-feeding, although a few survived for periods of 24–48 h. Feeding acitivity was greatly reduced after 50 min, but oviposition continued for up to 140 min. Before the female dryinid became quiescent, it sometimes rejected hosts suitable for oviposition, but only after it had handled several individuals during the same day.

While searching, female *G. sepsoides* periodically drum the substratum, or wave their antennae in the air. The rate of walking recorded at 19·5–21·5°C averaged 6·96 mm s^{-1}. Assuming that searching lasts for approximately three hours per day, the distances covered by females living for 14–21 days could amount to 223–334 m. While searching, the parasitoid reacts to movements of its prey, and reacts to hosts at a distance of 15 mm, pursuing them, or more rarely, jumping on them. Before jumping, the dryinid remains completely still for a few seconds, with its chelate forelegs held rigidly forwards. The hopper is then grasped at right angles to the longitudinal axis of the dryinid, with the chelae usually encircling the hind legs. The parasitoid may also restrain the host by applying its mandibles to the latter's abdominal tergites. The host may be lifted off the ground, or it may be held down on the substratum. Initially, the host may struggle, and a few individuals may escape, but the dryinid eventually curls its abdomen forwards and stings the host ventrally and paralyzes it. In feeding, the dryinid cuts the cuticle of the host's anterior abdominal tergites with its mandibles, sometimes inflicting large wounds which may extend over two segments. Subsequently, oviposition is accompanied by less intensive feeding or without any feeding at all. The eggs are laid immediately below the intersegmental membrane and their passage is sometimes clearly visible under a binocular microscope. In observations on 14 females, each in a separate container, 54 hosts were fed on, 54 fed on and oviposited in, and another 97 used for oviposition without feeding. Thus a dryinid that lived for 21 days could handle up to 95 hoppers. This is a minimum estimate, since one female lived for 27 days and was observed on alternate days only. It killed 50 and parasitized 42 hoppers. Thus, in its entire lifetime it could have handled 177 hoppers. After some laboratory tests, Strübing (1956) concluded that another *Gonatopus* species, *G. ombroides,* is a voracious feeder.

The second gonatopodine species observed by Waloff (1974) was *Pseudogonatopus distinctus*, which attacks Delphacidae. Its large, wingless females tended to drag their paralyzed hosts by their hind legs, sometimes for as long as 30 s. This behaviour was characteristic of this dryinid, but was observed only twice in the previous species. *P. distinctus* killed only 23% of its hosts, compared with 53% killed by *G. sepsoides*. Eleven *P. distinctus* females handled 4·3 hosts per day and lived for 12 days on average. Thus, this dryinid may kill or parasitize up to 90 delphacids during its lifetime.

A third gonatopodine, *Dicondylus bicolor*, was found by Waloff (1974) to be less voracious than the two preceding species, and as far as could be ascertained, it did not oviposit in any hosts upon which it fed. More detailed observations on this species have been made by Heikinheimo (1957) and Raatikainen (1967). In Britain this species is short-lived, whereas in Finland it is univoltine and lives for up to six weeks. Raatikainen found that on average each dryinid killed 2·7 *Javesella pellucida* nymphs per day. Thus, in addition to parasitism, this species may inflict mortality on 113 hosts by feeding throughout its lifetime.

The winged females of Anteoninae, like the wingless ones of Gonatopodinae, search for hosts by walking, not by flying. *Anteon pubicorne* sometimes chewed a small hole in the intersegmental membrane of its cicadellid host. It rarely killed hosts, since it normally fed only on the haemolymph which oozed from the puncture it had made in the host's integument. This was done while the female was ovipositing or otherwise handling the nymph. Sometimes the puncture was also used for oviposition. Becker (1975) noted similar behaviour in *Anteon ephippiger*. *Lonchodryinus ruficornis*, which also parasitizes Cicadellidae, was seen to bite at the base of the host's coxae, and imbibe the haemolymph that then exuded from the wound.

Although differences in the severity of host-feeding were noted among the three gonatopodine species studied by Waloff (1974), their behaviour differed significantly from that of the two anteonine ones. The Anteoninae inflicted only small wounds, and most of the hosts so attacked survived feeding.

Assuming that the predatory habits of some insect species preceded parasitic ones, the variations in behaviour associated with feeding and oviposition of gonatopodines, namely the use of a host individual either (1) for feeding only, or (2) for feeding and oviposition, or (3) for oviposition only, perhaps parallels one of the pathways whereby predators evolved into parasitoids. However, gonatopodines can be considered as specialized: the thoraces in apterous females are more reduced than in the other subfamilies, and together with dryininines their chelae show a more advanced structure, being larger and more complex (Richards, 1939). Therefore, it may be that their behaviour, which is intermediate between that of predators and parasitoids, is a secondary acquisition which accompanied the evolution of their chelae.

The host-feeding habit is widespread among hymenopteran parasitoids, having been recorded in 16 families in addition to Dryinidae (over 140 hymenopteran species in all) (Jervis and Kidd, 1986). Jervis and Kidd (1986) classified host-feeding behaviour according to whether (1) feeding is destructive (i.e. kills the host) or non-destructive, and (2) whether feeding accompanies oviposition (i.e. feeding is either concurrent or non-concurrent). Based on these criteria, the most common behaviour patterns among

Table 12
Host-feeding behaviour of Dryinidae.

Species	Behaviour	Reference
Lonchodryinus ruficornis	C–*ND*	Waloff (1974)
Anteon pubicorne	*C–ND*/NC–*ND*	Waloff (1974)
Anteon ephippiger	C–*ND*/NC–*D*	Becker (1975)
Echthrodelphax fairchildii	NC–*D*	Chandra (1980c)
Pseudogonatopus distinctus	C/NC–*D*	Waloff (1974)
P. flavifemur	NC–*D*	Chua and Dyck (1982)
P. nudus	NC–*D*	Chua and Dyck (1982)
Dicondylus bicolor	NC–*D*	Heikinheimo (1957), Raatikainen (1967), Waloff (1974)
D. dichromus	NC	Raatikainen (1961)
Gonatopus sepsoides	C–*D*/NC–*D*	Waloff (1974)
G. ombrodes	NC–*D*	Strübing (1956)
Richardsidryinus pyrillae	NC–*D*	Subba Rao (1957)

After Jervis and Kidd (1987).
The feeding puncture was invariably made with the mouthparts. C, concurrent; NC, non-concurrent; *D*, destructive; *ND*, non-destructive.

hymenopteran females are, in descending order: non-concurrent/destructive, concurrent/non-destructive, concurrent non-destructive/non-concurrent destructive, and concurrent non-destructive/non-concurrent non-destructive. The recorded host-feeding behaviour of twelve species of Dryinidae is summarized in Table 12. It is clear that for the majority of these species, the recorded pattern is incomplete; more detailed studies are therefore required before any meaningful generalizations about the occurrence and evolution of host-feeding behaviour in Dryinidae can be made.

Jervis and Kidd (1986) point out that host-feeding has been largely ignored by workers on parasitoid–host population dynamics. This is a serious oversight, since many parasitoids, including Dryinidae, kill significant numbers of hosts by feeding as well as by parasitism. Host-feeding may therefore have profound implications for parasitoid–host models. Collins *et al*. (1981) and Kidd and Jervis (unpubl.) have already found that host-feeding is likely to have a stabilizing influence on population interactions.

5. *Effects of Parasitism on Host Individuals*

The most generalized effect of dryinid parasitism on hosts is "parasitic castration", i.e. suppression or reduction of the internal reproductive organs, in both sexes. Also, parasitism in the nymphal stages frequently results in interference of development of external genitalia, resulting in the reduction of the aedeagus and styles. Parasitized females, for example, often have shorter ovipositors than unparasitized ones.

Parasitized individuals may also be paler than unparasitized ones, and this is especially noticeable in the more brightly coloured Typhlocybinae. Jervis (1978b) found that the most obvious visible effect of parasitism by *Aphelopus* species in adult hosts was depigmentation of the integument. In some Typhlocybinae, such as *Typhlocyba quercus* and *Kybos* species, the whole body including the wings may become depigmented. The characteristic red markings of *T. quercus* and deep green markings of *Kybos* may be replaced by either orange or yellow. Some endoskeletal structures may also be affected; for example, the apodemes of the second abdominal sternite of males can be much reduced or even absent. Whereas overall body colouration was affected in only a few species of Typhlocybinae, genital abnormalities and apodeme reduction occurred in a larger number of species.

An interesting effect of parasitism by dryinids was described by Müller (1960) in the males of *Euscelis incisus* and *E. distinguendus* parasitized by *Gonatopus sepsoides*. The aedeagus was significantly larger in parasitized individuals than in unparasitized ones, and parasitized males were larger as a whole. This contrasts with the usual reduction in genitalia caused by parasitoids. Müller suggests that reduction of body size occurs only if the dryinids attack very early nymphal instars, whereas parasitism of later nymphs results in delayed development which is accompanied by increase in growth, partly at the expense of gonad development. At least the male hosts on reaching the adult stage show a small but general increase in body size.

B. Pipunculidae

1. Systematic Position, Taxonomy and General Biology

The family Pipunculidae is the only group of parasitoid Diptera known to be associated with leafhoppers and planthoppers in Europe. Pipunculids are exclusively parasitic in Auchenorrhyncha (Table 13), attacking the nymphal and/or the adult stages. Females pounce on hosts, and using a large piercing ovipositor inject a single egg. Larvae develop singly, within the host's abdominal haemocoel. Depending on the species of pipunculid, and apparently also the size and species of host, the parasitoid completes its development in either the nymphal host, the adult host, or either of these stages. It consumes most or all host tissues except for the integument, and finally emerges by rupturing an intersegmental membrane. Pupation occurs either in soil or leaf litter, or on the host's food plant.

The systematics, morphology and palaeontology of Pipunculidae are reviewed by Aczél (1948). The other important monographs on the group are those of Coe (1966) and Hardy (1943).

Aczél (1954) considers pipunculids to be related to the Dolichopodidae, Platypezidae and Syrphidae which together form the series Aschiza. Except

Table 13
Hosts of Pipunculidae to family and subfamily level.

		Cicadellidae						
	Cercopidae	Idiocerinae	Macropsinae	Deltocephalinae	Typhlocybinae	Cixiidae	Delphacidae	Flatidae
Chalarus					×			
Verrallia	×		×					
Tomosvaryella				×				
Dorylomorpha				×				
Pipunculus		×[a]		×				
Cephalops				×[c]		×	×	
Eudorylas		×[b]		×				×[d]

From Coe (1966), revised by Benton (1975).
[a] Subramaniam (1922).
[b] Jervis (unpubl.).
[c] Albrecht (unpubl.).
[a] and [d] are non-European records.

for one species known from the Miocene, all recorded fossils are derived from Baltic amber (Oligocene).

The following works contain keys, descriptions or both, for the identification of adults: Coe (1966) (British fauna), Bankowska (1973) (Polish fauna), Lauterer (1981) (Czechoslovakian fauna), Albrecht (1979) (European *Dorylomorpha*), Kozánek (1981a,b), (Czechoslovakian *Pipunculus* and *Nephrocerus*) and Jervis (in prep.) (European *Chalarus*).

By associating reared adults with their puparial remains, a larval taxonomy can also be developed. Benton (1975), for example, found the structure of the posterior spiracular plate, prothoracic spiracles and cephalopharyngeal skeleton to be useful in the identification of second instar larvae and puparia of British Pipunculidae, and devised a key for the identification of several genera and some species. Jervis (1978b) likewise found that several species of *Chalarus* can be positively identified using larval characters alone.

Adult Pipunculidae have extraordinarily large compound eyes. These provide an extensive visual field necessary for the flies' characteristic hovering, often stationary flight. Field observations indicate that the foraging behaviour of females involves a large visual component (Benton, 1975; Huq, 1982; Jenkinson, 1903; Jervis, 1980a; Perkins, 1905c; Williams, 1918). Female *Verralia setosa*, for example, attempt to pierce birch buds and bud scales with their ovipositors, apparently mistaking the plant parts for nymphs of *Oncopsis flavicollis*, which are similar in size and colour (Benton, 1975).

Some species remove the host from the food plant and oviposit whilst in flight (Huq, 1982; May, 1979; Williams, 1918), whereas others leave the host *in situ* (Benton, 1975). Further interspecific differences have been recorded with respect to the host stage oviposited in. *Chalarus* species, for example, generally oviposit only in third, fourth and fifth instar nymphs (Jervis, 1980a), while *Pipunculus campestris, Eudorylas subterminalis* and *Tomosvaryella sylvatica* generally attack only second, third and fourth instar nymphs (Huq, 1982). *Cephalops semifumosus* attacks both nymphs and adults (Rothschild, 1964), while *Verrallia aucta*, like its North American relative *V. virginica* (Linnane and Osgood, 1977), oviposits only in adult Cercopidae (Coe, 1966; Whittaker, 1969). *V. aucta* females actively avoid the spittle that surrounds the host nymphs (Coe, 1966).

It has been suggested that female Pipunculidae are able to avoid superparasitism of hosts (Coe, 1966; Jenkinson, 1903; Williams, 1957). However, field-collected typhlocybid nymphs may contain up to seven offspring of *Chalarus*. Jervis (1980a) estimated that, on average, over 19% of *C. fimbriatus* larvae die as a result of larval competition arising from superparasitism. Therefore, it seems unlikely that females are capable of host discrimination. Such behaviour is, as far as is known, confined to hymenopteran parasitoids (Bakker *et al.*, 1985).

Further mortality of progeny may occur as a result of encapsulation by the host. Over 16% of eggs of *C. fimbriatus* were encapsulated by *Alnetoidia alneti* (Jervis, 1980a).

No information is available on the responses of foraging females to host density and spatial distribution.

When kept under laboratory or insectary conditions, adult pipunculids die within a few days (Jervis, 1978b; May, 1979; Rothschild, 1964). Females of *Chalarus* first contain ripe eggs 24 hours after emergence; there is therefore a preoviposition period. Adults of both sexes feed on honeydew in the field (Jervis, 1978b; Williams, 1918; Williams, 1957).

The female reproductive system is of a relatively simple type, with short lateral and median oviducts (Jervis, 1978b). The ovipositor, which consists of the modified seventh, eighth and ninth abdominal segments, incorporates a valve mechanism by which eggs are released from near the ovipositor tip following penetration of the host's integument. The ovipositor piercer bears numerous sensilla. The role they play in oviposition, if any, is unknown. Their form and pattern of distribution indicate that they may be involved in mechanoreception and chemoreception (Jervis, unpubl.).

Courtship and mating in Pipunculidae have been described for a number of species (Bristowe, 1950; Coe, 1966; Huq, 1982; Jervis, 1978b; Lauterer, 1981; May, 1979). All published accounts are far from complete, but a common picture of behaviour emerges. The male pursues and seizes the female in flight. The mating pair hover *in copula*, with the male positioned above the female, and may settle temporarily on vegetation. The male terminates copulation in flight, but the pair disengage while at rest. According to Huq (1982), a female may mate several times with the same male. Thereafter, she becomes unresponsive to further mating attempts. Huq also observed a female to mate between ovipositions.

Embryonic development follows egg deposition. The egg swells, apparently due to the intake of host fluids (Jervis, 1980a). Both the first and early second instar larvae have a large caudal vesicle, which is probably respiratory and excretory in function.

It is now generally agreed that there are only two larval instars. Descriptions of larvae and other immature stages are given by Aczél (1943), Benton (1975), Coe (1966), Huq (1982), Jervis (1980a), May (1979), Parker (1967), Rothschild (1964), Subramaniam (1922) and Whittaker (1969).

May (1979) showed that in *Cephalops curtifrons* the duration of the pupal stage is strongly temperature dependent. Males generally have a shorter pupal duration than females (Benton, 1975; Huq, 1982; Jervis, 1980a; May, 1979), and so tend to emerge earlier both in the insectary and in the field (Jervis, 1978b, 1980b; May, 1979). Benton (1975) also recorded statistically significant interspecific differences in pupal duration times.

2. *Phenological Studies*

Lauterer (1981, 1983) discusses the habitat preferences of pipunculid adults. Adults have been collected using a variety of methods: malaise traps (Benton, 1975; De Meyer and De Bruyn, 1984; Whittaker, 1969); aerial suction traps (Waloff, 1975; Whittaker, 1969); emergence traps (De Meyer and De Bruyn, 1984; Whittaker, 1969); nets (Benton, 1975; Waloff, 1975); and an aspirator (Jervis, 1980b). Some species have highly female- or male-biased field sex ratios (Jervis, 1980b; Rothschild, 1964).

Various methods have also been devised for rearing Pipunculidae from parasitized hosts (see Benton, 1975; Coe, 1966; Huq, 1982, 1984; Jervis, 1978a,b, 1980b; Strübing, 1957; Waloff, 1975), and Chandra (1978, 1980a) describes a method for rearing Dryinidae which may also prove suitable for Pipunculidae. So far only Huq (1982, 1984) and Strübing (1957) have succeeded in establishing breeding populations in the laboratory, and then only for a single summer season.

The phenologies of 28 European species are presently known (Table 14). The majority of species overwinter as diapausing pupae. Several however, overwinter as first instar larvae within hosts, resuming growth and development only when the latter come out of diapause. *Eudorylas obliquus* is unusual in that it overwinters both as pupae and larvae at the same locality (Silwood Park). In *Pipunculus campestris* the overwintering stages of English and German populations are apparently different (Table 14). Whether the phenology of this species actually does vary geographically is not known. However, given that voltinism in Auchenorrhyncha varies with latitude and photoperiod (Waloff and Solomon, 1973), intraspecific variation in parasitoid phenology is likely to occur.

Statistically significant interspecific differences in adult emergence time (i.e. flight period) have been recorded among coexisting species of Pipunculidae (Benton, 1975; Jervis, 1980c). For example, overwintering individuals of *Chalarus exiguus* emerge each year significantly later in the spring than those of *C. fimbriatus, C.* sp. A nr *spurius* and *C. pughi*. These temporal differences are correlated with differences in host relations, i.e. *C. exiguus*, which parasitizes a late-occurring host (*Alnetoidia alneti*), itself emerges late (Jervis, 1980c). Interspecific variability in emergence time, at least in *Chalarus*, is apparently a reflection of species differences in diapause development and/or postdiapause morphogenesis of the parasitoids.

Superimposed on these genetically determined interspecific differences are intraspecific differences which are likewise correlated with host relations. Thus, individuals of *C.* sp. A nr *spurius* derived from late-occurring host species emerge later in the season than those derived from early-occurring species (Jervis, 1980c). These host-associated intraspecific effects

Table 14
Voltinism and overwintering stages of European Pipunculidae.

Species	Voltinism	Overwintering stages	Remarks	References
Chalarus sp. nr *argenteus* Coe	Univoltine (Britain)	Pupa	Host univoltine	Jervis (1980b)
Chalarus sp. C nr *spurius* (Fallén)	Univoltine (Britain)	First instar larva in adults of *Empoasca vitis*	Host univoltine	Jervis (1980b): previously mentioned under name *C. parmenteri* Coe
Verrallia aucta (Fallén)	Univoltine (Britain, Belgium)	Pupa	Host univoltine	Whittaker (1969), Benton (1975), Waloff (1975), De Meyer and De Bruyn (1984)
Verrallia setosa Verrall	Univoltine (Britain, Belgium)	Pupa	Host univoltine	Benton (1975), De Meyer and De Bruyn (1984)
Pipunculus fonsecai Coe	Univoltine (Britain)	Pupa	Host univoltine	Benton (1975), Waloff (1980)
Pipunculus thomsoni Becker	Univoltine (Britain, Belgium)	?	Host univoltine	Waloff (1980), De Meyer and De Bruyn (1984)
Pipunculus zugmayeriae Kowarz	Univoltine (Britain)	Pupa	Host univoltine	Benton (1975), Waloff (1980)
Cephalops chlorionae Frey	*Univoltine* (Finland)	First instar larva in nymphs of *Chloriona glaucescens*	Host univoltine	Lindberg (1946)
Cephalops curtifrons Coe	Univoltine (Britain)	First instar larva in adults of *Stenocranus minutus*	Host univoltine	Benton (1975), May (1979), Waloff (1975, 1980)
Eudorylas fascipes (Zetterstedt)	Univoltine (Britain)	Pupa	Hosts bivoltine	Benton (1975), Waloff (1980)
Eudorylas fuscipes (Zetterstedt)	Univoltine (Finland)	Pupa	Host univoltine	Lindberg (1946)
Eudorylas zonatus Zetterstedt	Univoltine? (Britain)	?	None reared, but seasonal appearance suggests parasitoid is univoltine	Benton (1975), Waloff (1980)

Table 14 continued

Species	*Voltinism*	*Overwintering stages*	*Remarks*	*References*
Eudorylas zonellus Collin	Univoltine? (Britain)	?	None reared, but seasonal appearance suggests parasitoid is univoltine	Benton (1975), Waloff (1980)
Chalarus fimbriatus Coe	Bivoltine (Britain)	?	Hosts univoltine and bivoltine	Jervis (1980b)
Chalarus latifrons Hardy	Bivoltine (Britain)	Pupa	Known hosts univoltine; seasonal occurrence suggests parasitoid is bivoltine	Jervis (1980b)
Chalarus parmenteri Coe	Bivoltine? (Britain)	Pupa	Hosts univoltine and bivoltine; seasonal occurrence suggests parasitoid is bivoltine	Jervis (1980b)
Chalarus pughi Coe	Bivoltine (Britain)	Pupa	Known hosts bivoltine	Jervis (1980b)
Chalarus spurius (Fallén)	Bivoltine (Britain)	Pupa	Hosts bivoltine	Jervis (1980b), previously mentioned under name *C.* sp. B
Chalarus sp. A nr *spurius*	Bivoltine (Britain)	Pupa	Hosts univoltine and bivoltine	Jervis (1980b)
Dorylomorpha xanthopus (Thomson)	Bivoltine? (Belgium)	Pupa?	Host records and seasonal occurrence suggest parasitoid is bivoltine	De Meyer and De Bruyn (1984)
Tomosvaryella sylvatica (Meigen)	Bivoltine (Britain, Belgium, Germany)	Pupa	Hosts univoltine and bivoltine	Waloff (1975, 1980), De Meyer and De Bruyn (1984), Huq (1982)
Cephalops semifumosus (Kowarz)	Bivoltine (Britain, Belgium)	First instar larva in nymphs and adults of certain Delphacidae	Hosts univoltine and bivoltine	Rothschild (1966), Benton (1975), Waloff (1975, 1980), De Meyer and De Bruyn (1984)

continued

Table 14 continued.

Species	*Voltinism*	*Overwintering stages*	*Remarks*	*References*
Eudorylas obliquus Coe	Bivoltine (Britain)	First instar larva and pupa	Hosts bivoltine	Benton (1975), Waloff (1975, 1980)
Eudorylas obscurus Coe	Bivoltine (Britain)	Pupa	Hosts univoltine and bivoltine	Benton (1975), Waloff (1975, 1980)
Eudorylas subfascipes Collin	Bivoltine (Britain)	Pupa	Hosts bivoltine	Waloff (1975)
Tomosvaryella frontata (Becker)	Bivoltine, possibly trivoltine (France)	Pupa	Host multivoltine	Parker (1967)
Eudorylas subterminalis Collin	Facultatively trivoltine (Britain) Bivoltine (Germany)	Pupa	Hosts univoltine and bivoltine	Waloff (1975, 1980), Huq (1982)
Pipunculus campestris Latreille	Trivoltine (Britain) Bivoltine (Germany)	Pupa (Britain) first instar larva in adult *Euscelis incisus* (Germany)	Hosts univoltine and bivoltine	Benton (1975), Waloff (1980), Huq (1982)

are most probably a result of sensitivity to the host's physiology, i.e. prediapause conditions (Jervis, 1980c), rather than a result of disruptive selection for different "host races" (e.g. see Bush, 1975).

Although among Pipunculidae there is generally a close degree of temporal synchrony with hosts, the second generation of bivoltine *Chalarus* species in particular is rather poorly synchronized with the second generation of bivoltine hosts, because of the short development time of non-diapausing larvae (Jervis, 1980c). This asynchrony is thought to be the major cause of differences in percentage parasitism recorded between the two generations of certain host species (Jervis, 1980b,c). Variability in the degree of temporal synchrony between host and parasitoid populations leads to a non-random distribution of parasitism, which may have important consequences for parasitoid–host population dynamics, through the creation of a partial refuge (Hassell, 1986b).

Diapause has been artificially terminated, but with limited success, by subjecting pupae either to low temperatures for several weeks (Benton, 1975), or to long-day conditions (Huq, 1982).

3. *Host Specificity*

Data on host range have been obtained for over 52 European species of Pipunculidae (Appendix III). Only one species, *Cephalops obtusinervis*, has been reared from more than one subfamily of hosts. Rather surprisingly, it attacks hosts belonging to different families (Cicadellidae and Delphacidae: A. Albrecht, unpubl.). *Cephalops furcatus, Chalarus spurius, Ch. latifrons* and all *Verrallia* species reared so far are almost certainly narrowly polyphagous, while available evidence suggests that *Pipunculus fonsecai, P. zugmayeriae, Cephalops curtifrons, Chalarus exiguus* and *Ch.* sp. C nr. *spurius* are truly monophagous (Benton, 1975; Jervis, unpubl.).

The results of preliminary experiments carried out by Huq (1982) suggest that host specificity is at least partly the result of (a) discriminative oviposition behaviour, and (b) the differential effects of hosts on the survival of developing offspring, such that individuals derived from some host species do not emerge from the puparium.

Benton (1975) found insignificant inter- and intraspecific size differences among adult Pipunculidae reared from different species of host, adult size being strongly correlated with host size. Using regression analyses of morphometric data, he related the coefficient of variation in size of each pipunculid species to the number of host species attacked. It was shown that most of the variation in size could be explained in terms of host range; polyphagous species are more variable in size than monophagous ones, a reflection of the diversity in size among hosts. The coefficient of adult size variation can therefore be used as an index of the degree of host specificity. An understanding of size variation is also important because of the likely effects of a female's size on her fecundity, longevity and other components of fitness (see Waage and Ng, 1984).

4. *Effects of Parasitism on Host Individuals*

Auchenorrhyncha, particularly adults, may display various symptoms of parasitism by Pipunculidae (Chandra, 1980c; Giard, 1889; Huq, 1982; Jervis, 1978b; Lauterer, 1981; Lindberg, 1946; May, 1979; Morcos, 1953; Muir, 1918; Perkins, 1905c; Remane and Schultz, 1973; Ribaut, 1936; Rothschild, 1964; Waloff, 1980; Whittaker, 1969; Williams, 1957; Vidano, 1962; Yano *et al.*, 1985; Ylönen and Raatikainen, 1984). These include: depigmentation of the integument, genital abnormalities, reduced or increased body size, aberrations in wing venation, reduced size of apodemes, parasitic castration, and reduced locomotory ability. Parasitic castration is one of the most commonly recorded symptoms. The ovaries of females are poorly developed or even absent, as are the oviducts and accessory glands. The reproductive organs of males are similarly affected. Castrated males may nevertheless copulate with females and inseminate them (May, 1979) (see also Whittaker, 1969). Castration is most pronounced in those individuals that have been parasitized early in nymphal life. This applies also to many of the other symptoms of parasitism mentioned above. Host movement begins to be impaired after the parasitoid larva has reached the late first instar. Hosts containing parasitoids which are at earlier stages of development are able to walk or fly apparently normally (Jervis, 1978b; Waloff, 1980).

C. Strepsiptera

1. Systematic Position, Taxonomy and General Biology

Together with the other parasitoids of Auchenorrhyncha, Strepsiptera show many features of general biological interests. The main characteristics of this order are discussed by Kinzelbach (1971, 1978) and are summarized in Clausen (1962), Riek (1970) and Richards and Davies (1977).

As far as is known, all Strepsiptera are endoparasitic in other insects, for at least part of their life cycle. The sexes are strongly dimorphic, even in their endoparasitic larval stages. The adult male is free-living, but the abdomen of the neotenic, parasitic female is embedded in the host's haemocoel, only its fused head and thorax (forming the cephalothorax) protruding to the exterior. Exceptions to this are found in the primitive family Mengenillidae, females of which are free-living. Females of all Strepsiptera are viviparous, with hundreds, and in some species thousands, of eggs filling the body cavity, wherein they develop into minute, mobile larvae which in many species are able to jump. These "triungulinid" larvae are superficially similar to those of the beetle families Meloidae and Rhipiphoridae. In endoparasitic females there is a brood canal which is formed ventrally between the body of the female parasite and its last larval cuticle. The triungulinid larvae hatch within the female's body and reach the exterior by emerging through the aperture of the brood canal. Having become established in a host, they undergo hypermetamorphosis, moulting into grub-like, legless larvae which show sexual differentiation after the second moult.

Strepsipteran males are dark brown or black insects, ranging in size from 1·2 to 7·5 mm, the majority being between 2 and 3 mm long. They have conspicuous flabellate antennae, club-like forewings which are analogous to the hindwings of Diptera, and an enlarged metathorax bearing large, fan-shaped hindwings.

Interesting current work in Oxford, mainly by Kathirithamby and Spencer Smith, is beginning to shed light on some puzzling features of the Strepsiptera. For instance, the number of larval instars has always been uncertain, since all the estimates have been arbitrary and based solely on size increases.

Now, at least in *Elenchus tenuicornis* (Kirby), this problem has been resolved through the use of both scanning and transmission electron microscopy (Kathirithamby *et al.*, 1984). The triungulinid larvae undergo normal ecdysis, but the resulting grub-like larvae undergo two further apolyses, i.e. separation of the old cuticle from the epidermis without ecdysis. The increase in size between the second and the fourth instars occurs despite the retention of the exuviae, probably due to stretching and/or extension of the original convolutions in the cuticle. The occurrence of two apolyses in sequence without the shedding of exuviae is yet another unique feature of the Strepsiptera.

After extrusion, the male pupates *in situ* on the body of its host, and the fourth larval cuticle sclerotizes to form the puparium. In the females the pupal stage is suppressed (except in the free-living Mengenillidae), and the adult becomes sexually mature following extrusion of the cephalothorax through the host's cuticle.

The unusual emergence of the adult male of *Elenchus* has been described by Kathirithamby (1983). Prior to emergence, a membranous transparent sac appears on the ventral side of the head, between the eyes and below the antennnae and frons. Pulsations, by alternate inflations and deflations of the sac, rupture the puparium, and a cup-shaped cephalotheca separates off as the male emerges. In other words, a "ptilinum" helps to loosen the cephalotheca, and not the mandibles as was previously thought. The sac then collapses and is invaginated. It is analogous to the ptilinum of cyclorrhaphan Diptera, but differs in location.

After emergence, the short-lived male flies towards the female and copulates via her cephalothorax. Most male *E. tenuicornis* emerge in the morning, and they live for a few hours only. The same is true of *Halictophagus silwoodensis*, recently described by Waloff (1981). Raatikainen and Heikinheimo (1974) found that *E. tenuicornis* males occurred in the field for a period of six weeks in southern Finland, and that they began to fly when the daily sum of effective temperatures, i.e. those above 10°C, reached *ca.* 100 degree-days. Spencer Smith and Kathirithamby (1984) showed that the principal flight muscles of *Elenchus* are clearly of the asynchronous type with respect to the pattern of motor neural excitation, but that they differ strikingly from those of other insects by the small diameter of the fibres, the variable contour of the microfibrils, and the absence of tracheolar invaginations.

Male Strepsiptera, being so small, are rarely caught in aerial suction traps. Lewis and Taylor (1965) found only 15 male *E. tenuicornis* in a sample of 232,682 small, day-flying insects. Similarly, Raatikainen and Heikinheimo (1974) recorded only 61 male *E. tenuicornis* and 2 *Halictophagus ?curtisii* Dale in Curtis in netting apparatus set up in four latitudes in Finland (60°20′–*c.* 69°N) for periods varying between four and ten years.

Flight in males is, of course independent of dispersal, which in Strepsiptera can be achieved in two stages of the life history:

(1) First, the active, minute triungulinid larvae are released on to plants inhabited by their hosts. Included here is the apparent case of phoresy (the use of one animal species by another for the purposes of transport only) observed by Kathirithamby (1982, and in press) in rice fields of Malaysia, where some unstylopized individuals (stylopization is the term used to describe parasitism by Strepsiptera) of *Nilaparvata lugens* had larvae sitting on them. This suggests that phoresy itself could have been one of the steps in the evolution of parasitism by Strepsiptera. Of course, triungulinid larvae of Stylopidae, which parasitize Hymenoptera, depend on phoresy on adult hosts in order to reach the nests where they parasitize host larvae.
(2) Secondly, a more passive, but more long-range type of dispersal occurs by stylopized macropterous hosts, which are able to fly and so carry the parasites to new sites. Raatikainen (1972) caught numerous stylopized delphacids in his aerial nets in Finland, and his samples contained 4,543 *Javesella pellucida* (Fabricius), of which 773 (17%) bore visible *Elenchus tenuicornis*. Waloff (1973) recorded 4·9 and 1·6% stylopized *J. pellucida* in 323 first-generation and 258 second-generation specimens caught in aerial suction traps in Southern England.

Kinzelbach (1978) divides the Strepsiptera into two suborders, Mengenellidia and Stylopidia, and into nine families. Kathirithamby (pers. comm.) informs us that there are 436 species so far described from various parts of the world, and that there are a further 150 from Australasia alone. Many species are known from one sex only, since females in stylopized hosts are easier to collect than the minute, free living, very short-lived males. Recently, Drew and Allwood (1985) have described new stepsipteran species parasitizing fruit flies of the genus *Dacus* (Tephritidae), and have placed them in a new family, Dipterophagidae. However, Kathirithamby (in manuscript) has re-examined and redescribed these species, and considers that they belong to the Halictophagidae.

The Mengenellidia is considered to be the more primitive of the two suborders, and contains two families: Mengeidae, known only from fossil males in Baltic Amber, and Mengenillidae, which differ from other extant Strepsiptera in having free-living females and in parasitizing Thysanura.

Many authors, notably Crowson (1955), place strepsipterans among the Coleoptera, because the hind wings alone are used in flight, the abdominal tergites are more extensively sclerotized than the tergites, and their triungulinid larvae bear a strong resemblance to those of Meloidae and Rhipiphoridae. Also, a striking feature of the adult males and the free-living females is the absence of functional spiracles on the freely extended eighth abdominal segment, this being characteristic of the larger section of Coleoptera–Polyphaga. However, Kinzelbach (1971) considers that the relationship of Strepsiptera with other insects is uncertain and obscure, and that the palaeontological evidence contributes very little, since there are no pre-Tertiary fossils. Together with other authors, he concludes that the resemblance to Coleoptera may be due to convergent evolution.

2. *Host Specificity*

The known hosts of Strepsiptera, to both superfamily and family levels, are given in Table 15. They belong mainly to the Hemiptera and Hymenoptera–Aculeata, but also include Orthoptera, Dictyoptera and Diptera as well as Thysanura. Only two families, the Halictophagidae and Elenchidae, are known to parasitize Auchenorrhyncha in Europe. Their geographical distribution is shown in Fig. 4, and a list of their European hosts is given in Appendix IV. It is curious that most of the records of stylopization of Heteroptera occur outside Europe. Perhaps this is partially explained by the absence from Europe of Corioxenidae and Callipharixenidae. However, Halictophagidae, which occur throughout Europe, are known to have heteropteran hosts elsewhere in the world.

Among Halictophagidae, in those species of the genus *Halictophagus* that attack members of the superfamily Fulgoroidea (Tettigometridae, Delphacidae, Fulgoridae, Issidae) and the species of *Stenocranophilus* which are associated with Delphacidae only, host specificity appears to extend to host family level. Species of *Halictophagus* that attack cicadelloids (Cicadellidae) may have a somewhat narrower host range which extends to host subfamily level.

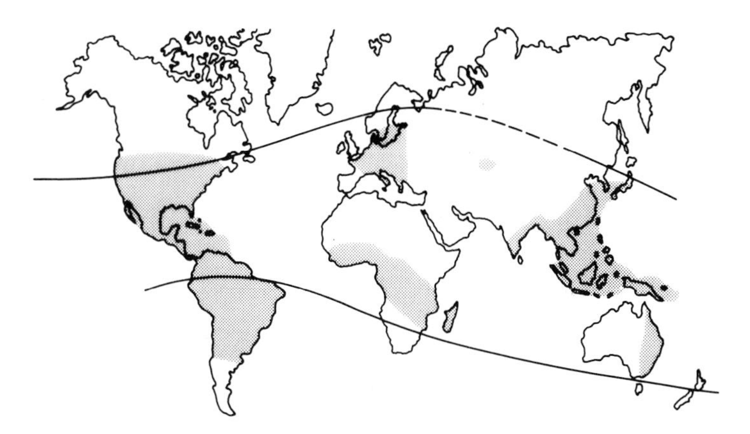

Fig. 4. World distribution of Elenchidae (—), and Halictophagidae (stipple) (Strepsiptera) (from Kinzelbach, 1978).

Table 15
Known hosts of Strepsiptera to superfamily or family levels.

Hosts			Strepsiptera								
Order	Suborder	Superfamily or family (* all superfamilies)	Mengeidae[1]	Mengenillidae	Corioxenidae	Halictophagidae	Callipharixenidae	Bohartillidae[2]	Elenchidae	Myrmecolacidae	Stylopidae
Thysanura	Zygentoma	Lepismidae		•							
Orthoptera	Ensifera	Tettigoniidae								•♀	
		Gryllidae								•♀	
		Gryllotalpidae								•♀	
	Caelifera	Tridactylidae				•					
Dictyoptera	Blattaria	Blattidae				•				•♀	
	Mantodea	Mantidae					•				
Hemiptera	Homoptera	Tettigometridae				•					
		Delphacidae				•			•		
		Fulgoridae				•			•		
		Dictyopharidae							•		
		Issidae				•					
		Flatidae							•		
		Ricaniidae							•		
		Eurybrachidae				•			•		
		Cercopidae				•					
		Membracidae				•					
		Cicadellidae				•			•		
	Heteroptera	Lygaeidae					•				
		Coreidae				•					
		Cydnidae			•						
		Scutellaridae				•	•				
		Pentatomidae			•	•	•				
Diptera	Nematocera	Phlebotomidae				•					
	Brachycera	Tabanidae							•		
	Cyclorrhapha	Platystomatidae							•		
		Tephritidae				•					
Hymenoptera	Apocrita	Mutillidae						•			
		Formicidae								•♂	
		Pompiloidea*									•
		Vespoidea*									•
		Sphecoidea*									•
		Apoidea*									•

Among Elenchidae, up to date, only one species, namely *Elenchus tenuicornis*, is known from Europe (Kinzelbach, 1978). It is said to attack about 60 species of Delphacidae, of which 39 are listed in Appendix IV.

3. *Effects on Individual Hosts*

Parasitism by Strepsiptera is usually referred to as stylopization. Stylopized hosts show varying degrees of reduction of internal and external genitalia and loss or suppression of sex-linked characters, and it has frequently been assumed that there is "sex reversal". This has been critically examined by Kathirithamby (1977a,b, 1979, 1982) in a number of delphacid and cicadellid species, and it has been shown that whereas stylopized individuals show abnormalities in sexual characters, there is no acquisition of characters of the opposite sex, i.e. no sex reversal as such. Kathirithamby (1977a) investigated very thoroughly the changes in primary sexual characters (i.e. internal genitalia), secondary characters (external genitalia used for copulation and oviposition) and tertiary characters (i.e. extragenital structures and organs) that occur in the delphacid *Javesella dubia* (Kirschbaum) when parasitized by *Elenchus tenuicornis*. She observed variations in the degree of damage resulting from stylopization, and these clearly depended on the time of establishment of the parasitoid. It was found that the internal sex organs in stylopized nymphs and adults of *J. dubia* were either reduced or absent. Rudiments of secondary sex organs were present but reduced in the nymphs, this reduction being more pronounced in adult males than in females, while the tertiary sexual characters or organs, such as colour and the tymbal apparatus (including the shield-shaped plate on the second abdominal tergite, together with the sternal apodemes) were absent in stylopized males. Also, the shape of the pale pygofer of males superficially resembles that of adult females, but is actually more like that of a pharate male. It may thus be concluded that development of the last few abdominal segments of stylopized males is suppressed at various points between the early and late stages of the pharate adult. Kathirithamby notes that stylopized male Delphacidae with pale colouration, peculiar shaped pygofers and lost tymbal organs have often been mistaken for females. Effects of stylopization by *Elenchus* on *Nilaparvata lugens* (Stål) and *Sogatella furcifera* (Horvath) in Malaysia have also been examined by Kathirithamby (1979, 1982), and some of her results are given in Table 16.

Kathirithamby (1977b) also examined the effects of stylopization on some species of Cicadellidae. In *Ulopa reticulata* parasitized by *Halictophagus*

Footnote to Table 15.

Data primarily from Kinzelbach (1978); also from Drew and Allwood (1985), Freytag (1985), Riek (1970), Richards and Davies (1977).

[1] Fossil ♂ in Baltic Amber.

[2] ♂♂ in Honduras, ♀♀ unknown.

Table 16
Effects of stylopization by *Elenchus* spp. on *Javesella dubia* and *Nilaparvata lugens*.

	Javesella			*Nilaparvata*		
Sex character	Lost	Lost/ suppressed	Unaffected	Lost	Lost/ suppressed	Unaffected
Primary						
Internal genitalia	+	+	–	+	–	–
Secondary						
External genitalia	–	+	–	–	+	–
Tertiary						
pigmentation ♂	–	+	–			
pigmentation ♀	–	–	–			
Shield-shaped plate on second abdominal tergite of adult ♂	+	–	–	–	+	–
Apodemes in adult ♂	+	–	–	–	+	–

From Kathirithamby (1979, 1982).

silwoodensis there was loss or suppression of internal sex organs, but the external genitalia showed much less change than in Delphacidae, although some stylopized females had much shorter ovipositors than unparasitized ones (males suffered mechanical damage to abdominal tergites when the larval strepsipterons emerged to pupate on the host). Kathirithamby concludes that whereas stylopized Delphacidae undergo morphological changes in their primary, secondary and tertiary sex organs and characters, this is far less pronounced in the three subfamilies of Cicadellidae so far examined, namely in Malaysian Deltocephalinae (*Nephotettix* spp.) and Cicadellinae (*Kolla* sp.) and British Ulopinae (*Ulopa reticulata*). It was also interesting that whereas in stylopized delphacids males showed greater suppression of external genitalia, females of the cicadellid *Ulopa* were more abnormal than males.

D. Other Parasitoids of Nymphs and Adults

1. Sarcophagidae

Auchenorrhyncha have recently been found to be parasitized by flesh-flies (Sarcophagidae). Soper *et al.* (1976) describe parasitism of adult male cicadas by *Colcondamyia auditrix* in Canada. The female parasitoids are attracted by the song of the male cicada, and rarely parasitize females.

Sarcophagid parasitism of cicada adults in Europe has so far not been recorded, but warrants close examination.

2. *Encyrtidae*

Arzone (1971) found that nymphs of the Buffalo Treehopper, *Bisonia stictocephala*, are attacked by an endoparasitic encyrtid, *Prionomastix morio* (Dalman), a rare but widespread wasp whose host was hitherto unknown. No information is available on its biology.

3. *Nematoda*

As far as can be ascertained, there are no records of nematode infections of Auchenorrhyncha in Europe. Infections probably occur, but are rare. Since nematodes kill their hosts, and locally can cause high mortality, they warrant discussion, although literature on them deals with non-European species of Auchenorrhyncha.

Poinar (1975), in his book on entomophagous nematodes, states that most nematode parasites of Hemiptera are accidental infections by Mermithidae, although some records from other nematode families are also known, Sperka and Freytag (1975) made a survey of 60,000 specimens of Auchenorrhyncha obtained from grassy pastures in Kentucky, and dissected out or reared from these more than 200 unidentified mermithid worms, the overall level of parasitism being 0·3%. Much higher infections of leafhoppers are reported from rice fields. The latter provide an ideal humid and wet environment for nematode transmission, as was pointed out by Greathead (1983). In contrast with the low infection levels in Kentucky, Yang (1982) recorded 69·7–71·0% mermithid infections in the brown rice planthopper, *Nilaparvata lugens*, 23–29% in *Sogatella furcifera* and 2·6–5·8% in *Laodelphax striatellus* in the Shanghai region of China. In Japan, nematodes caused up to 43% mortality in *N. lugens* and 70% in *S. furcifera* (Esaki and Hashimoto, 1930). In India, Manjunath (1978) found two species of *Hexamerus* in *N. lugens*, and noted that brachypterous hosts were more commonly parasitized than macropterous ones. The more common of the two *Hexamermis* species was also reared from *S. furcifera* and the cicadellid *Nephotettix nigropictus*.

Sperka and Freytag (1975) concluded that there was no apparent host specificity, and that the site and date of collection appeared more important than the species of host present. They found nematodes in six species of Cercopidae (including the European species *Philaenus spumarius* and *Aphrophora salicina*), in three species of Membracidae, in 28 species of Cicadellidae (including the European species *Psammotettix striatus* and *Alebra albostriella*), in four identified and several unidentified species of Delphacidae, in two species of Cixiidae, in one species of Dictyopharidae and one species of Issidae. They also noted that nematode parasitism seldom affected any external morphological features, including the genitalia, of their hosts. However, depigmentation and abdominal distension were apparent, and the internal genitalia of the host were either reduced or absent. Sperka and Freytag also found that nematodes and insect parasitoids simultaneously parasitized individual hosts. A few hosts infected with mermithids also contained either dryinid larvae or individuals of Strepsiptera. Moreover, in several specimens containing well developed nematodes, the strepsipteran females had fully developed triungulinid larvae. However, no mermithids were found in association with Pipunculidae.

IV. HYPERPARASITISM AND MULTIPLE PARASITISM OF AUCHENORRHYNCHA

A. Hyperparasitism

Pipunculidae and Dryinidae are themselves parasitized by a variety of Hymenoptera (Table 17). So far, only *Basalys* spp. and *Ismarus* spp. have been recorded in Europe. Lundbeck (1922) reared *B. erythropus* Kieffer from a puparium of *Dorylomorpha xanthopus* in Denmark, and Rothschild (1964) reared *B. fumipennis* Westwood from a puparium of *Cephalops semifumosus* in Britain. Females of both species appeared to have oviposited either in the pipunculid larva shortly after the latter had emerged from the secondary host, or in the puparium.

Unlike *Basalys* spp., *Ismarus* spp. are obligate hyperparasitoids. Four European species have been reared (Appendix V), among which a high degree of specificity is apparent. Although oviposition has not been observed, rearing data suggest that females oviposit in the larval dryinid while the latter is developing on the host leafhopper.

Keys to *Ismarus* spp. are provided by Hellén (1964) (Finland), Nixon (1957) (Britain) and Wall (1967) (Switzerland). Masner's (1976) revision of the genus should also be consulted.

Raatikainen (1961) reported a single case of parasitism of Dryinidae by a mite, *Achlorolophus gracilipes* (Kramer) (Erythraeidae). The latter was found inside(?) a young *Dicondylus dichromus* larva. *A. gracilipes* is also ectoparasitic on the dryinid's leafhopper host (*Javesella pellucida*) (Raatikainen, 1967).

Kinzelbach (1970) mentions records of nematode infections of Strepsiptera, but none of these refers to strepsipteran groups parasitizing Auchenorrhyncha.

B. Multiple Parasitism

Because of the difficulties associated with the identification of parasitoid larval stages to species level, no information has so far been obtained on

Table 17
Hyperparasitoids of Auchenorrhyncha.

Primary hosts	Hyperparasitoids[a]	References
	SECONDARY PARASITOIDS	
Pipunculidae	Diapriidae: *Basalys* spp.	Lundbeck (1922) Rothschild (1964)
	Eucoilidae: *Eucoila* sp.	Muir (1921)
	Eupelmidae: *Anastatus pipunculi* Perkins	Perkins (1906b)
	Encyrtidae: unidentified sp.	Perkins (1905d)
Dryinidae	Diapriidae: *Ismarus* spp.	Chambers (1955, 1981) Waloff (1975) Jervis (1979a) Freytag (1985)
	Ceraphronidae: *Ceraphron abnormis* Perkins	Swezey (1908)
	unidentified sp.	Chandra (1980c)
	Pteromalidae: unidentified sp.	Chandra (1980c)
	Aphelinidae: *Centrodora xiphidii* (Perkins)	Swezey (1908)
	Encyrtidae: *Chalcerinys* spp.	Perkins (1905a)
	Echthrodryinus spp.	Perkins (1906b)
	Echthrogonatopus spp.	Tryapitzyn (1964)
	Helegonatopus spp.	Freytag (1985)
	Hypergonatopus spp.	
	Ooencyrtus sp.	Bartlett (1939)
	TERTIARY PARASITOIDS	
Encyrtidae (*Helegonatopus*)	Eulophidae: unidentified sp.	Perkins (1906b)

[a] Greathead's (1983) statement that some Dryinidae are hyperparasitic is incorrect.

multiple parasitism (i.e. simultaneous parasitism of the same host individual by two or more parasitoid species) involving congeneric species. What little information has been obtained on multiple parasitism refers to the simultaneous occurrence of Dryinidae with Pipunculidae, and Dryinidae with Strepsiptera.

Multiple parasitism involving larvae of *Aphelopus* and *Chalarus* was frequently observed by Jervis (1980a), but the eventual outcome of these cases was not recorded. However, Buyckx (1948) found a second instar larva of *Chalarus* together with a third instar larva of *Aphelopus* in an adult typhlocybid, and observed that the pipunculid larva reached maturity and indirectly killed the *Aphelopus* larva (i.e. by killing the host). Since both larval instars of *Chalarus* possess sclerotized mouthparts and a caudal vesicle, whereas first to fourth instar larvae of *Aphelopus* possess no such structures, larvae of *Chalarus* would seem to be better equipped for destroying members of the rival genus. It therefore seems unlikely that the particular oviposition strategy employed by *Chalarus* females (i.e. attacking the last three nymphal instars of the host) has in any way been brought about by selection pressures arising from larval competition (Jervis, 1980a).

Heikinheimo (1957), Raatikainen (1967) and Abdul-Nour (1971) have observed multiple parasitism of hosts by Dryinidae and Strepsiptera. Both Heikinheimo (1957) and Raatikainen (1967) noted that individuals of *Javesella pellucida* may be simultaneously parasitized by *Dicondylus bicolor* and *Elenchus tenuicornis*. During a seven-year study, parasitism by *D. bicolor* alone varied between 0·1 and 9·9%, and that by *D. bicolor* together with *E. tenuicornis* between 0·1 and 2·9% (Raatikainen, 1967). Only 25% of larvae of the dryinid survived when the rival parasitoid was a male *Elenchus*, but survival was high when the strepsipteron was a female. Abdul-Nour (1971) in southern France recorded multiple parasitism involving the dryinid *Gonatopus sepsoides* and the strepsipteran *Halictophagus languedoci*. In seven hosts, the strepsipteron was male, in five it was female, and the other host bore two dryinid sacs and contained five strepsipterons (two second instar larvae and three developing males).

Part 2. ECOLOGICAL AND EXPERIMENTAL STUDIES

V. INTRODUCTION

Askew and Shaw (1986), in their recent discourse on parasitoid communities, give the following definition: "A community is a group of species having a high degree of spatial and temporal concordance, and in which the member species mutually interact to a greater or lesser extent".

Within each of the three groups of hosts here considered, namely the grassland leafhoppers (Cicadellidae), the grassland planthoppers (Delphacidae) and the arboreal Typhlocybinae (Cicadellidae), the member host species share a number of parasitoid species, and the assemblages of parasitoids attacking them can be regarded as three distinct parasitoid communities. These exophytic hosts and their associated parasitoids form intricate networks of interrelationships, not only because many of the parasitoids are polyphagous, but also because many of the host species support a number of commonly shared parasitoids. Figures 5 and 6 show that some cicadellid hosts are able to sustain as many as seven dryinid and nine pipunculid species. However, the composition of these parasitoid assemblages may vary between localities and probably also across the geographical range of the host species.

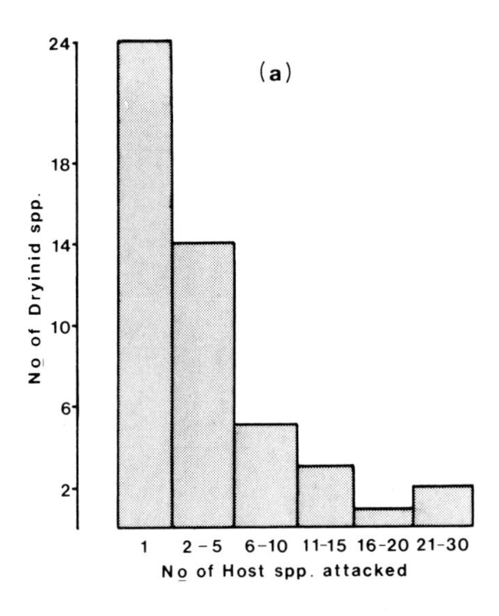

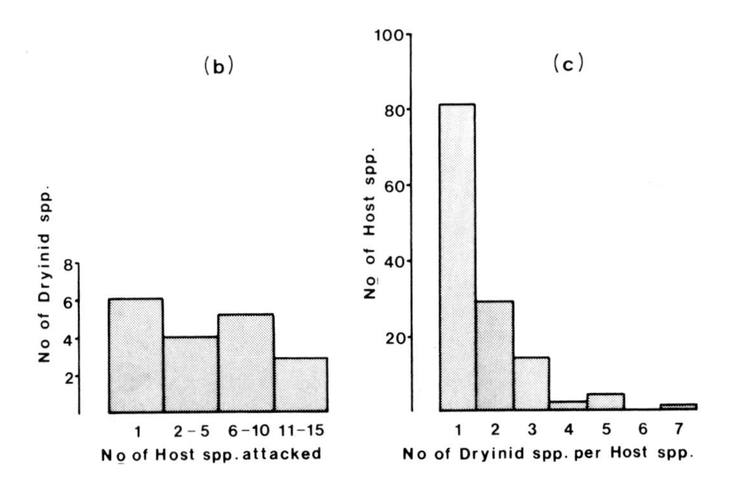

Fig. 5. European Dryinidae: (a) number of host species recorded per dryinid species; (b) same, in Britain; (c) number of dryinid species that a single host species is known to support.

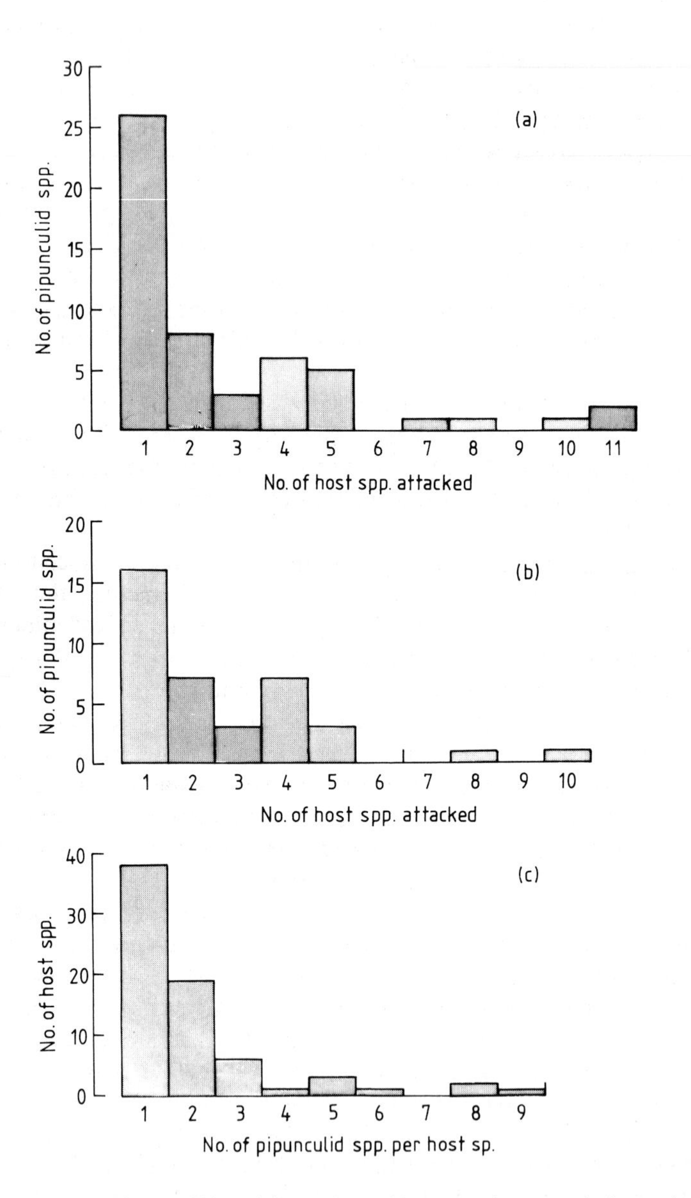

Fig. 6. European Pipunculidae: (a) number of host species recorded per pipunculid species; (b) same, in Britain; (c) number of pipunculid species that a single host species is known to support.

The assemblages of cicadellids and delphacids can be numerically abundant, and are species-rich. For example, in one hectare of acidic grassland in Silwood Park, 63 Auchenorrhyncha species were encountered, while on chalk grassland, Morris (1971) recorded 41 species. In other words, in some habitats the choice of hosts available to the parasitoids is extensive.

The parasitoid communities of Auchenorrhyncha are themselves comprised of constituent units or "components", each component consisting of the total of parasitoid species that attack a single host species within a defined habitat. Our term "component" corresponds to a "complex", as defined by Askew and Shaw (1986), but we find it more convenient to assign the word "complex" to the total number of parasitoid species belonging to the same family (e.g. complex of Dryinidae) that is found within a parasitoid community.

Since groups of parasitoid species utilize their hosts in a similar way, each component (and hence community) is made up of "guilds". Thus there is a guild of egg parasitoids, and another of nymphal–adult parasitoids. As the primary parasitoids may themselves be parasitized, there is yet another guild consisting of hyperparasitoids.

Leafhopper and planthopper communities, like others associated with exophytic hosts, appear to have a less complex structure than those associated with endophytic hosts, for example, in galls or leaf-mines (see Askew and Shaw, 1986); perhaps this is connected with the absence, or rarity, of facultative parasitoids in the former communities.

With the possible exception of the parasitoid communities associated with Aphidoidea, those associated with endophytic hosts are far better understood than those with exophytic hosts. In part, this can be explained by the ease with which the immature stages of parasitoids can be located and reared in galls and leaf-mines, and by the consequent certainty with which the parasitoid–host relationships can be discerned. The taxonomy of the nymphal stages of Auchenorrhyncha hosts is a subject for future research, although Vilbaste's (1968, 1982) keys to genera and certain species, Kathirithamby's (1973) key to 22 species of grassland leafhoppers, and Wilson's (1978) key to nymphs of woodland Typhlocybinae, as well as our own familiarity with the homopteran species discussed here, were a help. The taxonomy of the immature stages of the parasitoids is yet another subject of the future, and even the monographs on adult parasitoids, such as Olmi's (1984) on Dryinidae, and Coe's (1966) on Pipunculidae are still provisional, as many species remain to be described and the taxonomy of some genera is still in a confused state.

In Table 18 we have attempted to summarize three parasitoid communities associated with exophytic hosts best known to us; two located in acidic grassland sites at Silwood Park, and another located in deciduous woodland canopy in the area around Cardiff.

Table 18

Parasitoid communities associated with Auchenorrhyncha in deciduous woodlands in South Wales, and in acidic grassland in southern England.

(A) ARBOREAL TYPHLOCYBINAE

(1) Guild of egg parasitoids

Unknown

(2) Guild of nymphal–adult parasitoids

COMPLEX OF DRYINIDAE	COMPLEX OF PIPUNCULIDAE
Aphelopus atratus	*Chalarus argenteus*
A. melaleucus	*C. exiguus*
A. nigriceps	*C. fimbriatus*
A. serratus	*C. latifrons*
	C. parmenteri
	C. pughi
	C. sp. A nr. *spurius*
	C. sp. B nr. *spurius*

(3) Guild of hyperparasitoids

COMPLEX OF DIAPRIIDAE

Ismarus dorsiger (Diapriidae)

(B) GRASSLAND CICADELLIDAE

(1) Guild of egg parasitoids

COMPLEX OF MYMARIDAE	COMPLEX OF TRICHO-GRAMMATIDAE	COMPLEX OF APHELINIDAE
Anagrus holci	*Oligosita engelharti*	*Centrodora livens*
A. mutans[a]		
A. silwoodensis[a]		
Gonatocerus litoralis		
G. sp. nr *paludis*		

(2) Guild of nymphal–adult parasitoids

COMPLEX OF DRYINIDAE	COMPLEX OF PIPUNCULIDAE
Anteon ephippiger	*Dorylomorpha hungarica*
A. pubicorne	*D. xanthopus*
Lonchodryinus ruficornis	*Tomosvaryella kuthyi*
Gonatopus sepsoides	*T. palliditarsis*
	T. sylvatica
	Pipunculus campestris
	P. fonsecai
	P. thomsoni

Table 18 *continued*

(2) Guild of nymphal–adult parasitoids *continued*		
COMPLEX OF DRYINIDAE	COMPLEX OF PIPUNCULIDAE	
	T. zugmayeriae	
	Eudorylas fascipes	
	E. fuscipes	
	E. jenkinsoni	
	E. montium	
	E. obliquus	
	E. obscurus	
	E. ruralis	
	E. subfascipes	
	E. subterminalis	
(3) Guild of hyperparasitoids		
COMPLEX OF DIAPRIIDAE		
Ismarus rugulosus		
(C) GRASSLAND DELPHACIDAE		
(1) Guild of egg parasitoids		
COMPLEX OF MYMARIDAE	COMPLEX OF EULOPHIDAE	
Anagrus ensifer	*Tetrastichus mandanis*[b]	
(2) Guild of nymphal–adult parasitoids		
COMPLEX OF DRYINIDAE	COMPLEX OF PIPUNCULIDAE	COMPLEX OF ELENCHIDAE
Dicondylus bicolor	*Cephalops furcatus*	*Elenchus tenuicornis*
	C. curtifrons	
	C. perspicuus	
	C. semifumosus	
	C. subultimus	
	C. ultimus	
(3) Guild of hyperparasitoids		
COMPLEX OF DIAPRIIDAE		
Basalys fumipennis (Diapriidae)		

Community (A) occurred in broadleaved woodlands on the southern edge of the South Wales coalfields, close to Cardiff; (B) and (C) occurred in acidic grassland at Silwood Park, Berkshire.

[a] Hosts on *Juncus* and grasses.

[b] On *Juncus* only.

VI. MYMARIDAE, TRICHOGRAMMATIDAE AND EULOPHIDAE

A. Field Levels of Parasitism

Accurate estimates of percentage parasitism are very difficult to obtain. First, parasitized eggs cannot readily be distinguished from unparasitized ones until parasitoid larvae have reached the final instar; parasitism can therefore be seriously underestimated. Secondly, both hatched and unhatched host eggs are hard to detect in plant tissue. This, coupled with the fact that the larvae of some mymarid species (e.g. *A*. sp. nr *flaveolus*) take longer to develop than host embryos (Ôtake, 1968), may lead to serious overestimates. Thirdly, there are problems involved in identifying the eggs of different leafhopper species.

In order to overcome these difficulties, Ôtake (1967) devised a method involving the use of artificially infested plants containing eggs of known age. These plants are exposed in the field for a period of time, and are then returned to the laboratory and dissected (see Ôtake (1970b) for details of how to calculate estimates). This "trapping" method, or a similar technique, has been used by Ôtake (1967, 1970a,b, 1976, 1977), Becker (1975), Chandra (1980c), Fujimura and Somasundaram (1978, in Greathead (1983)), Lin (1974) and Miura *et al.* (1977). While it still does not provide realistic estimates of per cent parasitism, it nevertheless provides a more accurate means of obtaining comparative data on the field activity of parasitoids.

Levels of parasitism recorded in grassland leafhopper species at Silwood Park are shown in Table 19.

B. Functional Response to Host Density

The functional response of a parasitoid is the relationship between the number of hosts parasitized per unit time and host density. The slope of the response is the *per capita* searching efficiency (Hassell and Waage, 1984).

Moratorio (1977) investigated the functional responses of *Anagrus mutans* and *A. silwoodensis*. Both species show a typical type II (curvilinear) response *sensu* Holling (1959, 1966). Moratorio estimated for each species the two parameters of the response; a', the instantaneous searching efficiency, which determines the rate of increase of the response with host density, and T_h, the handling time, which determines the maximum attack rate. a' is similar for both species (*mutans* = 0·012, *silwoodensis* = 0·010), whereas T_h is considerably longer in *A. mutans* than in *A. silwoodensis* (0·66 h and 0·35 h, respectively). The two species therefore differ mainly in the number of hosts they parasitize at higher densities. Hassell (1982a,b) and Hassell and Waage (1984) point out that while the maximum attack rate is

Table 19
Egg mortality in four leafhopper and planthopper species at Silwood Park.

Leafhopper species	% Mortality in the field	% Parasitism	% Mortality due to *Tetrastichus*	Reference
S. minutus				
1969	59·1	15·9	—	May (1971)
1970	54·7	23·0	—	
C. anceps				
1953–54	48·2	—	6·8	Whalley (1956)
1954–55	42·0	0·1	6·3	
1959–60	27·1	0·3	13·1	Rothschild (1967)
1960–61	37.3	0·1	9·5	
1961–62	31·0	—	—	
C. viridis				
1969	61·7	9·1	12·0	Tay (1972)
1970	79·9	6·1	3·6	
1971	81·8	4·7	6·2	
M. sexnotatus				
1973–74	70·9–87·6	44·6	—	Becker (1975)

usually described in terms of handling time (i.e. non-searching time), it can equally well be the result of egg-limitation.

Moratorio (1977) found that *A. mutans* reared from eggs of *C. viridis* spent significantly less time ($P < 0{\cdot}01$) drilling than the smaller females reared from eggs of *D. hamata*, when both are presented with eggs of *D. hamata*. Whether adult size variation significantly affects the parasitoid's functional response—and whether it does so by affecting handling time and/or by affecting fecundity—is an intriguing problem awaiting investigation.

C. Spatial Variations in Levels of Parasitism

Recently, the question of how the percentage of hosts attacked by a parasitoid population varies spatially with host density (in particular, whether parasitism should be directly or inversely density dependent) has been addressed both empirically and theoretically (Chesson and Murdoch, 1986; Hassell, 1982b; Hassell *et al.*, 1985; Kidd and Mayer, 1983; Lessells, 1985; Morrison and Strong, 1980; Stiling and Strong, 1982b). The examples discussed in this context have included several *Anagrus* species (Seyedodeslami and Croft, 1980; Stiling, 1980a; Stiling and Strong, 1982b).

Stiling and Strong's (1982b) study revealed a tendency towards inverse density dependent parasitism within leaves and between leaves of the host plant, *Spartina alterniflora*, and also between isolated stands of *S. alterni-*

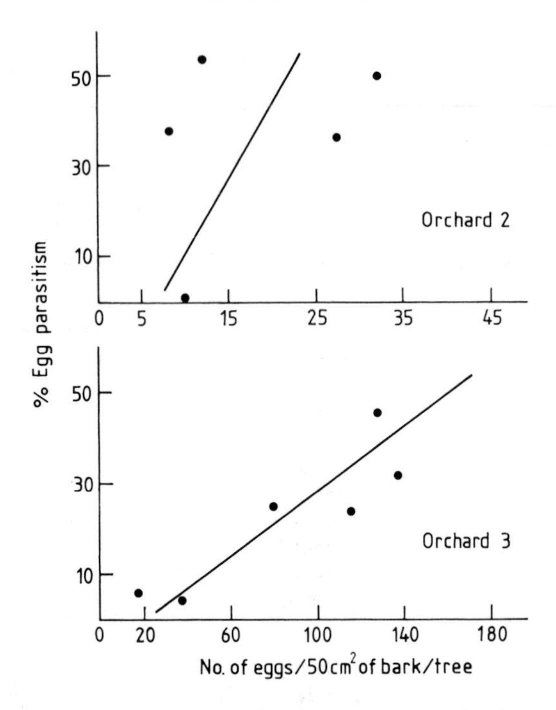

Fig. 7. Percentage parasitism of *Typhlocyba pomaria* eggs by *A. epos* between apple trees in two orchards in Grand Rapids, Michigan, USA, 1976: (a) $r = 0{\cdot}602$, $p < 0{\cdot}01$, $y = 6{\cdot}84 + 0{\cdot}301x$; (b) $r = 0{\cdot}863$, $p < 0{\cdot}01$, $y = 22{\cdot}05 + 2{\cdot}82x$. From Seyedodeslami and Croft (1980), Spatial distribution of overwintering eggs of the White Apple Leafhopper, *Typhlocyba pomaria*, and parasitism by *Anagrus epos*. *Environ. Entomol.* **9**, 624–28. Copyright © 1980 by Entomological Society of America. Reproduced by permission.

flora. On the other hand, Seyedodeslami and Croft's (1980) and Stiling's (1980a) studies revealed direct density dependent parasitism between trees and leaf pairs respectively (Figs. 7 and 8).

Two earlier studies also demonstrated direct density dependence: Becker's (1975) on *A. holci*, and Sasaba and Kiritani's (1972) on the trichogrammatid *Paracentrobia andoi*. Parasitism by *A. holci* varied significantly between groups of trap plants, while parasitism by *P. andoi* varied significantly between rice plant "stalks". Sasaba and Kiritani (1972) obtained data which suggested that there was also direct density dependent parasitism between rice paddy hills.

Lessells (1985) showed that three types of parasitism–host density relationship are theoretically possible in parasitoids that are maximizing their oviposition rate—not only direct density dependence, but also inverse density dependence and density independence. All of the above findings

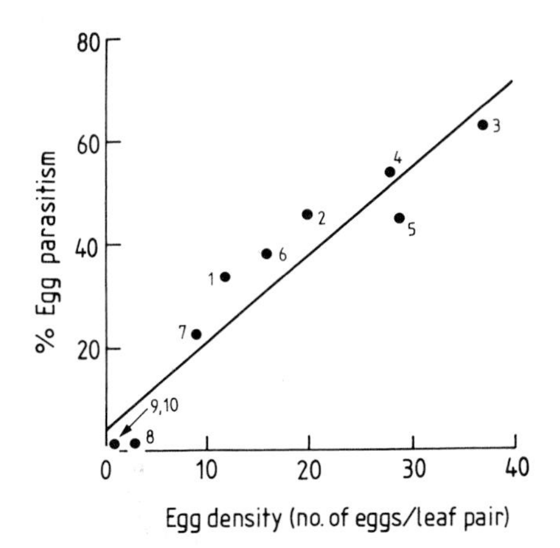

Fig. 8. Percentage parasitism of *Eupteryx* eggs by *Anagrus* between leaf pairs of *Urtica dioica* at Dowlais, South Wales, 1978: $r = 0{\cdot}94$, $p < 0{\cdot}001$, $y = 4{\cdot}02 + 1{\cdot}69x$; numbering of data points denotes position of leaf pair on stem (1, bottom pair; 10, apical pair). From Stiling (1980a), Competition and coexistence among *Eupteryx* leafhoppers (*Hemiptera: Cicadellidae*) occurring on stinging nettles (*Urtica dioica*). *J. anim. Ecol.* **49**, 793–805. Copyright © 1980 by British Ecological Society. Reproduced by permission.

therefore conform with theoretical expectations. A mechanistic explanation for them can be found in the balance between two counteracting processes (Hassell *et al.*, 1985); the spatial allocation of searching time by parasitoids in relation to host density per patch, and the degree to which exploitation is constrained by a relatively low maximum attack rate per parasitoid within a patch.

The inverse density dependent parasitism by *A. delicatus* could theoretically result from insufficient aggregation of searching time by female parasitoids in high density patches to compensate for any within-patch constraints on host exploitation imposed per parasitoid by time-limitation, egg-limitation or imperfect information on patch quality (Lessells, 1985). Conversely, the direct density dependent parasitism by *A. epos* and Stiling's *Anagrus* species could theoretically result from a degree of aggregation that more than compensates for any within-patch constraints per female on the exploitation of high density patches. However, Stiling and Strong (1982b) suggest that the inverse density dependent parasitism by *A. delicatus* could result from the disturbance caused to searching parasitoids by the periodic tidal inundation of *S. alterniflora* in its saltmarsh habitat. Evidence in support of their hypothesis comes from data on levels of parasitism within and between different plant parts. Percentage parasitism increases towards the apex of each leaf, and within plants towards the more apical (younger) leaves (Fig. 9). This is consistent with the finding that host eggs occurring at

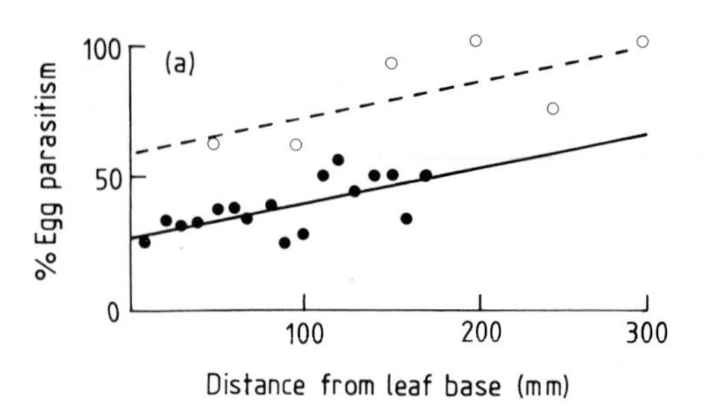

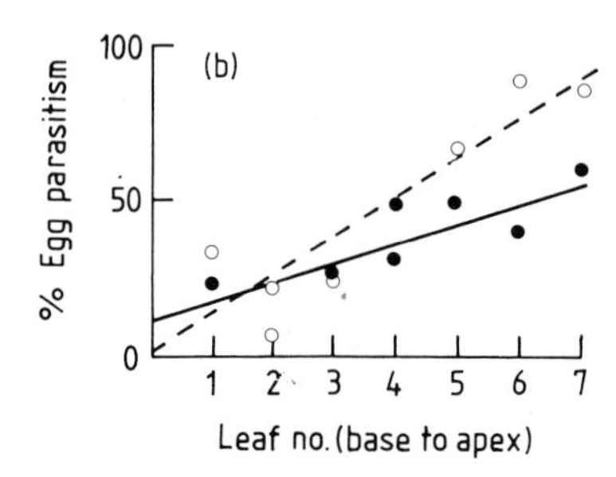

Fig. 9. Percentage parasitism of *Prokelisia marginata* eggs by *A. delicatus* (a) within leaves, and (b) between leaves of *Spartina alterniflora*, Oyster Bay, Florida, USA, 1980. ●, —, During winter; ○, -----, during summer. (a) $r = 0{\cdot}64$, $p < 0{\cdot}01$, $y = 27{\cdot}4 + 1{\cdot}24x$ in winter; $r = 0{\cdot}69$, $p < 0{\cdot}2$, $y = 59{\cdot}53 + 1{\cdot}30x$ in summer. (b) $r = 0{\cdot}92$, $p < 0{\cdot}01$, $y = 11{\cdot}56 + 6{\cdot}19x$ in winter; $r = 0{\cdot}91$, $p < 0{\cdot}01$, $y = 1{\cdot}36 + 12{\cdot}62x$ in summer. (From Stiling and Strong 1982b) Egg density and the intensity of parasitism, in *Prokelisia marginata* (Homoptera: Delphacidae). *Ecology*, **63**, 1630–35. Copyright © 1982 by Ecological Society of America. Reproduced by permission.

the tips of the leaf are exposed to the air for a longer time than those occurring at the bases (host egg densities are highest both in the basal portions of leaves, and in older leaves).

Unfortunately, there is at present insufficient information on the responses of female *Anagrus* to host patchiness to enable us to decide which of the above hypotheses best accounts for the observed distributions of parasitism.

Hassell (1984) and Chesson and Murdoch (1986) have shown that, contrary to conventional wisdom, inverse spatial patterns of parasitism can contribute to population stability. Whether direct or inverse relationships have the greater effect depends upon the characteristics of the host's spatial distribution (Hassell, 1985).

Aggregation of parasitoids in high host density patches increases the probability of encounters both between parasitoid females and between females and parasitized hosts. This mutual interference may result in a progressive decline in searching efficiency as parasitoid density increases, due to a reduction in searching time during and following each encounter (Hassell, 1978; Hassell and Waage, 1984; Waage and Hassell, 1982). However, Moratorio (1977) found no evidence of such a decline in either *A. mutans* or *A. silwoodensis* under laboratory conditions. Mutual interference, where it occurs in parasitoids, is a density dependent factor contributing to population stability (Hassell, 1978).

D. Temporal (Intergeneration) Variation in Parasitism

Egg mortality was found to be a key factor causing population changes in several grassland leafhopper species occurring at Silwood Park (see page 355). Unfortunately, there are insufficient data to allow us to determine whether or not temporal variations in parasitism are density dependent.

E. *Polynema striaticorne*, an Introduced Parasitoid Species

P. striaticorne was introduced in 1966 into the Piedmont Region of Italy from the United States, in an attempt to control populations of the Buffalo Treehopper, *Bisonia stictocephala*, a serious pest of grapevines (Vidano, 1966). The parasitoid was successfully established, and has since spread to many European countries, ranging from the Atlantic coast to the Black Sea coast (Vidano and Meotto, 1968). According to Vidano *et al.* (1985), *B. stictocephala* has been successfully controlled by *P. striaticorne*. Unfortunately, no detailed population data have so far been published on this example of classical biological control.

F. Impact of *Tetrastichus mandanis* on Host Populations

Overall field levels of mortality due to *T. mandanis*, recorded in *C. anceps* at Silwood Park, are shown in Table 19. Arzone (1974b) recorded 30% and 80% "parasitization" in the first and second generations respectively, of *C. dehneli*, at a site in the Piedmont region of Italy.

VII. DRYINIDAE

A. Effects of Dryinid Parasitism on Host Populations

Some examples of percentages of parasitism by Dryinidae are given in Table 20, but they do not always represent the true mortality inflicted by these parasitoids. For this, there has to be a full understanding of the predatory potential—or of its absence—in each dryinid species, and this knowledge is

Table 20

Mean per cent dryinid parasitism of adult hosts, except *Dicranotropis hamata*, where maximum parasitism is of fifth larval instar: limits in parentheses; n = number of generations.

(a) Locality and host species	1st generation		2nd generation		References
	n	Mean % parasitism and limits	n	Mean % parasitism and limits	
		BIVOLTINE SPECIES			
Silwood, Area I (s. England)					
Psammotettix confinis	4	15·8 (8·0–21·3)	4	4·6 (1·8–8·9)	Waloff (1975)
Jassargus distinguendus	4	4·6 (1·8–10·7)	4	1·5 (0–3·1)	Waloff (1975)
Adarrus ocellaris	4	6·9 (2·1–13·9)	4	2·3 (0–4·9)	Waloff (1975)
Silwood, Area II (s. England)					
Adarrus ocellaris	6	2·1 (0–4·9)	6	1·9 (0–3·9)	Waloff and Thompson (1980)
Dicranotropis hamata	5	32·4 (7·7–45·7)	4	25·8 (4·1–57·1)	Waloff and Thompson (1980)
Silwood, Oatfield (s. England)			2	0.4	Becker (1975)
Macrosteles sexnotatus					
Northern England					
Macrosteles viridigriseus	2	(1·2–2·5)	2	(0–2·5)	Khafagi (1986)
South of France					
Grassland cicadellids		2–10		(2–10)	Abdul-Nour (1971)
		UNIVOLTINE SPECIES			
Silwood, Area II (s. England)					
Diplocolenus abdominalis		$n = 6$	0·1 (0–0·4)		Waloff and Thompson (1980)
Elymana sulphurella		6	1·5 (0·2–3·6)		Waloff and Thompson (1980)
Recilia coronifera		6	2·6 (0–4·5)		Waloff and Thompson (1980)

USSR			
Oncopsis flavicollis (nymphs)	$n = 1$	30%	Ponomarenko (1986b)
Finland			
Javesella pellucida	$n = 7$	4·5 (0·2–11·1)	Raatikainen (1967)

(b) Percentages of dryinid parasitism in pest species in Rice (from Greathead, 1983)

	Nymphs	Adults
Nilaparvata lugens		
Philippines 1977	7·5	—
Thailand 1976–1977	2·1	0
India 1978	7·8	—
Sogatella furcifera		
Philippines 1977	4·9	—
Thailand 1976–1977	4·7	11·4
Nephotettix spp.		
Philippines 1977	0·2	—
Thailand 1976–1977	0·1	0·2

still rudimentary (page 309). Also, the early immature stages of dryinids can only be detected by dissection of hosts, but as yet there are no keys to the identification of dryinid larvae. Examples in Table 20 from Silwood Park, Berkshire in southern England were taken throughout the season and show that in the bivoltine host species dryinid parasitism tends to be higher in the first than in the second generation. The main dryinids attacking these grassland leafhoppers were *Lonchodryinus ruficornis, Anteon pubicorne* and *Gonatopus sepsoides*, and the difference in the degree of parasitism between the two host generations may be linked to the tendency of the potentially bivoltine *Lonchodryinus* and *Anteon* towards univoltinism and their early seasonal disappearance from the field.

In area II at Silwood, nymphs of *Dicranotropis hamata* were severely attacked by *Dicondylus bicolor*, and when percentage parasitism was plotted against nymphal density for nine successive generations, irregular, anti-clockwise cycling was revealed (Waloff and Thompson, 1980). This is suggestive of a delayed density dependent relationship, but since parasitism is not the key factor, the spiralling is obscured. However, a simple test indicated that generation time is involved in the parasitoid–host relationship, since the correlation coefficient between the number of nymphs recruited to generation t1 and the number parasitized in t1 is significant, as is that between recruits to t1 and the number parasitized in t2. The significance however, disappears by generation t3. Thus the effect of nymphal density in generation t1 on the numbers parasitized by *D. bicolor* persisted for two successive generations, but disappeared by the third.

The high levels of percentage parasitism in the above example are on the whole exceptional and local. Generally, they are much lower. For example, Abdul-Nour (1971) and Raatikainen (1967) recorded much lower levels in France and Finland respectively, although Ponomarenko (1986b) gives a figure of 30% parasitism of *Oncopsis flavicollis* nymphs by *Anteon jurineanum* in the USSR. For comparison, percentage parasitism of leafhopper pests of rice are also given in Table 20 (from Greathead, 1983).

Superparasitism in the European Dryinidae is also known and, for example, Jervis (1978b) recorded it in Typhlocybinae such as *Eupteryx aurata, Kybos* spp. and *Ribautiana ulmi*, some individuals of which bore two dryinid sacs, apparently of the same species. When the superparasitized hosts were kept alive, only one parasitoid larva completed its development. Grassland cicadellids bearing two sacs have also been seen by Waloff, but only on a few rare occasions. Thus, under certain conditions, intraspecific competition between dryinid larvae is a possibility which has not been explored. The best evidence for intraspecific competition comes from Khafagi (1986), who experimented with *Anteon pubicorne* and the leafhopper host *Macrosteles viridigriseus*. In the laboratory, superparasitized hosts carried two or three dryinid sacs each, but only one or two mature larvae emerged out of the hosts. Khafagi recorded 127 sacs on 59 hosts, from which 108 parasitoid larvae emerged. The resultant adult progeny were all male (96♂♂). Males reared from superparasitized leafhoppers were always smaller than the normal ones, i.e. those reared from hosts supporting only one parasitoid individual.

Recently, Chua *et al.* (1984) made detailed studies on *Pseudogonatopus flavifemur*, which is a major parasitoid of the brown rice planthopper, *Nilaparvata lugens*. They set out to assess the potential of this dryinid for the biological control of *N. lugens* and other pests of rice in the Philippines. In particular, they investigated the functional response of *P. flavifemur* to host density. It was found to be a sigmoid, i.e. type 3 *sensu* Holling (1966), response.

Chua *et al.* also investigated the response of females to spatial variations in host density, and found that females aggregated in areas of high host density. The authors therefore concluded that, on the basis of all their findings, *P. flavifemur* is a potentially important biological control agent, especially in programmes involving the repeated release of parasitoids during each generation of *N. lugens*. It is highly relevant that Hassell *et al.* (1977) and Hassell's analyses (1978) show that sigmoid functional responses can contribute markedly to stability in prey populations, as long as the predator populations remain relatively constant, and that both mutual interference and aggregation may have a stabilizing influence on a prey population (Hassell and May, 1974; Hassell and Waage, 1984).

VIII. PIPUNCULIDAE

A. Effects of Pipunculid Parasitism on Host Populations

Accurate estimates of percentage parasitism can be obtained only by dissection of individual hosts. Simply recording in a sample the number of hosts that have distended abdomens is likely to produce misleading results, as it is often impossible to distinguish reliably between gravid females and those whose abdomens are distended due to the presence of a pipunculid larva (Jervis, 1978a,b).

Percentage parasitism by Pipunculidae may reach high levels (Table 21). Both Waloff (1975) and Jervis (1980b) recorded marked differences in percentage parasitism between the two generations of bivoltine host species. In *Psammotettix confinis, Adarrus ocellaris* and *Euscelis* species, the tendency is for the second generation to be more heavily parasitized (Waloff, 1975). Waloff attributed this to bivoltinism of most of the Pipunculidae attacking these hosts, which contributes to a large numerical response by the parasitoids. In the Typhlocybinae studied by Jervis (1979d) the reverse

Table 21

Percentage parasitism by Pipunculidae of nine leafhopper species in British localities: 1 = first generation; 2 = second generation of bivoltine species).

	Philaenus spumarius	*Neophilaenus lineatus*	*Adarrus ocellaris*		*Jassargus distinguendus*		*Diplocolenus abdominalis*	*Arthaldeus pascuellus*		*Psammotettix confinis*		*Fagocyba cruenta*		*Alnetoidia alneti*
			1	2	1	2		1	2	1	2	1	2	
1964	38	31	—	—	—	—	—	—	—	—	—	—	—	—
1965	40	24	—	—	—	—	—	—	—	—	—	—	—	—
1966	35	23	—	—	—	—	—	—	—	—	—	—	—	—
1967	34	31	9·3	0	19·0	28·8	—	19·1	16·3	15·4	26·1	—	—	—
1968	—	—	0	23·9	23·5	19·7	—	20·3	9·3	14·4	26·6	—	—	—
1969	—	—	4·9	41·6	14·2	6·9	—	25·2	3·9	18·7	19·2	—	—	—
1970	—	—	7·9	42·6	15·9	23·9	—	13·5	10·0	12·9	19·3	—	—	—
1972	—	—	13·7	21·9	—	—	14·4	—	—	—	—	—	—	—
1973	—	—	5·5	25·9	—	—	6·3	—	—	—	—	—	—	—
1974	—	—	5·9	11·5	—	—	11·7	—	—	—	—	—	—	—
1975	—	—	14·1	21·8	—	—	15·1	—	—	—	3·8	—	—	—
1976	—	—	11·3	34·9	—	—	10·8	—	—	2·6	7·9	25·5	14·7	21·2
1977	—	—	21·1	39·0	—	—	12·4	—	—	12·7	34·2	—	—	—

Data from Jervis (1980b), Waloff (1975), Waloff and Thompson (1980), Whittaker (1969).

tendency was found. The major causal factor in this case was the temporal asynchrony of parasitoids and hosts referred to above.

The possible regulatory effects of Pipunculidae on host populations have been examined by Whittaker (1971a, 1973a) and Waloff (1975, 1980). Whittaker (1971a) showed that in a limestone grassland site in southern England (Upper Seeds, Wytham Woods), intergeneration mortality of adult *Neophilaenus lineatus* was more significantly density dependent when parasitism by *Verrallia aucta* was added to other mortality effects. In contrast, analysis of mortality in adult *Philaenus spumarius* at the same site revealed no density dependence (Whittaker, 1973b).

Whittaker (1973a) also examined the interaction between *V. aucta, N. lineatus* and *P. spumarius* at Upper Seeds. He found that in years when *Neophilaenus lineatus* was more abundant than *Philaenus spumarius, N. lineatus* was parasitized superproportionately by *Verrallia aucta*, while in years when *N. lineatus* was the less abundant host, it was parasitized subproportionately. Whittaker attributed this effect to either the slightly earlier emergence of *N. lineatus* compared with *P. spumarius*, or "apostatic" selection (page 358) by *V. aucta*. He suggested that each host species may act as a reservoir for *V. aucta* in years when the other host is less abundant, and that because of this, populations of the two species are more likely to be stable in each other's presence than when occurring alone.

Waloff (1975) and Waloff and Thompson (1980) investigated the dynamics of leafhopper populations in two grassland sites at Silwood Park. Using their data, percentage parasitism was plotted against log host density for seven species of Deltocephalinae; *Recilia coronifera, Adarrus ocellaris, Jassargus distinguendus, Diplocolenus abdominalis, Arthaldeus pascuellus, Psammotettix confinis* and *Elymana sulphurella*. Regression analyses of these data show that parasitism of *A. ocellaris* was inversely density dependent (Fig. 10) (although using a more critical test, plotting the k value of parasitism against host density, parasitism proved non-significant (Waloff, unpubl.)), and that parasitism of the other six species was density independent. Serial linkage of data points to produce time-sequence plots reveals no signs of delayed density dependence. There is therefore no evidence to suggest that, in the Silwood sites, Pipunculidae by themselves regulate host populations.

Kiritani *et al.* (1970) also were unable to find conclusive evidence for regulation of *Nephotettix cincticeps* populations by Pipunculidae in Japanese rice fields. Within each rice-growing season, in which four host generations developed, parasitism initially tended to be density dependent, but later tended to be inversely density dependent, the overall temporal pattern of parasitism being suggestive of the type of the anticlockwise spiralling observed in delayed density dependent relationships (Fig. 11).

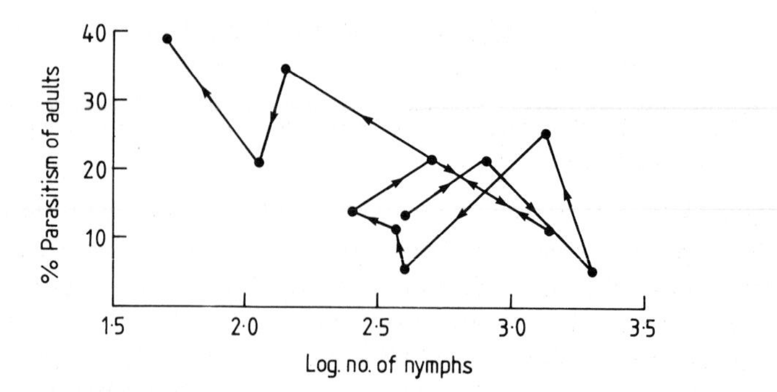

Fig. 10. Relationship between percentage parasitism of adults by Pipunculidae and density of third, fourth and fifth instar nymphs in *Adarrus ocellaris*, recorded over 12 generations at Silwood Park, 1972–1977. $y = -13{\cdot}54x + 54{\cdot}09, r = -0{\cdot}6, P < 0{\cdot}05$ (data from Waloff and Thompson, 1980).

Because pipunculids appear to be mainly visual hunters, they have been suggested as playing a role in the maintenance of the genetically based colour polymorphisms of some host species (Harper and Whittaker, 1976; Stiling, 1980c; Stewart, 1986; Whittaker, 1971b). Harper and Whittaker (1976) recorded frequency-dependent parasitism of certain morphs of adult *P. spumarius*, while Stiling (1980c) found that the black morphs of nymphal *Eupteryx urticae* contained disproportionately more eggs and larvae of *Chalarus spurius* than the orange morphs. However, the evidence in both cases remains inconclusive.

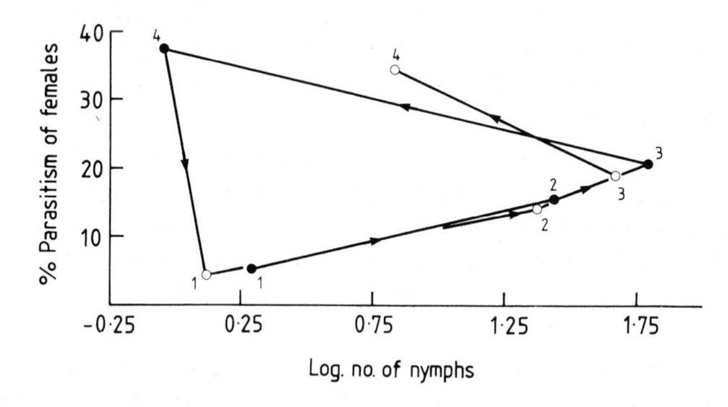

Fig. 11. Relationship between percentage parasitism of adult females by Pipunculidae and density of fifth instar nymphs in the Rice Green Leafhopper, *Nephotettix cincticeps*, recorded over four generations during each year in rice fields in Japan, 1966–1967. ●, 1966, ○, 1967; $y = 0{\cdot}196x + 19{\cdot}3$, $r = 0{\cdot}109$, $P > 0{\cdot}05$ (data from Kiritani *et al.*, 1970).

IX. STREPSIPTERA

A. Effects of Strepsipteran Parasitism on Host Populations, and Other Field Observations

As Freytag (1985) points out, very little work on the effects of Strepsiptera on host populations has been done so far. Numerous authors record levels of percentage parasitism, but the latter are very difficult to interpret unless they are linked to long-term population studies of the host species. Intensive work on rice planthoppers and leafhoppers has been done in the Far East, and is continuing (see Greathead's 1983 review), but much remains to be done elsewhere.

Detailed field investigations on the delphacid hosts of *Elenchus tenuicornis* were for several years carried out by Raatikainen and colleagues in Finland and other parts of Fennoscandia. *Javesella pellucida*, the most common harmful vector of crop viruses, received particular attention (Raatikainen, 1967). It was shown that, at least in some years, half of its populations on crops were stylopized. Data in Fig. 12 indicate that the decline of the host population after 1959 affected the abundance of the strepsipteron. Raatikainen considered that weather at the time of release of *Elenchus* triungulinid larvae greatly influenced their survival. This period was very dry in 1959 and again in 1963, and it is thought that many of these minute larvae desiccated before contacting their hosts. Raatikainen showed through experiments that survival, and hence searching time, of triungulinid larvae was closely linked with high temperatures and the associated fall in relative humidity. Moreover, insecticides were used on crops, mostly in 1959, and they must have killed a large number of parasitized hosts.

In their survey of the Strepsiptera of eastern Fennoscandia, Pekkarinen and Raatikainen (1973) collected 102 179 *J. pellucida* from different localities, of which 27% were stylopized.

Parasitism of Cicadellidae by *Halictophagus* species can reach even higher levels. For example, in southern France, Abdul-Nour (1971) recorded up to 68% stylopization of *Goldeus harpago* by *H. languedoci*, while in a small area of *Calluna vulgaris* in southern England *H. silwoodensis* parasitized 63% of *Ulopa reticulata* in 1969, 68% in 1970 and 65% in 1979 (Waloff, 1981).

Preliminary attempts at evaluating the importance of stylopization by *Elenchus tenuicornis* were made at Silwood Park (Waloff, 1975) in a 5104 m^2 area of grassland. Sixteen species of Delphacidae were recorded in this area, of which five, *Javesella pellucida, Hyledelphax elegantulus, Ribautodelphax angulosus, Kosswigianella exigua* and *Struebingianella dalei*, were common. These species are bivoltine and overwinter in their nymphal stages, which

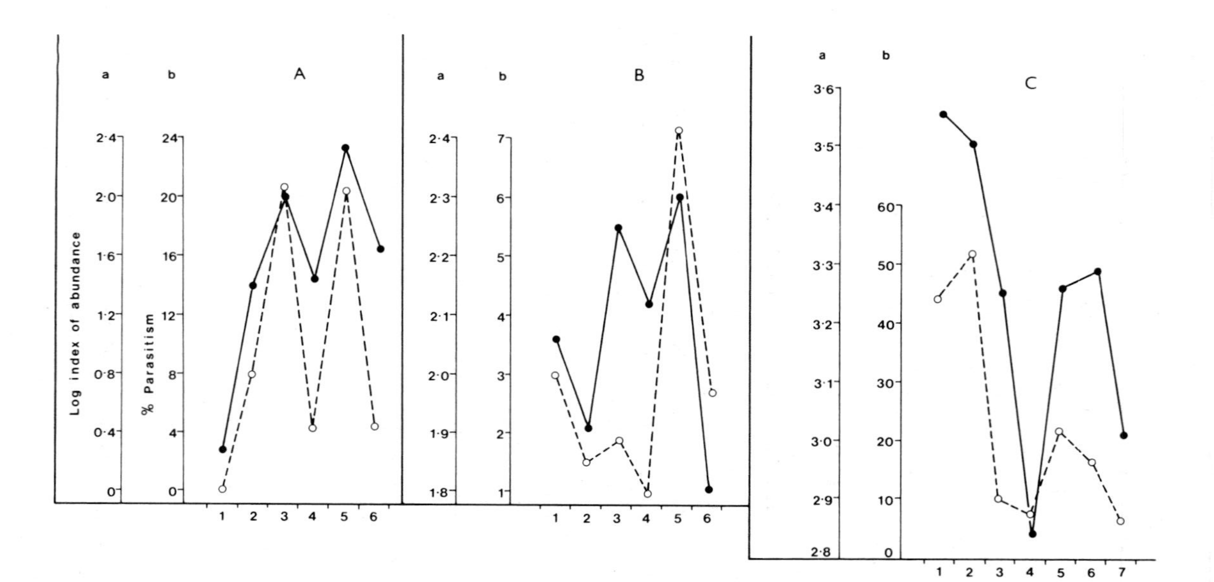

Fig. 12. (A) *Hyledelphax elegantulus* and (B) *Javesella pellucida*, in Silwood Park, Berkshire, England, (C) *J. pellucida* in Finland. ●—● in A and B = Log Index of abundance (from Waloff, 1975); in C = Log of total in samples (samples adjusted to same number per annum (from Raatikainen, 1967)). ○---○ in A, B and C is the percentage parasitism.

provide the winter reservoirs for *Elenchus* larvae. In a sequence of six generations the percentage stylopization tended to be proportional to the abundance of four of the five species (Waloff, 1975; and Fig. 12). Thus the populations of *Elenchus* and of its hosts expanded and contracted simultaneously, i.e. there was an immediate response of the parasitoid to host density. Since phenology and sequence of fluctuations through time are not identical among the delphacids, the non-specific triungulinid larvae tended to attack any potential host that was most readily encountered, and so the main reservoirs for the parasitoid changed from generation to generation. It is not impossible that *Elenchus* has a stabilizing effect on individual host populations, but when the numbers of the five common host species are pooled and the corresponding percentages of parasitism are calculated, the effect on the whole system shows signs of delayed density dependence (Waloff, 1980). Further studies and analyses would be of considerable interest.

An interesting aspect of stylopization, namely the apparent dependence of survival of the parasitoid on the choice of host plant made by the cicadellid hosts, emerged from Abdul-Nour's (1971) investigations on *Halictophagus languedoci* and its known hosts in southern France.

In typical Brachypodietum, with 70% *Brachypodium phaenicoides*, the trivoltine *H. languedoci* stylopized the most common hosts, *Adarrus taurus* and *Goldeus harpago*, the incidence and the degree of parasitism in these cicadellids depending on their phenology. Abdul-Nour (1971) calls this "Ecological Cycle 1".

In atypical Brachypodietum, with a 40% admixture of *Bromus erectus* there is a similar "Ecological Cycle 2" on *Arocephalus sagittarius* and *Jassargus obtusivalvis*. "Cycle 2" is supplemented by yet another cycle between three species of *Psammotettix*, i.e. by "Cycle 3", and also by "Cycle 1".

In mixed Brachypodietum with *Brachypodium phaenicoides, B. ramosum* and *B. erectus, Halictophagus* forms yet another "Ecological Cycle 4" on the predominant hosts in this habitat, i.e. on *Araldus propinquus* and *Adarrus geniculatus*. The extent of stylopization of these hosts was again supplemented by Cycles 1, 2 and 3.

However, in pure stands of either *Bromus erectus* or *B. ramosum*, the potential hosts, *Arocephalus sagittarius, Jassargus obtusivalvis, Adarrus geniculatus* and *Araldus propinquus*, are never parasitized, whereas in isolated tufts or islets of *Brachypodium phaenicoides* within these stands stylopized individuals are common. Abdul-Nour (1971) considered that nutritional regime was implicated, i.e. survival of *Halictophagus* depended on the type of food consumed by its hosts. It is, however, more likely that some physical or microclimatic features of the plant contribute to mortality of the triungulinid larvae.

Abdul-Nour (1971) set up laboratory experiments with adults of *Araldus propinquus* either on *B. phaenicoides* or *B. ramosum* and infected the two sets with *Halictophagus* triungulinid larvae. On *B. phaenicoides*, the parasitoids completed their development to the adult stage, whereas on *B. ramosum* there was no sign of stylopization, even in dissected hosts. Similar results were obtained in a parallel experiment using *Adarrus geniculatus* on *B. ramosum*. Further work on polyphagous host species is now needed.

X. HYPERPARASITOIDS

No information is available on levels of parasitism in European host populations. Chandra (1980c) reported between 30 and 50% parasitism of Dryinidae by both a ceraphronid and a pteromalid, in his study of rice hoppers in the Philippines.

XI. GENERAL DISCUSSION

A. Host Range Among Parasitoids

It is difficult to make any meaningful generalizations regarding host range, since the available data are far too incomplete. The host records for most species within the major primary parasitoid groups (Dryinidae, Pipunculidae, Auchenorrhyncha-associated Strepsiptera and Mymaridae—see Appendices) are based on very few individual rearings. In Mymaridae, there is the added uncertainty regarding the true identity of the parasitoids reared (page 295). Despite this, we have good reasons for concluding that in Dryinidae, Pipunculidae and Strepsiptera, the trend is towards polyphagy. In the former two groups, over half the species reared so far have been shown to attack more than one host species (Figs 5 and 6), and it is highly likely that a significant number of the apparently monophagous species will be shown to be polyphagous. Within the parasitoid fauna of Britain, as opposed to Europe as a whole, the same trend is apparent (Figs 5 and 6). Most importantly, polyphagous species outnumber monophagous ones within a local community; this applies to both the community associated with Auchenorrhyncha of acidic grassland, and the community associated with woodland canopy Typhlocybinae.

Although plant species is a major host habitat variable, particularly in woodland canopy, it appears that except for some monophagous species, few dryinids and pipunculids confine their attacks to hosts occurring on a single species of foodplant. Most attack hosts (monophagous and polyphagous) occurring on a variety of foodplants. Any host plant specificity that is observed among parasitoids of nymphs and adults is probably apparent rather than real. *Verralia setosa* for example, is host plant specific probably by virtue of the fact that its hosts (*Oncopsis flavicollis* and *O. subangulata*) are confined to birches (*Betula* spp.). Some species also show a lack of host habitat specificity. For example, five species of *Aphelopus* (*atratus, camus, melaleucus, nigriceps* and *serratus*) attack hosts both in woodland canopy and in open-field habitats (this contrasts with the apparently very marked habitat specificity of the mymarid, *Polynema woodi*).

Dryinidae and Pipunculidae, in common with other koinobionts (Askew and Shaw, 1986), parasitize a narrow range of host families. Only two species (*Cephalops obtusinervis* and *Dryinus collaris*), are known to parasitize members of more than one host family, Nevertheless, host taxon specificity in both parasitoid groups mainly extends to host subfamily level only. This is in contrast to Aphidiidae (Ichneumonoidea), among which the prevalent trend is towards parasitism of hosts belonging to one genus (Starý, 1981; Starý and Rejmánek, 1981).

B. Leafhopper–Parasitoid Population Dynamics

The greatest mortality in leaf- and planthopper populations in acidic grassland lies in their egg stage. In some species, e.g. *C. viridis* (Tay, 1972), *P. confinis* (Solomon, 1973), *D. hamata, A. ocellaris, D. abdominalis, E. sulphurella* and *R. coronifera* (Waloff and Thompson, 1980), egg mortality has been identified as the key factor, and numerically it forms the greatest component of the total mortality *K*. However, "egg mortality" is a composite term, which includes variations in fecundity per generation and death of eggs following oviposition. Finding eggs and identifying them with certainty in the large, natural Auchenorrhyncha communities is often precluded by the inordinate length of time it takes to do so. However, it is possible in some species, e.g. *C. viridis*, in which females oviposit into the stems of *Juncus,* or in *Stenocranus minutus*, in which the egg mass within the plant is covered externally by a whitish wax layer. It is also possible to use "trap" plants (see page 338), a technique used at Silwood Park by Becker (1975). Without further studies on the host egg stage, on carefully selected species, we cannot comment further on the role of egg parasitoids in the population dynamics of Auchenorrhyncha.

The egg, nymphal and adult stages of leafhoppers and planthoppers are contagiously distributed within the woodland canopy or open-field habitat. There is, in many species, a hierarchy of patchiness, the clumping of individual insects resulting from one or several of the following:

(1) Host plant specificity in oviposition behaviour by females (Claridge *et al.*, 1977; Claridge and Wilson, 1976, 1978; Stewart, 1983; Stiling, 1980b). This appears to be more important in some Auchenorrhyncha than in others. For example, Typhlocybinae and Macropsinae living on

trees, and Typhlocybinae living on herbaceous plants other than Gramineae, tend to be more specific than Cicadellidae and Delphacidae living in acidic grassland.

(2) Specificity in choice of egg site by ovipositing females (Claridge and Reynolds, 1972; Drosopoulos, 1977; Stiling, 1980b; Thompson, 1978; Waloff, 1979).

(3) The laying of eggs in discrete groups/masses (Ali, 1979; Bakkendorf, 1934; Drosopoulos, 1977; May, 1971; Rothschild, 1962; Tay, 1972).

(4) The tendency for nymphs and also adults to be feeding-site specific, behaviour which results in aggregation (Claridge *et al.*, 1981; McClure, 1974; Ôtake *et al.*, 1976; Q. Wei, pers. comm.).

Such patterns of host distribution are common among phytophagous insects, and are likely to be a major factor influencing the evolution of reproductive and behavioural strategies in parasitoids. It therefore comes as no surprise to find that females of an increasing number and variety of hymenopteran and dipteran parasitoids are being shown to respond to host patchiness by foraging in a non-random fashion; see van Alphen and Vet (1986) and Waage (1983) for reviews. Optimal foraging theory in particular predicts that foraging parasitoids will search so as to maximize the rate at which they find hosts suitable for oviposition. In foraging for patchily distributed hosts, optimal time allocation by parasitoids will involve concentration of search in higher density patches. The density dependent spatial variations in percentage parasitism at various levels of host patchiness by *Anagrus* (see page 339) suggest that females of at least some mymarid species may forage in this way, while the aggregative response demonstrated by Chua and Dyck (1982) and Chua *et al.* (1984) for *Pseudogonatopus flavifemur* suggests that female dryinids behave similarly.

Such behaviour, if it exists, has a potentially important regulatory influence on the dynamics of host-parasitoid interactions (Comins and Hassell, 1978), as does other non-random foraging behaviour (Hassell and May, 1974). Whether it and other potentially stabilizing behavioural components of parasitism (i.e. mutual interference and sigmoid functional response in *P. flavifemur*) play an important role in host population dynamics is unclear, since the evidence for between-generation density dependent parasitism by parasitoids in the Auchenorrhyncha considered here is either not available (e.g. for Mymaridae) or mostly equivocal. To summarize, among hosts attacked by Dryinidae and Pipunculidae, temporal changes in percentage parasitism have been shown to be density dependent in one species (*N. lineatus* parasitized by *Verrallia aucta*), probably inversely density dependent in one species (*A. ocellaris* parasitized by a complex of Pipunculidae), density independent in seven species (*P. spumarius, R. coronifera, J. distinguendus, D. abdominalis, A. pascuellus, P. confinis*, and *E. sulphurella* parasitized by Pipunculidae), and of an apparently delayed density dependent nature both in *D. hamata* parasitized by *Dicondylus bicolor*, and in *N. cincticeps* parasitized by Pipunculidae.

It is not known how actively or extensively the triungulinid larvae of Strepsiptera forage for hosts. Temporal changes in percentage parasitism were of an apparently delayed density dependent nature in *D. hamata, J. pellucida, H. elegantulus, R. angulosus* and *S. dalei*, when both the abundance of these five species and parasitism by *Elenchus tenuicornis* throughout six generations were pooled.

As Hassell (1986a,b) points out, while parasitoids clearly have the *potential* to regulate their host populations, evidence that they *actually do so* may be difficult to obtain. Detection of the regulatory ability of parasitoids may require much more detailed field studies than have been carried out to date. Hassell urges insect ecologists not only to construct life tables over as many generations as possible, but also to pay more attention to how births and deaths are distributed within the populations of each generation. Furthermore, he argues that such studies should be coupled, wherever possible, with manipulation experiments.

A particular difficulty in evaluating the possible regulatory effects of parasitoids on leafhopper populations, and ultimately on communities, is created by the highly labile nature of the host populations themselves (Waloff, 1980; Waloff & Thompson, 1980). This lability is a result of the considerable flight potential (mobility) of some species (e.g. *B. punctata, J. pellucida, M. parvicauda, Macrosteles* spp.). Consequently, populations may show rapid changes in their locations. Any stability that might be attained by these highly mobile populations, as the result of parasitism or other regulatory factors, will be detectable over a very wide area only.

It is possible that, due to the temporary nature of grassland-associated host populations, narrow host specificity (especially monophagy) among their parasitoids may have been selected against, due to the risk of local extinction over ecological time. Although some parasitoid stages (the eggs and larvae of Dryinidae and Pipunculidae, and the larvae and adults of Strepsiptera) may be transported by their hosts, parasitoid–host relationships must nevertheless be frequently disrupted as host populations shift from one location to another. This may account for the trend towards polyphagy among nymphal–adult parasitoids of grassland-associated hosts (although an alternative explanation may need to be sought for a similar trend seen among parasitoids attacking arboreal Typhlocybinae).

Polyphagous parasitoids, compared with monophagous ones, have a rather different dynamic relationship with their hosts. A broad host range will, for example, tend to buffer populations of such parasitoids from fluctuations in abundance of any one of their host species, producing interactions that are mainly uncoupled from a particular host population

(Hassell, 1978, 1986b). Polyphagous parasitoids may nevertheless cause density dependent mortality that is enough to regulate a host population. There is evidence that *Verrallia aucta*, which parasitizes *N. lineatus* and *P. spumarius*, may exert such an influence on populations of the former. As has been noted above (page 349), the proportion of *N. lineatus* attacked changes from less than to greater than expected as the proportion of *N. lineatus* available increases (Whittaker, 1971a). Whether this frequency dependent parasitism is the result of a behavioural response of the parasitoid, e.g. "switching" (Murdoch, 1969) = "apostatic selection" (Hubbard *et al.*, 1982) = "frequency dependent selection" (Clarke and O'Donald, 1964), is unclear, but whatever the cause, it can have a stabilizing effect. Interestingly, while a similar relationship may hold for populations of the alternative host species, *P. spumarius*, parasitism of the latter is density independent (Whittaker, 1973b).

In considering polyphagy, account may have to be taken of the differential effects of host species on the reproductive success (i.e. fitness) of polyphagous parasitoids. Examples of this are found in Mymaridae, in which fecundity is found to vary with host species (page 291), and it probably occurs also in Pipunculidae, in which adult size is correlated with host size (page 321). If a parasitoid attacks different-sized host species (clearly, many do), this is likely to have important consequences for parasitoid–host population dynamics.

A fundamental difference between arboreal and grassland populations of Auchenorrhyncha was earlier pointed out (page 355), that is, the tendency of the former to show a greater degree of host plant specificity. Adults of acidic grassland species are particularly mobile, and although they are confined to feeding on grasses (Gramineae), they move readily from one host plant species to another, depending on the nutritional state (e.g. amino acid content) of the host plant, which fluctuates both seasonally and annually (Prestidge and McNeill, 1983; McNeill and Southwood, 1978). One might therefore expect host population dynamics, and consequently host–parasitoid dynamics in the two communities, to be somewhat different. Unfortunately, to date, studies on parasitoid–leafhopper population dynamics have been almost exclusively confined to grassland hosts, with the result that no adequate data are available on arboreal communities (however, see Flanagan (1974)).

C. Interactions Between the Communities of Auchenorrhyncha and Those of Their Parasitoids

The whole system of Auchenorrhyncha communities and those of their parasitoids is held together by a network of interactions, many of which are surmised, but not yet ascertained, or understood. Some aspects of these interactions, such as the effects of host spatial distribution on behaviour of parasitoids and the possible importance of "switching", have already been discussed and will be only touched upon here.

Among the outstanding characteristics of this system are the species-richness of some of the host communities and the widespread polyphagy among the member species of the associated parasitoid communities. That is, many parasitoid species are able to utilize a number of host species, and conversely, many host species are able to sustain several parasitoid species from each of the complexes within a guild.

It is worth reiterating that recently Hassell (1986), in considering the potential of polyphagous and monophagous parasitoids in regulating host populations, has provided models for the interaction of both types of parasitoids with their hosts. He showed that, under certain conditions, polyphagous parasitoids may stabilize host populations, since mortality which they cause is directly density dependent, without the time delays which tend to produce oscillations in the populations.

The intricacies of the webs of multispecies interactions have been discussed by Hassell and Waage (1984), who speak of the "vertical" and the "horizontal", i.e. the two-dimensional structure of the webs. In the present context, the vertical component, which is made up of food chains, can be illustrated by the interactions on four trophic levels, e.g. the Diapriidae are hyperparasitic on Dryinidae, the latter are primary parasitoids of Cicadellidae, and these in turn feed on plant tissues, or sap of Gramineae. The horizontal dimension, at one trophic level, depends on species interactions of which little is known, and the vast problems of interspecific competition between parasitoids have yet to be studied. However, some adjustments made by both the parasitoid and the host communities towards partitioning of resources and coexistence are discernible. The homopteran hosts serve as oviposition sites and their haemocoels are the habitats within which the dryinid, pipunculid and strepsipteran larvae develop, i.e. they are the resources exploited by the parasitoid populations. Partitioning of resources takes place between the different parasitoid complexes, as well as within a particular complex. The first type can be illustrated by the following examples.

Both Dryinidae (*Aphelopus* spp.) and Pipunculidae (*Chalarus* spp.) oviposit in typhlocybine nymphs, but the females attack a different range of host instars. *Aphelopus* females lay their eggs in all five nymphal instars, whereas *Chalarus* females tend to lay in the last three only (Jervis, 1980c). Also, among the bivoltine grassland leafhoppers, the first host generations are more heavily attacked by dryinids than the second, whereas parasitization by the pipunculids is heavier in the second generation of hosts (Waloff, 1975; Waloff and Thompson, 1980). The difference in severity of dryinid attack on the two host generations can be explained by parasitoid phenology, and it also illustrates the second type of partitioning of resources, i.e. that within a single parasitoid complex. The most common dryinid species in the grassland study areas were *G. sepsoides, L. ruficornis* and *A. pubicorne*. The last two species, though potentially bivoltine, rarely have a second generation and even then tend to disappear from the field early in the season, having only the trivoltine *G. sepsoides* to exploit the host populations. The attacks by pipunculid species are usually heavier on the second generations of hosts, and this can perhaps be attributed to the absence of winter mortality and expansion of host populations and a large numerical response by the parasitoids. However, this is in direct contrast to what happens in the arboreal Typhlocybinae (see page 320; and Jervis, 1980b,c), where temporal asynchrony in the developmental rates of hosts and the non-diapausing *Chalarus* larvae results in low pipunculid parasitism of the second generation of hosts.

In the acidic grassland, the most abundant host species within a habitat tend to be heavily parasitized and to support a large number of parasitoid species. For example, in the south gravel area at Silwood Park, where *Psammotettix confinis* was the most abundant species, it supported three dryinid (*G. sepsoides, A. pubicorne* and *L. ruficornis*) and nine pipunculid species (*D. xanthopus, E. fascipes, E. jenkinsoni, E. obliquus, E. obscurus, E. subfascipes, P. campestris, T. sylvatica* and *T. kuthyi*). Similarly, *Adarrus ocellaris*, which predominated in another area, supported two dryinid and seven pipunculid species. As for Delphacidae, they suffered their heaviest parasitism from the strepsipteron *Elenchus tenuicornis*, and the percentages of parasitism per generation of the individual host species tended to be directly proportional to their abundance (see Fig. 12).

Another set of processes that holds this host–parasitoid system together and possibly influences host species coexistence, is the differential parasitism of alternative hosts, and possible switching by a single parasitoid species. Some evidence for this has been obtained by rearing out parasitoids from a series of host species, from the same area throughout the same generation (Waloff, 1975). Thus, alternate hosts of the dryinid *G. sepsoides* and the pipunculid *T. sylvatica* were *Psammotettix confinis, Jassargus distinguendus, Adarrus ocellaris* and *Arthaldeus pascuellus*, and those of the pipunculid *E. subterminalis*, the first three hosts and *Diplocolenus abdominalis*. This sharing of hosts creates a refuge from over-exploitation of any one host species. Creation of refuges, or of "enemy-free space" was one of the subjects of Lawton's (1986) contribution to the recent symposium on "Insect Parasitoids". Lawton considers that niches of member species of phytophagous insect communities are often influenced by their parasitoids, in attempts by the hosts to gain some enemy-free space, although he acknowledges that absolute safety from enemies is virtually impossible. He also points out that the more similar the host niches are, the more likely are the hosts to share polyphagous parasitoids, and this is highly relevant to the leafhoppers inhabiting grassland habitats.

Whether or not temporal and spatial refuges have evolved in response to pressure from natural enemies, they *do* exist, and some examples can be cited from the present studies. For instance, temporal escape from maximum parasitoid attack by means of asynchrony of development rates of *Chalarus* and the second generation of arboreal Typhlocybinae has already been mentioned. In grassland leafhoppers, the peaks of maximum abundance of the most common species of a community tend to form a time series (Waloff, 1979, 1980). Another, and striking example of temporal separation of peaks of abundance of coexisting "dominant" leafhoppers can be seen in Müller's (1978) studies in Thuringia. Similarly, Jervis (1980c), who examined the life histories of 29 species of arboreal Typhlocybinae, showed that there was a seasonal succession of these leafhoppers, which fell into five major groups in a time sequence. This temporal separation in abundance of species within the Auchenorrhyncha communities and the consequent spread of risk from enemies from extinction (Den Boer, 1968), must be a potent factor in the maintenance of balance between leafhoppers and their associated parasitoid communities. Refuges are provided to host species in their periods of low density, when the more abundant alternative hosts are available.

Examples of spatial, as opposed to temporal, "enemy-free space" are also known. Spatial variations in the levels of parasitism on the same host plant are well illustrated by Stiling and Strong's (1982b) work, discussed in the section on Mymaridae (see page 339). Another striking example was seen in Tay's (1972) work on *Cicadella viridis*. The cicadellid laid four fifths of its eggs in the lowest 150 mm of *Juncus* stems, where mymarid parasitism reached only 1·5%, whereas maximum parasitism of 20·2% occurred in stems 210–300 mm above ground level, where only a few eggs were laid.

Another interesting example of enemy-free space comes from Abdul-Nour's (1971) observations on the ecology of the strepsipteron *Halictophagus languedoci* in grassland in southern France (see page 353). A complex of leafhoppers was parasitized by the strepsipteron in *Brachypodium phaenicoides*, but the hosts escaped parasitism in stands of two other grasses, *B. ramosum* and *Bromus erectus*, even when parasitized individuals occurred in islets of *B. phaenicoides* within the stands. Reasons for this uneven distribution of parasitism are not clear, but its consequence is the provision of enemy-free space to the host species.

D. Parasitoids as Potential Biological Control Agents

As we have seen (Sections VI–IX), parasitoids may inflict high levels of mortality upon host populations, and have the potential to regulate the latter. They would therefore seem to be suitable candidates for the biological control of pest Auchenorrhyncha, and have, in fact, been employed with varying degrees of success against a variety of pest Cercopidae, Membracidae, Cicadellidae, Delphacidae and Flatidae throughout the World (for reviews see Clausen, 1978; Laing and Hamai, 1976; and Vidano *et al.*,

1985). Parasitoids are also currently being considered for use in the control of major pests such as *Empoasca fabae* in the United States (F. Herard, pers. comm.) and *Nilaparvata lugens* and other pest Auchenorrhyncha in Asia (Chua *et al.*, 1984; M. F. Claridge, pers. comm.).

Unfortunately, the problem with previous biological control programmes is that not only the failures but also the successes have been poorly documented, in terms of details of parasitoid–hopper population dynamics. It is therefore difficult to evaluate which factors have determined the outcome of a particular biological control attempt. In this chapter, we have drawn together information on a number of fundamental aspects of parasitoid biology and ecology. While it is clear that there remain many gaps in our knowledge, it is nevertheless our hope that we have succeeded in contributing to both a practical and a theoretical basis for the use of parasitoids in insect pest management.

ACKNOWLEDGEMENTS

We wish to thank the following, who kindly supplied unpublished information: A. Albrecht, M. F. Claridge, J. T. Flanagan, F. Herard, R. M. Khafagi, M. Moratorio and C. Vidano. We also wish to thank the following for useful discussions and criticism: N. A. C. Kidd, J. Kathirithamby, D. R. Lees, P. N. Ferns, A. J. A. Stewart and M. F. Claridge.

REFERENCES

Abdul-Nour, H. (1969). Une nouvelle espece de Strepsiptere, parasite de Jassidae (Hom. Auchen.): *Halictophagus languedoci* n. sp. *Ann. Soc. ent. France* **5**, 361–69.

Abdul-Nour, H. (1970). Les Strepsipteres, parasites d'Homopteres dans le Sud de la France. Description d'une nouvelle espece: *Halictophagus agalliae* n. sp. *Ann. zool. Ecol.* **2**, 339–44.

Abdul-Nour, H. (1971). Contribution a l'etude des parasites d'Homopteres Auchenorrhynques du Sud de la France: Dryinidae (Hymenopteres) et Strepsipteres. *These*, Acad. Montpellier, Univ. Sci. et Tech. Languedoc, 154 pp.

Aczél, M. (1948). Grundlagen einer Monographie der Dorylaiden (Dorilaiden-Studien 6). *Acta zool. lilloana* **6**, 5–168.

Aczél, M. (1954). Orthopyga and Camplopyga, new Divisions of Diptera. *Ann. ent. Soc. Am.* **47**, 75–80.

Ahlberg, O. (1925). Zikaden-Parasiten unter den Strepsipteren und Hymenopteren. *Medd. Centr. Fors. Jordbruks* **287**, 78–86.

Albrecht, A. (1979). Descriptions of seven new *Dorylomorpha* Aczel species from Europe (Diptera: Pipunculidae). *Ent. scand.* **10**, 211–18.

Ali, A. M. H. (1979). Biological investigations on the entomophagous parasites of insect eggs associated with *Juncus* species. *Ph.D. thesis*. University of Wales.

van Alphen, J. J. M. and Vet, L. M. (1986). An evolutionary approach to host finding and selection. In: *Insect Parasitoids*, 13th Symposium of the Royal Entomological Society of London (Ed. by J. K. Waage and D. J. Greathead). London: Academic Press. pp. 23–61

Annecke, D. P. and Doutt, R. L. (1961). The genera of the Mymaridae (Hymenoptera: Chalcidoidea). *S. Afr. Dept. Agric. Tech. Serv., Entomol. Mem.* **5**, 1–77.

Armstrong, T. (1936). Two parasites of the White Apple Leafhopper (*Typhlocyba pomaria* McA.). *Ann. Rep. Ontario Entomol. Soc.* **66**, 16–31.

Arzone, A. (1974a). Indagini biologiche sui parassiti oofagi di *Cicadella viridis* (L.) (Hem. Hom. Cicadellidae). I *Gonatocerus cicadellidae* Nik. (Hym. Mymaridae). *Annali Fac. Sci. agric. Univ. Torino* **9**, 137–60.

Arzone, A. (1974b). Indagini biologiche sui parassiti oofagi di *Cicadella viridis* (L.) (Hem. Hom. Cicadellidae). II *Oligosita krygeri* Gir. (Hym. Trichogrammatidae). *Annali Fac. Sci. agric. Univ. Torino* **9**, 193–214.

Arzone, A. (1974c). Indagini biologiche sui parassiti oofagi di *Cicadella viridis* (L.) (Hem. Hom. Cicadellidae). III *Polynema woodi* Hincks (Hym. Mymaridae). *Annali Fac. Sci. agric. Univ. Torino* **9**, 297–318.

Arzone, A. (1977). Reperti biologici ed epidemiologici su *Conomelus dehneli* Nast (Hom. Delphacidae), nuovo per l'Italia come un suo parassita, *Tetrastichus mandanis* Walk. (Hym. Eulophidae). *Boll. Zool. agric. Bach.* **14**, 5–16.

Askew, R. R. and Shaw, M. R. (1986). Parasitoid communities: their size, structure and development. In: *Insect Parasitoids*, 13th Symposium of Royal Entomological Society of London (Ed. by J. K. Waage and D. J. Greathead). London: Academic Press. pp. 225–264.

de Azevedo Marques, L. A. (1925). A cigarrhina nociva aos pomares (*Aethalion reticulatum* L.). *Chacaras e Quintaes* **32**, 33–7.

Bakkendorf, O. (1925). Recherches sur la biologie de *l'Anagrus incarnatus* Haliday, microhymenoptere parasite (accidental?) des oeufs de divers Agrionides. *Annls. Biol. lacustre* **14**, 249–70.

Bakkendorf, O. (1934). Biological investigations on some Danish hymenopterous egg-parasites. *Ent. Medd.* **19**, 1–134.

Bakkendorf, O. (1943). Report on an investigation of the local distribution of the components of the community *Chaetostricha pulchra* (Hym. Chalc.), *Tettigonia viridis* (Hem. Hom.) and *Juncus effusus* and *conglomeratus*. *Ent. Medd.* **23**, 31–6.

Bakker, K., van Alphen, J. J. M., van Batenburg, F. H. D., van der Hoeven, N., Nell, H. W., van Strien-van Liempt, W. T. F. H. and Turlings, T. C. J. (1985). The function of host discrimination and superparasitism in parasitoids. *Oecologia, Berlin* **67**, 572–76.

Balduf, W. V. (1928). Observations on the Buffalo Tree Hopper *Ceresa bubalus* Fabr. (Membracidae, Homoptera), and the bionomics of an egg parasite, *Polynema striaticorne* Girault (Mymaridae, Hymenoptera). *Ann. ent. Soc. Am.* **21**, 419–35.

Bankowska, R. (1973). Pipunculidae. *Pol. Tow. Ent.* **28**, 1–52.

Bartlett, K. A. (1939). A dryinid parasite attacking *Baldulus maidis* in Puerto Rico. *J. Agric. Univ. Puerto Rico* **22**, 497–98.

Becker, M. (1975). The biology and population ecology of *Macrosteles sexnotatus* (Fallén) (Cicadellidae, Hemiptera). *Ph.D. thesis*. University of London.

Benton, F. P. (1975). Larval taxonomy and bionomics of some British Pipunulidae. *Ph.D. thesis*. University of London.

Bentur, J. S., Sain, M. and Kalode, M. B. (1982). Studies on egg and nymphal parasites in rice planthoppers, *Nilaparvata lugens* (Stål) and *Sogatella furcifera* (Horvath). *Proc. Indian Acad. Sci. (Anim. Sci.)* **91**, 165–76.

Bristowe, W. S. (1950). Strange mating of *Pipunculus distinctus* Becker (Dipt; Pipunculidae). *Ent. mon. Mag.* **86**, 264.

Burks, B. D. (1979). Trichogrammatidae. In: *Catalog of Hymenoptera North of Mexico* (Ed. by K. V. Krombein, P. D. Hurd, D. R. Smith and B. D. Burks). Smithsonian Institution. Washington, DC: pp. 1033–43.

Bush, G. L. (1975). Modes of animal speciation. *Ann. Rev. Ecol. Syst.* **6**, 339–64.

Buyckx, E. J. E. (1948). Recherches sur un dryinide, *Aphelopus indivisus*, parasite de cicadines. *Cellule* **52**, 63–155.

Chambers, V. H. (1955). Some hosts of *Anteon* spp. (Hym., Dryinidae) and a hyperparasite *Ismarus* (Hym., Belytidae). *Ent. mon. Mag.* **91**, 114–15.

Chambers, V. H. (1981). A host for *Ismarus halidayi* Foerst. (Hym., Diapriidae). *Ent. mon. Mag.* **118**, 29.

Chandra, G. (1978). A new cage for rearing hopper parasites. *Int. Rice Res. Newslett.* **3**, 12.

Chandra, G. (1980a). Dryinid parasitoids of rice leafhoppers and planthoppers in the Philippines II. Rearing techniques. *Entomophaga* **25**, 187–92.

Chandra, G. (1980b). Dryinid parasitoids of rice leafhoppers and planthoppers in the Philippines. *Acta. Oecol.* **1**, 161–72.

Chandra, G. (1980c). Taxonomy and bionomics of the insect parasites of rice leafhoppers and planthoppers in the Philippines and their importance in natural biological control. *Philipp. Ent.* **4**, 119–39.

Chesson, P. L. and Murdoch, W. W. (1986). Aggregation of risk: relationships among host–parasitoid models. *Am. Nat.* **127**, 696–715.

Chu, Y. I. and Hirashima, Y. (1981). Survey of Taiwanese literature on the natural enemies of rice leafhoppers and planthoppers. *Esakia* **16**, 33–7.

Chua, T. H. and Dyck, V. A. (1982). Assessment of *Pseudogonatopus flavifemur* E. & H. (Dryinidae: Hymenoptera) as a biocontrol agent of the rice brown planthopper. *Proc. Int. Conf. Pl. Prot. in Tropics*, pp. 253–65.

Chua, T. H., Dyck, V. A. and Peña, N. B. (1984). Functional response and searching efficiency in *Pseudogonatopus flavifemur* Esaki & Hash. (Hymenoptera: Dryinidae), a parasite of rice planthoppers. *Res. Popul. Ecol.* **26**(1), 74–83.

Claridge, D. W., Derry, N. J. and Whittaker, J. B. (1981). The distribution and feeding of some Typhlocybinae in response to sun and shade. *Acta ent. fenn.* **38**, 8–12.

Claridge, M. F. and Reynolds, W. J. (1972). Host plant specificity, oviposition behaviour and egg parasitism in some woodland leafhoppers of the genus *Oncopsis* (Hemiptera Homoptera: Cicadellidae). *Trans. R. ent. Soc. Lond.* **124**, 149–66.

Claridge, M. F., Reynolds, W. J. and Wilson, M. R. (1977). Oviposition behaviour and food plant discrimination in leafhoppers of the genus *Oncopsis*. *Ecol. Ent.* **2**, 19–25.

Claridge, M. F. and Wilson, M. R. (1976). Diversity and distribution patterns of some mesophyll-feeding leafhoppers of temperate woodland canopy. *Ecol. Ent.* **1**, 231–50.

Claridge, M. F. and Wilson, M. R. (1978). Oviposition behaviour as an ecological factor in woodland canopy leafhoppers. *Ent. exp. appl.* **24**, 101–109.

Clarke, B. and O'Donald, P. (1964). Frequency-dependent selection. *Heredity* **19**, 201–206.

Clausen, C. P. (1940). *Entomophagous Insects*. McGraw-Hill, New York and London. 688 pp.

Clausen, C. P. (1978). *Introduced Parasites and Predators of Arthropod Pests and Weeds: a World Review*. Handbook No. 480. Washington, DC: United States Department of Agriculture.

Coe, R. L. (1966). Diptera: Pipunculidae. *Handbooks for the Identification of British Insects* 10 (2c). Royal Entomological Society of London. 83 pp.
Collins, M. D., Ward, S. A. and Dixon, A. F. G. (1981). Handling time and the functional response of *Aphelinus thomsoni*, a predator and parasite of the aphid *Drepanosiphum platanoidis*. *J. anim. Ecol.* **50**, 479–87.
Comins, H. N. and Hassell, M. P. (1979). The dynamics of optimally foraging predators and parasitoids. *J. anim. Ecol.* **48**, 335–51.
Craighead, F. C. (1921). Larva of the North American beetle *Sandalus niger* Knoch. *Proc. ent. Soc. Wash.* **23**, 44–8.
Crowson, R. A. (1955). *The Natural Classification of the Families of Coleoptera*. Nathaniel Lloyd, London. 187 pp.
Crowson, R. A. (1975). On a possible female of *Halictophagus curtisii* Dale (Col., Stylopoidea). *Ent. mon. Mag.* **111**, 62.
Currado, I. and Olmi, M. (1972). Dryinidae italiani: conoscenze attuali e nuovi reperti (Hymenoptera, Bethyloidea). *Boll. Mus. Zool. Univ. Torino* **7**, 137–76.
Debauche, H. R. (1948). Etudes sur les Mymarommidae et Mymaridae de la Belgique (Hymenoptera Chalcidoidea). *Mem. Mus. Hist. nat. Belg.* **108**, 1–248.
De Meyer, M. and De Bruyn, L. (1984). On the phenology of some Pipunculidae (Diptera) in Belgium. *Bull. Annls. Soc. R. belge. Ent.* **120**, 123–31.
Den Boer, P. J. (1968). Spreading of risk and stabilisation of animal numbers. *Acta biotheor.* **18**, 165–94.
Domenichini, G. (1966). Palaearctic Tetrastichinae. *Index of Entomophagous Insects*. Le Francois, Paris. 101 pp.
Donisthorpe, H. St. J. K. (1927). *British Ants, their Life-History and Classification*. Routledge, London.
Dout, R. L. (1961). The hymenopterous parasites of some Japanese leafhoppers. *Acta Hymenopt.* **1**, 305–14.
Doutt, R. L. and Nakata, J. (1965). Parasites for control of grape leafhopper. *Calif. Agric.* **19**, 3.
Doutt, R. L., Nakata, J. and Skinner, F. E. (1966). Dispersal of grape leafhopper parasites from a blackberry refuge. *Calif. Agric.* **20**, 14–15.
Doutt, R. L. and Viggiani, G. (1968). The classification of the Trichogrammatidae (Hymenoptera: Chalcidoidea). *Proc. Calif. Acad. Sci.* **35**, 477–86.
Dozier, H. L. (1932). Descriptions of new mymarid egg parasites from Haiti and Puerto Rico. *J. Dept. Agric. Puerto Rico* **16**, 81–91.
Drew, R. A. I. and Allwood, A. J. (1985). A new family of Strepsiptera parasitizing fruit flies (Tephritidae) in Australia. *Syst. Ent.* **10**, 129–34.
Drosopoulos, S. (1977). Biosystematic studies on the *Muellerianella* complex (Delphacidae, Homoptera, Auchenorrhyncha). *Meded. Landbouwhogesc. Wageningen* **77–14.** 1–33.
Esaki, T. and Hashimoto, S. (1930). Report on leafhoppers injurious to the rice plant and their natural enemies. *Pub. Kyushu Imp. Univ. Ent. Lab. Fukuoka* **2**, 1–29 (in Japanese).
Ewan, H. G. (1961). The Saratoga Spittlebug. *U.S. Dept. Agric. Tech. Bull.* No. 1250, 52 pp.
Fattig, P. W. (1955). The Cicadellidae or leafhoppers of Georgia. *Emory Univ. Mus. Bull. Georgia* **11**, 1–68.
Fenton, F. A. (1918). The parasites of leafhoppers, with special reference to the biology of Anteoninae. *Ohio J. Sci.* **28**, 177–212, 243–78, 285–96.
Flanagan, J. T. (1974). Population dynamics of the leafhopper *Alnetoidia alneti* (Dahlbom) and its effect on tree growth. *Ph.D. thesis*. University of Glasgow.
Flanders, S. E. (1950). Regulation of ovulation and egg disposal in parasitic Hymenoptera. *Can. Ent.* **82**, 134–40.

Freytag, P. H. (1985). The insect parasites of leafhoppers and related groups. In: *The Leafhoppers and Planthoppers* (Ed. by L. R. Nault and J. G. Rodriguez). New York: Wiley. pp. 423–67.

Ganin, M. (1969). Beitrage zur Erkenntnis des Entwicklungsgeschichte bei der Insekten. *Z. wiss. Zoll.* **19**, 381.

Giard, A. (1889). Sur la castration parasitaire des *Typhliocyba* par une larve d'hymenoptere (*Aphelopus melaleucus* Dalm.) et par une larve de Diptere (*Atelenevra spuria* Meig.). *C.r. hebd. Seanc. Acad. Sci., Paris* **109**, 708–10.

Gibson, G. A. P. (1986). Evidence of monophyly and relationships of Chalcidoidea, Mymaridae, and Mymarommatidae (Hymenoptera: Terebrantes). *Can. Ent.* **118**, 205–40.

Girault, A. A. (1916). Notes on North American Mymaridae and Trichogrammatidae (Hym.). *Entom. News, Philadelphia* **27**, 4–8.

Greathead, D. J. (1983). Natural enemies of *Nilaparvata lugens* and other leaf- and planthoppers in tropical agroecosystems and their impact on pest populations. *1st International Workshop on Leafhoppers and Planthoppers of Economic Importance*. Commonwealth Institute of Entomology. pp. 371–83.

Haeselbarth, E. (1979). Zur Parasitierung der Puppen von Forleule (*Panolis flammea* (Schiff.)), Kiefernspanner (*Bupalus pinarius* (L. J.)) und Heidelbeerspanner (*Boarmia bistortana* (Goeze)) in bayerischen Kiefernwalden. *Z. angew. Ent.* **87**, 311–22.

Hardy, D. E. (1943). Revision of the Nearctic Dorilaidae (Pipunculidae). *Kans. Univ. Sci. Bull.* **29**, 1–231.

Harper, G. and Whittaker, J. B. (1976). The role of natural enemies in the colour polymorphism of *Philaenus spumarius*. *J. anim. Ecol.* **45**, 91–104.

Hassan, A. I. (1939). The biology of some British Delphacidae (Homopt.) and their parasites with special reference to the Strepsiptera. *Trans. R. ent. Soc. Lond.* **89**, 345–84.

Hassell, M. P. (1978). *The Dynamics of Arthropod Predator–Prey Systems*. Princeton, NJ: Princeton University Press. 237 pp.

Hassell, M. P. (1982a). What is searching efficiency? *Ann. appl. Biol.* **101**, 170–175.

Hassell, M. P. (1982b). Patterns of parasitism by insect parasitoids in patchy environments. *Ecol. Ent.* **7**, 365–77.

Hassell, M. P. (1984). Parasitism in patchy environments: inverse density dependence can be stabilizing. *IMA J. Math. appl. Med. Biol.* **1**, 123–33.

Hassell, M. P. (1986a). Detecting density dependence *Trends Ecol. Evol.* **1**, 90–3.

Hassell, M. P. (1986b). Parasitoids and population regulation. In: *Insect Parasitoids*, 13th Symposium of the Royal Entomological Society of London (Ed. by D. J. Greathead and J. K. Waage). London: Academic Press. pp. 201–224.

Hassell, M. P., Lessells, C. M. and McGavin, G. C. (1985). Inverse density dependent parasitism in a patchy environment: a laboratory system. *Ecol. Ent.* **10**, 393–402.

Hassell, M. P. and May, R. M. (1974). Aggregation in predators and insect parasites and its effect on stability. *J. anim. Ecol. 43*, 567–94.

Hassell, M. P. and Waage, J. K. (1984). Host–parasitoid population interactions. *Ann. Rev. Ent.* **29**, 89–114.

Hassell, M. P., Lawton, J. H. and Beddington, J. R. (1977). Sigmoid functional responses by invertebrate predators and parasitoids. *J. anim. Ecol.* **46**, 249–62.

Hayat, M. (1974). Host-parasite catalogue of the egg-inhabiting aphelinid genera *Centrodora* Foerster, 1878, and *Tumidiscapus* Girault, 1911 (Hymenoptera, Chalcidoidea). *Polsk. Pismo. Entomol.* **44**, 287–98.

Heikinheimo, O. (1957). *Dicondylus lindbergi* sp. n. (Hym. Dryinidae) a natural enemy of *Delphacodes pellucida* (F.). *Ann. ent. fenn.* **23**, 77–85.

Hellén, W. (1964). Die Ismarinen und Belytinen Finnlannds. *Fauna fenn.* **18**, 1–68.
Holling, C. S. (1959). Some characteristics of simple types of predation and parasitism. *Can Ent.* **91**, 385–98.
Holling, C. S. (1966). The functional response of invertebrate predators to prey density. *Mem. ent. Soc. Can.* **48**, 1–86.
Hubbard, S. F., Cook, R. M. Glover, J. G. and Greenwood, J. J. D. (1982). Apostatic selection as an optimal foraging strategy, *J. anim. Ecol.* **51**, 625–32.
Huldén, I. (1984). Observations on an egg parasite of *Cicadella viridis* (Homoptera, Auchenorrhyncha). *Not. Ent.* **64**, 83–4.
Huq, S. B. (1982). A contribution to the biology of pipunculid-flies (Pipunculidae: Diptera). *Ph.D. thesis*. Free University, Berlin.
Huq, S. B. (1984). Breeding methods for Pipunculidae (Diptera), endoparasites of leafhoppers. *Int. Rice Res. Newslett.* **9**, 14–15.
Ison, C. H. (1959). Notes on the genus *Anagrus* (Mymaridae) with an account of rearing techniques. *J. Quekett. Mic. Cl., London* **4**, 221–30.
Jenkinson, F. (1903). *Verrallia aucta* and its host. *Ent. mon. Mag.* **39**, 222–23.
Jervis, M. A. (1978a). Homopteran Bugs. In: *A. Dipterist's Handbook* (Ed. by A. Stubbs and P. J. Chandler). Hanworth: Amateur Entomologist's Society. pp. 173–6.
Jervis, M. A. (1978b). Ecological studies on the entomophagous parasites of Typhlocybine leafhoppers. *Ph.D. thesis*. University of Wales.
Jervis, M. A. (1979a). Parasitism of *Aphelopus* species (Hymenoptera: Dryinidae) by *Ismarus dorsiger* (Curtis) (Hymenoptera: Diapriidae). *Ent. Gaz.* **30**, 127–9.
Jervis, M. A. (1979b). Courtship, mating and "swarming" in *Aphelopus melaleucus* (Dalman) (Hymenoptera: Dryinidae). *Ent. Gaz.* **30**, 191–3.
Jervis, M. A. (1980a). Studies on oviposition behaviour and larval development in species of *Chalarus* (Diptera, Pipunculidae), parasites of typhlocybine leafhoppers (Homoptera, Cicadellidae). *J. nat. Hist.* **14**, 759–68.
Jervis, M. A. (1980b). Life history studies of *Aphelopus* species (Hymenoptera: Dryinidae) and *Chalarus* species (Diptera: Pipunculidae), primary parasites of typhlocybine leafhoppers (Homoptera: Cicadellidae). *J. nat. Hist.* **14**, 769–80.
Jervis, M. A. (1980c). Ecological studies on the parasite complex associated with typhlocybine leafhoppers (Homoptera, Cicadellidae). *Ecol. Ent.* **5**, 123–36.
Jervis, M. A. (1986). New host records for *Aphelopus* (Hymenoptera: Dryinidae). *Ent. Gaz,* **37**, 37–38.
Jervis, M. A. and Kidd, N. A. C. (1986). Host-feeding strategies in hymenopteran parasitoids. *Biol. Rev.* **61**, 395–434.
Kathirithamby, J. (1971). Taxonomy development and morphology of the immature stages of cicadellidae. *Ph.D. Thesis,* University of London.
Kathirithamby, J. (1973). Key for the separation of larval instars of some British Cicadellidae (Hem. Homoptera). *Ent. mon. Mag.* **109**, 214–16.
Kathirithamby, J. (1977a). The effects of stylopisation on the sexual development of *Javesella dubia* (Kirschbaum) (Homoptera: Delphacidae). *Biol. J. Linn. Soc.* **10**, 163–79.
Kathirithamby, J. (1977b). Stylopisation in *Ulopa reticulata* (F.) (Homoptera, Cicadellidae). *Ent. mon. Mag.* **113**, 89–92.
Kathirithamby, J. (1979). The effects of stylopisation in two species of planthoppers in the Kirian District, West Malaysia (Homoptera: Delphacidae). *J. Zool. Lond.* **187**, 393–401.
Kathirithamby, J. (1982). *Elenchus* sp. (Strepsiptera: Elenchidae), a parasitoid of *Nilaparvata lugens* (Stål) (Homoptera: Delphacidae) in Peninsular Malaysia. *Proc. Int. Conf. Pl. Prot. in Tropics*, pp. 349–61.
Kathirithamby, J. (1983). The mode of emergence of the adult male *Elenchus*

tenuicornis (Kirby) (Strepsiptera: Elenchidae) from its puparium. *Zool. J. Linn. Soc.* **77**, 97–102.

Kathirithamby, J., Spencer Smith, D., Lomas, M. B. and Luke, B. M. (1984). Apolysis without ecdysis in larval development of a strepsipteran, *Elenchus tenuicornis* (Kirby). *Zool. J. Linn. Soc.* **82**, 335–43.

Kershaw, J. C. (1913). Recommendations for dealing with the Froghopper. *Dept. Agric. Trinidad & Tobago, Special Circ.* **9**, 1–10.

Khafagi, R. M. (1986). The biological relationships of *Macrosteles viridigriseus* (Homoptera) and its parasitoid *Anteon pubicorne* (Hymenoptera). *Ph.D. thesis*. University of Newcastle upon Tyne.

Kidd, N. A. C. and Mayer, A. D. (1983). The effect of escape responses on the stability of insect host–parasitoid models. *J. theor. Biol.* **104**, 275–87.

Kinzelbach, R. K. (1970). Nematoden bei *Mengenilla parvula* Silvestri. *Boll. Lab. Ent. Agr. "Filippo Silvestri" di Portici* **28**, 190–93.

Kinzelbach, R. K. (1971). Morphologische Befunde an Facheflugern und ihre phylogenetische Bedeutung (Insecta: Strepsiptera). *Zoologica Stutt.* **119**, 1–128, 129–256.

Kinzelbach, R. K. (1978). Strepsiptera. *Die Tierwelt- Deutschlands* **65**, 1–166.

Kiritani, K., Hokyo, N., Sasaba, T. and Nakasuji, F. (1970). Studies on population dynamics of the Green Rice Leafhopper, *Nephotettix cincticeps* Uhler: regulatory mechanism of the population density. *Res. Pop. Ecol.* **12**, 137–53.

Königsmann, E. (1978). Das phylogenetische system der Hymenoptera Teil 3: Terebrantes (unterordnung Apocrita). *Deutsch. Entomol. Z.* **25**, 1–55.

Kornhauser, S. J. (1919). The sexual characteristics of the membracid *Thelia bimaculata* (Feb.) I. External changes induced by *Aphelopus theliae* (Gahan). *J. Morph.* **32**, 531–635.

Kozánek, M. (1981a). Genus *Pipunculus* Latreille (Diptera, Pipunculidae) in Czechoslovakia. *Annot. zool. bot.* **142**, 1–16.

Kozánek, M. (1981b). Rod *Nephrocerus* Zetterstedt (Diptera, Pipunculidae) v Ceskoslovensku. *Biologia* **36**, 395–6.

Kryger, J. P. (1950). The European Mymaridae comprising the genera known up to *c*. 1930. *Ent. Medd.* **26**, 1–97.

Laing, J. E. and Hamai, J. (1976). Biological control of insect pests and weeds by imported parasites, predators and pathogens. In: *Theory and Practice of Biological Control* (Ed. by C. B. Huffaker and P. S. Messenger). New York: Academic Press. pp. 685–743.

Lauterer, P. (1981). Contribution to the knowledge of the family Pipunculidae of Czechoslovakia (Diptera). *Acta Mus. Moraviae Sci. nat.* **66**, 123–50.

Lauterer, P. (1983). Contribution to the knowledge of distribution and bionomics of some representatives of the family Pipunculidae in Central and Southern Europe. *Act. Mus. Moraviae. Sci. nat.* **68**, 131–8.

Lawton, J. H. (1986). The effect of parasitoids on phytophagous insect communities. In: *Insect Parasitoids,* 13th Symposium of the Royal Entomological Society of London, (Ed. by J. K. Waage and D. J. Greathead). London: Academic Press. pp. 265–287.

van Lenteren, J. C. (1981). Host discrimination by parasitoids. In: *Semiochemicals: their role in pest control* (Ed. by D. A. Nordlund, R. L. Jones and W. J. Lewis). New York: Wiley. pp. 153–79.

van Lenteren, J. C., Bakker, K. and van Alphen, J. J. M. (1978). How to analyze host discrimination. *Ecol. Ent.* **3**, 71–5.

Lessells, C. M. (1985). Parasitoid foraging: should parasitism be density dependent? *J. anim. Ecol.* **54**, 27–41.

Lewis, T. and Taylor, L. R. (1965). Diurnal periodicity of flight by insects. *Trans. R. ent. Soc. Lond.* **116**, 393–479.
Lin, K. S. (1974). Notes on some natural enemies of *Nephotettix cincticeps* (Uhler) and *Nilaparvata lugens* (Stål) in Taiwan. *J. Taiwan Agric. Res.* **23**, 91–115 (in Chinese, with English summary).
Lindberg, H. (1946). Die Biologie von *Pipunculus chlorionae* Frey und die Einwirkung von dessen Parasitismus auf *Chloriona*-Arten. *Acta Zool. Fenn.* **45**, 1–50.
Lindberg, H. (1950). Notes on the biology of dryinids. *Comment. Biol.* **10**, 1–19.
Linnane, J. P. and Osgood, E. A. (1977). *Verrallia virginica* (Diptera: Pipunculidae) reared from the Saratoga Spittlebug in Maine. *Proc. ent. Soc. Wash.* **79**, 622–23.
Lundbeck, W. (1922). Pipunculidae, Phoridae. *Diptera Danica* **6**, 1–447.
MacGill, E. I. (1934). On the biology of *Anagrus atomus* (L.): an egg parasite of the leaf-hopper *Erythroneura pallidifrons* Edwards. *Parasitology* **26**, 57–63.
Maillet, P. L. (1960). Sur le parasitisme d'oefs de le Cicadelle verte (*Cicadella viridis* L.), par un hymenoptere Mymaridae: *Anagrus atomus* (L.) forme *incarnatus* Hal. *Rev. Path. Veg.* **39**, 197–203.
Manjunath, T. M. (1978). Two nematode parasites of rice brown planthopper in Iran. *Int. Rice Res. Newslett.* **3**, 11–12.
Manjunath, T. M., Rai, P. S. and Gowda, G. (1978). Parasites and predators of *Nilaparvata lugens* in India. *Pans* **24**, 265–69.
Masner, L. (1976). A revision of the Ismarinae of the New World (Hymenoptera, Proctotrupoidea, Diapriidae). *Can. Ent.* **108**, 1243–66.
Matthews, M. J. (1986). The British species of *Gonatocerus* Nees (Hymenoptera: Mymaridae), egg parasitoids of Homoptera. *Syst. Ent.* **11**, 213–29.
May, Y. Y. (1971). The Biology and Population Ecology of *Stenocranus minutus* (Fabricius) (Delphacidae, Hemiptera). Ph.D. Thesis, University of London.
May, Y. Y. (1979). The biology of *Cephalops curtifrons* (Diptera: Pipunculidae), an endoparasite of *Stenocranus minutus* (Hemiptera: Delphacidae). *Zool. J. Linn. Soc.* **66**, 15–29.
McClure, M. S. (1974). Biology of *Erythroneura lawsoni* (Homoptera: Cicadellidae) and coexistence in the Sycamore leaf-feeding guild. *Environ. Entomol.* **3**, 59–68.
McKenzie, L. M. and Beirne, B. P. (1972). A grape leafhopper, *Erythroneura ziczac* (Homoptera: Cicadellidae), and its mymarid (Hymenoptera) egg-parasite in the Okanagan Valley, British Columbia. *Can. Ent.* **104**, 1229–33.
McNeill, S. and Southwood, T. R. E. (1978). The role of nitrogen in the development of insect/plant relationships. *Biological Aspects of Plant and Animal Coevolution* (Ed. by J. S. Harborne). London: Academic Press. pp. 77–98.
Miura, T., Hirashima, Y. and Wongsiri, T. (1979). Egg and nymphal parasites of rice leafhoppers and planthoppers, a result of field studies in Thailand in 1977. *Esakia* **13**, 21–44.
Moratorio, M. S. (1977). Aspects of the biology of *Anagrus* spp. (Hymenoptera: Mymaridae) with special reference to host–parasitoid relationships. *Ph.D. thesis.* University of London.
Morcos, G. (1953). The biology of some Hemiptera–Homoptera (Auchenorrhyncha). *Bull. Soc. Fouad 1er Entom.* **37**, 405–39.
Morris, M. G. (1971). Differences between invertebrate faunas in grazed and ungrazed chalk grassland. *J. appl. Ecol.* **8**, 37–52.
Morrison, G. and Strong, D. R. (1980). Spatial variations in host density and the intensity of parasitism: some empirical examples. *Environ. Entomol.* **9**, 149–152.
Muir, F. (1918). Pipunculidae and Stylopidae in Homoptera. *Ent. mon. Mag.* **54**, 137.

Muir, F. (1921). The Sugar Cane Leafhopper and its parasites in Hawaii. *Hawaii Planter's Rec.* **25**, 108–23.

Müller, H. J. (1960). Uber morphologische Folgen den Parasitierung von *Euscelis*-Mannchen (Homoptera–Auchenorrhyncha) mit Dryiniden-Larven. *Z. Morph. Okol. Tiere* **49**, 32–46.

Müller, H. J. (1978). Strukturanalyse der Zikadenfauna (Homoptera Auchenorrhyncha) einer Rasenkatena Thuringens (Leutratal bei Jena). *Zool. Jb. Syst. Bd.* **105**, 258–334.

Murdoch, W. W. (1969). Switching in general predators: experiments on predator specificity and stability in prey populations. *Ecol. Monogr.* **39**, 335–54.

Myers, J. G. (1922). Life-history of *Siphanta acuta* (Walk.), the Large Green Plant-hopper. *N.Z. Jl. Sci. Tech.* **5**, 256–63.

Myers, J. G. (1930). *Carabunia myersi* Watrst. (Hym., Encyrtidae), a parasite of nymphal froghoppers (Hom., Cercopidae). *Bull. ent. Res.* **21**, 341–51.

Nikol'skaya, N. M. (1951). A new species of *Gonatocerus* (Hymenoptera: Mymaridae) from eggs of the cicadellid *Cicadella viridis* (L.). *Entomol. Obozrenie* **16**, 315–34.

Nixon, G. E. J. (1957). Hymenoptera Proctotrupoidea Diapriidae subfamily Belytinae. *Handbooks for the Identification of British Insects* 8 (3d). Royal Entomological Society of London. 107 pp.

Norris, K. R. (1970). General biology. In: *The Insects of Australia* (Commonwealth Scientific and Industrial Research Organisation). Carlton, Victoria: Melbourne University Press. pp. 107–40.

Noyes, J. S. (1982). Collecting and preserving chalcid wasps (Hymenoptera: Chalcidoidea). *J. nat. Hist.* **16**, 315–34.

Olmi, M. (1984). A revision of the Dryinidae (Hymenoptera). *Mem. Am. Ent. Inst.* **37**(1), 1–946; **37**(2), 947–1913.

Orita, S. (1972). Some notes on *Lymaenon* sp. (Hymenoptera: Mymaridae), an egg parasite of Green Rice Leafhopper *Nephotettix cincticeps* Uhler (Homoptera: Cicadellidae) and its distribution in Hokuriku District. *Bull. Hokuriku Agric. Exp. Stat.* **14**, 122–127.

Ôtake, A. (1967). Studies on the egg parasites of the Smaller Brown Planthopper, *Laodelphax striatellus* (Fallén) (Hemiptera: Delphacidae). I. A device for assessing the parasitic activity, and the results obtained in 1966. *Bull. Shikoku Agr. Exp. Stat.* **17**, 91–103.

Ôtake, A. (1968). Studies on the egg parasites of the Smaller Brown Planthopper, *Laodelphax striatellus* (Fallén) (Hemiptera: Delphacidae). II. Development of *Anagrus* nr. *flaveolus* Waterhouse (Hymenoptera: Mymaridae) within its host. *Bull. Shikoku Agr. Exp. Stat.* **18**, 161–9.

Ôtake, A. (1969). Studies on the egg parasites of the Smaller Brown Planthopper, *Laodelphax striatellus* (Fallén) (Hemiptera: Delphacidae). III. Longevity and fecundity of *Anagrus* nr. *flaveolus* Waterhouse (Hymenoptera: Mymaridae). *Jap. J. Ecol.* **19**, 192–6.

Ôtake, A. (1970a). Studies on the Smaller Brown Planthopper *Laodelphax striatellus* (Fallén) (Hemiptera: Delphacidae). IV. Seasonal trends in parasitic and dispersal activities, with special reference to *Anagrus* nr. *flaveolus* Waterhouse (Hymenoptera: Mymaridae). *Appl. Ent. Zool.* **5**, 95–104.

Ôtake, A. (1970b). Estimation of parasitism by *Anagrus* nr. *flaveolus* Waterhouse (Hymenoptera, Mymaridae). *Entomophaga* **15**, 83–92.

Ôtake, A. (1976). Trapping of *Anagrus* nr. *flaveolus* Waterhouse (Hymenoptera: Mymaridae) by the eggs of *Laodelphax striatellus* (Fallén) (Hemiptera: Delphacidae). *Physiol. Ecol. Japan* **17**, 473–475.

Ôtake, A. (1977). Natural enemies of the Brown Planthopper. In: *The Rice Brown Planthopper* (Compiled by Food and Fertilizer Technology Centre for the Asian and Pacific Region), pp. 42–57. Tapei.

Ôtake, A., Somasundaram, P. H. and Abeykoon, M. B. (1976). Studies on populations of *Sogatella furcifera* Horvath and *Nilaparvata lugens* Stål (Hemiptera: Delphacidae) and their parasites in Sri Lanka. *Appl. Ent. Zool.* **11**, 284–94.

Parker, H. L. (1967). Notes on the biology of *Tomosvaryella frontata* (Diptera: Pipunculidae), a parasite of the leafhopper *Opsius stactogalus* on *Tamarix*. *Ann. ent. Soc. Am.* **60**, 292–95.

Patel, R. K. (1968). Records of natural enemies on *Sogatella furcifera* Horvth. *Indian J. Ent.* **30**, 321.

Peck, O., Boucek, Z. and Hoffer, A. (1964). Keys to the Chalcidoidea of Czechoslovakia (Insecta Hymenoptera). *Mem. Ent. Soc. Can.* **34**, 1–120.

Pekkarinen, A. and Raatikainen, M. (1973). The Strepsiptera of Eastern Fennoscandia. *Notulae Ent.* **53**, 1–10.

Perkins, R. C. L. (1905a). Leafhoppers and their natural enemies. I. Dryinidae. *Bull. Hawaiian Sug. Planter's Assoc. Exp. Stn.* **1**, 1–69.

Perkins, R. C. L. (1905b). Leafhoppers and their natural enemies. III. Stylopidae. *Bull. Hawaiian Sug. Planter's Assoc. Exp. Stn.* **1**, 90–111.

Perkins, R. C. L. (1905c). Leafhoppers and their natural enemies. IV. Pipunculidae. *Bull. Hawaiian Sug. Planter's Assoc. Stn.* **1**, 123–57.

Perkins, R. C. L. (1905d). Leafhoppers and their natural enemies. VI. Mymaridae, Platygasteridae. *Bull. Hawaiian Sug. Planter's Assoc. Stn.* **1**, 187–203.

Perkins, R. C. L. (1906a). Leafhoppers and their natural enemies. X. Dryinidae, Pipunculidae. *Bull. Hawaiian Sug. Planter's Assoc. Exp. Stn.* **1**, 483–99.

Perkins, R. C. L. (1906b). Leafhoppers and their natural enemies. VIII. Encyrtidae, Eulophidae, Trichogrammatidae. *Bull. Hawaiian Sug. Planter's Assoc. Exp. Stn.* **1**, 241–67.

Perkins, R. C. L. (1906c). Leafhoppers and their natural enemies. Introduction. *Bull. Hawaiian Sug. Planter's Assoc. Exp. Stn.* **1**, i–xxii.

Perkins, J. F. (1976). Hymenoptera Bethyloidea (excluding Chrysididae). *Handbks. Ident. Br. Insects* 5, (3a). Royal Entomological Society of London. 38 pp.

Pickles, A. (1932). Notes on the natural enemies of the Sugar-cane Froghopper (*Tomaspis saccharina* Dist.) in Trinidad, with descriptions of new species. *Bull. ent. Res.* **23**, 203–10.

Pickles, A. (1933). Entomological contributions to the study of the Sugar-cane Froghopper. *Trop. Agric.* **8**, 222–33.

Pierre, A. (1906). Biologie de *Tettigonia viridis* L. et de *Anagrus atomus* L., *Rev. Sci. Bourbonnais* **19**, 77–82, 117–21.

Poinar, G. O. (1975). *Entomophagous Nematodes*. Leiden.

Ponomarenko, N. G. (1968). On the biology of Dryinids (Dryinidae) parasites of leafhoppers. *Biul. Glavnovo Botan. Sada* **70**, 99–102.

Ponomarenko, N. G. (1971). Some peculiarities of development of Dryinidae. *Proc. 13th Int. Congr. Ent., Moscow 1968* **1**, 281–2.

Ponomarenko, N. G. (1975). Characteristics of larval development in the Dryinidae (Hymenoptera). *Ent. Obr.* **54**, 534–40.

Prestidge, R. A. and McNeill, S. (1983). Auchenorrhyncha–host plant interactions: leafhoppers and grasses. *Ecol. Ent.* **8**, 331–9.

Pruthi, H. S. (1937). Report of the Imperial Entomologist. *Sci. Rep. agric. Res. Inst. New Delhi* 1935–36, 123–37.

Raatikainen, M. (1961). *Dicondylus helléni* n. sp. (Hym., Dryinidae) a parasite of *Calligypona sordidula* (Stål) and *C. excisa* (Mel.). *Ann. Ent. Fenn.* **27**, 126–37.

Raatikainen, M. (1967). Bionomics, enemies and population dynamics of *Javesella pellucida* (F.) (Hom., Delphacidae). *Ann. agric. fenn. suppl.* **6**, 1–149.
Raatikainen, M. (1970). *Mesopolobus graminum* (Hardh) (Hym., Pteromalidae), its population dynamics and influence on *Javesella pellucida* (F.). *Ann. agric. fenn.* **9**, 106.
Raatikainen, M. (1972). Dispersal of leafhoppers and their natural enemies to oatfields. *Ann. agric. fenn.* **11**, 146–53.
Raatikainen, M. and Heikinheimo, O. (1974). The flying times of Strepsiptera males at different latitudes in Finland. *Ann. Ent. Fenn.* **40**, 22–25.
Regnier, R. (1921). Un ennemie du peuplier *Idiocerus populi* (Linné) Flor (Homop.) ou cicadelle du peuplier. *Ann. Epiphyties, Paris* **7**, 377–85.
Remane, R. and Schulz, K. (1973). Storung in der Ausbildung der ektodermalen weiblichen Genital-armatur im Zusammenhang mit parasitarer Kastration bei Zikaden der Gattung *Jassargus* Zachv. (Homoptera, Cicadellidae, Cicadelloidea). *Z. wiss. Zool.* **186**, 108–117.
Reynolds, W. J. (1975). Ecological investigations on some leafhoppers (Hemiptera: Homoptera, Cicadellidae) of woodland canopy. *M.Sc. thesis.* University of Wales.
Ribaut, H. (1936). Homopteres Auchenorhynches I (Typhlocybidae). *Faune de France* **31**, 1–228.
Richards, O. W. (1939). The British Bethylidae (*s.l.*) (Hymenoptera). *Trans. R. ent. Soc. Lond.* **89**, 185–344.
Richards, O. W. (1948). New records of Dryinidae and Bethylidae (Hymenoptera). *Proc. R. ent. Soc. Lond.* **23**, 14–18.
Richards, O. W. (1953). The classification of the Dryinidae (Hym.) with descriptions of new species. *Trans. R. ent. Soc. Lond.* **104**, 51–70.
Richards, O. W. and Davies, R. G. (1977). *Imm's General Textbook of Entomology*, Vol. 2. London. 1354 pp.
Riek, E. F. (1970). Strepsiptera. In: *The Insects of Australia* (Commonwealth Scientific and Industrial Research Organisation), Carlton, Victoria: Melbourne University Press. pp. 622–35.
Rothschild, G. H. L. (1962). The biology of *Conomelus anceps* (Germar) (Hemiptera: Delphacidae). *Ph.D. thesis.* University of London.
Rothschild, G. H. L. (1964). The biology of *Pipunculus semifumosus* (Kowarz) (Diptera: Pipunculidae), a parasite of Delphacidae (Homoptera) with observations on the effect of parasitism upon the host. *Parasitology* **54**, 763–69.
Rothschild, G. H. L. (1967). Notes on two hymenopterous egg parasites of Delphacidae (Hem.). *Ent. mon. Mag.* **103**, 5–9.
Sahad, K. A. (1982a). Biology and morphology of *Gonatocerus* sp. (Hymenoptera, Mymaridae), an egg parasitoid of the Green Rice Leafhopper, *Nephotettix cincticeps* Uhler (Homoptera, Deltocephalidae). I. Biology. *Kontyû* **50**, 246–60.
Sahad, K. A. (1982b). Biology and morphology of *Gonatocerus* sp. (Hymenoptera, Mymaridae), an egg parasitoid of the Green Rice Leafhopper, *Nephotettix cincticeps* Uhler (Homoptera, Deltocephalidae). II. Morphology. *Kontyû* **50**, 467–76.
Sahad, K. A. (1984). Biology of *Anagrus optabilis* (Perkins) (Hymenoptera, Mymaridae), an egg parasitoid of delphacid planthoppers. *Esakia* **22**, 129–44.
Sander, F. W. (1980). Beitrag zur Kenntnis des Mannchens von *Gonatopus sepsoides* Westwood). *Deutsch. Ent. Z.* **27**, 57–66.
Sasaba, T. and Kiritani, K. (1972). Evaluation of mortality factors with special reference to parasitism of the Green Rice Leafhopper, *Nephotettix cincticeps* Uhler (Hemiptera: Deltocephalidae). *Appl. Ent. Zool.* **7**, 83–93.

Schauff, M. E. (1984). The Holarctic genera of Mymaridae (Hymenoptera: Chalcidoidea). *Mem. Ent. Soc. Wash.* **12**, 1–67.
Seyedodeslami, H. and Croft, B. A. (1980). Spatial distribution of overwintering eggs of the White Apple Leafhopper, *Typhlocyba pomaria*, and parasitism by *Anagrus epos*. *Environ. Entomol.* **9**, 624–28.
Silvestri, F. (1918). Descrizione e notizie biologiche di alcuni Imenotteri Calcididi parasiti di uova di cicale. *Boll. Lab. Zoll. Gen. Agric. Portici* **12**, 252–65.
Silvestri, F. (1920). Contribuzione alla conoscenza dei parassiti delle ova del griletto canterio (*Oecanthus pellucens*, Scop., Orthoptera, Achetidae). *Boll. Lab. Zoll. Gen. Agrar. R. Scuola Sup. Agric. Portici* **14**, 219–50.
Silvestri, F. (1921). Notizie sulla cicale grigastra (*Tettigia orni* L.), sulla cicale maggiore (*Cicada plebeja* Scop.), sui loro parassiti e descrizione della loro larva neonata e della ninfa. *Boll. Lab. Zoll. Gen. Agric. Portici* **15**, 191–204.
Solomon, M. G. (1973). Ecological studies of grassland leafhoppers with special reference to *Psammotettix confinis* (Dahlbom) (Cicadellidae, Hemiptera). *Ph.D. thesis*. University of London.
Soper, R. S. Shewell, G. E. and Tyrell, D. (1976). *Colocondamyia auditrix* nov. sp. (Diptera: Sarcophagidae), a parasite which is attracted by the mating song of its host, *Okanagana rimosa* (Homoptera, Cicadidae). *Can. Ent.* **108**, 61–8.
Soyka, W. (1946). Beitrage zur klarung der Europaischen arten und gattungen der Mymariden (Hym.: Chalcidoid.). *Zbl. Gesamtgeb. Entomol.* **1**, 177–85.
Soyka, W. (1956). Monographie der *Polynema*-gruppe. *Abh. Zool. Bot. Ges. Wien* **19**, 1–115.
Spencer Smith, D. and Kathirithamby, J. (1984). Atypical "fibrillar" flight muscle in Strepsiptera. *Tissue and Cell* **16**, 929–40.
Sperka, C. and Freytag, P. H. (1975). Auchenorrhynchous hosts of mermithid nematodes in Kentucky. *Trans. Kentucky Acad. Sci.* **36**, 57–62.
Stáry, P. (1981). On the strategy, tactics and trends of host specificity evolution in aphid parasitoids (Hymenoptera, Aphidiidae). *Acta ent. bohemoslov.* **78**, 65–75.
Stáry, P. and Rejmánek, M. (1981). Number of parasitoids per host in different systematic groups of aphids: The implications for introduction strategy in biological control (Homoptera: Aphidoidea: Hymenoptera: Aphidiidae). *Ent. Scan. Suppl.* **15**, 341–51.
Stewart, A. J. A. (1983). Studies on the ecology and genetics of certain species of *Eupteryx* (Curt.); leafhoppers. *Ph.D. thesis*. University of Wales.
Stewart, A. J. A. (1986). Nymphal colour/pattern polymorphism in the leafhoppers *Eupteryx urticae* (F.) and *E. cyclops* Matsumura (Hemiptera: Auchenorrhyncha): spatial and temporal variation in morph frequencies. *Biol. J. Linn. Soc.* **27**, 79–101.
Stiling, P. D. (1980a). Competition and coexistence among *Eupteryx* leafhoppers (Hemiptera: Cicadellidae) occurring on stinging nettles (*Urtica dioica*). *J. anim. Ecol.* **49**, 793–805.
Stiling, P. D. (1980b). Host plant specificity, oviposition behaviour and egg parasitism in some leafhoppers of the genus *Eupteryx* (Hemiptera: Cicadellidae). *Ecol. Ent.* **5**, 79–85.
Stiling, P. D. (1980c). Colour polymorphisms in nymphs of the genus *Eupteryx* (Hemiptera: Cicadellidae). *Ecol. Ent.* **5**, 175–8.
Stiling, P. D. and Strong, D. R. (1982a). Parasitoids of the planthopper, *Prokelisia marginata* (Homoptera: Delphacidae). *Florida Entomol.* **65**, 191–2.
Stiling, P. D. and Strong, D. R. (1982b). Egg density and the intensity of parasitism in *Prokelisia marginata* (Homoptera: Delphacidae). *Ecology* **63**, 1630–5.
Strübing, H. (1956). *Neogonatopus ombrodes* Perkins (Hymenoptera, Dryinidae)

als Parasit au *Macrosteles laevis* Rb. (Homoptera–Auchenorrhyncha). *Zool. Beitrag.* **2**, 145–58.
Strübing, H. (1957). Ein Beitrag zur Biologie parasitischer Fliegen (Diptera–Pipunculidae). *Naturwiss.* **10**, 1–2.
Subba Rao, B. R. (1957). The biology and bionomics of *Lestodryinus pyrillae* Kieff. (Dryinidae: Hymenoptera), a nymphal parasite of *Pyrilla perpusilla* Walk; and a note on its role in the control of *Pyrilla. J. Bombay Nat. Hist. Soc.* **54**, 741–49.
Subba Rao, B. R. (1968). Descriptions of new genera and species of Mymaridae (Hymenoptera) from the Far East and the Ethiopian Region. *Bull. ent. Res.* **59**, 659–70.
Subba Rao, B. R. (1983). A catalogue of enemies of some important planthoppers and leafhoppers. *1st Int. Workshop on Leafhoppers and Planthoppers of Economic Importance.* Commonwealth Institute of Entomology 1983, pp. 385–403.
Subramaniam, T. V. (1922). Some natural enemies of Mango leaf-hoppers (*Idiocerus* spp.) in India. *Bull. ent. Res.* **12**, 465–7.
Swezey, O. H. (1908). On peculiar deviations from the uniformity of habit among chalcids and proctotrupids. *Proc. Hawaii ent. Soc.* **11**, 18–22.
Tachikawa, T. (1974). Hosts of encyrtid genera in the World (Hymenoptera: Chalcidoidea). *Mem. Fac. Col. Agric. Ehime Univ.* **19**, 185–204.
Tay, E. B. (1972). Population ecology of *Cicadella viridis* (*L.*) and bionomics of *Graphocephala coccinea* (Forster) (Homoptera: Cicadellidae). *Ph.D. thesis.* University of London.
Thompson, P. (1978). The oviposition sites of five leafhopper species (Hom. Auchenorrhyncha) on *Holcus mollis* and *H. lanatus. Ecol. Ent.* **3**, 231–40.
Tryapitzyn, V. A. (1964). Encyrtidae (Hymenoptera)—parasites of Dryinidae in U.S.S.R. *Zool. Zhurn.* **43**, 142–45 (in Russian, with English summary).
Tullgren, A. (1925). Om dvargstriten (*Cicadula sexnotata*, Fall.) och nagra andra ekonomiskt viktiga stritar. *Medd. Centralanst. försöks. jordbruks* **287**, 1–71 (in Swedish).
Vidano, C. (1962). Sulla alterata morfogenesi in Auchenorhichi parassitizzati e sulle sue interferenze speciographiche (Hemiptera Homoptera Jassidae). *Atti. Acad. Sci. Torino* **96**, 557–74.
Vidano, C. (1966). Introduzione in Italia di *Polynema striaticorne* Girault, parassita oofago di *Ceresa bubalus* Fabricius. *Boll. Soc. Entom. Italiana* **96**, 55–8.
Vidano, C. (1968). Precisione e labilita fenologiche di un Calcidoideo Mimaride. *Atti. Accad. Sci. Torino* **102**, 581–7 (in Italian, with English summary).
Vidano, C., Arzone, A. and Arno, C. (1985). Researches on natural enemies of viticolous Auchenorrhyncha. *Integrated Pest Control in Viticulture.* Expert's Group Meeting, Portoferraio, Italy, 26–28 Sept. 1985. pp 1–5.
Vidano, C. and Meotto, F. (1968). Moltiplicazione e disseminazione di *Polynema striaticorne* Girault (Hymenoptera Mymaridae). *Ann. Fac. Sci. Agr. Univ. Torino* **4**, 297–316.
Viggiani, G. (1969). Ricerche sugli Hymenoptera Chalcidoidea XX. Le specie paleartiche del genere *Lymaenon* Walker (Mymaridae) gruppo *Longicauda* (Enock), con descritzione di nuove specie. *Entomol. Bari* **5**, 37–50.
Vilbaste, J. (1968). Preliminary key for the identification of the nymphs of North European Homoptera Cicadinea. I. Delphacidae. *Ann. ent. Fenn.* **34**, 65–74.
Vilbaste, J. (1982). Preliminary key for the identification of the nymphs of North European Homoptera Cicadinea. II Cicadelloidea. *Ann. Zool. fenn.* **19**, 1–20.
Vungsilabutr, P. (1978). Biological and morphological studies of *Paracentrobia andoi* (Ishii) (Hymenoptera: Trichogrammatidae), a parasite of the Green

Leafhopper, *Nephotettix cincticeps* Uhler (Homoptera, Deltocephalidae). *Esakia* **2**, 29–51.

Waage, J. K. (1982). Sib-mating and sex ratio strategies in scelionid wasps. *Ecol. Ent.* **7**, 103–112.

Waage, J. K. (1983). Aggregation in field parasitoid populations: foraging time allocation by a population of *Diadegma* (Hymenoptera, Ichneumonidae). *Ecol. Ent.* **8**, 447–53.

Waage, J. K. and Godfray, H. C. J. (1985). Reproductive strategies and population ecology of insect parasitoids. In: *Behavioural Ecology, Ecological Consequences of Adaptive Behaviour*, 25th Symposium of the British Ecological Society. (Ed. by R. M. Sibly and R. H. Smith). Oxford: Blackwell. pp. 449–70.

Waage, J. K. and Hassell, M. P. (1982). Parasitoids as biological control agents—a fundamental approach. *Parasitology* **84**, 241–68.

Waage, J. K. and Ng, S. M. (1984). The reproductive strategy of a parasitic wasp. I. Optimal progeny and sex allocation in *Trichogramma evanescens*. *J. anim. Ecol.* **53**, 401–15.

Walker, I. (1979). Some British species of *Anagrus* (Hymenoptera: Mymaridae). *Zool. J. Linn. Soc.* **67**, 181–202.

Wall, I. (1967). Die Ismarinae und Belytinae der Schweiz. *Entomol. Abhandl.* **35**, 123–265.

Waloff, N. (1973). Dispersal by flight of leafhoppers (Auchenorrhyncha: Homoptera). *J. appl. Ecol.* **10**, 705–730.

Waloff, N. (1974). Biology and behaviour of some species of Dryinidae (Hymenoptera). *J. Ent.* **49**, 97–109.

Waloff, N. (1975). The parasitoids of the nymphal and adult stages of leafhoppers (Auchenorrhyncha; Homoptera). *Trans. R. ent. Soc. Lond.* **126**, 637–86.

Waloff, N. (1979). Partitioning of resources by grassland leafhoppers (Auchenorrhyncha, Homoptera). *Ecol. Ent.* **4**, 134–40.

Waloff, N. (1980). Studies on grassland leafhoppers (Auchenorrhyncha, Homoptera) and their natural enemies. In: *Advances in Ecological Research* (Ed. by A. Macfadyen), Vol. 11. London: Academic Press. pp. 81–215.

Waloff, N. (1981). The life history and descriptions of *Halictophagus silwoodensis* sp. n. (Strepsiptera) and its host *Ulopa reticulata* (Cicadellidae) in Britain. *Syst. Ent.* **6**, 103–13.

Waloff, N. and Thompson, P. (1980). Census data of populations of some leafhoppers (Auchenorrhyncha, Homoptera) of acidic grassland. *J. anim. Ecol.* **49**, 395–416.

Walter, G. H. (1983). "Divergent male ontogenies" in Aphelinidae (Hymenoptera: Chalcidoidea): a simplified classification and a suggested evolutionary sequence. *Biol. J. Linn. Soc.* **19**, 63–82.

Whalley, P. E. S. (1956). On the identity of species of *Anagrus* (Hym., Mymaridae) bred from leaf-hopper eggs. *Ent. mon. Mag.* **92**, 147–49.

Whalley, P. E. S. (1969). The mymarid (Hym.) egg-parasites of *Tettigella viridis* L. (Hem. Cicadellidae) and embryoparasitism. *Ent. mon. Mag.* **105**, 239–44.

Whittaker, J. B. (1969). The biology of Pipunculidae (Diptera) parasitizing some British Cercopidae (Homoptera). *Proc. R. ent. Soc. Lond.* **44**, 17–24.

Whittaker, J. B. (1971a). Population changes in *Neophilaenus lineatus* (L.) (Homoptera: Cercopidae) in different parts of its range. *J. anim. Ecol.* **40**, 425–43.

Whittaker, J. B. (1971b). The parasitization of colour polymorphs of *Philaenus spumarius* (L.) and of *Neophilaenus lineatus* (L.) (Homoptera) by a pipunculid. *Proc. 13th Int. Congr. Ent. Moscow 1968* **1**, 578–9.

Whittaker, J. B. (1973a). Density regulation in a population of *Philaenus spumarius* (L.) (Homoptera: Cercopidae). *J. anim. Ecol.* **42**, 163–72.

Whittaker, J. B. (1973b). Erratum, *J. anim. Ecol.* **42**, 829.

Williams, F. X. (1918). Some observations on *Pipunculus*, a fly which parasitizes the Cane Leafhopper at Pahala, Hawaii, February 11–April 25, 1918. *Hawaiian Planter's Rec.* **19**, 189–92.

Williams, J. R. (1957). The sugar-cane Delphacidae and their natural enemies in Mauritius. *Trans. R. ent. Soc. Lond.* **109**, 65–110.

Wilson, M. R. (1978). Descriptions and key to the genera of the nymphs of British woodland Typhlocybinae (Homoptera). *Syst. Ent.* **3**, 75–90.

Witsack, W. (1973). Zur Biologie und Okologie in zikadenelern parasitierender Mymariden der Gattung *Anagrus* (Chalcidoidea, Hymenoptera). *Zool. Jb. Syst. Bd.* **100**, 223–99.

Yang, Y. (1982). A preliminary study on the occurrence of rice planthopper nematodes. *Shanghai Agric. Sci. Tech.* **4**, 19–21.

Yano, K., Morakote, Satoh, M. and Asai, I. (1985). An evidence for behavioural change in *Nephotettix cincticeps* Uhler (Hemiptera: Deltocephalidae) parasitized by pipunculid flies (Diptera: Pipunculidae). *Appl. ent. Zool.* **20**, 94–6.

Ylönen, H. and Raatikainen, M. (1984). Uber die deformierung mannlicher kopulationsorgane zweien *Diplocolenus*-Arten (Homoptera, Auchenorrhyncha) beeinflusst durch parasitierung *Ann. ent. fenn.* **50**, 13–16.

APPENDICES

Appendices I–V list available rearing records for parasitoids of European Auchenorrhyncha.

Appendix I. Mymaridae, is based on records from Ahlberg (1925), Ali (1979), Arzone (1974a,c, 1977), Bakkendorf (1925, 1934), Becker (1975), Claridge and Reynolds (1972), Drosopoulos (1977), Hassan (1939), F. Herard (unpubl.), Huldén (1984), MacGill (1934), Maillet (1960), May (1971), Moratorio (1977), Morcos (1953), Nikol'skaya (1951), Pierre (1906), Raatikainen (1967), Regnier (1921), Reynolds (1975), Stiling (1980b), Tay (1972), Tullgren (1925), Vidano (1968), Vidano *et al.* (1985), Viggiani (1969), Whalley (1956) and Witsack (1973). * Introduced into Europe as part of a biological control programme.

Appendix II. Dryinidae, is based on records summarized in Olmi (1984), Freytag (1985), Jervis (1980c), Perkins (1976) and Waloff (1975, 1980), and also includes additional records by Chambers (1981), Jervis (1986) and F. Herard (unpubl.). Olmi's nomenclature is used throughout. Records of *Lonchodryinus ruficornis* from Typhlocybinae have been omitted, as we consider them to be questionable. * Other old and dubious records.

Appendix III. Pipunculidae, is based on records from A. Albrecht (unpubl.), Benton (1975), M. F. Claridge (unpubl.), F. Herard (unpubl.), Huq (1982), Jervis (1980c and unpubl.), Lauterer (1981, 1983), Parker (1967), C. Vidano (unpubl.) and Waloff (1975, 1980). Other records are taken from Coe (1966).

Appendix IV. Strepsiptera, is based on records from Kinzelbach (1978), Abdul-Nour (1969, 1970), Crowson (1975), Freytag (1985), Hassan (1939), Pekkarinen and Raatikainen (1973) and Waloff (1975, 1980). So far, all European parasitizations by Elenchidae have been attributed to *Elenchus tenuicornis*.

Appendix V. Diapriidae (*Ismarus* spp.), is based on the records given in the body of the table.

In Appendices I–IV, **numbers** indicate the **countries** in which the rearings took place: 1, Britain; 2, France; 3, Germany; 4, Finland; 5, Czechoslovakia; 6, Italy; 7, Spain; 8, Portugal; 9, USSR; 10, Netherlands; 11, Belgium; 12, Austria; 13, Greece; 14, Bulgaria; 15, Hungary; 16, Yugoslavia; 17, Poland; 18, Denmark; 19, Sweden.

Appendix I. Mymaridae

Hosts		*Anagrus atomus* (L.)	*A.* sp. nr. *atomus*	*A. bartheli* Tullgren	*A. ensifer* Debauche	*A. ensifer* Debauche var. E Walker	*A. holci* Walker	*A. incarnatus* Haliday	*A. mutans* Walker	*A. silwoodensis* Walker
MEMBRACIDAE	*Stictocephala bisonia* Kopp & Yonke									
CICADELLIDAE Cicadellinae	*Cicadella viridis* (L.)	● [1,2]	● [3]					● [1,2,3,6,18]	● [1]	● [1]
Idiocerinae	*Idiocerus confusus* Flor									
	I. distinguendus Kirschbaum									
	I. lituratus (Fallén)									
	I. populi (L.)									
	I. stigmaticalis Lewis									
	I. sp.									
Macropsinae	*Oncopsis alni* (Schrank)									
	O. flavicollis (L.)									
	O. subangulata (Sahlberg)									
	O. tristis (Zetterstedt)									
Aphrodinae	*Aphrodes* sp.									
Deltocephalinae	*Macrosteles sexnotatus* (Fallén)	● [19]					● [1]			
Typhlocybinae	*Empoasca vitis* (Göthe)		● [2]							
	Eupteryx aurata (L.)		● [1]							
	E. cyclops Matsumura		● [1]							
	E. urticae (Fabricius)		● [1]							
	Edwardsiana rosae (L.)	● [3]		● [19]						
	Hauptidia maroccana (Melichar)	● [1]								
	Zygina suavis Rey	● [6]								
DELPHACIDAE	*Anakelisia fasciata* Kirschbaum							● [3]		
	Stenocranus major (Kirschbaum)							● [3]		
	S. minutus (Fabricius)							● [3]		
	Conomelus anceps (Germar)				● [3]	● [1]		● [1,18]		
	C. dehneli Nast							● [6]		

A. stenocrani Walker	*A.* sp. "a"	*A.* sp. "b"	*A.* sp. "c"	*A.* sp. "c_3"	*A.* sp. "d"	*A.* sp. "e"	*A.* sp. "f"	*A.* sp. "g"	*A.* sp.	*Stethynium triclavatum* Enock	*Gonatocerus ater* Forster	*G. cicadellae* Nikol'skaya	*G. litoralis* (Haliday)	sp. nr. *maga* Girault	*G.* sp. nr. *paludis* (Debauche)	*G. tremulae* (Bakkendorf)	*G.* sp.	*Polynema atratum* Haliday	*P. bakkendorfi* Hincks	*P. euchariforme* Haliday	*P. striaticorne* Girault*	*P. vitripennis* (Forster)	*P. woodi* Hincks	*P.* sp. A	*P.* sp. B	*P.* sp. C	*P.* sp. D	*P.* sp.
																					●[6]							
	●[1]							●[1]			●[6]	●[6,9]											●[6]					●[4]
																								•[1]				
																											●[1]	
																									●[1]			
														●[2]		●[18]		●[18]										
																										●[1]		
																●[1]	●[6]			●[18]		●[1]						
																				●[1]								
																				●[1]								
																				●[1]								
																			●[1]									
													●[18]															
													●[1,19]		●[1]													
									●[2]	●[2]																		
●[1]																												
			●[1]	●[1]	●[1]																							

continued

Appendix I—*Continued*

Hosts	*Anagrus atomus* (L.)	*A.* sp. nr. *atomus*	*A. bartheli* Tullgren	*A. ensifer* Debauche	*A. ensifer* Debauche var. E Walker	*A. holci* Walker	*A. incarnatus* Haliday	*A. mutans* Walker	*A. silwoodensis* Walker
Ditropis pteridis (Spinola)									
Megamelus notula (Germar)							●[1]		
Mullerianella brevipennis (Boheman)									
M. fairmairei (Perris)				●[1]			●[1]	●[1]	
M. sp.									
Javesella pellucida (Fabricius)							●[4]		
Struebingianella lugubrina (Boheman)							●[3]		
unidentified sp. of Criomorphini							●[18]		
"unidentified Delphacidae"									

A. stenocrani Walker	A. sp. "a"	A. sp. "b"	A. sp. "c"	A. sp. "c_3"	A. sp. "d"	A. sp. "e"	A. sp. "f"	A. sp. "g"	A. sp.	*Stethynium triclavatum* Enock	*Gonatocerus ater* Forster	*G. cicadellae* Nikol'skaya	*G. litoralis* (Haliday)	sp. nr. *maga* Girault	*G.* sp. nr. *paludis* (Debauche)	*G. tremulae* (Bakkendorf)	*G.* sp.	*Polynema atratum* Haliday	*P. bakkendorfi* Hincks	*P. euchariforme* Haliday	*P. striaticorne* Girault*	*P. vitripennis* (Forster)	*P. woodi* Hincks	*P.* sp. A	*P.* sp. B	*P.* sp. C	*P.* sp. D	*P.* sp.
									●[1]																			
									●[10]																			
									●[10]																			
		●[1]					●[1]																					
						●[1]																						

Appendix II. Dryinidae

(II.1) Anteoninae with Cicadellidae as hosts

Hosts	*Lonchodryinus ruficornis* (Dalman)	*Anteon arcuetum* Kieffer	*A. brachycerum* (Dalman)	*A. ephippiger* (Dalman)	*A. flavicorne* (Dalman)	*A. fulviventre* Kieffer	*A. gaullei* Kieffer	*A. infectum* (Haliday)	*A. jurineanum* Latreille	*A. pubicorne* (Dalman)	*A. scapulare* (Haliday)
IDIOCERINAE											
Idiocerus sp.					●[1]						
I. confusus Flor					●[1,3]						
I. stigmaticalis Lewis					●[6]						
Populicerus albicans (Kirschbaum)					●[6]						
P. confusus (Flor)					●[1,3]						
P. laminatus (Flor)					●[1,9]						
P. populi (Linnaeus)					●[1]						
Rhytidodus decimusquartus (Schrank)		●[6]			●[6]						
Tremulicerus distinguendus (Kirschbaum)					●[6]						
JASSINAE											
Iassus lanio (Linnaeus)								●[1]			●[3*]
MACROPSINAE											
Oncopsis spp.			●[1]								
O. flaviocollis (Linnaeus)			●[1,9]						●[1,9]		
Macropsis sp.				●[3*]			●[3*]			●[3]	

Hosts	*Lonchodryinus ruficornis* (Dalman)	*Anteon arcuetum* Kieffer	*A. brachycerum* (Dalman)	*A. ephippiger* (Dalman)	*A. flavicorne* (Dalman)	*A. fulviventre* Kieffer	*A. gaullei* Kieffer	*A. infectum* (Haliday)	*A. jurineanum* Latreille	*A. pubicorne* (Dalman)	*A. scapulare* (Haliday)
DELTOCEPHALINAE											
Adarrus ocellaris (Fallén)	●[1,3]										
Arocephalus punctum (Flor)										●[1]	
Arthaldeus pascuellus (Fallén)	●[1]									●[1]	
Conosamus obsoletus (Kirschbaum)	●[1]										
Elymana sulphurella (Zetterstedt)	●[1]										
Euscelis incisus (Kirschbaum)	●[1]									●[1]	
Jassargus bisubulatus (Then)	●[6]										
J. distinguendus (Flor)	●[1]										
J. flori (Fieber)	●[1]										
J. obtusivalvis (Kirschbaum)	●[3*]										
Macrosteles sp.										●[1]	
M. laevis (Ribaut)	●[1,9]										
M. sexnotatus (Fallén)				●[1]						●[1]	
M. viridigriseus (Edwards)										●[1]	
Mocydia crocea (Herrich-Schaeffer)	●[1]			●[1]		●[1]				●[1]	
Psammotettix cephalotes (Herrich-Schaeffer)	●[1]										
P. confinis (Dahlbom)	●[1]				●[3*]					●[1]	
P. nodosus (Ribaut)	●[1]									●[1]	
P. striatus (Linnaeus)				●[9]							
Streptanus sordidus (Zetterstedt)	●[3*]									●[1]	

(II.2) Aphelopinae with Cicadellidae as hosts

Hosts	*Aphelopus atratus* (Dalman)	*A. camus* Richards	*A. melaleucus* (Dalman)	*A. nigriceps* Kieffer	*A. serratus* Richards
TYPHLOCYBINAE					
Aguriahana germari (Zetterstedt)			•[4]		
Alebra albostriella (Fallén)	•[1]				•[1]
A. wahlbergi (Boheman)	•[1]				
Alnetoidea alneti (Dahlbom)			•[1]		•[1]
Chlorita sp.		•[3]			
Edwardsiana sp.	•[1]		•[4]		•[1]
E. avellanae (Edwards)			•[1]		
E. bergmani (Tullgren)			•[4]		
E. crataegi (Douglas)	•[6]		•[1]		•[6]
E. flavescens (Fabricius)			•[1]		
E. geometrica (Schrank)			•[1]		•[1]
E. hippocastani (Edwards)	•[1]		•[1,2,9]		
E. lethierryi (Edwards)	•[1]		•[1,2]		•[1]
E. menzbieri Zachvatkin			•[4]		
E. plebeja (Edwards)			•[1]		
E. rosae (Linnaeus)	•[2,6]		•[1,2]		
Empoasca sp.			•[4]		
E. vitis (Göthe)	•[2]	•[2]	•[2]	•[2]	•[2]
Eupterycyba jucunda (Herrich-Schaeffer)					•[1]
Eupteryx aurata (Linnaeus)	•[1,2]				
E. cyclops (Matsumura)	•[1]				
E. mellissae Curtis	•[1]				
E. stachydearum (Hardy)	•[6]				
E. urticae (Fabricius)	•[1]				
Eurhadina concinna (Germar)				•[1]	

(II.2)—Continued

Hosts	*Aphelopus atratus* (Dalman)	*A. camus* Richards	*A. melaleucus* (Dalman)	*A. nigriceps* Kieffer	*A. serratus* Richards
Fagocyba sp.	● [1]				
F. cruenta (Herrich-Schaeffer)	● [1]		● [1,2,3,9]		● [1]
Fagocyba carri (Edwards)			● [1]		
Kybos smaragdula (Fallén)					● [1]
Lindbergina aurovittata (Doug)					● [1]
Linnavouriana decempunctata (Fallén)			● [4]		
Ossiannilssonola callosa (Then)			● [1]		
Ribautiana tenerrima (Herrich-Schaeffer)	● [1]				● [1]
R. ulmi (Linnaeus)	● [1]		● [1,2,9]		
Typhlocyba sp.	● [2]		● [2]		
T. bifasciata (Boheman)	● [1]				
T. quercus (Fabricius)	● [1]		● [1]		● [1]
Zygina sp.	● [1]				● [1]
Z. flammigera (Fourcroy)	● [2]		● [2]		

(II.3a) Gonatopodinae with Cicadellidae as hosts

Hosts	*Gonatopus*													*Tetradontochelys*	
	bernardi (Picard)	*bilineatus* Kieffer	*campestris* Ponomarenko	*distinguendus* Kieffer	*formicarius* Ljungh	*graecus* Olmi	*horvathi* Kieffer	*lunatus* Klug	*popovi* Ponomarenko	*rhaensis* Ponomarenko	*sepsoides* Westwood	*spectrum* (Snellen van Vollenh)	*striatus* Kieffer	*pedestris* (Dalman)	*pulicarius* (Klug)
APHRODINAE															
Aphrodes sp.													●[1]		
A. bicinctus (Schrank)		●[2]											●[1]		
DELTOCEPHALINAE															
Adarrus sp.								●[3]							
A. geniculatus (Ribaut)											●[2]				
A. ocellaris (Fallén)											●[1]				
A. taurus (Ribaut)				●[2]				●[2]			●[2]	●[2]			●[2]
Araldus propinquus (Fieber)				●[2]							●[2]	●[2]			
Arocephalus languidus (Fieber)											●[3]				
A. punctum (Flor)											●[1]				
A. sagittarius Ribaut											●[2]				
Arthaldeus pascuellus (Fallén)								●[4]			●[1]				
Cicadula sp.											●[1]				
"*Circulifer dubiosis*" (Matsumara)				●[2]											
Deltocephalus sp.					●[1]										
"*Deltocephalus pascuorum*"								●[3*]							
D. assimilis											●[12]				
Diplocolenus abdominalis (Fabricius)					●[6]						●[4,6]				
D. nigrifrons (Kirschbaum)					●[6]										
Doratura sp.						●[13]									
Elymana sulphurella (Zetterstedt)											●[1]				
Enantiocephalus cornutus (Herrich-Schaeffer)								●[3]							
Euscelis sp.											●[1,6]				
E. incisus (Kirschbaum)								●[3]			●[1,2,3,6]				

(II.3a)—Continued

Hosts	*Gonatopus*													*Tetradontochelys*	
	bernardi (Picard)	*bilineatus* Kieffer	*campestris* Ponomarenko	*distinguendus* Kieffer	*formicarius* Ljungh	*graecus* Olmi	*horvathi* Kieffer	*lunatus* Klug	*popovi* Ponomarenko	*rhaensis* Ponomarenko	*sepsoides* Westwood	*spectrum* (Snellen van Vollenh)	*striatus* Kieffer	*pedestris* (Dalman)	*pulicarius* (Klug)
Goldeus harpago Ribaut											●[2]				
Jassargus distinguendus (Flor)											●[1]				
J. flori (Fieber)											●[1]				
J. obtusivalvis (Kirschbaum)											●[2]				
J. repletus (Fieber)											●[14]				
Macrosteles sp.								●[4]							
M. laevis (Ribaut)														●[6]	
Mocuellus collinus (Boheman)					●[9]										
Opsius? stactogalus (Fieber)	●[2]														
Paluda elongata (Wagner)				●[2]							●[2]	●[2]			
Psammotettix alienaus (Dahlbom)					●[6]			●[6]			●[6]				
P. confinis (Dahlbom)					●[2*]	●[2]					●[1]				
P. nodosus (Ribaut)											●[1,2]				
P. notatus (Melichar)											●[2,6]				
P. putoni (Then)											●[2,6]				
P. striatus (Linnaeus)			●[9]	●[9]			●[9]	●[3]		●[9]	●[1,2,6]				
Recilia coronifera (Marshall)											●[1]				
Sorhoanus assimilis (Fallén)											●[1]				
S. xanthoneurus (Fieber)											●[1,12]				
Streptanus sordidus (Zetterstedt)											●[1]				
Thamnotettix sp.									●[9]						
T. confinis Zetterstedt											●[4]				
T. dilutior (Kirschbaum)		●[2]													
Turrutus socialis (Flor)											●[9]				

(II.3b) Gonatopodinae with Delphacidae as hosts

Hosts	*Agonatopoides solidus* (Haupt)	*Dicondylus bicolor* (Halliday)	*Dicondylus dichromus* (Kieffer)	*Donisthorpina pallida* (Ceballos)	*Echthrodelphax hortuensis* (Abdul-Nour)	*Haplogonatopus oratorius* (Westwood)	*Pseudogonatopus*: *P. albosignatus* (Kieffer)	*P. augustae* Currado & Olmi	*P. distinctus* (Kieffer)	*P. ligusticus* Currado & Olmi	*P. septemdentatus* (Sahlberg)	*P. rosellae* Currado & Olmi
Criomorphus albomarginatus Curtis		●[1,4]										
C. bicarinatus (Herrich-Schaeffer)		●[3]										
Delphacinus mesomelas (Boheman)		●[10]										
"*Delphax collinae*" (Boheman)		●[2]										
Dicranotropis divergens Kirschbaum												●[6]
Dicranotropis hamata (Boheman)		●[1,4,6,10]							●[1,3]			●[6]
Ditropis pteridis (Spinola)		●[1]							●[1]			
Gravesteiniella boldi (Scott)		●[1]							●[1]			
Hyledelphax elegantulus (Boheman)		●[1]							●[1]			
Javesella discolor (Boheman)									●[1]			
J. obscurella (Boheman)		●[4]										
J. pellucida (Fabricius)		●[1,4]							●[1,4]			
Kosswigianella exigua (Boheman)									●[1]			
Laodelphax striatellus (Fallén)	●[6]				●[6]	●[6]						

Hosts	*Agonatopoides solidus* (Haupt)	*Dicondylus bicolor* (Halliday)	*Dicondylus dichromus* (Kieffer)	*Donisthorpina pallida* (Ceballos)	*Echthrodelphax hortuensis* (Abdul-Nour)	*Haplogonatopus oratorius* (Westwood)	*Pseudogonatopus*					
							P. albosignatus (Kieffer)	*P. augustae* Currado & Olmi	*P. distinctus* (Kieffer)	*P. ligusticus* Currado & Olmi	*P. septemdentatus* (Sahlberg)	*P. rosellae* Currado & Olmi
Megadelphax sordidulus (Stål)		•[4]	•[4]		•[6]	•[6]						•[6]
Megadelphax sp.	•[6]							•[6]				
Metadelphax propinqua (Fieber)											•[9]	
Muellerianella fairmairei (Perris)				•[10]								
Muirodelphax aubei (Perris)	•[3]*											
Struebingianella dalei (Scott)									•[1]			
Ribautodelphax sp.		•[1]										
R. angulosus (Ribaut)									•[1]			
R. collinus (Boheman)		•[10]										
R. imitans (Ribaut)					•[2]		•[2]			•[2]		
R. pungens (Ribaut)		•[11]		•[10]								
Stiroma sp.		•[2]										
S. bicarinata (Herrich-Schaeffer)		•[4]										
Unkanodes excisa (Melichar)		•[3,4]	•[4]									
Xanthodelphax stramineus (Stål)				•[6]								

(II.4) Bocchinae and (II.5) Dryininae

	Bocchinae		Dryininae			
Hosts	*Bocchus europaeus* (Bernard)	*Mystrophorus formicaeformis* Ruthe	*Dryinus collaris* (Linnaeus)	*D. sanderi* Olmi	*D. terraconensis* Marshall	*Richardsidryinus corsicus* (Marshall)
CICADELLIDAE (Deltocephalinae)						
Deltocephalus sp.		• [1,3]				
Diplocolenus abdominalis (Fabricius)		• [9]				
DICTYOPHARIDAE						
Dictyophara europaea (Linnaeus)					• [2,3]	
CIXIIDAE						
Cixius contaminatus Flor.			• [1]			
C. nervosus (Linnaeus)			• [2,3]			
Tachycixius pilosus (Oliver)			• [3]			
ISSIDAE						
Caliscelis sp.	• [2]					
Hysteropterum flavescens (Ol.)						• [2]
H. latifrons				• [14]		
Issus coleoptratus (Geoffroy)			• [6]			

Appendix III. Pipunculidae (Part 1)

Hosts		*Chalarus argenteus* Coe	*C. exiguus* (Haliday)	*C. fimbriatus* Coe	*C. latifrons* Hardy	*C. parmenteri* Coe	*C.* sp. A nr. *parmenteri*	*C.* sp. B nr. *parmenteri*	*C. pughi* Coe	*C.* sp. nr. *pughi*	*C. spurius* (Fallén)	*C.* sp. A nr. *spurius*	*C.* sp. B nr. *spurius*	*C.* sp. C nr. *spurius*	*C.* sp. D nr. *spurius*	*Verrallia aucta* (Fallén)	*V. beatricis* Coe	*V. pilosa* Zetterstedt	*V. setosa* Verrall	*Tomosvaryella frontata* (Becker)	*T. kuthyi* (Aczél)	*T. pallidiarsis* (Collin)	*T. sylvatica* (Meigen)	*Dorylomorpha hungarica* (Aczél)	*D. maculata* Walker	*D. platystylis* Albrecht	*D. xanthopus* (Thomson)	*Pipunculus campestris* Latreille	*P. fonsecae* Coe
CERCOPIDAE																													
Aphrophorinae	*Philaenus spumarius* (L.)															●[1]													
	Neophilaenus lineatus (L.)															●[1]													
CICADELLIDAE																													
Idiocerinae	*Idiocerus populi* (L.)																												
Macropsinae	*Oncopsis alni* (Schrank)																●[1]												
	O. flavicollis (L.)																		●[1]										
	O. subangulata (Sahlberg)																	●[1]	●[1]										
	O. sp.																	●[1]											
Deltocephalinae	*Doratura stylata* (Boheman)																												●[1]
	Turrutus socialis (Flor)																												
	Adarrus ocellaris (Fallén)																						●[1]					●[1]	
	Jassargus distinguendus (Flor)																						●[1,3]						
	Diplocolenus abdominalis (Fabricius)																					●[1]	●[1]				●[1]	●[1]	
	Mocuellus collinus (Boheman)																												
	Sorhoanus assimilis (Fallén)																									●[5]			

continued

Appendix III (Part 1) continued.

Hosts	*Chalarus argenteus* Coe	*C. exiguus* (Haliday)	*C. fimbriatus* Coe	*C. latifrons* Hardy	*C. parmenteri* Coe	C. sp. A nr. *parmenteri*	C. sp. B nr. *parmenteri*	*C. pughi* Coe	C. sp. nr. *pughi*	*C. spurius* (Fallén)	C. sp. A nr. *spurius*	C. sp. B nr. *spurius*	C. sp. C nr. *spurius*	C. sp. D nr. *spurius*	*Verrallia aucta* (Fallén)	*V. beatricis* Coe	*V. pilosa* Zetterstedt	*V. setosa* Verrall	*Tomosvaryella frontata* (Becker)	*T. kuthyi* (Aczél)	*T. pallidtarsis* (Collin)	*T. sylvatica* (Meigen)	*Dorylomorpha hungarica* (Aczél)	*D. maculata* Walker	*D. platystylis* Albrecht	*D. xanthopus* (Thomson)	*Pipunculus campestris* Latreille	*P. fonsecae* Coe
Arthaldeus pascuellus (Fallén)																						●[1,3]				●[5]	●[1,3,5]	
Psammotettix confinis (Dahlbom)																				●[1]		●[1]				●[1]	●[1]	
P. kolosvarensis (Matsumura)																												
P. nodosus (Ribaut)																												
Graphocraerus ventralis (Fallén)																												
Conosanus obsoletus (Kirschbaum)																										●[1]	●[1]	
Euscelis incisus (Kirschbaum)																										●[1]	●[1,3,5]	
Streptanus marginatus (Kirschbaum)																								●[4]				
Doliotettix lunulatus (Zetterstedt)																												
Athysanus argentarius Metcalf																												
Mocydia crocea (Herrich-Schaeffer)																												
Thamnotettix confinis (Zetterstedt)																								●[4]				
Speudotettix subfusculus (Fallén)																								●[4]				
Cicadula albingensis Wagner																							●[3]					
C. flori (Sahlberg)																							●[5]				●[5]	
C. frontalis (Herrich-Schaeffer)																							●[5]					
C. quadrinotata (Fabricius)																							●[1]				●[1,3]	
C. saturata (Edwards)																							●[5]					
Elymana sulphurella (Zetterstedt)																											●[1]	

	athysanid nymphs																											
	Opsius stactogalus Fieber																			●[2]								
	Macrosteles laevis (Ribaut)																											●[1,3,5]
	M. sexnotatus (Fallén)																											●[1]
	M. variatus (Fallén)																											
	M. sp.																											
	Balclutha punctata (Fabricius)																											
Typhlocybinae	*Alebra albostriella* (Fallén)					●[1]							●[1]															
	Erythyria aureola (Fallén)											●[4]																
	Empoasca vitis (Göthe)						●[6]			●[2,7]	●[2]	●[2]		●[1,2,6]	●[2]													
	Kybos butleri (Edwards)				●[1]																							
	K. smaragdula (Fallén)				●[1]																							
	Eurhadina concinna (Germar)												●[1]															
	E. pulchella (Fallén)												●[1]															
	Eupteryx aurata (L.)								●[1]		●[1,2]																	
	E. cyclops Matsumura										●[1]																	
	E. melissae Curtis											●[1]																
	E. urticae (Fabricius)			●[1]					●[1]		●[1,2]																	
	Ribautiana tenerrima (Herrich-Schaeffer)											●[1]																
	R. ulmi (L.)											●[1]																
	Typhlocyba quercus (Fabricius)	●[1]				●[1]						●[1]																
	Fagocyba cruenta (Herrich-Schaeffer)					●[1]						●[1]																
	F. sp.											●[1]	●[1]															
	Edwardsiana bergmani (Tullgren)											●[4]																
	E. geometrica (Schrank)								●[1]			●[1]																
	E. hippocastani (Edwards)											●[1]																
	E. rosae (L.)											●[1]																
	E. sp.			●[1]					●[1]			●[1]																
	Alnetoidia alneti (Dahlbom)		●[1]	●[1,4]																								
	Zygina suavis Rey							●[6]																				

Appendix III. Pipunculidae (Part 2)

Hosts		P. thomsoni Becker	P. zugmayeriae Kowarz	P. spp.	Cephalops carinatus Verrall	C. chlorionae Frey	C. curtifrons Coe	C. furcatus (Egger)	C. oberon Coe	C. obtusinervis Zetterstedt	C. perspicuus de Meijere	C. semifumosus (Kowarz)	C. subultimus Collin	C. ultimus (Becker)	C. spp.	Eudorylas fascipes (Zetterstedt)	E. fuscipes (Zetterstedt)	E. fusculus (Zetterstedt)	E. jenkinsoni Coe	E. longifrons Coe	E. montium (Becker)	E. obliquus Coe	E. obscurus Coe	E. opacus (Zetterstedt)	E. ruralis Meigen	E. subfascipes Collin	E. subterminalis Collin	E. spp.
CERCOPIDAE																												
Aphrophorinae	*Philaenus spumarius* (L.)																											
	Neophilaenus lineatus (L.)																											
CICADELLIDAE																												
Idiocerinae	*Idiocerus populi* (L.)																											●[4]
Macropsinae	*Oncopsis alni* (Schrank)																											
	O. flavicollis (L.)																											
	O. subangulata (Sahlberg)																											
	O. sp.																											
Deltocephalinae	*Doratura stylata* (Boheman)																											
	Turrutus socialis (Flor)																										●[3]	
	Adarrus ocellaris (Fallén)															●[1]			●[1]			●[1]	●[1]			●[1]	●[1,3]	
	Jassargus distinguendus (Flor)																									●[1]	●[1]	
	Diplocolenus abdominalis (Fabricius)			●[4]																			●[1]				●[1]	●[4]

Mocuellus collinus (Boheman)																									●[3]	
Sorhoanus assimilis (Fallén)																										
Arthaldeus pascuellus (Fallén)														●[1]			●[1]			●[1]	●[1]				●[1]	
Psammotettix confinis (Dahlbom)														●[1]						●[1]	●[1]			●[1]	●[1,3]	
P. kolosvarensis (Matsumura)																									●[3]	
P. nodosus (Ribaut)																								●[1]	●[3]	
Graphocraerus ventralis (Fallén)		●[1]																								
Conosanus obsoletus (Kirschbaum)																										
Euscelis incisus (Kirschbaum)																							●[1,3]		●[3]	●[3]
Streptanus marginatus (Kirschbaum)																										
Doliotettix lunulatus (Zetterstedt)																										●[4]
Athysanus argentarius Metcalf			●[4]																							
Mocydia crocea (Herrich-Schaeffer)																		●[3]		●[1]						
Thamnotettix confinis (Zetterstedt)																										●[4]
Speudotettix subfusculus (Fallén)								●[4]					●[4]							●[3]						
Cicadula albingensis Wagner																										
C. flori (Sahlberg)																										
C. frontalis (Herrich-Schaeffer)																										
C. quadrinotata (Fabricius)														●[1]					●[3]	●[3]						
C. saturata (Edwards)																										
Elymana sulphurella (Zetterstedt)	●[1]														●[4]											
athysanid nymphs																				●[1]						
Opsius stactogalus Fieber																										
Macrosteles laevis (Ribaut)															●[3,17]											

continued

Appendix III (Part 2) continued.

Hosts		*P. thomsoni* Becker	*P. zugmayeriae* Kowarz	*P.* spp.	*Cephalops carinatus* Verrall	*C. chlorionae* Frey	*C. curtifrons* Coe	*C. furcatus* (Egger)	*C. oberon* Coe	*C. obtusinervis* Zetterstedt	*C. perspicuus* de Meijere	*C. semifumosus* (Kowarz)	*C. subultimus* Collin	*C. ultimus* (Becker)	*C.* spp.	*Eudorylas fascipes* (Zetterstedt)	*E. fuscipes* (Zetterstedt)	*E. fusculus* (Zetterstedt)	*E. jenkinsoni* Coe	*E. longifrons* Coe	*E. montium* (Becker)	*E. obliquus* Coe	*E. obscurus* Coe	*E. opacus* (Zetterstedt)	*E. ruralis* Meigen	*E. subfascipes* Collin	*E. subterminalis* Collin	*E.* spp.
	M. sexnotatus (Fallén)																●[1]											
	M. variatus (Fallén)																●[3]											
	M. sp.															●[3]												
	Balclutha punctata (Fabricius)																	●[4]						●[4]				
CIXIIDAE	*Tachycixius pilosus* (Olivier)							●[1]																				
DELPHACIDAE	*Stenocranus minutus* (Fabricius)						●[1,5]																					
	Chloriona glaucescens Fieber					●[4]																						
	Conomelus anceps (Germar)											●[1]		●[1]														
	Eurysa lineata (Perris)													●[3]														
	Ditropis pteridis (Spinola)								●[1]			●[1]	●[1]															
	Stiroma sp.				●[4]					●[4]					●[4]													
	Criomorphus williamsi China											●[1]		●[1]														
	Dicranotropis hamata (Boheman)				●[4]						●[1]	●[1]	●[1]	●[1]														
	Delphacodes venosus (Germar)											●[1]		●[1]														

Hyledelphax elegantulus (Boheman)											●[1]																
Javesella discolor (Boheman)											●[1]																
J. forcipata (Boheman)														●[4]													
J. pellucida (Fabricius)											●[1]																

Appendix IV. Strepsiptera

Family Halictophagidae

	Known host species	Family	Country where recorded
Genus *Halictophagus*			
H. agalliae Abdul-Nour	*Agallia consorbrina* Curtis	Cicadellidae	2
	A. laevis Ribaut		2
H. kuehnelti Hofender	*Fulgora europaea*	Fulgoridae	16
H. languedoci Abdul-Nour	*Adarrus geniculatus* Ribaut	Cicadellidae	2
	A. taurus Ribaur		2
	Arocephalus sagittarius Ribaut		2
	Araldus propinquus (Fieber)		2
	Goldeus harpago (Ribaut)		2
	Jassargus obtusivalvis (Kirschbaum)		2
	Psammotettix notatus (Melichar)		2
	P. putoni (Then)		2
	P. striatus (Linnaeus)		2

H. silwoodensis Waloff	*Ulopa reticulata* (Fabricius)		1
H. tettigometrae Silvestri	*Tettigometra concolor* Fieber	Tettigometridae	15
	T. impressiformis Mulsant & Rey		6
	T. impressopunctata Dufour		15
	T. obliqua Panzer		6
	T. pincta Fieber		6
Halictophagus spp.	*Eurysa lineata* (Perris)	Delphacidae	3*
	E. lurida Fieber		3*
	Eupelix cuspidata (Fabricius)	Cicadellidae	1
	Issus lavri	Issidae	16
Genus *Stenocranophilus*			
S. anomalocerus (Pierce)	*Laodelphax striatellus* (Fallén)	Delphacidae	15,17
S. quadratus	*Calligypona*, *Dicranotropis*, *Stenocranus* } Europe		

Appendix IV. Strepsiptera—*Continued*

Family Elenchidae

	Some known host species (all Delphacidae)	Country where recorded
Genus *Elenchus*		
E. tenuicornis (Kirby)	*Acanthodelphax denticauda* (Boheman)	4
	Chloriona glaucescens Fieber	2, 4, 15
	C. propinqua	15
	C. smaragdula (Stål)	4, 15
	C. unicolor (Herrich-Schaeffer)	4, 15
	Conomelus anceps (Germar)	1
	Criomorphus albomarginatus Curtis	1, 3, 15
	C. williamsi China	1
	Delphacinus mesomelas (Boheman)	15
	Dicranotropis hamata (Boheman)	1, 3, 4, 15
	D. carpathica Horvath	9, 15
	Ditropis pteridis (Spinola)	1
	Eurybregma nigrolineata Scott	1, 15
	Florodelphax leptosoma (Boheman)	3
	Hyledelphax elegantulus (Boheman)	1, 3, 15
	Javesella discolor (Boheman)	3, 4, 15

J. dubia (Kirschbaum)	1, 3, 4
J. forcipata (Boheman)	4, 5
J. obscurella (Boheman)	3, 4, 10
J. pellucida (Fabricius)	1, 3, 4, 5, 10, 15
Kosswigianella exigua (Boheman)	1, 3, 15
Megadelphax sordidula Stål	4, 10, 15
M. quadrimaculatus (Signoret)	1
Muellerianella brevipennis (Boheman)	10, 18
M. fairmairei (Perris)	1, 3
Muirodelphax denticauda (Boheman)	4
Paraliburnia adela (Flor)	3
Ribautodelphax albostriatus (Fieber)	3, 4
R. angulosus (Ribaut)	1
R. collina (Boheman)	3, 4
R. imitans (Ribaut)	2
Stegelytra putoni Rey	3
Stenocranus minutus (Fabricius)	1
Stiroma affinis Fieber	4
S. bicarinata (Herrich-Schaeffer)	3, 4, 10
Struebingianella dalei (Scott)	1
S. lugubrina (Boheman)	3, 4, 15
Xanthodelphax flaveolus (Flor)	3, 4, 15
X. stramineus Stål	4, 15

Appendix V. Hosts of European *Ismarus* spp.

	I. flavicornis (Thomson)	*I. dorsiger* (Curtis)	*I. halidayi* Forster	*I. rugulosus* Forster
Primary host	Unidentified; either *Anteon flavicorne* (Dalman) or *A. arcuatum* (Kieffer)[a]	Unidentified; probably *Aphelopus melaleucus* (Dalman)	An unidentified birch-associated *Anteon* sp. and *A. infectus* Haliday	Unidentified; either *Anteon lucidus* Haliday or *Lonchodryinus* sp.
Secondary host	Unidentified *Idiocerus* sp. (nymphs)	*Fagocyba cruenta* and *Ribautiana ulmi* (adults)	Unidentified *Oncopsis* sp. and *Iassus lanio* (nymphs)	*Streptanus sordidus* (adult)
Author	Chambers (1955)	Jervis (1979a)	Chambers (1955, 1981)	Waloff (1975)

[a] Chambers (1955) states that he reared both *Anteon flavicorne* and *A. flavicorne* var. *bensoni* Richards from *Idiocerus* spp. The latter dryinid has been synonymized with *A. arcuatum* Kieffer (see Olmi, 1985).

Index

Advances in Ecological Research Volumes 1–16

Cumulative List of Titles